ASTRONOMY AND ASTROPHYSICS LIBRARY

Series Editors: G. Börner, Garching, Germany
A. Burkert, München, Germany
W. B. Burton, Charlottesville, VA, USA and
   Leiden, The Netherlands
M. A. Dopita, Canberra, Australia
A. Eckart, Köln, Germany
E. K. Grebel, Heidelberg, Germany
B. Leibundgut, Garching, Germany
A. Maeder, Sauverny, Switzerland
V. Trimble, College Park, MD, and Irvine, CA, USA

Josef Kallrath · Eugene F. Milone

Eclipsing Binary Stars: Modeling and Analysis

Second Edition

Springer

Josef Kallrath
Department of Astronomy
University of Florida
Gainesville, FL 32611-2055
USA
josef.kallrath@web.de

Eugene F. Milone
Department of Physics & Astronomy
University of Calgary
2500 University Drive NW
Calgary, AB T2N 1N4
Canada
milone@ucalgary.ca

ISSN 0941-7834
ISBN 978-1-4419-0698-4   e-ISBN 978-1-4419-0699-1
DOI 10.1007/978-1-4419-0699-1
Springer New York Dordrecht Heidelberg London

Library of Congress Control Number: 2009928498

© Springer Science+Business Media, LLC 1999, 2009
All rights reserved. This work may not be translated or copied in whole or in part without the written permission of the publisher (Springer Science+Business Media, LLC, 233 Spring Street, New York, NY 10013, USA), except for brief excerpts in connection with reviews or scholarly analysis. Use in connection with any form of information storage and retrieval, electronic adaptation, computer software, or by similar or dissimilar methodology now known or hereafter developed is forbidden.
The use in this publication of trade names, trademarks, service marks, and similar terms, even if they are not identified as such, is not to be taken as an expression of opinion as to whether or not they are subject to proprietary rights.

Printed on acid-free paper

Springer is part of Springer Science+Business Media (www.springer.com)

Professori Dr. Hans Schmidt,

> magisto doctissimo ac clarissimo, patri et amico
> hoc opus votis optimis D.D.D.[1]

To Harlan Smith and Adriaan Wesselink,

> teachers extraordinaire, and, to my great fortune,
> my graduate advisors.[2]

---

[1] Josef Kallrath.
[2] Eugene F. Milone.

Foreword

Have you ever stopped at a construction project on the way to your office and the day's astrophysics? Remember the other onlookers – folks just enjoying the spectacle, as we all do in following developments away from our areas of active work? We are excited and thrilled when the Hubble Space Telescope discovers an Einstein Cross, when the marvelous pulsars enter our lives, and when computer scientists put a little box on our desk that outperforms yesterday's giant machines. We are free to make use of such achievements and we respect the imagination and discipline needed to bring them about, just as onlookers respect the abilities and planning needed to create a building they may later use. After all, each of us contributes in our own areas as best as we can.

In addition to the serious onlookers there will be passersby who take only a casual look at the site. They may use the building later, but have little or no interest in its construction and give no thought to the resources needed to bring it to completion. Upon arriving at work, those persons write astronomy and astrophysics books at various levels, in which they must say something about close binary stars. Usually a page or two will do, and the emphasis is on the MLR (mass, luminosity, radius) data obtained only from binaries. The role of binaries in stellar evolution also may be awarded a page or so, perhaps meshed with binaries being homes of black holes and neutron stars. We live in an era of ever more applied research, with national priorities set by the interests and judgments of select committees. Consequently, most authors tell us the answer to one central question: What have binaries done for us lately? Well, of course, binaries are alive and well as sources of fundamental information on many fronts. But what of the fun, intrigue, and beauty of close binaries?

However, I do not want to be hard on the generic text authors because I remember my initial reaction to binaries. A fellow graduate student (later a very accomplished researcher) was doing a binary star project and I could not imagine why he was so interested. He tried to explain it all, but it just was not working: subject = nonexciting. However, I soon was into binaries anyway, just due to being surrounded by binary star work (yes, with appalling conformity). Time went by and then something happened to turn the view around – it was Su Shu Huang's work on ϵ Aurigae and also on β Lyrae. Here was pure distilled cleverness and insight. Huang looked at problems that had been examined exhaustively by several of the most celebrated astrophysicists and saw things that had eluded everyone. Suddenly binaries were locales where mystery could transform into understanding if one looked in the right way. But where does one learn to do this sort of thing, in a course? Not at most schools. Can one learn from a book? Well, there are books on binary stars, but they mainly serve as repositories of formulas, derivations, and diagrams, and some follow the ideas of only one person or "school." A few books give recipes for procedures developed by their authors, usually without providing insight. There has long been a need for a book that takes a wide view of binary star models and their interface with observations, and that is the goal set by Josef Kallrath and Eugene F. Milone, who together have broad experience in binary star models and observations. Their creation has conscientious coverage throughout most of the "models versus observations" field. It can guide interested persons into the overall field and be a helpful companion as they explore new examples, such as in the *initial approach* (what is going on?), a *settling-in stage* (is it a standard situation or are there complications?), *getting up to speed* (developing intuition and extracting maximum information), and finally *evaluation of results*.

Examine the Contents to see a variety of topics not found in the few preceding books in this general area. Here we find extensive treatment of history, terminology, observational methods, accuracy, binary models *from the ground up*, system morphology, a sense of where things are going, perspectives for long-range development, guides to exploration of the literature, and even philosophy. Although not all important categories of binaries are covered, nor are all individual binaries of special interest, the coverage in this first edition is remarkable. Protest marches for inclusion of symbiotics, ultra-compact X-ray binaries, etc., in future editions may well be successful. For now the emphasis is on more general considerations. We see a balance between *hands-on* and *automated* analysis. Extreme *hands-on* advocates typically get things roughly right and can recognize novel features but fail to extract all available information. Extreme advocates of the *automated* school can reach optimal solutions for standard cases but miss anything new (there is more to astrophysics than parameter adjustment). We need to navigate between these extremes.

The names of some luminaries of the binary star field may seem to be underrepresented, for example, those in structure and evolution. Should we not be reading more about Eggleton, Kippenhahn, Lucy, Ostriker, Paczynski, Plavec, Taam, van den Heuvel, Webbink, etc.? But remember that it is a book about direct representation of observations through models and must be kept to a reasonable size, and there are excellent books on structure and evolution.

Although the scope is limited to models, observations, and related mathematics, there is something here for everyone. Thus we learn that the Kolmogorov–Smirnoff test is not, after all, a way to distinguish vodkas. And there are binaries for everyone. Game players will like the one that stays in eclipse 90% of the time and comes out for only 10% (*PK Boo*). Gadget afficionados prefer the remote paging device, *b Per*. We have a thing to play in *TV Cet* and a place to stay in *HO Tel*. And then there is the only star with a question mark in its name (*Y Sex?*). So peruse the book, learn from it, and enjoy close binaries. If you happen to find some MLR data along the way, so much the better.

R. E. Wilson

Preface to the Second Edition

Di, coeptis ... *adspirate meis*　　　　　　　　(*Gods, aid my undertaking*)
Ovid (43 B.C.–A.D. 17), *Metamorphoses* 1,2-4

The second edition arose from the authors' and the publisher's observations that 10 years after the first edition a new edition is needed to cover the impressively long list of new physical features and analyzing methods in eclipsing binary (*EB*) star analysis. Direct distance estimation through *EB* analysis is one of the highlights. Complete derivation of the ephemerides and third-body orbital parameters from light and radial velocity curves is another. Incorporation of interpolation-based approximations to stellar atmospheres has become common practice. Limb-darkening coefficients do not need to be entered explicitly but are locally computed as function of temperature, gravity, wavelength, and chemical composition.

EB research has made great contributions to stellar astrophysics for over a century, e.g., resolution of the Algol paradox, insights into the physics of cataclysmic variables, and improved understanding of W UMa stars. Bright *EB*s can be observed and analyzed for orbital and physical properties to high accuracy with even modest equipment. The advent of larger telescopes and powerful instrumentation also allows analysis and distance estimations of *EB*s in Local Group galaxies even as far as M31 and M33. Large telescopes also allows the observation and study of eclipsing very-low-mass stars, brown dwarfs and planets, and even planets in *EB*s. The detection of extra-solar planets by transit methods is a field not entirely outside the *EB* research where *EB* techniques have been used successfully.

The Kepler mission[1] (cf. Koch et al. (2006)) launched on March 6th, 2009, the GAIA mission to be launched in 2012 or the *Large Synoptic Survey Telescope* (LSST; http://www.lsst.org/), in discussion for after 2015, will add a new challenge to the field: The analysis of large numbers of *EB* light curves from surveys. Finally, enhanced or completely new software is available for *EB* research. PHOEBE (Sect. 8.2) is an example of a platform-independent *EB* software with an attractive graphical user interface.

---

[1] Updated details are at http://www.kepler.arc.nasa.gov.

The proceedings of IAU Symposium No. 240 (2006) [entitled *Binary Stars as Critical Tools & Tests in Contemporary Astrophysics*, edited by Hartkopf et al. (2007)] provide an excellent overview on state-of-the-art and ongoing activities in close binary research. They briefly review major advances in instrumentation and techniques, new observing and reduction methods, and discuss binary stars as critical tools and tests for studying a wide variety of astrophysical problems. *Tools of the Trade and the Products they Produce: Modeling of Eclipsing Binary Observables* edited by Milone et al. (2008) is another source highlighting recent advances. We see strong enhancements both in physics and in *EB* software:

- additional physical features
   1. an alternative method to derive a binary's ephemeris;
   2. effects of third bodies on light curves and radial velocity curves;
   3. *EB*s with intrinsically variable components;
   4. stellar atmosphere approximation functions;
   5. direct distance estimation;
   6. color indices as indicators of individual temperatures;
   7. spectral energy distribution as independent data source; and
   8. main sequence constraints;
- enhanced programs and new software;
- techniques for analyzing large numbers of light curves, and;
- *EB*s in extra-solar planet research.

We largely retain the structure of the first edition. Some sections have been added to the existing chapters, especially the *Eclipsing Binary Guide for Researchers in Other Fields* in Chap. 1. What is now called Chap. 5 hosts most of the new material. Chapters 5, 6, and 7 of the first edition are now Chaps. 6, 7, and 8 of this second edition.

Last but not least: The book is now part of Springer's *Astronomy and Astrophysics Library Series* which both indicates and acknowledges its wider relevance for astronomy and astrophysics.

Gainesville, FL, US Josef Kallrath
Calgary, AB, Canada Eugene F. Milone

Preface to the First Edition

Di, coeptis ... *adspirate meis* (*Gods, aid my undertaking*)
Ovid (43 B.C.–A.D. 17), *Metamorphoses* 1,2-4

This book arose from the realization that light curve modeling has not had a full expository treatment in 40 years. The last major exposition was that of Russell & Merrill (1952), and that treatment dealt exclusively with the Russell–Merrill spherical star model and with the process of light curve rectification. The present work may be the first comprehensive exposition of the merits of the major modern light curve analysis methods, notwithstanding the pioneer efforts of many investigators beginning with Kopal (1950), who again described mainly his own efforts. The need for a sourcebook and didactic presentation on the subject was recognized in the course of planning a conference on light curve modeling which was held in Buenos Aires and Cordoba in July–August, 1991. The proceedings of the Argentina meetings (Milone 1993) review the current state of light curve modeling methods and focus on special topics but do not give a general review. The only previous meeting devoted exclusively to the topic of comparative light curve modeling methods was IAU Colloquium No. 16 in Philadelphia in September 1971.

Although there is an extensive literature on eclipsing binary research, the paucity of instructional materials in the area of light curve analysis is striking. The graduate student, the researcher, or the advanced amateur astronomer must struggle through journal articles or conference proceedings and must read between the lines to glean the details of the modeling process. Most of the didactic books on light curve modeling date to the presynthetic light curve era. The monograph by Russell & Merrill (1952) is one major example and that by Kopal (1959) is another. Yet the need is acute: As with many other areas of science, the computer revolution has given many astronomers, amateur as well as professional, the tools to attempt light curve solutions. In this work, we provide a suitable background for the new modeler, a useful source book for the experienced modeler, and a springboard for the development of new modeling ideas. For the Wilson–Devinney approach in particular, we elaborate on some of the subtle details that determine the success or failure of light curve computation.

Methods of analyzing eclipsing binary data involve

1. the specification of an astrophysical model;
2. the selection of an algorithm to determine the parameters; and
3. the estimation of the errors of the parameters.

The focus of the book is primarily on models and algorithms rather than on model applications and individual binaries. Nevertheless, models and algorithms are illustrated through investigations of individual stars. The review by Wilson (1994) deals with the intuitive connection between binary star models and light curves and the historical development of the field. Here we concentrate on the mathematical formulation of models and the mathematical background of the algorithms. We present a self-contained, elementary treatment of the subject wherever possible. The book is written for the reader who is familiar with the basic concepts and techniques of calculus and linear algebra. As an aid to the exploration of higher-level material, a brief introduction into the theory of optimization and least squares methods is provided in Appendix IV. Besides presenting the physical and mathematical framework, we have tried to present the material so that it is not far from actual implementation.

The book consists of four major parts: Introductory material (Chaps. 1 and 2); the physical and mathematical core (Chaps. 3 and 4); practical approaches (Chaps. 5 and 6); and the authors' views on the structure of future light curve programs (Chap. 7). The appendices provide further mathematical details on specific topics and applications and point to other sources. The structure of the book should assist readers who are taking their first steps into the field to get a sound overview and also experienced researchers who are seeking a source book of formulas and references.

Chapter 1 gives a nonmathematical overview of the field. In particular, the issues of what can be derived from eclipsing binary stars and why these data are relevant to astrophysics in general are considered. Here we introduce the general concept of equipotential surfaces. Because an eclipsing binary analyst needs understanding of observational data, some background on the database and methods of data acquisition is necessary. Therefore, in Chap. 2 we review observational methods relevant to data analysis. Chapter 3 contains a general approach to light curve modeling. The chief concern is the solution of the direct problem: computing light curves, radial velocity curves, and other observables for a given set of parameters. To this end, a mathematical framework is presented for the relevant astrophysics. The inverse problem of the determination of eclipsing binary parameters is discussed in Chap. 4. As in Chap. 3, we present a formal approach that may serve as a platform for further developments in data analysis. Besides attending to critical issues in light curve analysis, the formulation allows various methods to be related and discussed from a common point of view. Chapter 5 gives characteristics of existing light curve programs and coding details for some of them. Here, the purpose is to provide the new light curve analyst with an overview to explore concepts relevant to the field, including the astrophysics of particular programs and how they work. Because the Wilson–Devinney program is the most frequently used light curve analysis tool, Chap. 6 discusses special eclipsing binary cases analyzed with the Wilson–Devinney program and related programs. In order not to overburden Chaps. 5 and 6, most of the practical details of the Wilson–Devinney program are collected in Appendix D.1. Chapter 7 summarizes ideas and strategies for building light curve programs and previews the coming decade. It is intended as a stimulant to further eclipsing binary research and light curve program development.

Some topics that are covered only briefly will have more extensive treatment in some later edition. A first group covers further model extensions requiring time rather than phase as input quantity, e.g., variable periods, apsidal motion, eclipsing binaries with intrinsically variable components, spot migration, circumstellar flows. Other extensions will focus on particular types of binaries: cataclysmic variables, symbiotic stars and other red-giant binaries, W Serpentis stars, high, low, and ultralow mass X-ray binaries, binaries with atmospheric eclipses, and individual strange objects such as ε *Aurigae*, β *Lyrae*, and *Cygnus X-1*.

Gainesville, FL, US
Calgary, AB, Canada

Josef Kallrath
Eugene F. Milone

References

Hartkopf, W. I., Guinan, E. F., & Harmanec, P. (eds.): 2007, *Binary Stars as Critical Tools and Tests in Contemporary Astrophysics*, No. 240 in Proceedings IAU Symposium, Dordrecht, Holland, Kluwer Academic Publishers

Koch, D., Borucki, W., Basri, G., Brown, T., Caldwell, D., Christensen-Dalsgaard, J., Cochran, W., Dunham, E., Gautier, T. N., Geary, J., Gilliland, R., Jenkins, J., Kondo, Y., Latham, D., Lissauer, J., & Monet, D.: 2006, The Kepler Mission: Astrophysics and Eclipsing Binaries, *Ap. Sp. Sci.* **304**, 391–395

Kopal, Z.: 1950, The Computation of Elements of Eclipsing Binary Stars, *Harvard Observatory Monograph* **8**, 1–181

Kopal, Z.: 1959, *Close Binary Systems*, Chapman & Hall, London

Milone, E. F. (ed.): 1993, *Light Curve Modeling of Eclipsing Binary Stars*, Springer, New York

Milone, E. F., Leahy, D. A., & Hobill, D. W.: 2008, *Short-Period Binary Stars: Observations, Analyses, and Results*, Vol. 352 of *Astrophysics and Space Science Library*, Springer, Dordrecht, The Netherlands

Russell, H. N. & Merrill, J. E.: 1952, The Determination of the Elements of Eclipsing Binary Stars, *Princeton. Obs. Contr.* **26**, 1–96

Wilson, R. E.: 1994, Binary-Star Light-Curve Models, *PASP* **106**, 921–941

Acknowledgments

Robert E. Wilson[1] provided unstinting help in the clarification and exposition of his model and program and in carefully reviewing drafts of the present volume. Dirk Terrell[1] provided a great deal of help in explicating and coding details of the light curve modeling improvements. Andrew T. Young[2] provided valuable discussions concerning photometry. It is a pleasure to thank A. M. Cherepashchuk[3] for translating an important segment of his work into English for inclusion in the chapter on what we call "The Russian School" of light curve modeling.

Albert P. Linnell,[4] Stephen Mochnacki,[5] and Petr Hadrava[6] provided detailed material on their light curve programs. Dirk Terrell provided a description of his Windows PC interface to the Wilson–Devinney program. David H. Bradstreet[7] made `Binary Maker` 2.0 and 3.0 available to us; his program helped us to produce many of the pictures in this book.

We are grateful to a number of colleagues who read the manuscript critically and made many valuable suggestions, especially Horst Drechsel,[8] Paul Etzel,[9] Hilmar Duerbeck,[10] Horst Fichtner,[11] Jason McVean,[12] Petr Hadrava, Edward Olson,[13]

---

[1] Department of Astronomy, University of Florida, Gainesville, Florida, USA.

[2] Astronomy Department, San Diego State University, San Diego, California, USA.

[3] Sternberg Astronomical Institute, Moscow, Russia.

[4] Department of Physics and Astronomy, Michigan State University, East Lansing, Michigan, USA.

[5] Department of Astronomy, University of Toronto, Toronto, Ontario, Canada.

[6] Astronomical Institute, Academy of Sciences of the Czech Republic, Ondřejov, Czech Republic.

[7] Department of Physical Sciences, Eastern College, St. Davids, Pennsylvannia, USA.

[8] Universitätssternwarte Bamberg, Bamberg, Germany.

[9] Astronomy Department, San Diego State University, San Diego, California, USA.

[10] WE/OBSS, Brussels Free University (VUB), Brussels, Belgium.

[11] Fakultät für Physik und Astronomie, Ruhr-Universität Bochum, Bochum, Germany.

[12] Department of Physics and Astronomy, University of Calgary, Calgary, AB.

[13] Department of Astronomy, University of Illinois, Urbana, USA.

Johannes P. Schlöder,[14] Klaus Strassmeier,[15] Richard Walker,[16] and Robert E. Wilson. Parts of the manuscript were also read by Bernd-Christoph Kämper[17], Hilde Domgörgen,[18] and Olaf Evers.[24] Bernd-Christoph Kämper, Horst Fichtner, and Eric Leblanc,[19] have been very helpful in checking out references.

We are grateful also to a number of colleagues who provided illustrations: Fred Babott,[20] Horst Drechsel, Hilmar W. Duerbeck, Jason MvVean, Jaydeep Mukherjee,[21] Colin D. Scarfe,[22] Dirk Terrell, and Robert E. Wilson. Finally, we thank Michael Alperowitz[23] and Robert C. Schmiel[24] for careful checking of the ancient Greek and Latin proverbs and historical quotations and to Christine Bohlender[18] and Norbert Vormbrock[18] who helped in the production of postscript graphics.

The second edition benefited again strongly from Robert E. Wilson's encouraging support, extensive discussions, and personal notes on his light curve program. The time spent with Bob and the hospitality in his house was also a personal enjoyment. Walter Van Hamme[25] helped to clarify various points on the stellar atmosphere option added to the Wilson–Devinney program. Andrej Prša[26] provided additional explanations on his software package PHOEBE and created Fig. 8.4 for this book. Hilmar Duerbeck,[27] Roland Idaczyk[28], and Steffen Rebennack[29] read the manuscript carefully, identified subtle problems, and made valuable suggestions. Steve Howell[30] reviewed the photometric material in Chaps. 1 and 2 and made numerous helpful suggestions. Petr Hadrava read parts of the manuscript and corrected minor problems. Steffen Rebennack supported the LaTeX-embedded production of the figures. Springer's editing team was very constructive and our main editor, Harry Blom, patiently allowed this book to become mature and reach its current state. Finally, Alex Jack helped with proofreading.

---

[14] Interdisziplinäres Zentrum für wissenschaftliches Rechnen (IWR) der Universität Heidelberg, Heidelberg, Germany.

[15] Universitätssternwarte Wien, Vienna, Austria.

[16] US Naval Observatory, Flagstaff Station, Flagstaff, Arizona, USA.

[17] Universitätsbibliothek Stuttgart, Stuttgart, Germany.

[18] BASF SE, Ludwigshafen, Germany.

[19] DAO, Herzberg Institute of Astrophysics, Victoria, BC

[20] Rothney Astronomical Observatory, University of Calgary, Calgary, AB.

[21] Department of Astronomy, University of Florida, Gainesville, Florida, USA.

[22] Department of Physics and Astronomy, University of Victoria, Victoria, British Columbia, Canada.

[23] Seminar für klassische Philologie der Universität Heidelberg, Germany.

[24] Department of Greek and Latin, University of Calgary, Calgary, AB.

[25] Department of Physics, Florida International University, Miami, FL 33199, USA.

[26] Villanova University, Dept. Astron. Astrophys., 800 E Lancaster Ave, Villanova, PA 19085, USA.

[27] WE/OBSS, Brussels Free University (VUB), Brussels, Belgium.

[28] Ngaio, Wellington 6035, New Zealand.

[29] Center for Applied Optimization, University of Florida, Gainesville, FL.

[30] WIYN, NOAO, Tucson, AZ.

Contents

Foreword vii

Preface to the Second Edition ix

Preface to the First Edition xi
    References xiii

Acknowledgments xv

List of Figures xxvii

Journal Abbreviations xxxi

Acronyms and Abbreviations xxxiii

Mathematical Nomenclature and Symbols, Physical Units xxxv

Part I Introduction

1 Introduction 3
  1.1 Eclipsing Binaries and Other Variable Stars 3
      1.1.1 Eclipsing Variables 5
           1.1.1.1 Algols 5
           1.1.1.2 β Lyrae 6
           1.1.1.3 W Ursae Majoris or W UMa 6
      1.1.2 Pulsating Variables 7
      1.1.3 Eruptive Variables 8
  1.2 Overview of the Problem 10
      1.2.1 Why Binary Stars Are Important 10
           1.2.1.1 Visual Double Stars 11
           1.2.1.2 Spectroscopic Binaries 13
           1.2.1.3 Eclipsing Binaries 14

		1.2.2	Phenomenological Classification of Eclipsing Binary Light Curves	15
		1.2.3	Morphological Classification of Eclipsing Binaries	18
		1.2.4	What Can Be Derived from Eclipsing Binaries	22
		1.2.5	Why Data Derived from Eclipsing Binaries Are Important	22
	1.3	The History of Light Curve Modeling		23
		1.3.1	The Pioneers – The Age of Geometry	23
		1.3.2	The Age of Computational Astrophysics	24
		1.3.3	Determining Astrophysical Parameters	25
		1.3.4	Later Generations of Light Curve Models	25
		1.3.5	Astrophysical Problems Solved by Light Curve Methods	26
	1.4	EB Guide for Researchers in Other Fields		27
		1.4.1	Eclipsing Binaries and Standard Candles	27
		1.4.2	Eclipsing Binaries in ExtraSolar Planet Research	28
		1.4.3	Nomenclature: Primary and Secondary Component	29
		1.4.4	Where Are the Radii?	29
		1.4.5	Precession and Apsidal Motion	29
		1.4.6	Looking for Eclipsing Binary Standard Software	30
		1.4.7	Analytic Techniques and Numerical Analysis	30
	1.5	Selected Bibliography		30
	References			31

2 The Database and Methods of Data Acquisition 37
   2.1 Photometry 37
      2.1.1 Photoelectric Photometry 37
      2.1.2 Two-Star Photometers 39
      2.1.3 Photoelectric Observations 41
      2.1.4 Imaging Data 43
      2.1.5 Photometric Data Reduction 44
      2.1.6 Significance of Cluster Photometry 47
   2.2 Spectroscopy 48
      2.2.1 Radial Velocities 51
      2.2.2 Spectrophotometry 53
      2.2.3 Line-Profile Analysis 55
   2.3 Polarimetry 57
   2.4 Magnetometry 59
   2.5 Doppler Profile Mapping 60
   2.6 Advice to Observers 60
   2.7 Eclipsing Binary Data from Surveys 62
   2.8 Terminology: "Primary Minimum" and "Primary Star" 65
   2.9 Selected Bibliography 66
   References 67

Part II Modeling and Analysis

3 A General Approach to Modeling Eclipsing Binaries 75
- 3.1 System Geometry and Dynamics 77
  - 3.1.1 Coordinates and Basic Geometrical Quantities 77
  - 3.1.2 Dynamics and Orbits 82
    - 3.1.2.1 Circular Orbits 85
    - 3.1.2.2 Eccentric Orbits 85
  - 3.1.3 Spherical Models 89
  - 3.1.4 Ellipsoidal Models.................................... 92
  - 3.1.5 Roche Geometry and Equipotential Surfaces 95
    - 3.1.5.1 Circular Orbits and Synchronous Rotation 96
    - 3.1.5.2 Circular Orbits and Asynchronous Rotation 98
    - 3.1.5.3 Eccentric Orbits and Asynchronous Rotation 101
    - 3.1.5.4 Approaches Including Radiation Pressure 103
  - 3.1.6 Binary Star Morphology 109
- 3.2 Modeling Stellar Radiative Properties 114
  - 3.2.1 Gravity Brightening................................... 115
  - 3.2.2 Stellar Atmosphere Models 119
  - 3.2.3 Analytic Approximations for Computing Intensities 119
  - 3.2.4 Center-to-Limb Variation 120
  - 3.2.5 Reflection Effect 123
  - 3.2.6 Integrated Monochromatic Flux 128
- 3.3 Modeling Aspect and Eclipses 128
- 3.4 Sources and Treatment of Perturbations 131
  - 3.4.1 Third Light 131
  - 3.4.2 Star Spots and Other Phenomena of Active Regions 132
  - 3.4.3 Atmospheric Eclipses 136
  - 3.4.4 Circumstellar Matter in Binaries 137
    - 3.4.4.1 Gas Streams 139
    - 3.4.4.2 Gas Stream in the *VV Orionis System* 140
    - 3.4.4.3 Disks and Rings 142
    - 3.4.4.4 Stellar Winds................................ 144
    - 3.4.4.5 Attenuating Clouds 144
- 3.5 Modeling Radial Velocity Curves 145
- 3.6 Modeling Line Profiles...................................... 150
- 3.7 Modeling Polarization Curves 152
- 3.8 Modeling Pulse Arrival Times 155
- 3.9 Self-Consistent Treatment of Parallaxes........................ 156
- 3.10 Chromospheric and Coronal Modeling......................... 157
- 3.11 Spectral Energy Distribution 158
- 3.12 Interstellar Extinction 159
- 3.13 Selected Bibliography 160
- References... 161

Contents

4 Determination of Eclipsing Binary Parameters 169
    4.1 Mathematical Formulation of the Inverse Problem 169
        4.1.1 The Inverse Problem from the Astronomer's Perspective 173
            4.1.1.1 The Input Database 173
            4.1.1.2 General Problems of Nonlinear Parameter Fitting .. 174
            4.1.1.3 Special Problems of Nonlinear Parameter Fitting in Light Curve Analysis...................... 175
            4.1.1.4 On the Use of Constraints 178
            4.1.1.5 Assignment of Weights 179
            4.1.1.6 Simultaneous Fitting 182
    4.2 A Brief Review of Nonlinear Least-Squares Problems............. 183
        4.2.1 Nonlinear Unconstrained Least-Squares Methods 184
        4.2.2 Nonlinear Constrained Least-Squares Methods 185
    4.3 Least-Squares Techniques Used in Eclipsing Binary Data Analysis .. 186
        4.3.1 A Classical Approach: Differential Corrections 187
        4.3.2 Multiple Subset Method and Interactive Branching 190
        4.3.3 Damped Differential Corrections and Levenberg–Marquard Algorithm 190
        4.3.4 Derivative-Free Methods............................... 191
            4.3.4.1 The Simplex Algorithm 192
            4.3.4.2 Powell's Direction Method 196
            4.3.4.3 Simulated Annealing 197
        4.3.5 Other Approaches 198
    4.4 A Priori and A Posteriori Steps in Light Curve Analysis 198
        4.4.1 Estimating Initial Parameters 198
        4.4.2 Criteria for Terminating Iterations 202
        4.4.3 The Interpretation of Errors Derived from Fitting 204
        4.4.4 Calculating Absolute Stellar Parameters from a Light Curve Solution 205
            4.4.4.1 The Complete Data Case...................... 206
            4.4.4.2 The Incomplete Data Case 209
    4.5 Suggestions for Improving Performance 210
        4.5.1 Utilizing Symmetry Properties.......................... 211
        4.5.2 Interpolation Techniques............................... 211
        4.5.3 Surface Grid Design 213
        4.5.4 Analytic Partial Derivatives 214
        4.5.5 Accurate Finite Difference Approximation................ 216
    4.6 Selected Bibliography 216
    References... 217

5 Advanced Topics and Techniques 221
    5.1 Extended Sets of Observables and Parameters.................... 221
        5.1.1 Inclusion of Absolute Parameters in Light Curve Analysis... 222
        5.1.2 Determining Individual Temperatures 224

Contents   xxi

		5.1.2.1	Temperature Estimations 224
		5.1.2.2	Color Indices as Individual Temperature Indicators . 226
		5.1.2.3	Both Temperatures from Absolute Light Curves ... 228
	5.1.3	Traditional Distance Estimation 230
	5.1.4	Direct Distance Estimation 231
	5.1.5	Main-Sequence Constraints 233
	5.1.6	Intrinsic Variability of Eclipsing Binaries' Components 234
5.2	Multiple Star Systems and their Dynamics 235
	5.2.1	Third-Body Effects on Light and Radial Velocity Curves 235
	5.2.2	Ephemerides Derived from Whole Light Curves and Radial Velocity Curves... 238
5.3	Analyzing Large Numbers of Light Curves 241
	5.3.1	Techniques for Analyzing Large Numbers of Light Curves .. 241
	5.3.2	The Matching Approach 242
		5.3.2.1	Solving Linear Regression Problems 243
		5.3.2.2	Generation and Storage of the Archive Curves 243
	5.3.3	The Expert Rule Approach 244
	5.3.4	Simplified Physical Models 244
	5.3.5	Artificial Neural Networks 245
5.4	Extrasolar Planets 245
	5.4.1	General Comments About Substellar Objects.............. 247
	5.4.2	Methods to Find "Small"-Mass Companions 247
		5.4.2.1	Astrometry Variations 248
		5.4.2.2	Direct Imaging and Spectroscopy................ 248
		5.4.2.3	Radial Velocity Variations of the Visible Component 249
		5.4.2.4	Gravitational Lensing......................... 249
		5.4.2.5	Transit Eclipses 250
		5.4.2.6	Indirect Effects: O–C Variation.................. 251
		5.4.2.7	Effects on Disks 251
	5.4.3	Star–Planet Systems and Eclipsing Binary Models 251
		5.4.3.1	Comparing Stars, Brown Dwarfs, and Planets 251
		5.4.3.2	Transit Geometry and Modeling Approaches 252
		5.4.3.3	Representing Planets in the WD Model 254
		5.4.3.4	HD 209458b: Transit Analysis of an ExtraSolar Planet.......................................255
		5.4.3.5	The OGLE-TR-56 Star Planet System 257
5.5	Selected Bibliography 258
References.. 258

Part III Light Curve Programs and Software Packages

6 Light Curve Models and Software 265
    6.1 Distinction Between Models and Programs 265
    6.2 Synthetic Light Curve Models................................ 266

		6.2.1	The Russell–Merrill Model and Technique 266
		6.2.2	The "Eclipsing Binary Orbit Program" EBOP 273
		6.2.3	The Wood Model and the WINK program 277
	6.3	Physical Models: Roche Geometry Based Programs 277
		6.3.1	Binnendijk's Model 278
		6.3.2	Hadrava's Program FOTEL 278
		6.3.3	Hill's Model 279
		6.3.4	Linnell's Model 279
		6.3.5	Rucinski's Model 282
		6.3.6	Wilson–Devinney Models 282
			6.3.6.1 The 1998 Wilson–Devinney Model 282
			6.3.6.2 New Features in the 1999–2007 Models 287
	6.4	Cherepashchuk's Model 289
	6.5	Other Approaches 295
		6.5.1	Budding's Eclipsing Binary Model 295
		6.5.2	Kopal's Frequency Domain Method 295
		6.5.3	Mochnacki's General Synthesis Code, GENSYN 297
		6.5.4	Collier–Mochnacki–Hendry GDDSYN Spotted General Synthesis Code 297
		6.5.5	Other Spot Analysis Methods 298
	6.6	Selected Bibliography 298
	References 299

7	The Wilson–Devinney Program: Extensions and Applications 305
	7.1 Current Capabilities of WDx2007 306
	7.2 Atmospheric Options 307
	7.2.1 Kurucz Atmospheres in WDx2007 307
	7.2.2 Kurucz Atmospheres in WD 309
	7.3 Applications and Extensions 310
	7.3.1 The Eclipsing X-Ray Binary *HD 77581/Vela X-1* 311
	7.3.2 The Eclipsing Binaries in *NGC 5466* 312
	7.3.3 The Binary *H235* in the Open Cluster *NGC 752* 315
	7.3.4 The Field Binary *V728 Herculis* 317
	7.3.5 The Eclipsing Binaries in *M71* 317
	7.3.6 The Eclipsing Binaries in 47 Tuc 319
	7.3.7 The Well-Studied System *AI Phoenicis* 320
	7.3.8 HP Draconis 321
	7.3.9 Fitting of Line Profiles 324
	7.4 The Future 324
	7.4.1 "The Future" as Envisioned in 1999 324
	7.4.2 The Future (as Seen in 2009) 326
	References 327

8 Light Curve Software with Graphical User Interface and Visualization ... 331
- 8.1 Binary Maker ... 331
- 8.2 PHOEBE ... 335
- 8.3 NIGHTFALL ... 337
- 8.4 Graphics Packages ... 337
- References ... 339

9 The Structure of Light Curve Programs and the Outlook for the Future ... 341
- 9.1 Structure of a General Light Curve Analysis Program ... 341
  - 9.1.1 Framework of the Light Curve Models ... 342
  - 9.1.2 Framework to Embed Least-Squares Solvers ... 343
- 9.2 Procedural Philosophies ... 344
- 9.3 Code Maintenance and Modification ... 345
- 9.4 Prospects and Expectations ... 346
- References ... 348

Part IV Appendix

A Brief Review of Mathematical Optimization ... 351
- A.1 Unconstrained Optimization ... 351
- A.2 Constrained Optimization ... 356
  - A.2.1 Foundations and Some Theorems ... 356
  - A.2.2 Sequential Quadratic Algorithms ... 359
- A.3 Unconstrained Least-Squares Procedures ... 360
  - A.3.1 Linear Case: Normal Equations ... 361
  - A.3.2 The Linear Case: An Orthogonalization Method ... 362
  - A.3.3 Nonlinear Case: A Gauß–Newton Method ... 364
- A.4 Constrained Least-Squares Procedures ... 367
- A.5 Selected Bibliography ... 368

B Estimation of Fitted Parameter Errors: The Details ... 369
- B.1 The Kolmogorov–Smirnov Test ... 369
- B.2 Sensitivity Analysis and the Use of Simulated Light Curves ... 370
- B.3 Deriving Bounds for Parameters: The Grid Approach ... 371

C Geometry and Coordinate Systems ... 373
- C.1 Rotation of Coordinate Systems ... 373
- C.2 Volume and Surface Elements in Spherical Coordinates ... 374
- C.3 Roche Coordinates ... 378
- C.4 Solving Kepler's Equation ... 379

D	**The Russell–Merrill Model**	381
	D.1 Ellipticity Correction in the Russell–Merrill Model	381
E	**Subroutines of the Wilson–Devinney Program**	385
	E.1 ATM – Interfacing Stellar Model ATMospheres	386
	E.2 ATMx – Interfacing Stellar Model ATMospheres	386
	E.3 BBL – Basic BLock	386
	E.4 BinNum – A Search and Binning Utility	387
	E.5 BOLO – Bolometric Corrections	387
	E.6 CofPrep – Limb-Darkening Coefficient Preparation	387
	E.7 CLOUD – Atmospheric Eclipse Parameters	387
	E.8 CONJPH – Conjunction Phases	388
	E.9 DGMPRD – Matrix–Vector Multiplication	388
	E.10 DMINV – Matrix Inversion	388
	E.11 DURA – Constraint on X-Ray Eclipse Duration	388
	E.12 ELLONE – Lagrangian Points and Critical Potentials	388
	E.13 FOUR – Representing Eclipse Horizon	390
	E.14 FOURLS – Representing Eclipse Horizon	390
	E.15 GABS – Polar Gravity Acceleration	391
	E.16 JDPH – Conversion of Julian Day Number and Phase	391
	E.17 KEPLER – Solving the Kepler Equation	391
	E.18 LC and DC – The Main Programs	391
	E.19 LCR – Aspect Independent Surface Computations	392
	E.20 LEGENDRE – Legendre Polynomials	392
	E.21 LIGHT – Projections, Horizon, and Eclipse Effects	393
	E.22 LimbDark – Limb Darkening	394
	E.23 LinPro – Line Profiles	394
	E.24 LUM – Scaling of Polar Normal Intensity	394
	E.25 LUMP – Modeling Multiple Reflection	394
	E.26 MLRG – Computing Absolute Dimensions	396
	E.27 MODLOG – Handling Constraints Efficiently	396
	E.28 NEKMIN – Connecting Surface of Over-Contact Binaries	396
	E.29 OLUMP – Modeling the Reflection Effect.......................	396
	E.30 OMEGA* – Computing $\Omega(r)$	404
	E.31 PLANCKINT – Planck Intensity	405
	E.32 READLC* – Reading Program Control Parameters	405
	E.33 RING – The Interface Ring of an Over-Contact Binary	405
	E.34 RanGau – Generation of Gaussian Random Numbers	405
	E.35 RanUni – Generation of Uniform Random Numbers	405
	E.36 ROMQ – Distance Computation of Surface Points	405
	E.37 ROMQSP – Distance Computation of Surface Points	406
	E.38 SIMPLEX* – Simplex Algorithm	406
	E.39 SinCos – Surface Grid Sine and Cosines	406
	E.40 SQUARE – Building and Solving the Normal Equations..........	406

E.41 SPOT – Modeling Spots...406
E.42 SSR* – Computation of Curves and Residuals 407
E.43 SURFAS – Generating the Surfaces of the Components 407
E.44 VOLUME – Keeping Stellar Volume Constant 407
References...407

F Glossary of Symbols 411

Index 417

List of Figures

1.1	Classes of light curves	16
1.2	U, V, and infrared light curves of *Algol*	18
1.3	Projections of equipotential Roche surfaces	19
1.4	Roche potential and shape of a detached binary system	20
1.5	Roche potential and shape of a semi-detached binary	21
1.6	Roche potential and shape of an over-contact binary	21
2.1	RADS instrument on the RAD's 41-cm telescope	40
2.2	RADS differential photometry	40
2.3	CCD image frame of the globular cluster *NGC 5466*	48
2.4	Spectra of early spectral-type eclipsing binaries	49
2.5	Radial velocity curve of *AI Phoenicis*	51
2.6	Cross-correlation functions for *RW Comae Berenices*	52
2.7	An R_J light curve taken by the Baker–Nunn patrol camera	63
3.1	Definition of a right-handed Cartesian coordinate system	78
3.2	Definition of spherical polar coordinates	78
3.3	Surface normal and line-of-sight	79
3.4	Plane-of-sky coordinates I	80
3.5	Plane-of-sky coordinates II	81
3.6	Orbital elements of a binary system	83
3.7	Relationships between phase and orbital quantities	84
3.8	The orbit of *HR 6469* with $e = 0.672$	86
3.9	True anomaly and eccentric anomaly	87
3.10	Schematic light curve in the spherical model	90
3.11	Angular momentum vectors of orbital and stellar rotation	99
3.12	Nonsynchronous rotation	100
3.13	*Inner* radiation pressure effects	105
3.14	Effects of full radiation pressure	108
3.15	Lagrangian points L_1^p and L_2^p in the *BF Aurigae* system	109
3.16	The double-contact binary *RZ Scuti*	113
3.17	Center-to-limb variation	121
3.18	Light variation caused purely by the reflection effect	124
3.19	Light curves with albedo varying from 0 to 1	127
3.20	Geometrical condition for visible points	129

3.21	Projected plane-of-sky distance	130
3.22	Sun spots	133
3.23	Trajectories in a binary system	141
3.24	Gas streams in *VV Orionis*	142
3.25	Disk formation in *SX Cassiopeiae*	143
3.26	Modeling of the Rossiter effect in *AB Andromedae*	147
3.27	Radial velocity curves for point masses	148
3.28	Variation in the intrinsic line profile	151
3.29	Line profiles and Rossiter effect	151
3.30	Polarization curves of *SX Cassiopeiae*	154
3.31	He II spectroheliogravarPhic image of the Sun	158
4.1	Sum of the squared residuals versus parameters	175
4.2	Sum of the squared residuals versus parameters	176
4.3	Contour plot of $\sum wr^2$ versus T and i	177
4.4	Contour plot of $\sum wr^2$ versus q and Ω	178
4.5	Noise contributions in various regimes of star brightness	180
4.6	The geometry of the Simplex algorithm	193
4.7	Flow chart of the Simplex algorithm	195
4.8	Geometry of contact times	200
5.1	Planet transiting in front of a star	253
5.2	The star has radius R	253
5.3	The HST transit light curve and best-fit model	255
5.4	The radial velocity curve and best fit model	256
6.1	The relation among δ/r_g, $k = r_s/r_g$, and α	268
6.2	Geometry of the secondary eclipse of *RT Lacertae*	272
6.3	The infrared light and color curves of *RT Lacertae*	273
6.4	Basic eclipse geometry	290
6.5	More complex models: The X-Ray system *SS433*	294
7.1	*HD 77581* light curve and radial velocities, Vela X-1 pulse delays	312
7.2	Light curves of *NH31* in the globular cluster *NGC 5466*	313
7.3	Light curve of the over-contact system *H235*	316
7.4	Radial velocities of the over-contact system *H235*	316
7.5	Cluster magnitude diagram of *M71* with isochrones	318
7.6	B primary minimum of the *HP Draconis* light curve	322
7.7	B secondary minimum of the *HP Draconis* light curve	323
7.8	Fitting line profiles	323
8.1	The limit of an eclipse of an over-contact system. Created with the help of BM2 (Bradstreet 1993)	332
8.2	Shallow eclipses. Created with the help of BM2 (Bradstreet 1993)	333
8.3	Deep eclipses. Created with the help of BM2 (Bradstreet 1993)	334
8.4	A screen-shot of the PHOEBE graphical user interface	336
8.5	High-angle, three-dimensional analysis of the residuals	338
8.6	Low-angle, three-dimensional analysis of the residuals	338
B.1	Standard deviation of the fit versus mass ratio	371
C.1	Rotation of Cartesian coordinate systems	374

List of Figures

C.2	Derivation of the surface element	378
E.1	Structure of subroutine BBL	387
E.2	The WD main program: LC	391
E.3	The WD main program: DC	392
E.4	Subroutine LCR	393
E.5	Geometry of the reflection effect 1	397
E.6	Geometry of the reflection effect 2	398
E.7	Geometry of the reflection effect 3	399
E.8	Geometry of the reflection effect 4	400
E.9	Geometry of the reflection effect 5	402
E.10	Geometry of the reflection effect 6	403

Journal Abbreviations

ACM Trans. Math. Software	ACM Transactions on Mathematical Software
Acta Astron.	Acta Astronomica
AJ	The Astronomical Journal
Ann. Rev. Astron. Astrophys.	Annual Review of Astronomy and Astrophysics
ApJ	The Astrophysical Journal
ApJ Suppl	The Astrophysical Journal Supplement Series
Ap. Sp. Sci.	Astrophysics and Space Science
A&A	Astronomy and Astrophysics
A&A Suppl.	Astronomy and Astrophysics Supplement Series
Astron. Nachr.	Astronomische Nachrichten
Astron. Rep.	Astronomical Report
BAAS	Bulletin of the American Astronomical Society
Bull. Math. Biol.	Bulletin of Mathematical Biology
CMDA	Celestial Mechanics and Dynamical Astronomy
Comp. J.	The Computer Journal
J. Comp. Phys.	Journal of Computational Physics
Mem. R. Astron. Soc.	Memoirs of the Royal Astronomical Society
MNASSA	Monthly Notices of the Royal Astronomical Society of Southern Africa
MNRAS	Monthly Notices of the Royal Astronomical Society
Quart. Appl. Math.	Quarterly of Applied Mathematics
Observatory	The Observatory
PASP	Publications of the Astronomical Society of the Pacific

Sov. Astron.	*Soviet Astronomy*
Space Sci. Rev.	*Space Science Reviews*
Vistas	*Vistas in Astronomy*
Zeitschr. f. Astrophys.	*Zeitschrift für Astrophysik*

Acronyms and Abbreviations

ως συνελώς ειπεῖν   (*to say it short*)

APT	automatic photometric telescope
CCD	charge coupled device
CCF	cross-correlation function
CDM	conjugate direction method
CLV	center-to-limb variation
CMD	color-magnitude diagram
DLS	damped least squares
EB	eclipsing binary
HST	Hubble Space Telescope
IAU	International Astronomical Union
IPMs	interior point methods
JDN	Julian Day Number
MACHO	massive compact halo object
MSC	main sequence constraint
NDE	Nelson–Davis–Etzel; usually used to refer to the NDE model
NLP	nonlinear programming
OGLE	optical gravitational lensing experiment
PMT	photomultiplier tube
RADS	Rapid Alternate Detection System
RV	radial velocity
SB1	single-lined spectroscopic binary
SB2	double-lined spectroscopic binary
SED	spectral energy distribution
SI	physical units according to the Système International
SQP	sequential quadratic programming
WD	Wilson–Devinney; used to refer to the Wilson–Devinney model or program
w.r.t.	with respect to
ZAMS	zero-age Main Sequence

Mathematical Nomenclature and Symbols, Physical Units

"What's in a name?"   Shakespeare: *Romeo and Juliet*, ii, 2

A few general rules are observed: vectors are marked as bold characters, *e.g.*, **x**, **n**, or **r**. The product **a** · **b** of two vectors **a**, **b** $\in \mathbb{R}^n$ is always understood as the scalar product $\mathbf{a}^T \mathbf{b} = \sum_{i=1}^{n} a_i b_i$. Matrices are indicated with sans serif font, *e.g.*, A. The list below gives our mathematical symbols and operators.

$\mathbb{R}^n$	the n-dimensional vector space of real (column) vectors with n components
∇	gradient operator $\nabla := \nabla_x = \frac{\partial}{\partial \mathbf{x}} = \left(\frac{\partial}{\partial x_1}, \ldots, \frac{\partial}{\partial x_n}\right)^T$ applied to a scalar-valued function f
$:=$	defines the quantity on the left side of an equation by the term on the right side of the equation
$\equiv$	quantity on the left side of the equation is set identically to the term on the right side of the equation
$\doteq$	indicates approximation, as in a first-order Taylor series expansion
$\mathbf{x}^T$	the transposed vector, $\mathbf{x}^T := (x_1, \ldots, x_n)$ is a row vector
$\mathbf{e}_i$	unit vector associated with the ith coordinate axis
$\mathbb{1}$	identity matrix of appropriate dimension
$\mathcal{M}(m, n)$	set of matrices with m rows and n columns
$\forall$	it represents "for all"

Superscripts indicate attributes of a quantity, e.g., L^{bol}, a bolometric luminosity. Subscripts are used for indexing and counting. The subscript j is always used to refer to one of the binary components. To avoid confusion with the symbols R and L, the radii and luminosities of the binary components are written in calligraphic style $\mathcal{R}$ and $\mathcal{L}$ if they are in absolute (or solar) units. Symbols used in this book are listed in Appendix F.

Throughout this book we mostly use SI units. However, when referring to original papers or figures, CGS or even special units such as Å cannot be avoided. Where possible we give only generic physical dimensions of quantities, such as mass, length, time, energy.

Part I
Introduction

Chapter 1
Introduction

In this chapter we first identify eclipsing binaries (*EBs*) as a class of variable stars, which we discuss generally. We then sketch the importance of *EBs* for the determination of fundamental stellar data. We discuss the conditions under which *EB* light curve data enable astronomers to derive stellar masses and other parameters. Finally, we provide the foundation for understanding *EB* models based on equipotential surfaces.

Although originally classified phenomenologically (namely, from light curve appearance), *EBs* are now understood on the basis of much firmer physics. The improved understanding has led to a morphological basis for classification.

1.1 Eclipsing Binaries and Other Variable Stars

Οπου ακούς πολλά κεράσια, βάστα μικρό καλάτι
(*Don't get overwhelmed, and be cautious*)

Variable stars are stars that vary in apparent brightness with time. In fact, *all* stars are variable at some level of precision, over some timescale. In astronomy there are three basic timescales:

- *dynamic* (the time it would take for a star to collapse under gravity if radiation and particle pressure were removed), typically tens of minutes;
- *thermal* (the time to exhaust its stored thermal energy), typically millions of years (as the energy is depleted by the luminosity); and
- *nuclear* (the time to exhaust its nuclear energy), typically, billions of years.

The relevant timescales for variable stars are between the dynamic and the thermal, but certainly much closer to the dynamic.

In fact the term "variable star" is usually reserved for stars that vary in brightness by some detectable amount over the interval of the observations. We have no prehistoric record of such events, but we certainly have ancient records. See, for example, Kelley & Milone (2005, esp., Chap. 5) for an extensive summary. In the recent past (50 years or so), the observational precision has been of order 0.01 magnitude or more. At present, photometry has, in principle if not usually in practice, improved by an order of magnitude, and at the level of millimagnitudes, most stars will appear variable. For example, Howell et al. (2005) found in a survey of the galactic cluster NGC 2301 that 56% of 4000 stars were variable at an amplitude of 0.002 magnitude or greater, the precision limit of the survey for the brightest 5 magnitudes of the survey. To keep our present exposition within reasonable bounds, for present purposes, for the most part we will stick to the more classical limit to define a "variable star," namely a star with brightness variation of >1% or so and over timescales of millennia or less (down to seconds or less). More specifically, in the wider literature variable stars have been held to be variable if they vary in optical wavelengths (~ 0.35 to $<1.0\,\mu m$) over intervals of decades or less; cf. Hoffmeister et al. (1985).

There are many fine monographs on the subject of variables, both generally and specifically. We will mention the latter under the appropriate group. The most recent general summary of which we are aware is by Sterken (1997). Other general works which are somewhat more dated but still offer interesting insights include Hoffmeister et al. (1984/1985); Petit (1985); Strohmeier (1972); and Glasby (1969).

Like the animal, vegetable, and mineral categories in the "Twenty Questions" parlor game of some decades back, variable stars are classically assigned to one of three main categories:

- "geometric" variables;
- "pulsating" variables; and
- "eruptive" variables.

In another classification scheme, a broader distinction was made between "extrinsic" and "intrinsic" variables, with "geometric" variables considered "extrinsic," and the other two "intrinsic." We shall discuss geometric, pulsating, and eruptive variables in sequence.

A "geometric" variable varies not due to its own physical behavior but because of changing aspect,[1] i.e., the viewable part of a star changes with time. This category includes the *EBs* and also examples where the eclipse is due to a disk or a planet. It can also include pulsars, which vary mainly because of rotation, and spotted stars in the sense that the spots cause light modulation over a rotation period; spots usually do not last for decades, but there are exceptions [e.g., *RW Com*, cf. Milone et al. (1980)]. Finally, a cataclysmic variable (see below) may have a large "hump" (due to a hot spot) in its light curve which may be asymmetric due to eclipse by the companion star. Thus the degree to which geometric effects cause the observed variation will differ with the type of system.

---

[1] Aspect means the appearance of an object as viewed from a given direction.

1.1.1 Eclipsing Variables

Eclipsing variables are periodic (that is, the cycle of variation repeats relatively reliably). This broad grouping was historically divided into three phenomenological classes according to the appearance of the light curves: Algols, β Lyrae systems, and W Ursae Majoris systems. The characteristics of these light curve types are discussed in the following subsections.

1.1.1.1 Algols

The prototype is β Persei, also known as Algol. In visible passbands, the striking characteristics are approximately constant light outside eclipse and minima that fall and rise abruptly and occupy only a small fraction of the full light curve, typically less than $\sim$15% for each minimum. Typically, the longer the period, the shorter the fraction of light curve taken up by eclipse. The periods range from days to weeks or more in length. Usually such light curves suggest little interaction between components. This is often but not generally true because the shapes of light curves in optical bands can be misleading. For example, where the depths of the two minima are very different, the temperatures of the component stars are different,[2] and the hotter, bluer star may dominate the light from the system. The light curve may rise near secondary minimum, indicating a "reflection effect," actually a reprocessing of the hotter stars' radiation as it impinges on the atmosphere of its companion, increasing the cooler star's luminosity in the irradiated area – best seen around the secondary minimum. If it were not for the secondary minimum, which may be shallow or even absent in optical passbands, the reflection effect would peak at the phase of mid-secondary minimum. The effect is especially noticeable if the cooler star is significantly larger. If looked at in infrared passbands, the cooler star will contribute relatively more to the combined light and may resemble a β Lyrae-type light curve (see below). In extreme cases, the redder component is so highly evolved that it may be filling its Roche lobe and sending a stream of material toward its companion. This is, in fact, the case with Algol itself (cf. Chen & Reuning (1966)). Not all systems with Algol-like light curves will be in this state, however; the components are often two similar stars and not so close to each other that they are distorting each other's shapes. When the stars are far apart, their shapes may be approximated by spheres. It suffices to note that the spherical approximation would not be adequate for all cases. We discuss further this class of eclipsing variable under the *EB* designation "EA" in the section below. As outlined in Sect. 3.1.6 binary systems in which components are well within their Roche lobes are called "detached" systems and those in which one component fills its Roche lobe are called "semi-detached."

---

[2] As we will learn later, this conclusion is valid only for circular orbits (actually, true Algols are interacting and are likely to have only circular orbits) and stars of similar size.

1.1.1.2 β Lyrae

The prototype gives its name to the class. The light curve continuously varies across the cycle of variation, and the minima occupy a fairly large proportion of the cycle. The periods are typically days, but when giants or supergiants are involved, the period may be much longer. The important thing is not the cycle length or the scale of the system, but the size of the stars relative to the size of the orbit. The continuous variation of light is partially due to the changing aspects of the stars as they rotate, classically known as the "ellipsoidal variation."

The relative depths of the minima indicate the temperature difference between components; redder passbands tend to show less different depths. The light curves give the (correct) impression that the stars are interacting gravitationally. In fact, the stars are undergoing tidal distortions and their shapes reflect this distortion. Roche geometry is generally used to accurately model these systems' properties. This class of eclipsing variable is further discussed under the *EB* type "EB" in Sect. 1.2.2.

1.1.1.3 W Ursae Majoris or W UMa

The prototype is an eclipsing binary with period less than a day, characteristic of the class. Like β Lyrae stars, the light curve varies continuously, but the depths of the minima are usually similar, but rarely exactly identical. Systems that exhibit these light curves are thought to arise from binaries in physical contact, not through a stream, but through an actual neck of material that bridges the small distance between the inward pointing edges of the components. Such systems are known as "over-contact" or if just barely touching, as "contact" systems. Although some astronomers use the term *contact* to refer to both contact and over-contact systems, here we will use the term exclusively in its narrower meaning. Roche geometry is used generally for the accurate modeling of the components of these systems. There are two subclasses of WUMa systems, about which capable astronomers argue endlessly: A-type and W-type systems. In A-type systems the more massive star is larger and hotter; in W-type systems, the more massive star is larger but cooler than its companion. Although both types of systems may exhibit asymmetries in light curves, the W-type tends to exhibit more of this sort of behavior. There may be a difference in depth of up to 0.1 magnitude. As well, there may be a difference in brightness between the maxima, a phenomenon sometimes referred to as the "O'Connell effect," which is quantitatively defined as

$$\Delta m = m_{II} - m_{I} \qquad (1.1.1)$$

where *m* refers to the magnitude and the subscripts I and II refer to the maxima following the primary and secondary minimum, respectively. This means that Δm is positive if maximum I is brighter (has a smaller magnitude) than maximum II. The O'Connell effect may be found in many close binary systems of several types, but quite often in W-type UMa systems; cf. Davidge & Milone (1984) for a discussion of contributive causes. This class of eclipsing variables is further discussed under *EB* type "EW" in Sect. 1.2.2.

1.1.2 Pulsating Variables

Pulsating variables undergo variations in radius due to intrinsic variation of temperature and pressure. They may be strictly periodic, as in RR Lyrae stars or Cepheid variables, or merely cyclic, as in RV Tauri or Mira variables. The period of variation may be very rapid – minutes for some high-temperature variables – to years for the Miras. The General Catalogue of Variable Stars (Kholopov 1985) lists the following types: α Cygni; β Cephei; Cepheids; W Virginis; δ Scuti; Irregular; Mira; PV Telescopii; RR Lyrae; RV Tauri; Semiregular; SX Phoenicis; and ZZ Ceti; most of them have assigned subtypes, which we omit here. Of greatest interest to those outside the variable star community are the RR Lyrae variables, with approximately constant luminosities, and the Cepheids, with luminosities that increase with period. Such stars are considered to be "standard candles," and may be used to determine the distance to any ensemble in which they are found. RR Lyrae stars are giant stars that have periods of about half a day. Cepheids are supergiant stars that have periods from 1 to tens of days. Both are found in the field and in globular clusters, but RR Lyrae stars are much more common. There are two types of Cepheids, the classical Cepheids, members of Population I that are younger and are associated with the galactic plane, and Population II Cepheids, found in the galactic halo and in globular clusters. The realization that there are two types of Cepheids by Walter Baade led to a revised (primarily, extragalactic) distance scale (by a factor of 2). Closely related to the Cepheids and RR Lyrae objects are the δ Scuti stars (so designated by Harlan J. Smith 1955) and their globular cluster-resident cousins, the SX Phoenicis stars. These are subgiants or dwarfs (luminosity classes IV and V, respectively) that have periods of pulsation that are typically small fractions of a day. All three groups are found in the "instability strip" on the Hertzsprung–Russell diagram (luminosity or absolute magnitude versus spectral type, stellar temperature or color index: basically brightness plotted against color), where no stable stars are found. The Cepheids, being the most luminous, lie uppermost, the RR Lyrae stars below them, and the δ Scuti stars lie the lowest, straddling and just above the Main Sequence, the locus of all stars powered by the fusion of hydrogen in their cores, where stars spend most of their lives. Pulsating stars display a $P\sqrt{\rho}$ relation, that is, the product of the period and the square root of the density is a constant for a particular group of pulsating stars:

$$P\sqrt{\rho} = \sqrt{\frac{1}{\kappa G}}, \qquad (1.1.2)$$

where κ is a dimensionless constant and G is the gravitational constant. Thus, the shorter period stars are more dense, in accord with their luminosity classes. A pulsating star may be pulsating at its fundamental frequency, but sometimes it pulsates in the first or even second overtone (like a whistle that is blown too hard). Delta Scuti stars may exhibit pulsations at many periods at the same time; these arise from nonradial pulsation modes, in which some zones of the star are expanding, while others are contracting. Stellar seismometry deals with the many modes of oscillation that can be found in most stars, including the Sun. Of course, in the case of the Sun the oscillations are of very low amplitude (as well as very numerous).

1.1.3 Eruptive Variables

Eruptive variables involve violent transient events (explosions). These may vary from small flares (on stars of the main sequence) to "flare stars" (where the energies may be as much as 10^3 times greater) to complete destruction of the star (supernovae). The GCVS (Kholopov 1985) lists the following types of variable stars in the "eruptive" category: Stars with brightness increase due to "violent processes and flares" but they include also "shell events or matter outflow" in stellar winds with possible interaction with interstellar matter. As a consequence, this categorization includes many slower phenomena as well as stars undergoing very rapid changes. The GCVS types belonging to this category include

- FU Orionis variables (typically 100-fold brightening over intervals of months, followed by constancy or slow decline over years to decades. Possibly associated with the T Tauri evolution stage "Orion variables". A reflecting nebula spectrum is always seen, and emission lines are seen at outburst);
- γ Cassiopeiae variables [Rapid rotators, these objects are hot, emission line stars (Be stars), and Doppler-shifted spectral lines indicate rapid outflow. Their light variation may be as much as 1.5 magnitudes];
- Orion variables (irregular variables associated with nebulosity. Some are found to be T Tauri stars, an early stage in stellar evolution; some are rapidly varying; spectral types distinguish subcategories);
- R Coronae Borealis variables [described as both eruptive and pulsating, these objects suffer fading and gradual recovery over intervals of months; the variation is thought to involve ejections of carbon (soot, basically)];
- S Doradus variables (high-luminosity stars characterized by envelope ejections; P Cygni and η Carinae are members of this class. The latter, at least, has been identified as an Asymptotic Giant Branch star undergoing thermal pulses, a late stage in stellar evolution);
- UV Ceti variables (typically late-type dwarf stars undergoing flares with rise times of seconds and recoveries of minutes of duration; a related group is associated with "Orion variables". In clusters, these objects are simply known as "flare stars");
- Wolf–Rayet variables (very hot stars with irregular light variations).

In addition, the GCVS identifies RS Canum Venaticorum variables among the eruptive variables. These objects are typically giant or subgiant interacting stars

1.1 Eclipsing Binaries and Other Variable Stars

characterized by strong magnetic field interactions and active chromospheric regions, causing quasi-cyclic variations in light curve shape outside of eclipse (cf. Hall (1975)). As these are enhanced forms of phenomena associated with all late-type, main sequence stars (such as the Sun) there is disagreement about whether or not RS CVn-like systems deserve a separate group designation. In this book, we prefer to call systems that exhibit enhanced active regions merely as "RS CVn-like."

The GCVS has another category in which the more violent members of eruptive variables are placed; the "cataclysmic" variables, described as having outbursts produced by thermonuclear processes either on the surface or in the interior. The name "cataclysmic variable" usually means a slightly different type of object to much of the variable star community; it involves a white dwarf and a cooler, less evolved star that has filled its inner Lagrangian surface (see Sect. 3.1.6) resulting in a semi-detached binary star system. The system is further characterized by a stream of material feeding either directly on to, or into a disk surrounding, the white dwarf. As the material in the disk loses angular momentum, it spirals onto the white dwarf's surface. This is, in fact, a nova or even supernova, waiting to happen, but it is (usually!) not happening (yet!). See Warner (1995) for a definitive discussion of CVs and Hellier (2001) for practical details. In any case, the GCVS category includes novae of various types, namely "fast," "slow," or "very slow," depending on the rate of decline from maximum outburst), "recurrent" (i.e., having been seen to recur), or "nova-like" (having spectra resembling novae at minimum light); and the supernovae. Supernovae types I and II are associated with stellar populations II and I, respectively. Population I stars are relatively young, and supernovae from this population are thought to include at least some single massive stars that have undergone catastrophic collapse due to the formation of iron in an endothermic reaction in their cores. The removal of energy in this process causes a deficiency in the pressure so that the weight of the overlying layers cannot be sustained, resulting in a massive implosion and catastrophic explosion as the imploding material bounces off the highly compressed core, dispersing the atmosphere in the surrounding interstellar medium.

Objects called "novae" ("new" stars) have been known since ancient times, but supernovae, the far more luminous phenomena, are newly recognized. Walter Baade recognized that some novae were extraordinarily luminous and called these "Hauptnovae," (or "chief novae"; others used the terms "giant novae," or "more luminous novae," or even "super-novae" to describe them) and he and Fritz Zwicky are said to have coined the term "supernovae" (see Osterbrock (2001)). The relative energy of the outburst ranges from $\sim 10^{44}$ for novae to $\sim 10^{48}$ ergs for supernovae. It is now known that novae are the products of mass exchange in highly evolved binary star systems, involving a white dwarf as the recipient star. Some supernovae (type Ia in particular) are similar. However, in the case of novae, the system survives, and the nova phenomenon may recur. In the case of supernovae, there are drastic changes to the exploding star, and a recurrence may not be possible. Three known outcomes of this catastrophic event are

- the core of the star collapses to form a neutron star, an object so dense that electrons and protons are forced into neutrons. Rotating neutron stars are seen as pulsars;
- the core collapses into a black hole, an object so dense that light cannot escape its overwhelmingly strong gravitational field within a certain distance of its center;
- the star dissipates into a rapidly expanding debris cloud without a core remnant.

This concludes our brief summary of variable stars. We now concentrate on eclipsing systems and their treatment.

1.2 Overview of the Problem

Coup d'oeil   (*As at one glance: A brief survey*)

We begin with a discussion of the importance of binary stars and give a historical overview of the types of binaries and of the modeling and analysis of *EB* light curves and conclude with a summary of the nomenclature and symbols we use throughout.

1.2.1 Why Binary Stars Are Important

Binary stars are important, first, because they are numerous. Latham et al. (1992, p. 140) conclude that the frequency of spectroscopic binaries detected in the galactic halo is not significantly different from that in the disk, despite differences in kinematic properties and chemical composition. The observed frequency is approximately 20%; the actual frequency is higher because many binaries remain undetected. In the solar neighborhood, where we have the benefit of proximity so that proper motion variations can be detected, the frequency is more than 50% – and several stars are in fact multiple systems.

The second reason why binaries are important is that they are the primary source of our knowledge of the fundamental properties of stars. For example, the direct determination of the *mass* of any astronomical object requires measurable gravitational interaction between at least two objects (galaxy–galaxy, star–star, star–planet, planet–satellite). In galaxy–galaxy interactions, the distances and separations are so large that no detectable motion on the plane of the sky is possible. In star–planet interactions the objects contrast so greatly in brightness that outside the solar system only the highest possible – and until recently rarely attained – precision can resolve the objects. Typically in the latter case, only the star's motions are detectable, and the properties of that star must be assumed, mainly on the basis of binary star studies, in order to deduce the properties of the planet. In star–star interactions, the variations in position and velocity caused by orbital motion are detectable for a wide range of stellar separations and up to at least a factor of 5 in brightness. It is often the case that *both* stars may be studied in any of several ways, depending on their distances, brightnesses, and motions. Other basic properties of stars and of the systems they constitute can be determined through analysis of observational data, depending

1.2 Overview of the Problem

on the observational technique by which the interaction is studied. The four main types of binaries described by the observational technique are visual, astrometric, spectroscopic, and *EB* systems. We discuss each type in turn.

1.2.1.1 Visual Double Stars

For visual double stars, true binary star systems (as opposed to purely optical doubles) in which both components are visible and resolvable in a telescopic eyepiece, it is possible to determine the component masses M_1 and M_2. They are derived from Kepler's third law (3.1.62) and the moment equation $a_1 M_1 = a_2 M_2$, where a_1 and a_2 are the semi-major axes of the absolute orbits of the components about a common center of mass.[3] The derivable orbital elements include the size or semi-major axis a and shape of the relative orbit and the inclination i of the plane of the orbit against the plane of the sky. Because, however, in most cases[4] only the angular semi-major axes can be determined in this way, parallax measurements are needed to establish linear values. Due to the limited accuracy of parallax measurements, this method has been restricted to the near-solar neighborhood, within about 30 parsecs; however, high spatial resolution surveys have improved the situation. The Hipparcos space astrometry mission (1989–1993) acquired median astrometric accuracies of ~ 0.001 arc-sec, and the resulting catalogue contains 12,195 detected double or multiple star systems. Such a nearby sample of stars may suffer from selectivity effects. Most of the stars of this sample have spectral types later than F5, for example. Nevertheless, it is a valuable sample because it enables us to calibrate stellar luminosities, which is the basis for all standard candles of all the types of stars thus studied.

If only one component is visible, because the other is too faint and/or is too close to its brighter companion to be separated through telescopic resolution alone, gravitational effects may help us to prove that the system is a binary. Such a system, in which an orbital motion is detected by astrometric methods, is called an *astrometric* binary. The faint companion may be nominally resolvable but hidden in the glare of the bright component. *Sirius B* is such a star: The much smaller and fainter component of the "Dog Star", the "Pup" was first observed visually by Alvan G. Clark in 1862, but Sirius had been recognized by 1844 to be a binary on the basis of its proper motion variability discovered by Friedrich Wilhelm Bessel (1845). For a fine discussion of the extraction of data from astrometric binaries in general and of the Sirius system in particular, we recommend Aitken (1964) and Lindenblad (1970).

Another interesting type of astrometric binary is presented by cases where the components are so close that they are, or have been until recently, unresolvable.

---

[3] Unfortunately, in only a few cases is it possible to measure the semi-major axes a_1 and a_2 of the absolute orbits separately. In most cases, only the relative orbit and its semi-major axis $a = a_1 + a_2$ can be determined.

[4] There are a few cases of spectral–visual binaries which also give the absolute value of a.

Resolving power, or the ability to resolve fine detail, can be described mathematically by

$$\Delta = 1.22\frac{\lambda}{D}, \qquad (1.2.3)$$

where Δ is the minimum angular separation in radians, D is the aperture of the telescope, and λ is the wavelength in the same units. This quantity is, in fact, the central radius of the diffraction disk or *Airy disk*, the central portion of the diffraction image or *Airy figure*. See Couteau (1981, p. 32) for a lucid discussion. Adaptive optics[5] makes use of a "reference star" to achieve sub-arc-second seeing within a small region of the field of view known as the *isoplanatic patch*, within which atmospheric fluctuations are correlated to ~ 1 rad. The actual isoplanatic patch is a few arc-secs on the sky, typically. This technique permits ground-based telescopes to achieve an order of magnitude improvement in resolution. If other contributions to the "seeing budget" can be minimized as well, the resolution can approach the theoretical (angular) resolving power of the telescope. Adaptive optics are necessary to overcome the effect of atmospheric seeing; in space, instrumental resolution is the limiting condition. The repaired Hubble Space Telescope (*HST*), for example, has an effective resolution of about 0.05 arc-sec, permitting direct viewing of both the separation and the rough surface details of the Pluto–Charon system. Direct angular measurements of some of the largest of the sky's bright stars are now possible.

A number of less direct but more effective techniques also permit high angular resolution:

- **Lunar occultations**: The edge of the Moon occasionally occults a star or stellar system within the maximum range of its declination: about $\pm 28°$. Analysis of the resulting diffraction pattern intensities can determine binary star separations and even the diameters of stars down to about 0.001 arc-sec.
- **Phase interferometry**: Around 1920, Michelson (1920); Michelson & Pease (1921); and Pease (1925) determined the sizes of bright red giant stars with the help of a phase interferometer mounted on the 100-in. telescope at Mt. Wilson. The practical limit to angular resolution with this method was about 0.01 arc-sec and was set by two factors: mechanical flexure of the interferometer arm and atmospheric seeing. The arm bore two mirrors which were the equivalents of Young slits, and whereas a length of 25 ft was successful, an attempt at 50 ft was not. More recent work in this area has been done by groups in France (beginning with A. Labeyrie in 1974), a group at the US Naval Observatory (Flagstaff, Arizona), and at JPL (beginning with Shao and Staelin in 1979), among others.
- **Aperture synthesis**: Several modern groups have succeeded in using arrays of telescopes separated by up to 100 m and improved equipment to produce higher quality in resolution and stability and to extend the interferometry to two dimensions. The availability of new autocorrelation methods to combine fringes from separate telescopes permits the determination of binary separations

---

[5] For a comprehensive review of adaptive optics, see Beckers (1993).

1.2 Overview of the Problem

even from sites which are nonoptimal both astrometrically and photometrically [see Baldwin et al. (1996) for a description of work by the Cambridge Optical Aperture Synthesis Telescope (COAST) group].

- **Intensity interferometry**: Brown et al. (1974a, b) measured the diameters of blue stars with an intensity interferometer at a multiple telescope observatory at Narrabri, Australia, beginning in the 1950s. The technique involves determining the correlation between the light received by several collectors (in the Narrabri case, 6.5 m incoherent dishes). See Brown (1968) for a basic review of these techniques as well as the use of lunar occultations.
- **Speckle interferometry**:[6] Speckle observations involve the determination of pattern parameters in which the atmosphere acts as a diffusing screen [see Schlosser et al. (1991, pp. 119–123), for a simple but clear exposition]. The method has proven very fruitful for visual binary work. By the mid-1990s, the CHARA group at Georgia State University, led by Harold A. McAlister, made more than 40,000 speckle observations of more than 7,300 stars or systems, and many more observations were being carried out (Mason et al. 1996).

Long-baseline interferometry permits the resolution of many spectroscopic binaries. As is the case for all well-determined visual binaries, coupled with high-precision radial velocity data, the parameters can yield all the geometric elements of the orbits. To a certain degree, the relative brightness of the components can also be obtained and in combination with photometry can provide a distance. An excellent example of such collaboration can be found in the work of Scarfe et al. (1994) and Van Hamme et al. (1994). Interferometric observations from space offer many advantages, among them a spectral range from the far-ultraviolet to the far-infrared. This means the possibility of observations of objects such as protostar binaries which radiate in the far-infrared. The techniques of long-baseline optical and infrared interferometry were reviewed by Shao & Colavita (1992).

A calibration of stellar surface brightness making use of the measured sizes of stars was carried out first by Wesselink (1969). This information source can be useful in several ways, e.g., initial values for radii (given a spectroscopic estimate of luminosity and photometric color index) for light curve modeling might be obtained from his Fig. 2. This plot shows the radii of stars superimposed on a color-magnitude array. A short catalogue of derived stellar sizes was compiled by Wesselink et al. (1972).

1.2.1.2 Spectroscopic Binaries

The detection and analysis of spectroscopic binaries is not subject to geometrical resolution limits as are angular measurements. With sufficient light gathering power, it is possible to investigate spectroscopic binaries even in nearby galaxies and to derive the luminosity ratio and mass ratio.

---

[6] Although *speckle interferometry* is called interferometry, the reader should be aware that the concept is completely different from the others kinds of interferometry mentioned above. Whereas the latter involves a certain base line, speckle interferometry is rather a correlation method.

The luminosity ratio, i.e., the relative luminosities of the component stars, can be derived from spectra alone, by a method developed by Petrie (1939) at the Dominion Astrophysical Observatory. The determination of the mass ratio is more difficult. It can be derived only spectroscopically under favorable conditions, namely where the components have similar luminosities (say within a factor of 5). In that case, a *radial velocity curve*[7] may be observed for each component; both radial velocity curves enable us to compute the spectroscopic mass ratio. Note that spectroscopists usually define the mass ratio as the more massive over the less massive star.[8] Such a system is called a *double-lined spectroscopic binary* (SB2). If only one component can be observed spectroscopically, the system is called a *single-lined spectroscopic binary* (SB1). In this case, a useful quantity defined in (4.4.34) and known as the *mass function* can still be obtained which, according to (4.4.38), provides a lower bound on the sum of masses and gives a lower bound on the unobserved mass in any case (because the observed star cannot have mass less than zero).

The calculation of masses and radii requires the inclination, i, which cannot be found from spectroscopic data alone. In the SB1 case, the mass ratio also is not known. If i is sufficiently large [9] and the separation of the components is sufficiently small, the binary appears as an *EB*.

1.2.1.3 Eclipsing Binaries

A variable star observer measures a time-dependent flux, the display of which versus time or *phase* (the repeated foldings of the time into the period of variation) is known as the *light curve*. The acquisition and reduction of photometric observations will be discussed in Sect. 2.1. *EB*s establish a special class of variables stars. For the nomenclature and classification of variable stars we refer the reader to the book *Light Curves of Variable Stars* by Sterken and Jaschek (1997) and Wilson (2001). Whereas eruptive, pulsating, rotating, and cataclysmic variables are said to be intrinsic variables caused by different physical mechanisms, *EB*s are extrinsic variables requiring models including both astrophysics and geometry.

As we have indicated, an eclipsing variable is a binary system whose orbital motion is in a plane sufficiently edge-on to the observer for eclipses to occur. The smaller the orbit relative to the sizes of the stars, the greater the likelihood of eclipses. For a special subgroup of *EB*s (so-called over-contact binaries, with a

---

[7] A radial velocity curve is a plot of the star's velocity component toward (or away from) the observer versus time or orbital phase (essentially the fraction of an orbital cycle). See Fig. 3.27 for an example and Sect. 2.2 for details.

[8] See Sect. 2.8 for terminology concerning stars 1 and 2, as well as the uses of the term "primary" in referring to components and eclipses.

[9] The inclination i is defined such that an edge-on orbit has $i = 90°$.

common envelope) eclipses may occur, although perhaps not perceptibly, even if the inclination is as small as 35°. Illustrations demonstrating the visibility of eclipses at low inclinations for an over-contact system are in Sect. 8.1 (Figs. 8.1, 8.2, and 8.3). These binaries usually have orbital periods of less than 10 days and in most cases less than 1 day. Among the exceptions are some rare cases of hot and/or developed systems. The longest period *EB* known at present is ε *Aurigae* [see, for instance, Caroll et al. (1991)] with an orbital period of 27.1 years. According to Kepler's third law this binary has an orbit relatively large[10] compared to the sizes of the components. Historically, considerations concerning the likelihood of eclipses lead to a connection between *EB*s and "close binaries." In the early days, a "close binary" was defined as a binary with component radii not small compared to the stars' separation. This definition was later replaced by a more physical definition related to the evolution of the components by Plavec (1968), which we discuss at the end of Sect. 1.2.3.

EB studies often involve the combination of photometric (light curve) and spectroscopic (mainly, radial velocity curve) data. Analysis of the light curve yields, in principle, the orbital inclination and eccentricity, relative stellar sizes and shapes, the mass ratio in some cases, the ratio of surface brightnesses, and brightness distributions of the components among other quantities. If radial velocities are available, the masses and semi-major axis may also be determinable. Many other parameters describing the system and component stars may be determined, in principle, if the light curve data have high enough precision and the stars do not differ greatly from the assumed model. The prediction of the information content of particular light curves has been a major topic of concern in binary star studies; the exposition of this topic is an important component of the present work also.

1.2.2 Phenomenological Classification of Eclipsing Binary Light Curves

Examples of prototypical light curves are shown in Fig. 1.1. They correspond to the classical categories, discussed above, of "Algol," "β Lyrae," and "W UMa" light curves, also known as *EA*, *EB*, and *EW* light curves, respectively.

The EA light curves are typically almost flat-topped, suggesting that effects due to the proximity of the components are small, with a large difference between the depths of the two minima. Indeed, in some wavelengths the secondary minimum may be undetectable, and there may be an increase in light near the expected phase of secondary minimum due to the "reflection effect."

---

[10] Note that ε *Aurigae* is enormously large. It contains an F0 supergiant with an estimated radius, depending on the distance, between about 100 and 227 $R_\odot$.

(a) Synthetic "Algol" type light cure (V band).

(b) Synthetic "β Lyrae" type light cure (V band).

(c) Synthetic "W UMa" type light cure (V band).

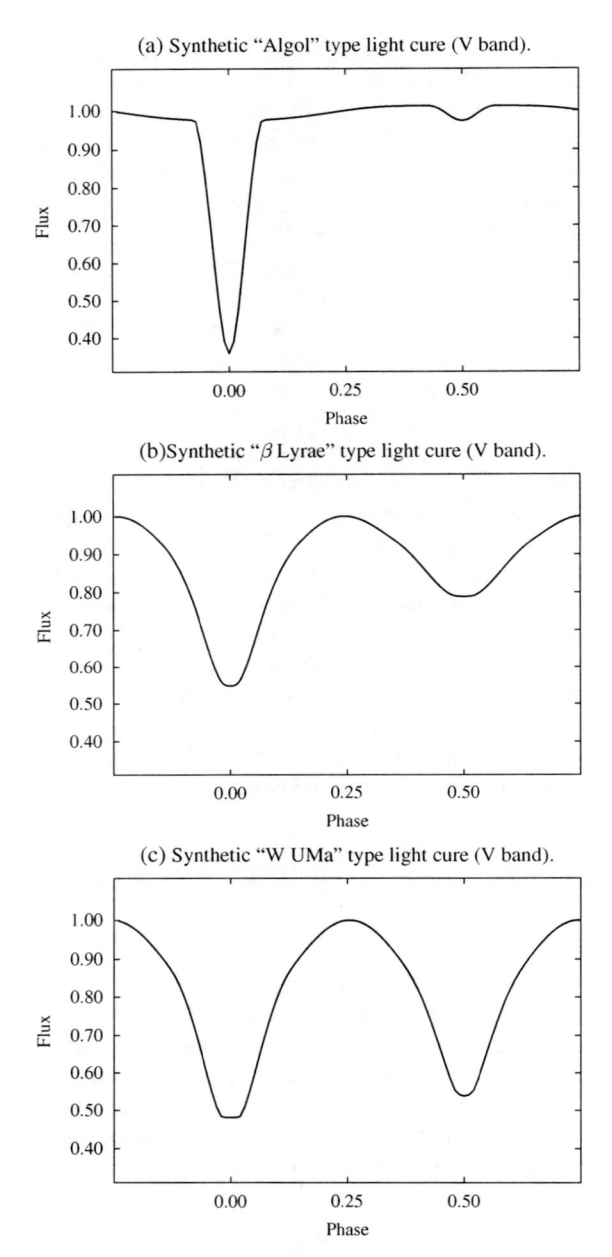

Fig. 1.1 (over) Classes of light curves. (**a**) Shows a synthetic "Algol"-type light curve (*V band*). It has been produced using the parameter file *algolv.bmd* from the Binary Maker 2.0 examples collection (Bradstreet, 1993). (**b**) Shows a synthetic "β Lyrae"-type light curve (*V band*) and has also been produced with Binary Maker 2.0 (Bradstreet, 1993) using the parameters for *RU Ursae Minoris* given in the *Pictorial Atlas* (Terrell et al. 1992,(p. 107)). (**c**) Shows a synthetic "W UMa"-type light curve (*V band*). It has been produced using the parameter file *abandb.bmd* from the Binary Maker 2.0 examples collection (Bradstreet, 1993)

The EB light curves, on the other hand, are continuously variable (the "ellipsoidal variation"[11]), characteristic of tidally distorted components, and with a large difference in depths of minima indicating components of quite different surface brightness.

Finally, the EW (or W UMa) light curve is also continuously variable, but with only a small difference in the depths of the minima. The variation outside the eclipse in the latter two types is indeed due to proximity effects (mainly the tidally distorted shapes of the stars), but the EB light curves arise from detached[12] or semi-detached binaries, whereas EW systems are over-contact.[13] The expression "EA light curve," on the other hand, is somewhat misleading. Judged by the light curve, the system may *look* undistorted, but only in light from the visible (or, as infrared astronomers refer to it, the "optical") part of the spectrum. In the infrared, for example, *Algol* itself presents a continuously variable light curve and a fairly deep secondary minimum (Fig. 1.2). This reveals quite clearly that the bluer, hotter component in the system is relatively small and undistorted, and its radiation enhances the bright inner face of its companion.

These considerations show the value of treating all aspects of the light of the system in light curve analysis, and not only their geometric characteristics. Unfortunately, in many cases the system geometry has been the only goal of light curve investigations.

As studies of *Algol* itself show, *EB* analysis is a formidable astrophysical task (see Sect. 1.3.5 for further examples). The field includes radiation physics and sometimes hydrodynamics. It borrows methods from celestial mechanics, thermodynamics, and other branches of physics. Physical models are required for radiation transport in the components' atmospheres and for the dynamic forces controlling the stellar mass distributions.

---

[11] The expression *ellipsoidal variation*, or less correct *oblateness effect*, is more generally used in the context of the Russell–Merrill model (see Sect. 6.2.1), where the shape of the light curve is modeled as due to ellipsoidal stars. The term *oblateness* should be reserved for rotational but not for tidal distortion.

[12] The expressions "detached," "semi-detached," and "over-contact" arise from morphological classification of binaries (Sects. 1.2.3 and 3.1.6). Detached systems have separated stars. Semi-detached systems are still separated but one component fills its critical lobe.

[13] Sometimes, these systems are called *contact system*; in this book we reserve this term only for the case in which both components fill their critical lobes exactly. This special mathematical case seems not to occur in real binaries. More generally, in the *over-contact systems* both stars overfill their inner Lagrangian surfaces and establish a common envelope. Such systems can exist for astronomically significant times only if the orbits are circular and the components rotate synchronously.

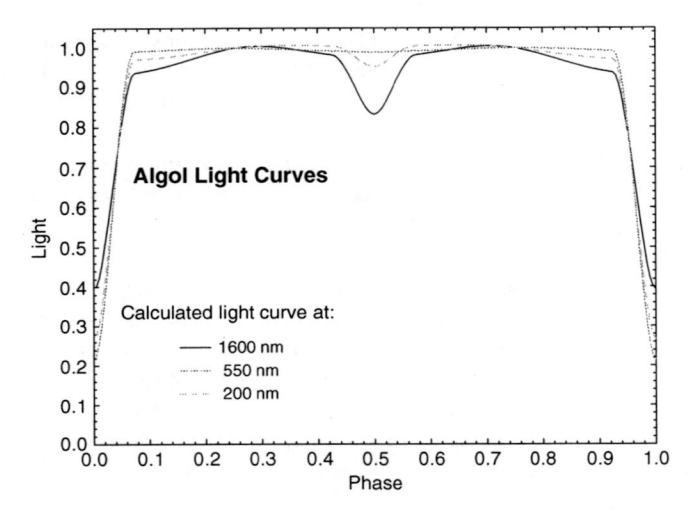

Fig. 1.2 U, V, and infrared light curves of *Algol*. The plot has been produced with Binary Maker 2.0. The parameters are taken from the *Pictorial Atlas* (Terrell et al. 1992, p. 239) and from Kim (1989)

1.2.3 Morphological Classification of Eclipsing Binaries

The dynamic forces controlling the stellar mass distributions involve the effects of rotation, tides, and noncircular orbits. For an introductory-level discussion of all these effects, see Wilson (1974). Fortunately, tidal forces produce circular orbits and synchronous rotation in many interacting binaries. A detailed and excellent analysis of the tidal evolution in close binary systems is provided by Hut (1981). The orbital period of a synchronous rotator in a circular[14] orbit is the same as the rotation period. We will discuss only synchronous rotation in this section.

Another physical simplification reduces the mathematical complexity: Although the stars may be relatively large and considerably distorted, they attract one another nearly as if their entire masses were concentrated into mass points at their centers. Therefore, only two forces need to be considered in the circular orbit and synchronous rotation case:

1. gravitational attractions of two mass points and
2. the centrifugal force due to the rotation of the entire binary system about its center of mass.

Given that both gravitational and centrifugal forces are time-wise constant for corotating matter, we can expect to find solutions for static configurations in the corotating frame. A somewhat similar problem was solved by the French mathematician É. Roche (1820–1883) in the nineteenth century (Roche, 1849, 1850). The basic

---

[14] Note that eccentric binaries tend to have their angular rotations locked at the orbital angular rate at the periastron (Hut, 1981).

1.2 Overview of the Problem

concept for understanding the solutions of that problem is equipotential surfaces (briefly, equipotentials). These are surfaces on which the sum of rotational and gravitational energy per unit mass is constant. On these level surfaces, also called "Roche surfaces," the component of the force vector tangential to these surfaces vanishes, i.e., the local force vector is everywhere normal to them. The Roche surfaces are indeed the static surfaces we are interested in: They corotate with the orbital motion of the binary. Binary component surfaces are now modeled as equipotentials (similarly on Earth, where ocean and lake surfaces follow equipotentials). The force perpendicular to the surface is different for different equipotentials and also varies as a function of location on a particular surface unless this surface is a sphere. In the vicinity of the Earth, equipotentials due to the combined gravitational forces of Earth and Moon are almost spheres. A family of binary system equipotentials projected onto their orbital plane is illustrated in Fig. 1.3. One point in this figure is called the Lagrangian point L_1^p, after the French mathematician J. L. Lagrange (1713–1765). For a corotating test particle at L_1^p the gravitational and rotational forces balance, so the particle feels no force. The Roche surface[15] passing through L_1^p consists of two ovoid surfaces called the *Roche lobes* of the components. The two ovoids touch at L_1^p.

1. detached systems (Fig. 1.4), if neither component fills its Roche lobe;
2. semi-detached systems, if one component fills its Roche lobe, and the other does not; and
3. over-contact systems, if both components exceed their Roche lobes.

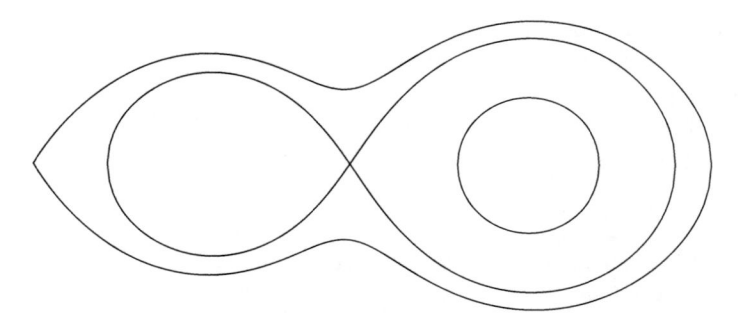

Fig. 1.3 Projections of equipotential Roche surfaces. The plot, showing equipotential Roche surfaces projected onto the orbital plane, was produced with `Binary Maker` 2.0 for a binary system with mass ratio $q = 0.5$ and Roche potentials $\Omega_1 = 5$, $\Omega_2 = 3$. The outer curve corresponds to the surface passing through L_2^p. The next inner one passing through L_1^p (point where lines cross) represents the Roche lobes for both stars. Both curves depend only on the mass ratio. Finally, the inner near, circular curves are the stars corresponding to the above given potentials and mass ratio

---

[15] In more general models, including eccentric orbits and asynchronous rotation, the expression Roche surface and Roche lobe will be replaced by *critical surface* and *critical lobe*.

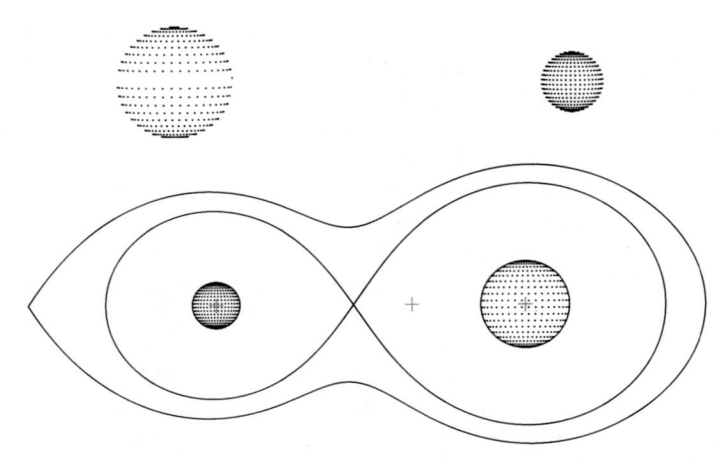

Fig. 1.4 Roche potential and shape of a detached binary system. The plot has been produced using Binary Maker 3.0 and the *YZ Cassiopeiae* parameter set provided in the examples collection (Bradstreet & Steelman, 2004). ($f_1 < 0$, $f_2 < 0$)

The interpretation and physical properties (such as stability) associated with the morphological classes introduced above is discussed in further detail in Sect. 3.1.6. The form of a component is closely related to the *contact parameter*, or sometimes, *fill-out factor*, f, which measures the degree of lobe filling [Chap. 3, definition (3.1.101)].

If eccentric orbits and nonsynchronous rotation are considered, then additional configurations besides the one discussed above may occur [see Chap. 3 or Wilson (1979, 1994)].

There is some correspondence between the morphological classification based on the Roche lobes and the phenomenological classification presented in the previous section:

Algol-type light curves $\Rightarrow$ semi-detached systems (Fig.1.5),
W UMa-type light curves $\Rightarrow$ over-contact systems (Fig. 1.6).

Note that the phenomenological classification of the β Lyrae-type light curve has no morphological counterpart. Sometimes, β Lyrae-type light curves are produced by detached systems, sometimes by semi-detached systems, and sometimes also by systems having marginal over-contact; see the *Binary Stars Pictorial Atlas* (Terrell et al. 1992), for example. However, there are semi-detached binaries that are not Algols (e.g., cataclysmic variables) and over-contact binaries that are not W UMa's (e.g., over-contact binaries like *TU Muscae*).

Having this concept and basic understanding of Roche potentials, it is possible to give a physically useful definition of *close binaries* following Plavec (1968, p. 212): Close binaries are those systems in which a component fills its critical Roche lobe

1.2 Overview of the Problem

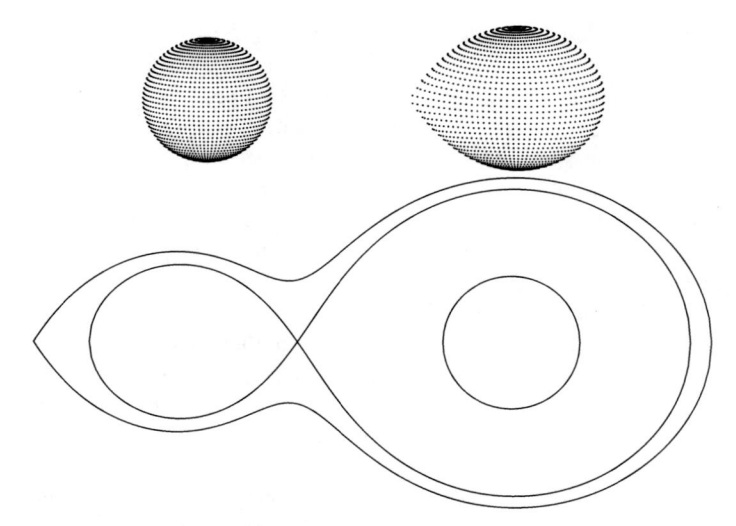

Fig. 1.5 Roche potential and shape of a semi-detached binary. The plot has been produced with `Binary Maker` 2.0 using the *Algol* parameter set in the examples collection (Bradstreet, 1993). The primary component fills its Roche lobe ($f_1 = 0$, $f_2 < 0$)

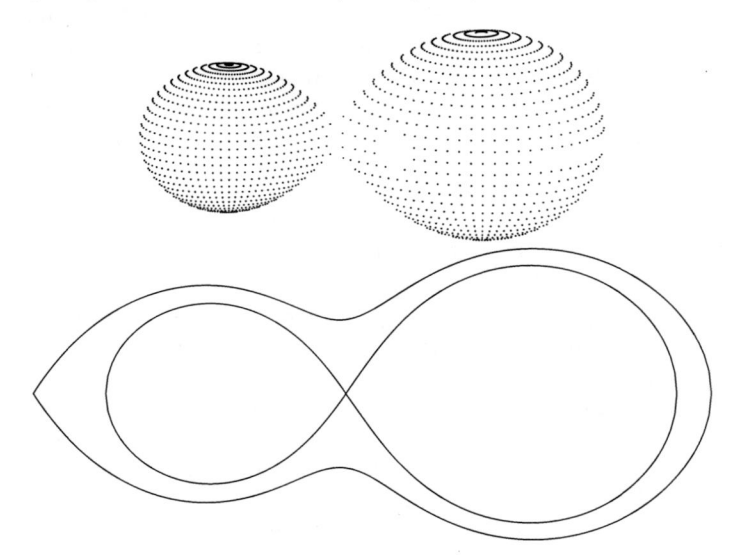

Fig. 1.6 Roche potential and shape of an over-contact binary. The plot has been produced with `Binary Maker` 2.0 for the *TY Bootis* parameter set in the examples collection (Bradstreet, 1993). Both components exceed their Roche lobes ($0 < f_{1,2} \leq 1$, $f_2 = f_1$)

at some stage of its evolution. Prior to this evolutionary definition the custom was to define a close binary as one in which the dimensions of at least one component are of the same order of magnitude as the separation.

1.2.4 What Can Be Derived from Eclipsing Binaries

The analysis of photometric light curves alone cannot provide absolute dimensions of binary stars or their orbits. The reason is a scaling property with respect to the relative orbital semi-major axis a: If all linear dimensions are increased by a certain factor, the associated light curve changes can be canceled by shifting the binary to a larger distance. A light curve can provide the orbital inclination and, among other parameters, relative quantities such as the radii in units of a, ratio of luminosities, stellar figures, and perhaps the photometric mass ratio.

Radial velocity curves can provide the mass ratio and the scaling factor a in physical units if the inclination is known from another kind of observation. With a and the period P known, the masses can be found unambiguously from (4.4.14) and (4.4.16). Similarly, i must be known to derive orbital dimensions [cf. (4.4.14, 4.4.15, 4.4.16 and 4.4.17)] from a radial velocity curve. Combining these rules, we find the following:

The full determination of absolute eclipsing binary parameters requires both a light curve and a radial-velocity curve for each component. *EB*s are informative objects because they allow photometry and spectroscopy to be combined effectively. Eclipsing, double-lined systems are rare but very valuable. If the data quality is high and the binary configuration is well conditioned, we have a fundamental source of information about sizes, masses, luminosities, and distances or parallaxes of stars. Many other parameters can be determined from precise light curve data if the configuration fulfills certain requirements, for example, by having complete eclipses. Because such stars may be found over the full range of ages, they also tell how stars evolve – at least in binary stars.

1.2.5 Why Data Derived from Eclipsing Binaries Are Important

The early- and mid-age evolution of a star depends almost uniquely on its mass and its initial chemical composition. Therefore, in order to test stellar structure and stellar evolution theories, it is desirable to have as many accurate masses and other star parameters as possible. In addition, these data help to improve our understanding of such exotic objects as X-ray[16] binaries, novae, and Wolf–Rayet stars. Unfortunately, despite much progress, far too few accurate masses are available, especially for stars of early (O and B) spectral type (Popper 1980). These very hot stars are important in order to understand the upper main sequence. They are of special interest because they undergo mass loss due to stellar winds. Knowledge of age and composition is basic to understanding the evolution of a star. Such information is sometimes available for members of star clusters. A great wealth of knowledge can be gained whenever binaries in clusters can be successfully analyzed. The cluster membership

---

[16] X-ray binaries are interacting close binary systems which contain a neutron star or a black hole [Krautter (1997)]. They are discovered on the basis of their strong X-ray emission which is of the order of 10^{28}–10^{31} W.

links age with mass, luminosity, and radius of each component, and if the chemical composition is known, this potent combination allows detailed testing of stellar evolution theory.

1.3 The History of Light Curve Modeling

Natura non facit saltum   *(Nature makes no leap)*
after Aristotle (384–322 B.C.)

1.3.1 The Pioneers – The Age of Geometry

*EB*s play a special role in binary research. Henry Norris Russell (1887–1957), one of the most distinguished astronomers, spoke about the "Royal Road of Eclipses" (Russell 1948b). Traveling this road entails the decoding of the messages encrypted in the light curves of eclipsing variables. As a rule, a light curve is determined by geometric effects due to eclipses and by physical proximity effects between the components. In the past, light curves were "rectified" in order to get rid of the "ellipticity" and "reflection" effects and such other perturbations from the light curve as could be modeled by a truncated Fourier series [see Russell (1948a); Russell & Merrill (1952)]. In doing so, a triaxial ellipsoid model was transformed into a spherical model, with which the light curve solution could be obtained in a straightforward way from tables or with the aid of nomographs. Computer programs based on rectifiable models were developed by Jurkevich (1970), who investigated the suitability to machine coding of a number of existing light curve approaches, including two of Kopal's rectifiable models, and by Proctor & Linnell (1972). However, the underlying assumptions of rectifiable models rarely hold in reality. As a rule, fully accurate, reliable solutions could be expected only for well-separated, detached systems. Systematic deviations could be observed for semi-detached and especially for over-contact systems. For over-contact systems in particular, the solutions were almost always misleading if not completely wrong. This is illustrated by the case of *TY Bootis*, an over-contact system which several modelers have tackled. Table 1.1 summarizes the results.

The photometric data by Carr (1972) were analyzed with the Russell–Merrill method by Carr. The same data were used in an analysis (listed as WD) using the Wilson–Devinney program, today the most frequently used model and program in the *EB* community, hereafter abbreviated the WD model or WD program and further described in Sect. 6.3.6. A new data set was also analyzed with the Wilson–Devinney method (Milone et al. 1991). For an explanation of the parameter notations in the first column see the Symbols' List in Appendix F. The square-bracketed values for the absolute radii in column 2 were computed on the basis of Carr's values for $r_s = R_s/a$ and $r_g = R_g/a$ and the WD-determined value of a for the Carr data [r_s and r_g are the radii of the "smaller" and "greater" stars in

Table 1.1 Analysis of *TY Bootis* by several approaches

Parameter	Carr (1972)[17]	WD-1	WD-2
$a/R_\odot$	...	2.304 (14)[18]	2.318 (12)
e [19]	0	0	0
i	$73°\!.7$	$78°27\,(12)$	$77°\!.50\,(8)$
$q = M_2/M_1$	...	2.153 (10)	2.138 (7)
$\Omega_1 = \Omega_2$	...	5.374 (16)	5.370 (8)
T_1/K [17]	...	5524	5834 (150)
$\Delta T_1/K$	...	310 (5)	365 (5)
$R_1/R_\odot$	[0.89]	0.75 (1)	0.71 (1)
$R_2/R_\odot$	[0.84]	1.05 (2)	1.02 (1)
$k = r_s/r_g$	0.94 (3)	0.71 (2)	0.70 (2)
l_{1B}	0.570	0.414 (2)	0.430 (1)
l_{1V}	0.555	0.394 (1)	0.417 (1)

the Russell–Merrill model (Sect. 6.2.1)]. The classical analysis, although carefully carried out, failed to yield the correct configuration let alone the correct ratio of sizes for the system. To be sure, the lack of radial velocity data did not help Carr's analysis, but the Russell–Merrill method does not permit these data be used in any rigorous way in the analysis itself. The WD analysis is essentially confirmed by Rainger et al. (1990) who used Hill's LIGHT2 program (see Sect. 6.3.3). However, even today, many intractable *EB* cases, especially those involving systems with thick disks, variable gas streams, atmospheric eclipse phenomena, and associated transient emission features, have not been satisfactorily solved with modern methods. And yet they were bravely tackled with the Russell–Merrill technique, because nothing else was available. Some of those cases still provide vigorous challenges to any light curve modeling code.

1.3.2 The Age of Computational Astrophysics

Significant progress was made in the early 1970s. Models and programs were developed to compute (synthetic) light and velocity curves directly. Such models and programs were based on spherical stars, treated in EBOP, the *Eclipsing Binary Orbit Program* [Nelson & Davis (1972), Etzel (1981), and Etzel (1993)]; ellipsoidal geometry, treated in WINK developed by Wood (1971) and newer versions of EBOP. Lucy (1968), Hill & Hutchings (1970), Wilson & Devinney (1971), and Mochnacki & Doughty (1972a, b) produced models and programs based on Roche

---

[17] Ellipses in this column indicate parameters not derived by Carr. Entries in the adjacent following column were determined from Carr's (1972) data with the WD program by Milone et al. (1991) and Milone (1993, p. 197–199).

[18] As elsewhere in this volume, a value in parentheses following a quantity specifies the uncertainty in that quantity in units of the last decimal place. The uncertainty given is the mean standard error (m.s.e.), or standard deviation, unless indicated otherwise.

[19] Assumed and unadjusted.

geometry. Lucy's was probably the first attempt at direct calculation of light curves; it was limited to over-contact systems describable by a single value of the potential. Only bolometric light curves were computed and effects of mutual irradiation were neglected. Hill and Hutchings provided an early calculation of irradiation effects, assuming a spherical primary for *Algol*. These new approaches permitted the computation of light curves on the basis of complex physical models describing the dynamic forces controlling the stellar mass distributions and the radiation transport in the components' atmospheres. Physical models based on equipotentials and Roche geometry are implemented in the Wilson–Devinney program [Wilson & Devinney (1971), Wilson (1979)] and in LIGHT2 [Hill & Rucinski (1993) and citations contained therein]; see Section 6.3 for more references of the Roche model-based programs. Inversely, the development of physical models and programs for *EB*s led to least-squares determinations of light curve parameters. The first use of least-squares for a physical light curve model was by Wilson & Devinney (1971, 1972, 1973); the next was Lucy (1973). The computational implementation of Roche models coupled with least-squares analyses really started the age of computational astrophysics in *EB* research.

1.3.3 Determining Astrophysical Parameters

The analysis of photometric and spectroscopic data of *EB*s has been performed during the last 35 years by means of synthetic light curves from which parameters such as the mass ratio q, inclination i, and temperature T_2, among other quantities, have been derived with the help of algorithms capable of solving nonlinear least-squares problems. The procedures in use differ significantly, depending on the physical model for the shapes of the stars and the method of solving the (nonlinear) least-squares problem. Three methods used to obtain the light curve solution are *Differential Corrections* [Wilson & Devinney (1971, 1972), Lucy (1973), Napier (1981)]; a derivative-free determination through the use of the *Simplex Algorithm* [Kallrath & Linnell (1987)]; and *Levenberg–Marquardt*-type schemes also known as *Damped Differential Corrections* [Hill (1979), Kallrath et al. (1998)]. The physical parameters derived from the photometric data may depend on the model but should not depend on the least-squares solver.

1.3.4 Later Generations of Light Curve Models

The physical models discussed in this book in greatest detail are those developed by Wilson & Devinney (1971) and later versions by Wilson (1979, 1990, 1998, 2003, 2007). For a brief review of the models we recommend Wilson (1994). Readers interested in the Nelson–Davis–Etzel model should consult Etzel & Leung (1990). Chapter 5 also describes the models and programs by Hill & Rucinski [Hill (1979), Hill & Rucinski (1993)], by Linnell (1984, 1993), by Hadrava (1997, 2004), and others.

The Wilson–Devinney model has been selected as our principal analysis research tool both for its intrinsic virtues and because of its widespread popularity. Its usage has increased to the point that it is used for the majority of light curve analyses performed at the present time (McNally 1991, p. 485), Milone (1993). And last but not least, because it is expandable in the sense that new astrophysical features can be incorporated as the field progresses. Here we mention a few examples: radial velocities (Wilson & Sofia 1976), star spots [Milone et al. (1987), Kang & Wilson (1989)], Kurucz atmospheres (Milone et al. 1992b), line profiles (Mukherjee et al. 1996), radiation pressure effects (Drechsel et al.1995), parameter estimation methods (Kallrath, 1987, 1993), and improvement of convergence by using the Levenberg–Marquardt algorithm (Kallrath et al. 1998).

1.3.5 Astrophysical Problems Solved by Light Curve Methods

Several important astrophysical problems have been solved with major help from light curve solution methods, e.g., the *Algol Paradox* [cf. Pustylnik (2005) for a historical review], the structure of W UMa stars, bolometric albedos of convective envelopes, and undersized subgiants [cf. Wilson (1994)]. Progress in understanding intriguing binaries such as ε Aurigae[20] and β Lyrae has been made by including gas streams and disks in light curve modeling (see Sect. 3.4.4.1).

The improvement of light curve solution methods has contributed to our understanding of physical processes in stars. The earliest work by Russell was applied immediately to the determination of absolute parameters of stars, the precision of which improved as analytic techniques kept pace with observational techniques.

A breakthrough occurred with the introduction of Roche geometry. An example of improved astrophysical understanding through *EB* light curve analysis is the successful modeling of W UMa stars as over-contact systems. These very abundant binaries are excellent laboratories for convection in stars. Their fast orbital motion makes them attractive candidates for gravitational wave astronomy. In the early days these objects, as all *EB*s, were modeled as ellipsoids. The problem was that light curve solutions found detached configurations[21] but W UMas have long been known to be main sequence objects with mass ratios much different from unity. Yet the components have very nearly equal surface temperature as shown both by light curves and spectra. Individual main sequence stars of unequal mass cannot have equal surface temperatures, so Kuiper (1941, 1948) argued that they must be over-contact binaries with energy exchange. This is because energy exchange is not

---

[20] Apparently ε Aurigae's variability was first noticed in the eclipse of 1821 by Johann Fritsch, who seems not to have published the discovery but just passed it along in some way. The first quantitatively observed eclipse was that of 1848, with pre-eclipse observations at least back to 1846. The 1848 observations by Argelander seem not to have been published until 1903 (*Astron. Nachr.* Vol. 164, p. 83) by Ludendorff. The early history of this star is discussed by M. Güssow (1936, Veröff. Univ. Sternwarte Berlin-Babelsberg, Vol. 11, No. 3).

[21] Note that ellipsoidal models could, in principle, produce solutions with overlapping ellipsoids.

possible in detached systems. W UMas are well suited to equipotential representation; isomorphism with the Roche model is excellent, which is not true of an ellipsoidal representation. The gravity effect is also important and nicely taken care of as the surface potential gradient. The overall result has been that many inconsistencies and strange results were eliminated by Roche equipotential models [cf. Lucy (1968), Mochnacki & Doughty (1972a, b), Wilson & Devinney (1973), and Lucy (1973)].

The successful modeling of Algol systems as semi-detached gave quantitative reinforcement to the already accepted solution of the *Algol paradox*: The hotter, more massive primaries were clearly main sequence stars, but the less massive secondaries had radii much too large to be on the main sequence (i.e., they were evolved subgiants or giants). This finding contradicted the well-accepted picture that more massive stars evolve faster than less massive stars (see page 137 for the resolution of the Algol paradox). For a still profitable discussion of Algols, their history, evolution, relation to other binaries, and circumstellar environment refer to Batten (1989). The next stage in the study of Algols is to understand the evolution subsequent to mass transfer episodes through observation of binaries with major circumstellar mass flows. Examples might be the disk-enshrouded binary β *Lyrae* [see Hubeny & Plavec (1991) and Sects. 3.4.4.1 and 3.4.4.3], the unusual binary *V356 Sagittarii* with its opaque ring of recently transferred matter (Wilson & Caldwell 1978), *KU Cygni* with its thick, dusty accretion disk (Olson 1988), *AX Monocerotis* with scattering clouds in its environment (Elias et al. 1997), and many symbiotic stars.

1.4 EB Guide for Researchers in Other Fields

Professional researchers and amateur astronomers from outside the *EB* field can benefit from *EB* analysis techniques. A need may arise because an astrophysically interesting object, say a pulsating star, is in an *EB*, or because of geometric similarity to *EB* problems, as in extrasolar planetary transits. X-ray binary researchers also encounter *EBs* from time to time. *EB* research provides many models and software packages.

Some models and software are physically and geometrically simple, whereas others are sophisticated and allow insertion or revision of physics. Some provide a user-friendly interface, others do not. None serve all needs without modification. This target provides a few hints for those who approach the field the first time and are puzzled about what they read in various publications.

1.4.1 Eclipsing Binaries and Standard Candles

Larger telescopes and powerful instrumentation enable analysis of faint *EBs* in clusters, or even Local Group galaxies such as the Large and Small Magellanic Clouds, M31 (Ribas et al. 2005) and M33 [cf. Hilditch et al. (2005), or Bonanos et al.

(2006a, b)]. These *EB*s can provide direct distance determinations to the host objects [cf. Ribas et al. (2004)] as long as one can obtain the required data, i.e., radial velocity curves and at least one light curve. Although *EB*s allow distance estimations very accurately in favorable cases, they are not really standard candles as are Cepheid variables, RR Lyrae variables, and supernovae. *EB*s are rather individual objects that can provide distances to objects around them if the binary model assumed is correctly chosen. With this proviso, if the universe were to contain only one *EB* and be otherwise empty, one could derive the binary's distance.

1.4.2 Eclipsing Binaries in ExtraSolar Planet Research

Extrasolar planet research has similarities with *EB* studies in the sense that similar data, light, and radial velocity curves are used. A star–planet (or other low-luminosity object) system, with transits and radial velocities for the star only, is in many respects analogous to a single-lined spectroscopic and detached *EB*. As the number of detected transiting planets increases (on July 1, 2009, the *Extrasolar Planet Encyclopedia*[22] listed 59 transiting planets), *EB* analyzing methods become more and more important to extrasolar planet researchers. In favorable cases they can give the mass ratio, inclination, as well as period, rate of period change, semi-major axis, stellar, and planetary radius.

As in a single-lined binary, the mass ratio q cannot be determined from the velocity curve, one can proceed as follows: The mass of the star, $\mathcal{M}_s$, must be assumed, for instance, based on its spectral characteristics, from evolutionary models, e.g., assuming the star is on the main sequence. As for inclinations i between $80°$ and $90°$, $\sin i$ and $\sin^3 i$ vary only between 0.98 and 1, and 0.94 and 1, resp., and as q is usually small, e.g., $q \leq 10^{-3}$, Kepler's third law

$$P^2 = \frac{4\pi^2 a^3}{G\mathcal{M}_s(1+q)} \qquad (1.4.1)$$

enables us to derive a good estimate of the semi-major axis, a, from the period P. Initially, adopting a reasonable value for the planetary mass, $\mathcal{M}_p$, q follows as $q = \mathcal{M}_p/\mathcal{M}_s$.

Limb-darkening law and coefficients can be taken from Van Hamme (1993). The star's rotational period is most likely much smaller than P and requires the rotation parameter F_1 to be set to values much smaller than unity. As the time of minima in planetary transits may not be well defined, it is an option to use time as an independent parameter and fit the ephemeris parameters (reference epoch T_0, period P, and possibly rate of period change dP/dt) together with inclination i, a and systemic velocity, V_γ, and radii (or Roche potentials). As the triple (P, a, q) inserted in (1.4.1) might give different values for the stellar mass than the pregiven value $\mathcal{M}_s$, some iterations are necessary. Examples are given in Sect. 5.4.3.3.

---

[22] URL http://exoplanet.eu/catalog-transit.php.

1.4.3 Nomenclature: Primary and Secondary Component

It can be very frustrating to see in the literature the terms *primary* and *secondary* component used in contrasting ways among photometrists and spectroscopists. Section 2.8 tries to cover this problem and to give some helpful orientation. Related to this problem is also the *definition of zero phase*. Light curve observers associate zero phase with the primary minimum whereas for radial velocity curve astronomers the time of periastron passage sometimes appears to be more suitable.

1.4.4 Where Are the Radii?

Most *EB* models and programs nowadays do not use radii as adjustable parameters but rather the Roche potentials Ω_1 and Ω_2 which cover the full range of morphological types from detached, semi-detached, to over-contact binaries as described in Sect. 3.1.6 . The star–planet systems in extrasolar planet research are detached systems. For those, the relative radii and Roche potentials are coupled by $r_i \sim 1/\Omega_i$.

1.4.5 Precession and Apsidal Motion

Unfortunately the term "precession" often is used unadvisedly not only in the binary star literature but also in other areas. For example, many textbooks and public documentaries speak of the "precession" of planet Mercury's orbit in regard to a well-known prediction of General Relativity Theory (GRT). However, the described GRT phenomenon is orbit *rotation*, not precession. Orbit rotation (apsidal motion) is rotation *within* the orbit's own plane, so only one plane is involved, whereas precession involves two planes and may be described in terms of (conical) motion of one plane's normal around the other's. So although precession and orbit rotation may arise in a common context and perhaps have some common physics, they are geometrically distinct. Sometimes the term *nodal precession* is used for true precession and *precession* (unqualified) for orbit rotation. That does make a distinction if the adopted meaning is made clear, but why use confusing terminology? With such dual terminology, one might refer to precession as true precession, but then there would be an extra unnecessary and conflicting terminology. Precession need not refer to an orbit but can be precession of a gyroscope, including precession of the Earth's equatorial plane (equivalently precession of its rotation axis). We recommend taking a page from a physics text, where precession refers to phenomena such as gyroscopic precession that necessarily involve two planes.

As outlined in Sect. 5.2.1, a third body in a binary system causes apsidal motion *accompanied* by precession of the orbital rotation axis. Both effects result from the rotation of the binary's orbital frame around the barycenter and are effective on the same timescale.

1.4.6 Looking for Eclipsing Binary Standard Software

The *EB* literature is full of different models and programs, and until recently there has been no commonly accepted standard model or data reduction program. The Wilson–Devinney (WD) model is the one most frequently used, but probably not everyone in the community of *EB* researchers would agree that the WD model is a de facto standard.

1.4.7 Analytic Techniques and Numerical Analysis

The *EB* field is sometimes advanced on this. Selecting appropriate increments for computing numerical derivatives is an issue. A crucial part of the numerical analysis is also the use of proper weights and the computation of standard errors of the estimated parameters. The interpretation of the standard errors requires great care if parameters are strongly correlated.

1.5 Selected Bibliography

Utilia et delectabilia   (*Useful and delightful*)

This section is intended to guide the reader to recommended books or articles on variables stars and *EB*s.

- *The Binary Stars* by Aitken (1935, first edition) and its revised 1964 version by Jack T. Kent – a classic, and a good source of the physics and mathematics in the treatment of the astrometry and radial velocities in visual double stars studies.
- *Properties of Double Stars* by Binnendijk (1960) – another classic work, treating also *EB*s.
- *Binary and Multiple Systems of Stars* by Batten (1973). A good general introduction to binaries.
- Readers interested in Algol binaries are referred to the book *Algols* by Batten (1989).
- *Interacting Binaries* by Sahade & Wood (1978). A treatment of the types of binary stars close enough to each other to affect each other's shape and evolution. The treatment, primarily from a historical point of view, is dated but still of great interest.
- *Interacting Binary Stars* by Pringle & Wade (1985). An excellent composite treatment by a number of experts in the field treating both the physical states and the evolution of interacting binaries.
- *Cataclysmic Variable Stars* by Warner (1995). The definitive work on these highly interacting binary stars.
- For the nomenclature and classification of variable stars we refer the reader to the book *Light Curves of Variable Stars* by Sterken and Jaschek (1997), and in particular to Chap. 1 in this reference. It presents a wealth of typical light and color curves to allow identification, together with a detailed and up-to-date description of each subclass.

- A brief review and phenomenological approach to *EB*s is provided in Sterken (1997, Chap. 6). This chapter, besides Algol, β Lyrae, and W UMa systems, also discusses a more special group: The RS Canum Venaticorum-type systems.
- *Binary Stars – A Pictorial Atlas* by Terrell et al. (1992) contains parameters, light curves, and three-dimensional views of about 335 *EB*s.
- The textbook *Introduction to Close Binary Stars* by Hilditch (2001) provides a thorough introduction to binary stars as well as related aspects in stellar astrophysics, stellar structure and evolution, and observational astrophysics.
- *Resolving the Algol Paradox and Kopal's Classification of Close Binaries with Evolutionary Implication* is an interesting historical write-up by Pustylnik (2005) on one of the greatest contributions of the *EB* research to astrophysics.
- *Astrophysics of Variable Stars* by Sterken & Aerts (2006) provides well-prepared review contributions on data sources of variable stars, binary stars and *EB*s, and also stellar pulsation.
- The proceedings of IAU Symposium No. 240 (2006) under the title *Binary Stars as Critical Tools & Tests in Contemporary Astrophysics* edited by Hartkopf et al. (2007) provide an excellent overview on state-of-the-art and ongoing activities in close binary research. They review major advances in instrumentations and techniques, new observing techniques and reduction methods and discuss binary stars as critical tools and tests for studying a wide variety of important astrophysical problems.
- *Understanding Variable Stars* by Percy (2007) provides a basic exposition of variable stars for college students with some background in astronomy and for active amateur astronomers. It discusses both the history of the subject and the properties of each of the variable star groups.
- *An Introduction to Close Binary Stars* by Hilditch (2001) is a somewhat advanced treatment of binary stars, intended for upper level under-graduate students and graduate students, this book deals with the full range of interacting variables, from X-ray binaries and cataclysmic variables to over-contact systems. It discusses both the structure and the evolution of these objects. Its concluding chapter discusses image reconstruction.
- *Brightest Diamond in the Night Sky* by Holberg (2007) is dedicated to the historical and astronomical importance of the Sirius system, a binary star involving the brightest star in the sky after the Sun and a white dwarf, the compact core of a more massive companion that evolved more quickly than its bright, white companion.

References

Aitken, R. G.: 1964, *The Binary Stars*, Dover Publications, Philadelphia, PA, 3rd edition
Baade, W.: 1944, The Resolution of Messier 32, NGC 205, and the Central Region of the Andromeda Nebula, *ApJ* **100**, 137–146
Baldwin, J. E., Beckett, M. G., Boysen, R. C., Burns, D., Buscher, D. F., Cox, G. C., Haniff, C. A., Mackay, C. D., Nightingale, N. S., Rogers, J., Scheuer, P. A. G., Scott, T. R., Tuthill, P. G., Warner, P. J., Wilson, D. M. A., & Wilson, R. W.: 1996, The First Images from an

Optical Aperture Synthesis Array: Mapping of Capella with COAST at Two Epochs, *A&A* **306**, L13–L16

Batten, A. H. (ed.): 1973, *Binary and Multiple Systems of Stars*, Pergamon Press, Oxford, UK

Batten, A. H. (ed.): 1989, *Algols*, Kluwer Academic Publishers, Dordrecht, Holland

Beckers, J. M.: 1993, Adaptive Optics for Astronomy: Principles, Performance, and Applications, *Ann. Rev. Astron. Astrophys.* **31**, 13–62

Bessel, F. W.: 1845, Über Veränderlichkeit der Eigenen Bewegung der Fixsterne, *Astron. Nachr.* **22**, 145–167

Binnendijk, L.: 1960, *Properties of Double Stars*, University of Pennsylvannia Press, Philadelphia, PA

Bonanos, A. Z., Stanek, K. Z., Kudritzki, R. P., Macri, L., Sasselov, D. D., Kaluzny, J., Bersier, D., Bresolin, F., Matheson, T., Mochejska, B. J., Przybilla, N., Szentgyorgyi, A. H., Tonry, J., & Torres, G.: 2006b, The First DIRECT Distance to a Detached Eclipsing Binary in M33, *Ap. Sp. Sci.* **304**, 207–209

Bonanos, A. Z., Stanek, K. Z., Kudritzki, R. P., Macri, L. M., Sasselov, D. D., Kaluzny, J., Stetson, P. B., Bersier, D., Bresolin, F., Matheson, T., Mochejska, B. J., Przybilla, N., Szentgyorgyi, A. H., Tonry, J., & Torres, G.: 2006a, The First DIRECT Distance Determination to a Detached Eclipsing Binary in M33, *ApJ* **652**, 313–322

Bradstreet, D. H.: 1993, Binary Maker 2.0 – An Interactive Graphical Tool for Preliminary Light Curve Analysis, in E. F. Milone (ed.), *Light Curve Modeling of Eclipsing Binary Stars*, pp 151–166, Springer, New York

Bradstreet, D. H. & Steelman, D. P.: 2004, *Binary Maker 3.0*, Contact Software, Norristown PA, 19087.

Brown, R. H.: 1968, Measurement of Stellar Diameters, *Ann. Rev. Astron. Astrophys.* **6**, 13–18

Brown, R. H., Davis, J., & Allen, L. R.: 1974a, The Angular Diameters of 32 Stars, *MNRAS* **167**, 121–136

Brown, R. H., Davis, J., Lake, R. J. W., & Thompson, R. J.: 1974b, The Effect of Limb-Darkening on Measurements of Angular Size with an Intensity Interferometer, *MNRAS* **167**, 475–483

Caroll, S. M., Guinan, E. F., McCook, G. P., & Donahue, R. A.: 1991, Interpreting Epsilon Aurigae, *ApJ* **367**, 278–287

Carr, R. B.: 1972, Photoelectric UBV Photometry of TY Bootis, *AJ* **77**, 155–159

Chen, K.-Y. & Reuning, E. G.: 1966, Infrared Photometry of beta Persei., *AJ* **71**, 283–296

Couteau, P.: 1981/78, *Observing Visual Double Stars; Original Title: L'Observation des Étoiles Doubles Visuelles*, MIT Press; original publisher Flammarion, Cambridge, MA, A. H. Batten translated this 2nd edition

Davidge, T. J. & Milone, E. F.: 1984, A Study of the O'Connell Effect in the Light Curves of Eclipsing Binaries, *ApJ Suppl.* **55**, 571–584

Drechsel, H., Haas, S., Lorenz, R., & Gayler, S.: 1995, Radiation Pressure Effects in Early-Type Close Binaries and Implications for the Solution of Eclipse Light Curves, *A&A* **294**, 723–743

Etzel, P. B.: 1981, A Simple Synthesis Method for Solving the Elements of Well-Detached Eclipsing Systems, in E. B. Carling & Z. Kopal (eds.), *Photometric and Spectroscopic Binary Systems*, pp. 111–120, D. Reidel, Dordrecht, Holland

Etzel, P. B.: 1993, Current Status of the EBOP Code, in E. F. Milone (ed.), *Light Curve Modeling of Eclipsing Binary Stars*, pp. 113–124, Springer, New York

Etzel, P. B. & Leung, K.-C.: 1990, Synthesis Methods for Eclipsing-Binary Light Curves, in I. Ibanoglu (ed.), *Active Close Binaries*, pp. 873–879, Kluwer Academic Publishers, Dordrecht, Holland

Glasby, J. S.: 1969, *Variable Stars*, Vol. 50 of *International Series of Monographs on Natural Philosophy*, Harvard University Press, Cambridge, MA

Hadrava, P.: 1997, *FOTEL 3 – User's Guide*, Technical report, Astronomical Institute of the Academy of Sciences of the Czech Republic, 25165 Ondrejov, Czech Republic

Hadrava, P.: 2004, FOTEL 4 – User's guide, *Publications of the Astronomical Institute of the Czechoslovak Academy of Sciences* **92**, 1–14

References

Hall, D. S.: 1975, On the period variations and light curve changes in RS Canum Venaticorum, *Acta Astronomica* **25**, 215–224

Hartkopf, W. I., Guinan, E. F., & Harmanec, P. (eds.): 2007, *Binary Stars as Critical Tools and Tests in Contemporary Astrophysics*, No. 240 in Proceedings IAU Symposium, Dordrecht, Holland, Kluwer Academic Publishers

Hellier, C.: 2001, *Cataclysmic Variable Stars: How and Why they Vary*, Springer-Praxis

Hilditch, R. W.: 2001, *An Introduction to Close Binary Stars*, Cambridge University Press, Cambridge, UK

Hilditch, R. W., Howarth, I. D., & Harries, T. J.: 2005, Forty Eclipsing Binaries in the Small Magellanic Cloud: Fundamental Parameters and Cloud Distance, *MNRAS* **357**, 304–324

Hill, G.: 1979, Description of an Eclipsing Binary Light Curve Computer Code with Application to Y Sex and the WUMA Code of Rucinski, *Publ. Dom. Astrophys. Obs.* **15**, 297–325

Hill, G. & Hutchings, J. B.: 1970, The Synthesis of Close-Binary Light Curves. I. The Reflection Effect and Distortion in Algol, *ApJ* **162**, 265–280

Hill, G. & Rucinski, S. M.: 1993, LIGHT2: A light-curve modeling program, in E. F. Milone (ed.), *Light Curve Modeling of Eclipsing Binary Stars*, pp. 135–150, Springer, New York

Hoffmeister, C., Richter, G., & Wenzel, W.: 1984, *Veränderliche Sterne*, Springer-Verlag, Heidelberg, 2nd edition

Hoffmeister, C., Richter, G., & Wenzel, W.: 1985, *Variable Stars*, Springer-Verlag, Heidelberg, German to English translated 2nd edition

Holberg, J. B.: 2007, *Sirius – Brightest Diamond in the Night Sky*, Springer, New York

Howell, S. B., VanOutryve, C., Tonry, J. L., Everett, M. E., & Schneider, R.: 2005, A Search for Variable Stars and Planetary Occultations in NGC 2301. II. Variability, *PASP* **117**, 1187–1203

Hubeny, I. & Plavec, M. J.: 1991, Can Disk Model Explain β Lyrae?, *AJ* **102**, 1156–1170

Hut, P.: 1981, Tidal Evolution in Close Binary Systems, *A&A* **99**, 126–140

Jurkevich, I.: 1970, Machine Solutions of Light Curves of Eclipsing Binary Systems, in A. Beer (ed.), *The Henry Norris Russell Memorial Volume*, Vol. 12 of *Vistas*, pp. 63–116, Pergamon Press, Oxford, UK

Kallrath, J. & Linnell, A. P.: 1987, A New Method to Optimize Parameters in Solutions of Eclipsing Binary Light Curves, *Astrophys. J.* **313**, 346–357

Kallrath, J., Milone, E. F., Terrell, D., & Young, A. T.: 1998, Recent Improvements to a Version of the Wilson–Devinney Program, *ApJ Suppl.* **508**, 308–313

Kang, Y. W. & Wilson, R. E.: 1989, Least-Squares Adjustment of Spot Parameters for Three RS CVn Binaries, *AJ* **97**, 848–865

Kelley, D. H. & Milone, E. F.: 2005, *Exploring Ancient Skies – An Encyclopedic Survey of Archaeoastronomy*, Springer, New York

Kholopov, P. N. (ed.): 1985, *General Catalogue of Variable Stars*, "NAUKA" Publishing House, Moscow, 4th edition

Kim, H.-I.: 1989, BV Light Curve Analysis of Algol, *ApJ* **342**, 1061–1067

Krautter, J.: 1997, X-Ray Binaries, in C. Sterken & C. Jaschek (eds.), *Light Curves of Variable Stars – A Pictorial Atlas*, Cambridge University Press, Cambridge, UK

Kuiper, G. P.: 1941, On the Interpretation of Beta Lyrae and Other Close Binaries, *ApJ* **93**, 133–177

Kuiper, G. P.: 1948, Note on W Ursae Majoris Stars, *ApJ* **108**, 451–452

Latham, D. W., Mazeh, T., Torres, G., Carney, B. W., Stefanik, R. P., & Davis, R. J.: 1992, Spectroscopic Binaries in the Halo, in A. Duquennoy & M. Mayor (eds.), *Binaries as Tracers of Stellar Evolution*, pp. 139–144, Cambridge University Press, Cambridge, UK

Lindenblad, I.: 1970, Relative Photographic Positions and Magnitude Difference of the Components of Sirius, *AJ* **75**, 841–847

Lucy, L. B.: 1968, The Light Curves of W Ursae Majoris, *ApJ* **153**, 877–884

Lucy, L. B.: 1973, The Common Convective Envelope Model for W Ursae Majoris Systems and the Analysis of their Light Curves, *Ap. Sp. Sci.* **22**, 381–392

Mason, B. D., Hartkopf, W. I., Gies, D. R., McAlister, H. A., & Bagnuolo, W. G.: 1996, Binaries in Clusters and the Field: First Results of a Speckle Survey of O Stars, in E. F. Milone & J.-C.

Mermilliod (eds.), *The Origins, Evolution, and Destinies of Binary Stars in Clusters*, pp. 40–43, A.S.P. Conference Series, Provo, UT

McNally, D. (ed.): 1991, *Reports on Astronomy Symposium on the Theory of Computing*, No. XXIA in Close Binary Stars, The Netherlands, IAU

Michelson, A. A.: 1920, On the Application of Interference Methods to Astronomical Measurements, *ApJ* **51**, 257–262

Michelson, A. A. & Pease, F. G.: 1921, Measurement of the Diameter of α Orionis with the Interferometer, *ApJ* **53**, 249–259

Milone, E. F. (ed.): 1993, *Light Curve Modeling of Eclipsing Binary Stars*, Springer, New York

Milone, E. F., Chia, T. T., Castle, K. G., Robb, R. M., & Merrill, J. E.: 1980, RW Comae Berenices I. Early Photometry and UBV Light Curves, *ApJ Suppl.* **43**, 339–364

Milone, E. F., Groisman, G., Fry, D. J. I. F., & Bradstreet, H.: 1991, Analysis and Solution of the Light and Radial Velocity Curves of the Contact Binary TY Bootis, *ApJ* **370**, 677–692

Milone, E. F., Wilson, R. E., & Hrivnak, B. J.: 1987, RW Comae Berencis. III. Light Curve Solution and Absolute Parameters, *ApJ* **319**, 325–333

Mochnacki, S. W. & Doughty, N. A.: 1972a, A Model for the Totally Eclipsing W Ursae Majoris System AW UMa, *MNRAS* **156**, 51–65

Mochnacki, S. W. & Doughty, N. A.: 1972b, Models for Five W Ursae Majoris Systems, *MNRAS* **156**, 243–252

Mukherjee, J. D., Peters, G. J., & Wilson, R. E.: 1996, Rotation of Algol Binaries – A Line Profile Model Applied to Observations, *MNRAS* **283**, 613–625

Napier, W. M.: 1981, A Simple Approach to the Analysis of Eclipsing Binary Light Curves, *MNRAS* **194**, 149–159

Nelson, B. & Davis, W. D.: 1972, Eclipsing-Binary Solutions by Sequential Optimization of the Parameters, *ApJ* **174**, 617–628

Olson, E. C.: 1988, Photometry of Long-Period Algol Binaries: IV. KU Cygni and its Thick, Dusty Accretion Disk, *AJ* **96**, 1439–1446

Osterbrock, D. E.: 2001, Who Really Coined the Word Supernova? Who First Predicted Neutron Stars?, in *Bulletin of the American Astronomical Society*, Vol. 33 of *Bulletin of the American Astronomical Society*, p. 1330

Pease, F. G.: 1925, Measurement of the Spectroscopic Binary Star Mizar with the Interferometer, *Proc. Nat. Acad. Sci. USA* **11**, 356–357

Percy, J. R.: 2007, *Understanding Variable Stars*, Cambridge University Press, Cambridge, UK, Understanding variable stars/John R. Percy. Cambridge: Cambridge University Press, 2007. xxi, 350 p. : ill. ; 26 cm. ISBN: 9780521232531 (hbk.)

Petit, M.: 1985, *Variable Stars*, John Wiley & Sons, New York

Petrie, R. M.: 1939, The Determination of the Magnitude Difference between the Components of Spectroscopic Binaries, *Publ. Dom. Astrophys. Obs.* **7**, 205–238

Plavec, M. J.: 1968, Mass Exchange and Evolution of Close Binaries, *Adv. in Astron. Astrophys.* **6**, 201–278

Popper, D. M.: 1980, Stellar Masses, *Ann. Rev. Astron. Astrophys.* **18**, 115–164

Pringle, J. E. & Wade, R. A.: 1985, *Interacting Binary Stars*, Cambridge University Press, Cambridge, UK

Proctor, D. D. & Linnell, A. P.: 1972, Computer Solution of Eclipsing-Binary Light Curves by the Method of Differential Corrections, *ApJ Suppl.* **24**, 449–477

Pustylnik, I. B.: 2005, Resolving the Algol Paradox and Kopal's Classification of Close Binaries with Evolutionary Implications, *Ap. Sp. Sci.* **296**, 69–78

Rainger, R. P., Hilditch, R. W., & Bell, S. A.: 1990, The Contact Binary System TY Bootis, *MNRAS* **246**, 42–46

Ribas, I., Jordi, C., Vilardell, F., Fitzpatrick, E. L., Hilditch, R. W., & Guinan, E. F.: 2005, First Determination of the Distance and Fundamental Properties of an Eclipsing Binary in the Andromeda Galaxy, *ApJ Letters* **635**, L37–L40

Ribas, I., Jordi, C., Vilardell, F., Giménez, Á., & Guinan, E. F.: 2004, A Program to Determine a Direct and Accurate Distance to M31 from Eclipsing Binaries, *New Astronomy Review* **48**, 755–758

References

Roche, Éd.: 1849, La Figure d'une Masse Fluide Soumise à l'Attraction d'un Point Éloigné (Premiere Partie), *Academie des Sciences et Lettres de Montpellier, Mémoires de la Section des Sciences* **Tome Premier (1847–1850)**, 243–262

Roche, Éd.: 1850, La Figure d'une Masse Fluide Soumise à l'Attraction d'un Point Éloigné (Seconde Partie), *Academie des Sciences et Lettres de Montpellier, Mémoires de la Section des Sciences* **Tome Premier (1847–1850)**, 333–348

Russell, H. N.: 1948a, Idealized Models and Rectified Light Curves for Eclipsing Variables, *ApJ* **108**, 388–412

Russell, H. N.: 1948b, The Royal Road to Eclipses, in *Centennial Papers*, Vol. 7 of *Harvard Observatory Monographs*, pp. 181–209, Harvard College Observatory, Cambridge, MA

Russell, H. N. & Merrill, J. E.: 1952, The Determination of the Elements of Eclipsing Binary Stars, *Princeton. Obs. Contr.* **26**, 1–96

Sahade, J. & Wood, F. B.: 1978, *Interacting Binaries*, Pergamon Press, Oxford, UK

Scarfe, C. D., Barlow, D. J., Fekel, F. C., Rees, R. F., Lyons, R. W., Bolton, C. T., McAlister, H. A., & Hartkopf, W. I.: 1994, The Spectroscopic Triple System HR 6469, *AJ* **107**, 1529–1541

Schlosser, W., Schmidt-Kaler, T., & Milone, E. F.: 1991, *Challenges of Astronomy*, Springer, New York

Shao, M. & Colavita, M. M.: 1992, Long-Baseline Optical and Infrared Interferometry, *Ann. Rev. Astron. Astrophys.* **30**, 457–498

Smith, H. J.: 1955, Low-Luminosity Intrinsic Variables with Periods Less than 0.2 Day, *AJ* **60**, 179–180

Sterken, C. & Aerts, C. (eds.): 2006, *ASP Conf. Ser. 349: Astrophysics of Variable Stars*

Sterken, C. & Jaschek, C. (eds.): 1997, *Light Curves of Variable Stars – A Pictorial Atlas*, Cambridge University Press, Cambridge, UK

Strohmeier, W.: 1972, *Variable Stars*, Vol. 50 of *International Series of Monographs on Natural Philosophy*, Pergamon Press, Oxford

Terrell, D., Mukherjee, J. D., & Wilson, R. E.: 1992, *Binary Stars – A Pictorial Atlas*, Krieger Publishing Company, Malabar, FL

Van Hamme, W.: 1993, New Limb-Darkening Coefficients for Modeling Binary Star Light Curves, *AJ* **106**, 2096–2117

Van Hamme, W., Hall, D. S., Hargrove, A. W., Henry, G. W., Wasson, R., Barksdale, W. S., Chang, S., Fried, R. E., Green, C. L., Lines, H. C., Lines, R. D., Nielsen, P., Powell, H. D., Reisenweber, R. C., Rogers, C. W., Shervais, S., & Tatum, R.: 1994, The Two Variables in the Triple System HR 6469 = V819 Her: One Eclipsing, One Spotted, *AJ* **107**, 1521–1528

Warner, B. (ed.): 1995, *Cataclysmic Variable Stars*, Cambridge University Press, Cambridge, UK

Wesselink, A. J.: 1969, Surface Brightnesses in the U, B, V System with Applications on Mv and Dimensions of Stars, *MNRAS* **144**, 297–311

Wesselink, A. J., Paranya, K., & DeVorkin, K.: 1972, Catalogue of Stellar Dimensions, *A&A Suppl.* **7**, 257–289

Wilson, R. E.: 1974, Binary Stars – A Look at Some Interesting Developments, *Mercury* pp. 4–12

Wilson, R. E.: 1979, Eccentric Orbit Generalization and Simultaneous Solution of Binary Star Light and Velocity Curves, *ApJ* **234**, 1054–1066

Wilson, R. E.: 1990, Accuracy and Efficiency in the Binary Star Reflection Effect, *ApJ* **356**, 613–622

Wilson, R. E.: 1994, Binary-Star Light-Curve Models, *PASP* **106**, 921–941

Wilson, R. E.: 1998, *Computing Binary Star Observables (Reference Manual to the Wilson–Devinney Program*, Department of Astronomy, University of Florida, Gainesville, FL, 1998 edition

Wilson, R. E.: 2001, Variable Stars, in P. Murdin (ed.), *Encyclopedia of Astronomy and Astrophysics*, Vol. 4, pp. 3424–3433, Institute of Physics Publishing (Nature Publishing Group), Bristol, Philadelphia (London)

Wilson, R. E.: 2007, Close Binary Star Observables: Modeling Innovations 2003–2006, in W. I. Hartkopf, E. F. Guinan, & P. Harmanec (eds.), *Binary Stars as Critical Tools and Tests in Contemporary Astrophysics*, No. 240 in Proceedings IAU Symposium, pp. 188–197, Kluwer Academic Publishers, Dordrecht, Holland

Wilson, R. E. & Caldwell, C. N.: 1978, A Model of V356 Sagittarii, *ApJ* **221**, 917–925

Wilson, R. E. & Devinney, E. J.: 1971, Realization of Accurate Close-Binary Light Curves: Application to MR Cygni, *ApJ* **166**, 605–619

Wilson, R. E. & Devinney, E. J.: 1972, Addendum and Erratum to 1971 paper, *ApJ* **171**, 413

Wilson, R. E. & Devinney, E. J.: 1973, Fundamental Data for Contact Binaries: RZ Comae Berenices, RZ Tauri, and AW Ursae Majoris, *ApJ* **182**, 539–547

Wood, D. B.: 1971, An Analytic Model of Eclipsing Binary Star Systems, *AJ* **76**, 701–710

Chapter 2
The Database and Methods of Data Acquisition

The intention here is to summarize aspects of observational astronomy relevant to light curve acquisition and modeling. Thus, we discuss passband[1] profiles of observational data because some codes no longer consider the observations to be monochromatic fluxes. Passbands are finite wavelength-width stretches of the electromagnetic spectrum. Sources of errors in the observational data are discussed because light curve analysis codes can have weighting which is light-level dependent. Such errors depend to a large extent on observational techniques, so these too need to be described. The incorporation of Doppler profile analysis techniques in light curve codes requires high-resolution spectrophotometric data with excellent signal-to-noise ratios to extract profile information. The same can be said for magnetometry. Polarization data require another dimension of information in the form of a series of position angle measurements.

2.1 Photometry

Fiat lux et lux facta est (And God said, "Let there be light,"
Genesis i:3   and there was light)

2.1.1 Photoelectric Photometry

Photoelectric photometry until the last decade or so has been and arguably still is the most precise and accurate means of obtaining flux measurements in optical astronomy. There are a number of reasons for this situation. Under most circumstances, the best possible precision for a source of given brightness is obtained when shot noise, the Poisson statistical variation in the photon count, is the dominant source[2] of uncertainty. Under such circumstances, the precision is said to be "photon limited"

---

[1] Some astronomers prefer the term *bandpass*.

[2] This statement is nearly correct, but if a bright star would have a seeing noise of less than 1%, this could change.

and the error is proportional to $\sqrt{N}$; when variation in the background dominates, the observation is said to be "background limited" (for further discussion of sources of noise in astronomical photometry and their effects on light curve analysis, see Section 4.1.1.5). Photoelectric photometers have been capable of higher precision than imaging devices, basically because CCDs (see Sect. 2.1.4) originally were not designed as optical devices but as electronic switches; consequently their properties have not been conducive to photometry of the highest accuracy. See Young et al. (1991) for a fuller discussion. PMTs can be used to observe brighter stars, even if neutral density filters are used to attenuate the light; bright stars create challenges for CCDs because of either saturation or image profile wing effects. Notwithstanding this situation, CCD detectors have improved greatly over the past 20 years, and the efforts of meticulous observers and analysts have been able to achieve remarkable precision. Moreover, further improvements are being made, and, indeed, have been made, as Howell et al. (2003) and Johnson et al. (2009) demonstrate. These papers discuss the use of orthogonal transfer CCDs. In the latter paper one of these CCDs is used to achieve a precision of 0.47 millimagnitudes in the observtaion of an extra-solar planet transit eclipse. A fine general source for CCD astronomy is Howell (2000). See, e.g., Gilliland et al. (1993) for an application of CCD photometry to astroseismology. Finally, CCDs have major advantages in doing photometric studies of multiple faint objects in crowded fields. See Howell et al. (2005) for an example.

A photoelectric photometer consists of an individual phototube or solid state detector which generates a current when light falls on it. In a photomultiplier, the current may be accelerated from the photocathode surface onto a succession of other surfaces (called dynodes) at monotonically increasing potentials. The accelerating voltages cause increasing electron ejection across each successive dynode stage, creating an increasing cascade of electrons and providing gains of up to 10^7. Particular passbands are defined by colored glass or thin-metal-coated interference filters in combination with the spectral sensitivity of the detector. In practice, the actual net instrumental passbands are usually not known very well and transformations are achieved by observing a list of "standard" stars with well-established magnitudes and color indices in the standard system to be emulated. Examples of broad passbands (of order 100 nm) are the *UBVRI* of the Johnson system [see Johnson & Morgan (1953) and Johnson (1966)] and R_C and I_C bands of the Cousins (1976, 1978) system; examples of intermediate passbands ($\leq\sim$30 nm) are the *uvby* of the Strömgren system (Strömgren 1966). In these examples, the letters indicate the color of the transmitted light in an almost intuitive way: ultraviolet, blue, visual, red, and infrared for Johnson–Cousins, but ultraviolet, violet, blue, and yellow for the Strömgren system. There are many other systems, created to study specific astrophysical problems. These include the Geneva (Hauck 1968), Vilnius (Straižys 1975),Walraven, DDO, and Washington systems (Canterna 1976). See Bessell (1979) for intercomparisons and examples of other cases, and Moro & Munari (2000) for a catalogue of many photometric systems, extended in Fiorucci & Munari (2003). This work has been updated and appears at this writing at the url http://ulisse.pd.astro.it/Astro/ADPS/Systems/index.html. In addition to these,

2.1 Photometry

however, spectrophotometric data can be integrated over as large a passband as necessary to provide data for light curve analysis over virtually any wavelength range for which adequate atmospheric models exist. Spectrophotometers, however, suffer from scattered light problems and are difficult to calibrate. A much more serious difficulty for Earth-bound observations is the problem of obtaining adequate comparison star spectrophotometry to deal with atmospheric extinction (see Sect. 2.2.2).

A calibrated load resistor can convert the small phototube current to a voltage drop which can be recorded by analog or digital means, or the pulses of cascaded electrons from single photon impacts on the photocathodes can be counted. The pulse-counting technique has a drawback, viz., the problem of pulse overlap or coincidence from the brighter stars. Generally detection systems are unable to distinguish between simultaneous or nearly simultaneous photon arrivals, so that a coincidence or "dead time" correction must be applied to the observed count rate. Nevertheless, because of its potentially high measurement precision and the convenience of data handling and storage, pulse counting is the most common type of photoelectric photometry registry now in use. Note that the capability to provide large numbers of simultaneously observed comparison stars is an advantage of CCDS, if the PSF across the frame is uniform and unvarying. If they are not, great care is required in the image processing to ensure that they do not actually degrade the photometry. Howell et al. (1988) and Howell (1992) demonstrates how to carry out differential CCD photometry.

2.1.2 Two-Star Photometers

A serious practical limitation to good photometry is a nonconstant sky. The sky is variable in both transparency and brightness. The latter may be due to atmospheric emissions (aurora, airglow) or to the reflection of city lights from clouds and haze. Therefore, it is a major advantage to be able to use a nearby comparison star to obtain differential light curves if the observations can be obtained at a faster rate than the sky undergoes variation. For this reason, two-star (more generally, "two-channel") photometers have become important for photometric work. The visual polarizing photometer used at Harvard College Observatory since the 1870s and at Princeton Observatory from about 1911 onward and the analogue electronic version, the Walraven photometer used in South Africa in the 1950s, were pioneering efforts and produced many useful light curves. Today there are many variants of these photometers which have been developed at different observatories. Most depend on having separate light path detectors, or electronics for the different channels. The one with which we are most familiar does not. It is the *Rapid Alternate Detection System* (RADS), used at the University of Calgary's Rothney Astrophysical Observatory since about 1981 (Milone et al. 1982). This system (Figs. 2.1 and 2.2) employs a single pulse-counting detector and a swiveled secondary mirror which is driven by the dial-in settings of a function generator. The amplitude of the throw, the

Fig. 2.1 RADS instrument on the RAD's 41-cm telescope. Shown is the controller for the RADS. It consists of a function generator which controls the successive positions of the secondary mirror within a cycle involving successive settings on the program star, sky near the comparison star, the comparison star, and the sky near the program star

duty cycle, and the delay time for mirror settling can be entered separately for each of four positions. The delay time depends on the aperture, because a smaller aperture requires more stability for the image as the mirror ringing dies down. The sum of

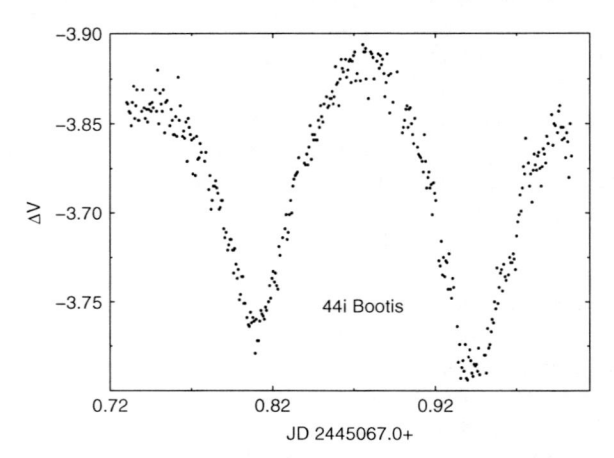

Fig. 2.2 RADS differential photometry. Differential V RADS light curve *44i Bootis* on JD 2445067 published by Robb and Milone (1982)

the duty cycles and time delay determines the chopping frequency, which may be as high as 20 or 30 Hz; the system is usually operated closer to 1 Hz, however, because of the overhead caused by image motion and the delays at each station. The chopping line of the mirror may be rotated to coincide with the line between two stars. Two of the channels are usually consigned to observe the sky near the stars (near cities, "cloudy skies" usually also means "bright skies," so the flux spatially and temporally near the stars must be sampled also). The pulse-counting electronics are gated to the position of the mirror, and auxiliary controls permit the programming of filter changes. RADS works for stars which are separated by up to 45 arc-min. It thus may be superior to imaging devices with typical fields of view of a few arc-minutes for observations of objects in sparse areas of the sky, because it is highly desirable to have stars of similar spectral energy distribution and brightness to avoid systematic effects. Schiller & Milone (1990) discuss a study in which single-channel photoelectric and CCD photometry were carried out simultaneously on similar sized telescopes at McDonald Observatory of a star (the δ Scuti star *DY Herculis*) in a sparse CCD frame. The photoelectric light curve was far superior if only because the small chip size allowed no bright comparison star to be imaged simultaneously.

Two-star systems such as RADS are insensitive to first-order extinction, and even to "second-order" extinction (many astronomers prefer to use the more strictly correct term "color-dependent extinction") if the comparison star is carefully matched in spectral distribution. Milone & Robb (1983) discuss the practical use of the system and demonstrate its effectiveness.

2.1.3 *Photoelectric Observations*

Photoelectric data may be obtained in the form of direct current, voltage, or pulse counts, but pulse counting became the data acquisition method of choice for photoelectric photometry. Recording digital observations is now almost always done by computers. Two types of data are sought: Standard star data and program star data. "Standard stars" are those which help to define the photometric system being used and their observation will provide the standardization needed for the program star data. The brightness of the "program star" must be determined in the passbands of the standard system and its observations need to be transformed to that standard system. The process of standardization, which we discuss in detail below, is important because observations corrected only for terrestrial atmospheric extinction are written in a kind of private code, effectively, and are subject to misinterpretation or miscomprehension until decoded. If the program star is variable, the observations must be gathered in intervals which permit good time resolution. Usually such observations are made in conjunction with constant-light "comparison" and "check" stars. The comparison star is observed before and after the variable star and should be sampled at frequent intervals. It provides the first-order extinction correction, and if it is relatively well matched in color to the variable star, the brightness difference between them should be independent of the color effects of extinction, again, to first order.

Multiple comparison stars are especially valuable when the variable star amplitude is very low, as in many δ Scuti variables, a class of short-period pulsating stars. In any case, observation of a second comparison star is a good idea. The comparison stars must be observed frequently and over a relatively long range of time; consequently, if one of them turns out to be variable, the other comparison star (the check star) will save the day. The variable star observer measures a time-dependent flux,[3] the display of which against time or *phase* (the repeated foldings of the time into the period of variation) is known as the *light curve*. The reduction and standardization of photometric observations will be discussed in Sect. 2.1.5.

In *EB* light curves, the photometric phase is the decimal fraction of the cycle of variation, with zero phase often set at mid-primary minimum. This is the photometric usage; spectroscopists may use the instant of passage through a node of the orbit (one of the two points marking the intersection of the orbit with the plane of the sky) as zero phase. In eclipsing variable star work, orbital phase may be computed as

$$\Phi = \Phi_s + frac\left\{\frac{\tau - E_0}{P}\right\}, \quad \tau := t + \delta t, \tag{2.1.1}$$

where Φ_s denotes a constant offset which often will simply be zero, $frac\{x\}$ denotes the decimal part of x, the time of mid-observation, t, is best expressed in the continuously increasing Julian Day Number and decimal fraction thereof; δt is the heliocentric correction, the correction for the difference between light travel time of the starlight to the Earth and that to the Sun; τ is the Heliocentric Julian Date; E_0 is the epoch or instant of an adopted time of minimum; and P is the orbital period. Typical precision in P is $\sim 0\overset{d}{.}0001$, or better. Both E_0 and P are determined from a series of times of minimum light, and the precision increases with the range in time over which the observations are obtained (assuming that the period and epoch are constant and there are no very large gaps in the record). Each individual time of minimum requires a careful set of observations and several methods are available for the determinations [see, e.g., Ghedini (1982)]. The heliocentric correction depends on the relative locations on the sky of the star and the Sun (and therefore the solar date and the coordinates of the star). It may be computed[4] or interpolated in the tables of Landolt & Blondeau (1972). If there are poorly determined or no eclipses, radial velocities may provide the best ephemeris. For eccentric orbit binaries, the phasing is more complicated as discussed in Sect. 3.1.2.2.

---

[3] Usually the magnitude differences, $\Delta m_k = m^* - m_k$, relative to a comparison star, are measured, but in light curve analysis we prefer normalized flux. Therefore the conversion $I_k = 10^{-0.4(\Delta m_k - \Delta_0)}$ is applied to all measurements k where Δ_0 is chosen such that the maximum normalized flux is about unity.

[4] See Henden & Kaitchuck (1982) or Duerbeck & Hoffmann (1994) for computational formulas and examples.

2.1 Photometry 43

2.1.4 Imaging Data

An important advantage of the charge coupled device (CCD) is that it can detect many sources simultaneously, as can a photographic plate. However, the CCD response is more nearly linear and is linear over a much larger range of signal than is that of photographic emulsion, and detection of faint sources is possible even in the presence of bright sources. CCDs exceed photographic plates in linearity, quantum efficiency, and in dynamic range and can exceed PMTs in these properties as well as in sky sampling. The disadvantages are the extensive image processing and computer storage required for image frames (although improvements in computer storage have greatly alleviated the latter problem). There is also a standardization problem arising from having a large number of detectors, each with its own spectral sensitivity, which also varies with the direction of the incoming illumination. The image processing, first, requires the determination and removal of the bias structure of the chip (a kind of zero-point determination); second, the determination and subtraction of "dark current"; third, the flat-fielding or sensitivity determination of each pixel (the sensitivity correction of each image frame can be achieved through division by the normalized flat for that particular passband); fourth, the identification and removal of cosmic ray hits from the image frame; and fifth, the subtraction of sky background from the light distribution of each star image on each frame and the conversion of the integrated corrected flux to magnitudes. Then the reduction process to determine and correct for atmospheric extinction and to transform the extinction-corrected observed brightness of each image to a standard system must be carried out. The size of each image file is $n_x \times n_y$, where $n_{x,y}$ is the number of pixels in a line or column. Thus a 2048-square chip with information digitized into 16 bits (giving $2^{16} = 65,536$ levels) or 2 bytes has more than 8 MB ($2048^2 \times 2$) of information. Each image-processing step generates another file, and a doubling of the file size may occur when floating point numbers are generated for the pixel values.

Because of its dependence on chip temperature, the bias structure in the image frame must be determined with high precision through statistics of many bias frames, close to the time of the program object observation. Just reading out a chip changes its temperature, so the procedure cannot be perfect, although the effect usually is not great.

Sensitivity calibration requires "flats," or frames exposed to a uniform light source. The accuracy of the flat-fielding procedure relies on the uniformity of the illuminating source. There are three sources in common use. The first is the twilight sky, which varies spatially and temporally, but matches the colors of blue cluster stars better than does an illuminated screen in the dome, which is a second source. The dome screen is usually illuminated by a projector with filter slides to block longer wavelength light. It is difficult to illuminate the screen strictly uniformly so that even though it is greatly out of focus to the detector, the screen may not illuminate the chip uniformly either. A variety of this type of source involves lamps shining through a translucent diffusing screen directly into the telescope. A third source can be a selected field in the night sky – a field devoid of many stars. By a procedure known as "dithering," the moving of the telescope by small increments several times, the star images which are in the frame will be displaced along the chip. The taking of the median of each pixel over several such frames can eliminate remaining stars. In addition, there are several important checks and procedures which precise photometry demands. "Dark current" must be determined for a variety of exposure times. Cosmic ray "hits" must be removed through direct viewing of the image frames or through an automated procedure.

Image processing packages, such as can be found in the general purpose packages IRAF or MIDAS, are used to accomplish many of these steps: The determination of magnitude, and the subtraction of bright star images, or of columns of charge which may "leak" from bright, saturated sources. Revealing fainter images may be done with other packages, such as DAOPHOT, DOPHOT, or, if only a few profiles are to be measured, ROMAFOT in MIDAS. The aim of these packages is to produce magnitudes of stars (with uncertainties) for each of the image frames. Typically there are two methods of obtaining magnitudes: Through the determination of the stellar profile parameters (and the application to the program stars on the frame); and from the summation of the flux through a series of apertures centered on the stars. The extrapolation to large radius can be carried out with auxiliary programs, principally DAOGROW. The processes are discussed by Stetson (1998), the originator of DAOPHOT.

Once all these steps are accomplished, equal care has to be paid to the reduction and standardization of the data, requirements to which both photoelectric and CCD photometry must adhere. For modern methods to treat CCD data, see Howell (2006).

2.1.5 Photometric Data Reduction

In area images, one presumably has nonvariable stars on the frame; if they are not saturated they can be used for extinction determination by the *Bouguer* extinction method. The raw magnitude of each star of a CCD frame (after bias level subtraction and division by a normalized flat have been applied to the frame) may be plotted against airmass. Alternatively, a suitable mean magnitude of all (nonvariable) stars brighter than a given limit may be used. If the colors of those stars are not known, the latter technique begs questions about the color terms, so it is better to use the values from a relatively few, well-determined comparison stars for this purpose. Commonly used equations of condition are

$$\Delta := m_\lambda - k'' X c = m_{\lambda_0} + k'_\lambda X \qquad (2.1.2)$$

and

$$c - k'' X c = c_0 + k'_c X, \qquad (2.1.3)$$

2.1 Photometry

where c is the color index, k' and k'' are the first- and "second-order" extinction coefficients, X is the airmass, and the subscript 0 indicates outside-atmosphere (airmass = 0) magnitude or color index.

The least-squares solution for two unknowns, k'_λ and m_{λ_0}, and n observed data points may be obtained for each comparison star:

$$k'_\lambda = \frac{n \sum \Delta_i X_i - \left(\sum X_i\right)\left(\sum \Delta_i\right)}{n \sum X_i^2 - \left(\sum X_i\right)^2}, \quad (2.1.4)$$

$$m_{\lambda_0} = \frac{\left(\sum \Delta_i\right)\left(\sum X_i^2\right) - \left(\sum X_i\right)\left(\sum \Delta_i X_i\right)}{n \sum X_i^2 - \left(\sum X_i\right)^2}. \quad (2.1.5)$$

Similar expressions hold for the color index coefficient, k'_c and c_0. The method assumes that the second-order coefficient is constant and is known. See Schlosser et al. (1991, p. 129) for a detailed discussion of this method. Further refinements have been discussed by Sterken & Manfroid (1992). One of these involves the combinations of data from different nights to find the extra-atmosphere magnitudes and colors (assuming that the local system is stable from night to night).

A second method is referred to as the Hardie extinction method (Hardie 1962). It involves observations of groups of stars with a range of color indices at both low and high airmass. The equations of condition are

$$dm_\lambda - dm_{\lambda_0} - k''d(Xc) = k'_\lambda dX \quad (2.1.6)$$

and

$$dc_\lambda - dc_{\lambda_0} - k''d(Xc) = k'_c dX, \quad (2.1.7)$$

where the differential quantities refer to difference between star pairs at high and low airmass. The pairs must be of similar color index. The least-squares result for the one-unknown problem is

$$k'_\lambda = \frac{\sum\{[(dm_\lambda - dm_{\lambda_0})_i - k''d(Xc)_i]dX_i\}}{\sum(dX)_i^2} \quad (2.1.8)$$

with a similar expression for k'_c. This method works well with a cluster of stars which can be observed at different times of night and therefore at different airmasses, or with stars in two or more Harvard Selected Areas which have been systematically observed and standardized by Landolt [cf. Landolt (1983)].

The techniques are applicable to both photoelectric and CCD photometry. With meticulous care, relative photometric precision of the order of 2–3 millimagnitudes is possible, but absolute photometry is not so easily accomplished to this degree of accuracy. Transformation to a standard system is an important procedure which astronomers ignore to their peril. The result can be disastrous, especially for the combination of data from different observers. Failure to transform adequately is akin to writing a research paper in one's private code. Hardie's (1962) treatment of transformations from the local system (magnitudes and color indices transformed from raw instrumental values to outside the atmosphere values) involves the assumption of a linear relationship between the local and the standard systems. The relationships are as follows:

$$m_{std} = m_0 + \varepsilon c_{std} + \zeta_\lambda \qquad (2.1.9)$$

and

$$c_{std} = \mu c_0 + \zeta_c. \qquad (2.1.10)$$

The ε coefficient is usually small with an absolute value less than 0.1 for well-matched systems. For μ, we expect a number near 1, typically in the range between 0.9 and 1.1. ζ in each case is the zero point. If the filters and detector are similar to those used to establish the standard system, and no other major transmission element or exotic reflecting surface has been introduced in the light path, equations (2.1.9) and (2.1.10) are not bad approximations. However, the linearity of any set of transformations is always open to question [see Young (1974)]. Even when the local system is designed to approximate the standard system closely, there are occasions when it may fail to do so. Milone et al. (1980) discuss a situation where the zero points were time dependent. Apparently a broken heater wire for the fused quartz window of the PMT chamber[5] led to a slow build up of frost on the window which resulted in a linear decrease in sensitivity of the photometer over time.

The solutions for the two unknowns for each equation are readily found. For the magnitude equations of condition, (2.1.9), they are

$$\varepsilon_\lambda = \frac{n \sum [(\Delta m) c_{std}] - \sum (\Delta m) \sum c_{std}}{n \sum c_{std}^2 - \sum c_{std} \sum c_{std}}, \qquad (2.1.11)$$

$$\zeta_\lambda = \frac{\sum (\Delta m) \sum c_{std}^2 - \sum [(\Delta m) c_{std}] \sum c_{std}}{n \sum c_{std}^2 - \sum c_{std} \sum c_{std}}, \qquad (2.1.12)$$

where the quantity $(\Delta m) = m_{std} - m_0$. For the color equations of condition, they are

$$\mu_c = \frac{n \sum [c_0 c_{std}] - \sum c_0 \sum c_{std}}{n \sum c_0^2 - \sum c_0 \sum c_0}, \qquad (2.1.13)$$

---

[5] The chamber in which the photomultiplier tube in a photoelectric photometer is located. It is usually made of μ-metal and acts like a Faraday cage against the ambient magnetic field of the Earth, so that the latter does not affect the streaming of electrons across the dynodes. The window through which light enters is usually of fused quartz to permit ultraviolet radiation to pass. It must be heated to avoid dew and/or frost accumulation.

2.1 Photometry

$$\zeta_c = \frac{\sum c_{std} \sum c_0^2 - \sum [c_0 c_{std}] \sum c_{std}}{n \sum c_0^2 - \sum c_0 \sum c_0}. \quad (2.1.14)$$

With the extinction and transformation coefficients and zero points, the raw differential magnitudes and color indices may be transformed to the standard system. Similar techniques can be carried out in infrared astronomy, but standardization has been problematic [see Milone (1989)]. New infrared passbands to minimize the effects of water vapor extinction and improve transformability have been proposed by the IAU Working Group on Infrared Astronomy (Young et al. 1994). The near-IR portion of this suite, namely *iZ, iJ, iH*, and *iK*, has been fabricated and tested (Milone & Young 2005, 2007, 2008).

2.1.6 Significance of Cluster Photometry

CCD photometry has proven its value in cluster studies. Observations of faint stars at or fainter than the main sequence turn-off in many globular clusters are being made for the first time.

Two powerful techniques for determining ages of star clusters are the following:

1. The fitting of isochrones to a cluster's color magnitude diagram (*CMD*) (see Fig. 7.5). An important test for the accuracy of this fitting is the agreement of the isochrones with the *CMD* locations of stars of known masses. This method is model dependent and requires accurate isochrones to be successful. An example of the method is provided in Milone et al. (2004).
2. The comparison of the sizes of evolved components with model predictions. This technique is not entirely independent of the first because evolutionary predictions are needed for this technique also, and the radius is a function of the luminosity. Substantial numbers of variables and nonvariables may be imaged in both direct and multi-object-spectrograph imaging.

Variable stars in clusters can provide independent assessments of cluster distance and permit checks on luminosity calibrations. Figure 2.3 shows what can be achieved. A group at the University of Calgary is analyzing *UBVI* data obtained with the SDSU Mt. Laguna Observatory 1-m telescope equipped with a back-illuminated, thinned 800 × 800 chip. Averages of the brightest nonsaturated, nonvariable stars in the field define an artificial "comparison star" for each frame. Typical (m.s.e.) uncertainties in each observation are about $0.^m05$ for the faintest ($V = 19.^m5$) and about $0.^m005$ for the brightest ($V = 14^m$) variables. The study of variable stars in clusters can provide fundamental information about both the ensemble and the individual stars. The techniques are described in numerous sources, e.g., in Milone (2003) and Milone et al. (2004). Consult Howell et al. (2005) for an example of the use of CCD techniques to study an open cluster.

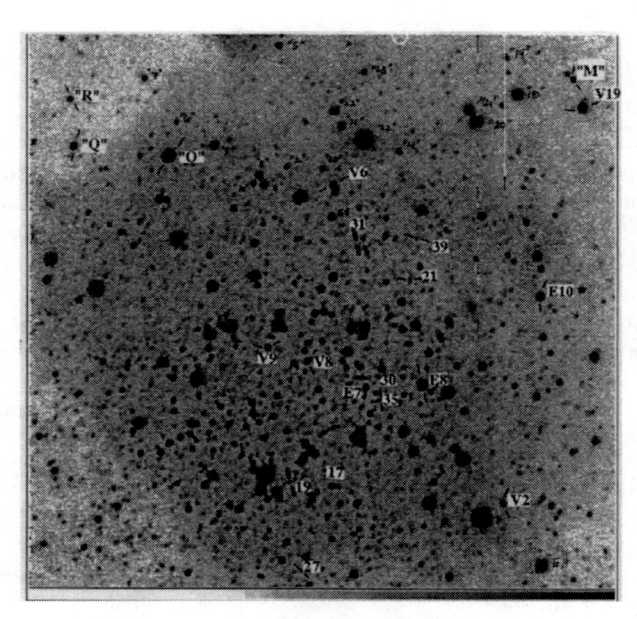

Fig. 2.3 CCD image frame of the globular cluster *NGC 5466*. This *I* (infrared) passband CCD image frame was obtained at Mt. Laguna by E. F. Milone. *NGC 5466* contains at least three eclipsing binary blue stragglers

2.2 Spectroscopy

Multum in parvo   *(Much in little)*

An important difference between photometry and spectroscopy is spectral resolution, defined as

$$R := \lambda/\Delta\lambda, \qquad (2.2.1)$$

where $\Delta\lambda$ is the smallest discernible unit of bandwidth or the smallest wavelength-resolution element of the instrument (photometer or spectrograph) which is being used. In broad-band photometry, $R \approx 10$, in intermediate-band, $R \approx 20-50$, and in narrow-band photometry, $R \leq 500$ typically. Spectroscopy implies that the details of the spectrum are important and $R > 1000$ usually, depending on the application and spectrum range. In highest resolution spectroscopy, $R \geq 10^5$. Such resolution is required for fine spectral analysis and for the highest precision radial velocity studies, e.g., extrasolar planet research.

There are differences in aims, instrumentation, techniques, and emphases of various kinds between photometry and spectroscopy, in addition to the resolution difference. We consider later the special case of spectrophotometry, where the purposes are closer to those of photometry. Spectral details are important for several very

2.2 Spectroscopy

different kinds of studies, namely classification, line profile analysis, and radial velocity determination.

Spectral classification, a useful way to estimate temperature and luminosity, usually is done at relatively low spectral resolution. The criteria are the relative strengths of features which may be (possibly blended) absorption lines of various atomic and molecular species. Spectral lines vary in strength with temperature, which partly determines the relative populations of atoms in the lower states of the transition represented by the absorption line. Figure 2.4 shows the spectra of some binaries with high temperatures. Plots of relative flux versus wavelength are in Jacoby et al. (1984), who provide a useful "library of stellar spectra", covering a large range of spectral types. The analysis of spectral line profiles, on the other hand, requires

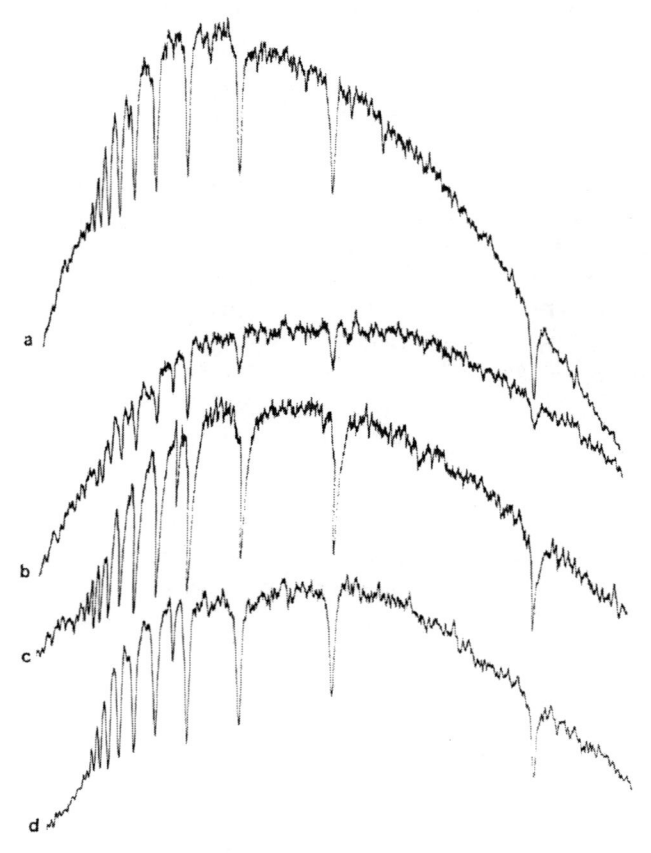

Fig. 2.4 Spectra of early spectral-type eclipsing binaries. (**a**) *CX Canis Majoris*, (**b**) *TU Crucis*, (**c**) *AQ Monocerotis*, and (**d**) *DQ Velorum*, from Fig. 2 in Milone (1986)

the highest resolution to determine abundances, observe Zeeman splitting (through magnetic fields), and model atmosphere parameters. Because this topic is beyond our present scope, we refer the reader to detailed accounts by Aller (1963), Mihalas (1965, 1978), and Gray (1992).

The measurement of a Doppler shift in a spectral line[6] permits the determination of a radial velocity for that spectral line and may be carried out at intermediate resolutions. Higher resolutions are desirable for precise radial velocity work. The resolution is determined by the size of a resolution element of the detector, the scale of the spectrum at the detector, the projected slit width, and various optical characteristics of the spectrograph. The resolution defined by equation (2.2.1) depends on the wavelength and the size of the smallest useful spectral element. That size is determined by the dispersion properties of the grating. The grating resolution must at least be matched by the detector in order to be realized. The spectral scale is usually described in terms of the linear dispersion, or, more commonly, the *reciprocal linear dispersion*,

$$\frac{d\lambda}{ds} = \frac{w \cos i'}{mf}, \qquad (2.2.2)$$

where w is called the "width" of a rule or groove of the grating (more correctly, it is the separation of one groove from another), i' is the diffracted angle at the grating, m is the order of the spectrum, and f is the focal length of the spectrograph camera which converges the (parallel) light emerging from the (plane) grating onto the detector at the output focal plane. The dispersion is also determined by the grating properties used to obtain the spectrum. The number of grooves, N, and the width of the grooved part of the grating, W, are related to w by the relation

$$w = W/N. \qquad (2.2.3)$$

Equation (2.2.3) may be substituted into equation (2.2.2) in order to eliminate w. Finally, reflection gratings are usually *blazed*, so that the maximum reflected energy goes into the diffracted order of interest. For normal incidence at the grating, the blaze angle is achieved for wavelength

$$\lambda_b = \frac{w}{m} \sin 2\alpha_b, \qquad (2.2.4)$$

where α_b is the blaze angle, the difference between the normals of the grating and the grooved surface.

---

[6] Note that the radial velocities of spectral lines may be affected by blending with those due to circumstellar matter.

2.2.1 Radial Velocities

High resolution is essential for spectral analysis, whereas low resolution may be quite sufficient for spectral classification. To be useful in binary star modeling, radial velocities as displayed in Fig. 2.5 must have a high level of precision and accuracy. Red shifts for distant galaxies are very large, for example, and there may be a paucity of (identified) lines in the visible region, so precisions of tens and even hundreds of km/s may not be considered too low. In binary star or pulsating variable star work, the radial velocity precision must be much higher to yield fundamental stellar data, such as the mean radii of pulsating stars or the masses of eclipsing system components. At present writing, typical radial velocity uncertainties are of order 1 km/s. Some very high precision has been achieved by specialized techniques, however; the use of absorption cells to ensure stability in conjunction with high-resolution spectrographs on large telescopes [see, e.g., Campbell et al. (1988), who used a hydrogen fluoride absorption cell] provided a means to detect very low-mass companions to nearby stars. The best that had been achieved to 1995 was $\sim$10 m/s; in 1998 it was $\sim$4 m/s (Kuerster et al. 1994). This improvement has been the main reason for the burst of discoveries of planetary systems of other stars beginning in 1995. See Butler et al. (1996) for a summary of these radial velocity improvements, and the prospects of further improvement.

Spectra can be measured directly or cross-correlated against a reference spectrum. The process of cross-correlation involves a method of systematic digital comparison between two spectra, in order to obtain the difference between the two radial velocities. Simkin (1974) first applied the technique to astronomy (in her case, to galaxy spectra). Typically, the Fourier transform of the linearized and rectified

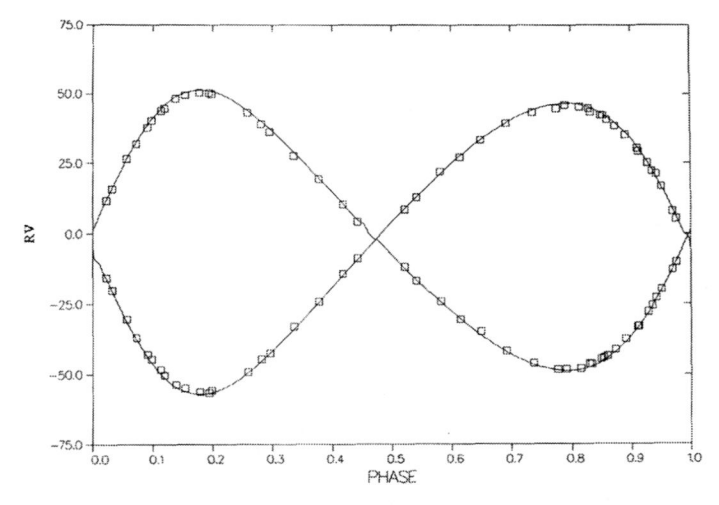

Fig. 2.5 Radial velocity curve of *Al Phoenicis*. This figure [Fig. 3 of Milone et al. (1992), adapted from Andersen et al. (1988)] of the long-period totally eclipsing system shows the observed and computed radial velocity curves

spectrum of the program star along with the conjugate transform of the reference spectrum is calculated, and the product of the transforms is evaluated. The cross-correlation function (*CCF*) can be described as the Fourier transform of this product. The reference spectrum may be that of a standard star with well-established radial velocity, a spectrum of the program star itself (at a specified phase), or a synthetic spectrum. Especially useful software programs, REDUCE and VCROSS, were developed to reduce and analyze stellar spectra by Hill et al. (1982) and Hill (1982). Wilson et al. (1993), working with this software, used for the δ *Scuti* star *EH Librae* a reference spectrum of the same star obtained at a phase where the expansion was close to zero. This work achieved relative velocity precisions of a few hundreds of meters per second. The velocity difference was obtained from the spectral shift which produced the peak of the resulting CCF. When the program star is an *EB*, the CCF typically has two peaks, as in Fig. 2.6. Additional complications may be present, as we note below and in Sect. 2.2.3, but cross-correlations have measured radial velocities in many cases that were considered unpromising only a decade earlier, at least to the level of uncertainty of $\sim$10 km/s. Beginning in the 1980s [cf. McLean (1981)], cross-correlation techniques have proven crucial to our understanding of contact and other short-period interacting systems. A somewhat different method, involving a spectral "broadening function" (*BF*) and a modified singular value decomposition technique (*SVD*), has been used in recent years to obtain radial velocities by Rucinski (2002) and references therein. The broadening function basically transforms a narrow-lined standard star spectrum into a rotationally broadened one, more suitable for comparison with a short-period *EB*. The templates are standard star spectra and the method works best if they are of the same spectral types as the component stars of the binary. In fact, three-component systems

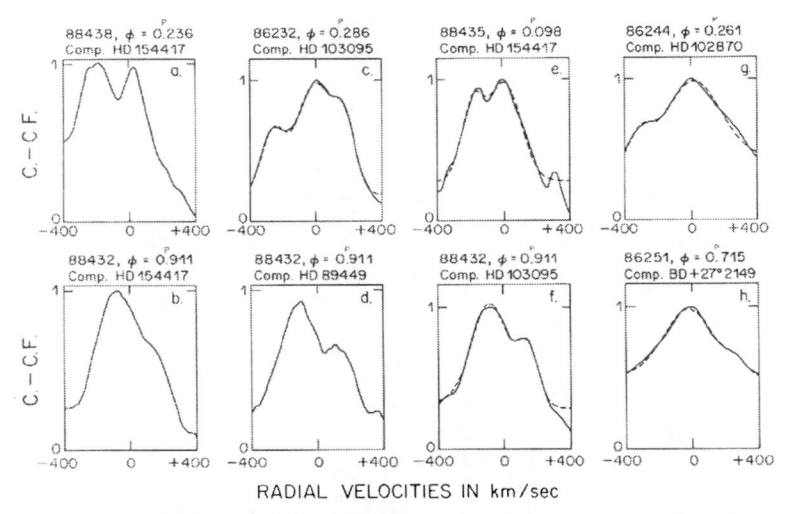

Fig. 2.6 Cross-correlation functions for *RW Comae Berenices*. This figure is Fig. 1 in Milone et al. (1984, p. 110)

2.2 Spectroscopy

can be handled as well. The coefficients of the polynomial that defines a particular *BF* are determined by least squares, with Rucinski's (1992, 1999) *SVD* treatment. Higher RV precision is achieved by Rucinski in the cases where both *BF* and *CCF* methods are tried.

Whatever detection method is used, standards need to be intercompared because different techniques may have systematic differences. The IAU "*Reports*" lists the status of standard stars and indicates those suspected of variability. An up-to-date list can be found in the *Astronomical Almanac* for the current year. For bright stars, high-resolution techniques have yielded very small errors of measurement such that planets are being sought and detected with increasing confidence. However, the very high precision obtained for this purpose is unlikely to be achieved soon for short-period interacting *EB*s because of their rotational broadening, line blending, and stellar activity.

2.2.2 Spectrophotometry

Suppose we are observing from space so that atmospheric extinction problems are absent. If we can accomplish the necessary standardizations of effective passbands, the combination of spectroscopy and photometry can provide a rich bounty for light curve analysis. A large number of narrow passbands can greatly improve the radiative modeling of stars, because the wavelength-dependent parameters are adjustable for each passband and together provide strong leverage for the determination of the temperatures and for any thermal perturbations projected onto the disk surfaces. In addition, the weight of the nonwavelength-dependent parameters is increased by virtue of the large number of light curves. Suppose now that we attempt to do similar observations from Earth. As for broad- or intermediate-band photometry, the spectrophotometric narrow-band photometry can be carried out relative to comparison stars, but corrections must be applied for differential extinction, and practically, it is very hard to do the comparison star observations.

The photometric precision requisite for high-quality narrow-band light curves is possible only if the photon flux is high enough. For a precision of 1% (neglecting other noise sources), we require a signal-to-noise ratio

$$S/N = \sqrt{\bar{n}} \qquad (2.2.5)$$

of 100 so that the mean count $\bar{n}$ needs to be

$$\bar{n} \approx 10,000. \qquad (2.2.6)$$

If the resolution element is of order 1 nm, only bright stars can be observed with intermediate-sized telescopes for reasonable length ($\tau \leq 30$ min) exposure times, because the signal-to-noise ratio, $S/N \propto \sqrt{\tau}$. In practice, some compromise between size of telescope and spectral resolution must be made. Schlosser et al. (1991, pp. 206–207) demonstrate the effects of increasing the observing time on the S/N of spectrophotometric data.

Phase-dependent spectrophotometry has been carried out by Etzel (1988, 1990), who demonstrates its value for determining flux distributions and therefore color temperatures of components in totally eclipsing systems.

A promising technique to disentangle the separate spectral distributions of the component stars from the composite spectrum of an *EB* is discussed by Simon & Sturm (1994). The method of spectra disentangling (Hadrava, 1995) combines simultaneously the splitting of the spectra observed at different orbital phases into spectra of individual components of the spectroscopic binary, the measurement of radial velocities at each phase and the solution of orbital elements. The first part of this problem for known radial velocities may be solved, e.g., by the method of tomography separation. The second part is the aim of the cross-correlation method, for which template spectra of the components are needed to find their shifts in the observed spectra. Finally, the third part involves the solution of the radial-velocity curves. The direct fitting of observed spectra by the best component spectra and the orbital parameters (or individual radial velocities) is thus a more reliable and less laborious procedure. The method could be generalized to get still more information from the line-profile variations.

KOREL is a code developed by Hadrava (2004) for spectra disentangling using Fourier transforms. It allows application of the method of "relative line photometry," i.e., to find the variations of line strengths. From the beginning in 1997 KOREL takes into account up to five components in a hierarchical structure of a multiple[7] stellar system. One of them (e.g., the widest one) can be identified with the telluric lines, which can be separated from the observed spectra in this way and can yield a check (or correction) of the proper wavelength scales of individual exposures. Several regions around important spectral lines can be solved simultaneously. The numerical method of the solution is based on the fact that the modes of Fourier transforms of component spectra are multiplicative factors at the complex exponentials corresponding to Doppler shifts in the Fourier transformed space of frequency logarithms. They can thus be calculated by the least-squares method, whereas the orbital parameters (or radial velocities) can be fitted, e.g., by the Simplex method. The technique is said to work on artificial spectra down to a signal-to-noise ratio of 10. An important requirement to identify which part of the continuum originates from each star to be able to scale properly equivalent widths of the lines in the decomposed spectra is that the light ratio in the range of the spectra must be known with high precision – either from lightcurve analysis or from the spectra themselves

---

[7] Another code, SPSYN by Barden and Huenemoerder dating back to Barden (1985), also supports the analysis of triple systems and is based on *Fast Fourier Transforms*.

2.2 Spectroscopy

if the separate features of the spectra can be discerned (at quadratures,[8] say) and the components are not too dissimilar in temperature and luminosity. The pure determination of orbital parameters is reliable without this condition as the light ratio can be estimated from fitting the decomposed spectra by model spectra. The requirement that the light ratio does not vary with phase tends to limit the technique to widely separated eclipsing (or noneclipsing) binaries, although the authors indicate that the limitations of the constant light ratio assumption is overcome by the variable line strength factors implemented in KOREL of 1997. A more difficult but not impossible task is to overcome the requirement that there be no variation of line-profile shape. To disentangle any variability requires the model, for instance, of Cepheid pulsations, to be fitted with some free parameters. Light-ratio variability is disentangled by the line-strength factors, variations of shape need more sophisticated models.

2.2.3 Line-Profile Analysis

Lines in stellar spectra are images of the input slit on the output plane of the spectrograph. However, even after allowance for the finite width of the input slit and instrumental diffraction, it is seen that spectral lines are broadened by a number of mechanisms originating in the star itself. The most basic of these mechanisms is the *natural broadening* that arises because the atoms' energy levels have probability distributions. According to the *Heisenberg Uncertainty Principle*, the product of uncertainties in position and in momentum and the product of uncertainties in time and in energy are of the order of Planck's constant, h:

$$\Delta x \Delta p \geq h, \quad \Delta t \Delta E \geq h, \quad h = 6.62608 \cdot 10^{-34} \text{ Js}. \tag{2.2.7}$$

Because the energy of a photon is equal to the difference between two energy levels of the radiating or absorbing atom, the frequency or wavelength of the photon will be affected accordingly. In particular, an atom absorbs from the continuum radiation at a wavelength which may be slightly different from the most probable one. The combination of all such absorptions produces an absorption profile.

A second source is *collisional broadening*. This broadening is due to perturbations of the energy level of the atom because of either of two effects. First, the passage of an atom in the vicinity of an atom undergoing an absorption or emission of a photon changes both the kinetic energy of the disturbing atom and the energy of the photon. This is known as *collisional damping*. A second effect is due to the

---

[8] At quadratures each component's spectrum may be seen, as in the rare case of visual binaries with elongation about an arcsec. However, the latter would have long periods and small radial velocity amplitude and interferometry would provide a better light ratio. The individual spectra could be obtained during eclipses, as Simon & Sturm (1994) did to prove that their method works effectively.

effects of the electromagnetic fields of nearby charged particles. This mechanism is known as *Stark broadening*; in most stars, the first effect is the more important except in the lines of hydrogen and of some helium lines. The denser the stellar atmosphere, the stronger these perturbations and the broader the spectral lines. This effect is important in the strong lines and is relatively strong in dwarf stars, which are more compact, than in giants and supergiants, which are much less dense. If the number of absorbers is small, the broadening is mainly due to thermal Doppler broadening (discussed next). As the number of absorbers increases, the line core saturates and the "damping wings," due to a combination of natural and collisional damping, begin to dominate. The *curve of growth*[9] of spectral lines can be deduced from high-resolution spectra, and the abundances found.

A third major broadening source is thermal *Doppler broadening*. In the high-temperature environment of stars, the line-of-sight motions will be Doppler shifted relative to both the radiation upwelling from below and to the observer. The result is again a displacement of the absorption from line center for many photons, and a finite width. Because the motions of the atoms depend on the temperature of the stellar atmosphere, Doppler broadening depends on the temperatures of the regions in which the absorptions occur. This is the dominant source of line broadening for weak lines in the spectra of slowly rotating stars.

A discrete splitting of the energy levels of certain species of atoms occurs in the presence of magnetic fields. This is called *Zeeman splitting*. The number of components and their relative strengths differ from line to line; the lines are also polarized (see Sect. 2.3) in this process. Basically the splitting is proportional to the magnetic field strength, and inversely proportional to the mass of the atom. The typical pattern is a triplet, but each component of the triplet may itself be split into multiple components. The outside members of the triplet (or triplet groups) are called the σ-component(s) and the inside ones, the π-component(s). The *longitudinal* effect is seen if the σ-components are circularly polarized (in opposite directions, with left-handed circular polarization producing the higher frequency component) and the π-component absent; this occurs when the line-forming region is viewed along the direction of the magnetic field. The *transverse* effect is seen if the σ-components are linearly polarized perpendicular, and the π-component linearly polarized parallel, to the direction of the magnetic field. When there are many noncoherent local magnetic fields associated with the line-producing region, a blend of these effects can be expected, blurring the Zeeman components and causing a net broadening of the line.

Mass motions on stellar surfaces (micro- and macro-turbulence, large-scale circulations, and stellar rotation, for example) contribute to line broadening through Doppler shifts. The effect of star rotation on spectral lineprofiles of *EB*s (Mukherjee et al. 1996) can be observed during eclipses in the form of the Rossiter effect, where

---

[9] The term *curve of growth* refers to a graph of line strength versus the effective number of absorbing atoms [cf. Aller (1963)].

one of the two observed sets of radial velocities varies with the masking of the velocity contributions of the eclipsed part of a star (see Fig. 3.28). In binary systems of very short period, rotational broadening can degrade radial velocity accuracy and even prevent measurement, especially for stars near spectral types A and earlier. In such stars, the spectrum is dominated by very broad hydrogen and helium lines, but even the weak lines of other elements are spread out into a characteristic dish-shaped profile. "Blending" of lines compounds the problem. It is all the more remarkable, therefore, that over-contact systems containing stars of early spectral types (i.e., high-temperature stars) have been analyzed at all, let alone with good precision. With cross-correlation and related techniques being used regularly, the situation is expected to improve even more.

2.3 Polarimetry

Cetera desunt   *(The remaining (parts) are lacking)*

Polarization of the light of a binary component was predicted by Chandrasekhar (1946a, b). While searching for a confirmation of the prediction, Hall (1949a,b, 1950) and Hiltner (1949a,b) discovered the polarization of starlight due to scattering by interstellar dust. Additional sources of polarization in interacting binaries are

- light of one component scattered at the surface of the other;
- starlight scattered by a circumstellar disk, stream, or other locus of concentrated gas;
- thermal bremsstrahlung in the stellar environment (electron scattering in gas flows or in coronae); or,
- nonthermal bremsstrahlung (from flares);
- electron scattering in high-temperature atmospheres; and
- magnetic surface fields.

Observationally, the light observed through different rotations of a polarizing analyzer is measured at certain angles and corrected for instrumental polarization. The basic parameters are the following:

1. The *fractional polarization*,

$$P = \frac{\mathcal{F}_{max} - \mathcal{F}_{min}}{\mathcal{F}_{max} + \mathcal{F}_{min}}, \quad (2.3.1)$$

where $\mathcal{F}$ refers to the observed flux or power through the polarizing filter (or polarizing *analyzer*), whereas the subscripts indicate maximum or minimum transmission of the flux at particular angular settings of the analyzer, 90° apart.

2. The *position angle* of maximum transmission, θ.

The polarization can also be given in terms of the weighted means of the Stokes quantities,[10] Q and U. Concisely expressed,

$$Q = P\cos 2\theta, \quad U = P\sin 2\theta. \tag{2.3.2}$$

Much astrophysical information about the nature of scattering disks and electron scattering envelopes around stars can be obtained in this way [see Wilson (1993)].

Although there are many polarimetric publications on *EB*s, most have been in the form of surveys rather than time-wise variation. Surveys can be very useful in identifying candidate stars, but tell us little about specific polarimetric behavior. Of course, polarimetry requires much brighter stars (or larger telescopes) than does photometry, and also much more sophisticated instrumentation. Still, polarimetric curves have been published for interesting close binaries but they typically have serious shortcomings. For example, the observational difficulties mentioned above result in rather small numbers of data points being collected. Attempts to compensate for this problem lead to folding of polarimetric curves on the orbital period, yet most polarimetric phenomena are not strictly periodic so that phased polarimetry has very limited usefulness.

Pioneering work in astronomical polarimetry was carried out by James Kemp (1927–1988), who was the first to discover circular polarization in the continuum of a white dwarf, GJ 742 (Kemp et al. 1970), and the first (Kemp et al. 1983) to discover the limb polarization in an *EB* (Algol, in fact) that was predicted by Chandrasekhar (1946a, b).

In recent years, much progress has been made in this field. Work by J. Landstreet and his group in Canada and by J-F. Donati and his associates in France has been carried to a high level of precision through instruments such as ESPaDOnS (Echelle SpectroPolarimetric Device for the Observation of Stars) in use at the Canada-France-Hawaii Telescope. The device is designed to measure polarization in spectral lines and permits the sky spectrum to be measured simultaneously. The exposures cover nearly the entire range 370–1050 nm, at resolutions of 68,000 and 80,000, depending on observing mode. With this instrument, stars as faint as 14th magnitude have been studied. The instrument must still take four exposures to produce the four Stokes quantities. However, there is another concern. In addition to the extensive observing time, painstaking care must be taken to keep the target object centered in the observing aperture, because systematic drift to one edge of the aperture can introduce continuum polarization; moreover, determination of the amount of polarization requires measurement of the difference in intensity between two beams emerging from the Wollaston prism, and these beams are transmitted by fiber optics to the spectrograph, and guiding errors may effect the two beams

---

[10] Usually referred to as the "Stokes parameters." R. E. Wilson has pointed out that these quantities are measured directly. They are better called "Stokes quantities" to avoid confusion with parameters derived from light curve analysis.

differently. It has been estimated that instrumental scatter due to these effects can amount to 1 percent. When S/N requirements are 500 or more to achieve the goal of exploring polarization of weak sources as a function of phase, to permit magnetic field mapping, this level of uncertainty can be a concern. On the modeling side, polarimetric curves have had to be "rectified" to treat unwanted effects, as in the old light curve analysis days. Such analyses required removal of those effects from the observations, rather than expansion of the theory to reproduce them, as in normal scientific practice. Another shortcoming has been the use of artificial fitting functions (such as Fourier series) rather than direct representation in terms of a definite model. Finally, the majority of polarimetry papers have contained no published observations but only graphical presentations, so data have not been available in suitable form for improved analysis. Hopefully, practices will improve in the future.

Another reason for lack of adequate polarimetry studies may have been that until recently there have been few modeling programs which made use of polarization data. This is no longer the case, and hopefully, investigators will pay this important field more attention. Closer collaboration between model developers and observers of polarimetric data would also improve the situation. Observers should strive for good coverage not only in phase but also in time (see Sect. 2.6, item 5).

2.4 Magnetometry

Embarras des richesses   (*Embarrassment of riches*)

George E. Hale (1908) discovered magnetic fields in sunspots. Despite failures to detect a general solar field, he foresaw the possibility of detecting magnetic fields in other stars, a feat which was accomplished in 1946. A catalogue of stars showing large Zeeman effects (implying the existence of magnetic fields as high as 5000 Gauß) was published by H. W. Babcock (1958). The technique (Babcock, 1962) involves separating the Zeeman components with the help of a polarization analyzer; the Zeeman splitting is different for the perpendicular and parallel components of the magnetic field. High-precision spectrophotometry and thus large telescopes are required. The stars with strong magnetic fields are mostly of early type (the "magnetic variables" are typically anomalous A-type stars), although large magnetic fields have been found in F-type stars also. In solar and cooler stars polarization is not easily measured, because the circular components of opposite polarity tend to cancel when localized dipolar fields add together. Marcy (1984) succeeded in measuring magnetic fields in 19 of 29 G and K stars examined through a technique which involved linear polarization components. He also found evidence for magnetic fields in bright RS CVn-type stars. To be fully useful for light curve modeling, the technique must be coupled to a mapping process. This has already been performed for polarization variations connected to radial velocity fields, as we note in the next section.

The comprehensive analysis of the eclipsing magnetic binary *E1114+182* by Biermann et al. (1985) shows what data might be obtained from magnetic binaries. This binary is the first eclipsing *AM Herculis* binary system and the shortest eclipsing cataclysmic variable known. *Einstein* X-ray, optical photometry and spectrophotometry, linear polarimetry, and radio emission data enter the analysis and provide tight information on the physical and geometrical status of this binary system.

2.5 Doppler Profile Mapping

De proprio motu   (*of its own motion*)

Armin Deutsch (1958), William Wehlau (1967), and, more recently, John Rice (1996) and references therein, and other observers have applied line profile analysis to locate dark regions on single, rotating stars with strong magnetic fields. Goncharsky et al. (1982, 1983) used Doppler profile measuring techniques to analyze Ap stars, and Vogt et al. (1987) and Strassmeier (1994) used it to map cool spot regions on RS CVn-type binaries. The idea is that a spotted region will cause a depression in the continuum flux from that region of the star. If the region can be associated with a particular velocity field, and thus with a wavelength shift in the profile, analysis of the line profile for dips (absorptions due to dark spots) or bumps (emission due to bright spots) can then help to locate the longitude with respect to the central meridian of the spotted region. The method prefers fast rotating stars because it requires[11] $v \sin i \geq 20$ km/s [see, for instance, Strassmeier (1997, Sect. 5.3)] in order to be effective, a condition that does not usually hold in stars of spectral types F, G, and K (Gray 1988, Chap. 7, p. 21). However, Strassmeier & Rice (1998) succeeded even in analyzing *EK Draconis* with $v \sin i \geq 17.5$ km/s. Gray (1988, Chap. 7, p. 23) suggests an additional method of determining the longitude placement of "star patches" using profile asymmetries. *Stellar tomography* is a term that has been used to describe the use of high-resolution spectral profiles to explore the velocities of components of Algol and other binary star systems that have circumstellar material. One of the most successful of the codes of which we are aware is SHELLSPEC, developed by Jan Budaj and Mercedes Richards. The simultaneous use of WD-type codes with this type of software tool may prove invaluable for future investigations. In fact Miller et al. (2007) seems to do just this. Details about SHELLSPEC may be found at the url http://www.astro.sk/ budaj/shellspec.html.

2.6 Advice to Observers

Docendo discimus   (*We learn by teaching*)

This section contains suggestions to observers of *EB*s to help improve the database and its subsequent analysis.

---

[11] Note that v denotes the speed of rotation.

2.6 Advice to Observers

1. Mid-range to long-period binaries ($P > 5^d$, but especially $P > 50^d$) nee vations of all kinds. The $P > 50^d$ systems with giant or bizarre cor may have interesting light curves even if they do not eclipse. The longer period binaries are nearly unexplored territory. They are perfectly suitable for APTs, (robotic) automatic photometric telescopes [cf. Milone et al. (1995) or Strassmeier et al. (1997)].
2. Infrared light curves have been neglected relative to optical light curves. Infrared light curves are especially needed for binaries with large temperature differences between the two components. Observations over at least one full cycle are critically important, especially if an optical light curve can be observed simultaneously. The main problem in the infrared has been the lack of a single set of standard passbands to which observations that are made from any observing site could be transformed. This is because the original Johnson JHKLMNQ passbands were not designed to fit cleanly within the Earth's atmospheric windows. Subsequent observers dealt with the problem by redesigning passbands suitable for their own observing sites. These, too, were not optimum for sites with different water vapor content. Consequently, there have been several generations of such passbands produced, and transformations between infrared passbands of different generations are particularly prone to systematic errors. The difficulty is that atmospheric water vapor absorption produces curvature in the extinction curve between 0 and 1 air- (actually water vapor) mass, an effect named after Forbes (1842), and this curvature may differ from hour to hour as well as night to night, let alone from season to season. Even differential light curves may be affected by systematic as well as random noise depending on the distance of comparison from target stars and the data sampling frequency. Beginning in the late 1980s (Milone 1989), a new approach was undertaken by IAU Commission 25. An infrared working group (IRWG) was set up to design a new of passbands that are optimally placed in the atmospheric windows. The result by Young et al. (1994) is a new set of infrared passbands (iZ, iJ, iH, iK, iL, iL', iM, iN, in, iQ) are transformable to a higher precision than are all previous passbands. The near infrared set (iZ, iJ, iH, iK) have been field tested and found useful for any site at which photometry can be carried out, with superior S/N and greater insensitivity to water vapor than nearly all previous infrared passbands (Milone & Young 2005). With such passbands, precise light curves may be achievable, and with that precision, more precise and accurate parameters made be determined.
3. Some "observational" papers that do not list any observations are being published. Advances in interpretation are such that 5 or 10 years after publication, the observations may remain the only worthwhile part of a paper. Graphs of observations and phased observations (without the absolute time information) are no substitute for actual data. Nor are promises of availability a substitute for published numbers. CD ROMs, data archives (such as the CDS in Strasbourg), and World Wide Web pages now make it easy to publish most kinds of observations. We should consider the likely long-term permanency of the repository to be selected. The commissions 27 (Variable Stars) and 42 (Close Binary Systems)

of the International Astronomical Union (IAU) might also support the archiving of variable-star data (Sterken & Jaschek, 1997, p. 2).
4. Certain binaries with active mass flow need to be followed continuously over at least several orbits. Such objects are suitable targets for APTs.
5. Concerning polarimetry, it is very important to observe over several consecutive orbits and to document the absolute time instead of only phases. Although light curves are nearly periodic, polarization curves mainly reveal transient events. They usually show only a jumble if folded in phase space. Whereas most light curves are published and nonpublication is (or should be) the exception, it is the other way around for polarization curves – usually they are not published and soon lost.
6. X-ray binaries constitute a new candidate class for precise light curve analysis. These objects provide X-ray duration constraints and arrival times of X-ray pulses. Here again, it is very important to publish actual observations, including absolute time information. As is shown in Sect. 7.3.1, X-ray binaries can provide excellent data for simultaneous fitting (see Sect. 4.1.1.6) of multiband light curves, optically determined radial velocities, and pulse arrival times – potentially a remarkable bounty of separate kinds of information.
7. Spectrophotometry provides an even greater potential bounty, and, in principle, thousands of light curves. Important requirements are that the spectra must be free of scattered light effects and similar comparison star spectra must be available (both conditions are rarely met). Consequently, precise spectrophotometry is rare. Such data need to be carefully processed so that the resulting light curves have the requisite precision. Data may be binned in order to improve precision, at the expense of spectral resolution, but the advantages of many multiwavelength light curves still obtain – the chief of which are the radiative properties of the components.

2.7 Eclipsing Binary Data from Surveys

Many surveys have been conducted and those that have high resampling rates have produced variable star discoveries. One such survey known to us is that being conducted with the Baker–Nunn Patrol Camera (BNPC) of the University of Calgary's Rothney Astrophysical Observatory. This f/1 instrument as currently configured has a 4096×4096 chip in an FLI CCD camera as its detector and captures more than 19 square degrees of the sky on a single exposure. M. Williams has used the instrument to detect more than 30 variable stars in the range 11–15 magnitude (in a passband approximately equivalent to Johnson's R) in a single sky field. Of these, 24 are eclipsing variables. A fitted R_J light curve is shown in Fig. 2.7. The analysis was carried out with the WD package including both simplex and damped least-squares options. The fitted curve is for a semi-detached model.

Much wider surveys have been carried out in searches for gravitational lensing (OGLE for Optical Gravitational Lensing Experiment), which is monitoring the

2.7 Eclipsing Binary Data from Surveys

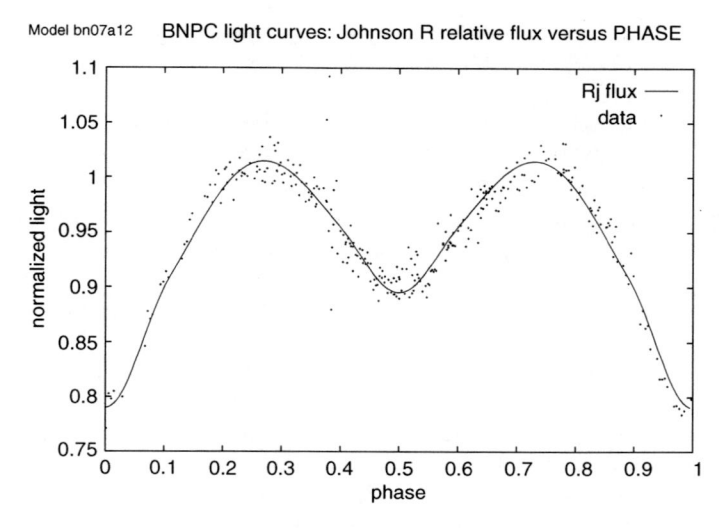

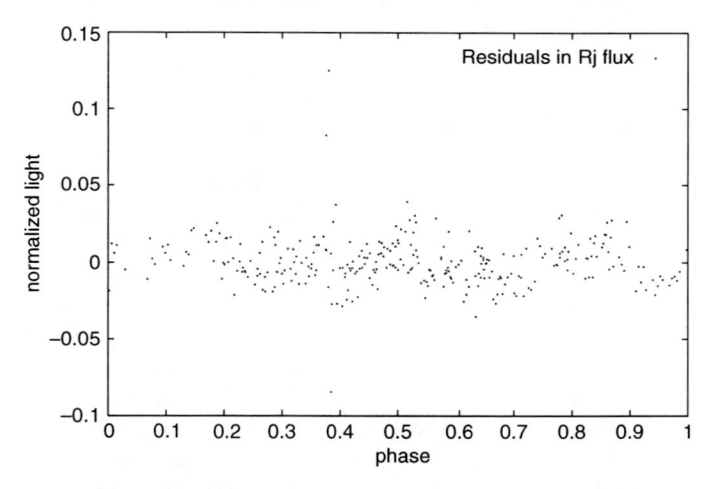

Fig. 2.7 An R_J light curve of an eclipsing binary observed with the Baker–Nunn patrol camera. The binary has been modeled as a semi-detached system

galactic bulge, and has produced hundreds of variables, includes tens of candidates for planetary transits, and even some cases of lensing, apparently by Jovian-mass planets. From 2001, the survey has been carried out from Chile with the 1.3-m Warsaw telescope equipped with an 8k MOSAIC camera consisting of eight SITe 2048 × 4096 chips (8192 × 8192 pixels) with a scale of 0.26 arcsec/pixel, giving a field of view of 35′ × 35′ on the sky.

A similar survey is aimed at finding MACHOs (Massive Astrophysical Compact Halo Objects) as components of the dark matter in galactic halos. Although variable stars have been discovered in this survey in large numbers (at least 611 *EB*s, for example; cf. Alcock et al. 1997), one of the more exciting results of this survey is that dark objects observed as lenses may have masses near 0.5 solar-masses, typically. Such a mass coupled with low luminosity suggests white dwarfs; cf. (Bennett et al. 2005) or Chabrier et al. (1996). If this assessment is accurate, 50% of the dark halo may be composed of these objects, and if so, they may be five times as numerous as main sequence stars. However, this high a white dwarf contribution to the dark matter halo has been questioned, for example, by Torres et al. (2008); other results have emerged from this (MACHO) large survey, not least of which has been a new determination for the distance of the Large Magellanic Cloud (Alcock et al. 2004).

An important space survey that yielded impressive results was the Hipparcos–Tycho mission. The Hipparcos (for HIgh Precision PARallax COllecting Satellite) astrometry mission and the accompanying Tycho two-passband photometric instrument discovered more than 8000 variable stars, including *EB*s. HIPPARCOS itself obtained 13 million observations of 118,000 stars and recorded them with high photometric precision (median precision of 0.0015 magn. for $m_H < 9$ or better), whereas TYCHO observed 1 million stars and obtained, on average, ~18% more observations per star, but to a lower precision. The mission concluded in 1993, after 4 years of operation. The Hipparcos and Tycho Catalogue (Perryman et al. 1997) lists 11,597 possible variable star detections, of which 8237 were new. But further data mining produced 2675 more candidate variable star discoveries (Koen & Eyer 2002).

A survey space mission that holds much promise for variable star discoveries is the GAIA[12] mission described in Munari (2003) (GAIA was originally named for Global Astrometric Interferometer for Astrophysics, and although no longer applicable, the acronym has been kept for continuity purposes.) GAIA is to be launched in 2012, by current estimates. This mission is to contain astrometric, spectrophotometric, and spectroscopic instruments (Perryman 2002). It is expected to measure positions of a billion stars, of which 18 million are expected to be variable (Eyer & Cuypers 2000), among them at least 1 million *EB*s. The radial velocity spectrometer, with resolution $R \approx 11,200$, will yield the third dimension of the kinematic motion for all stars brighter than 15th magnitude. The instrument configuration is intended to produce kinematic, brightness, luminosity, temperature, metallicity, and extinction discrimination for the stars of the Milky Way, and perhaps also, to some extent, the other galaxies of the local group. As currently planned, it has changed from initially planned photometry that involved 5 broad and 14 narrow passbands ideally designed to produce much of this discrimination (Jordi et al. 2004a, b) to spectrophotometry. Thus, to replicate the photometry, integrations need to be carried out over the spectral energy distribution. Even with lower expected precision for astrometric and radial velocity determination, the results should nevertheless greatly

---

[12] http://www.rssd.esa.int/index.php?project=GAIA&page=index provides detailed information on GAIA.

enhance our knowledge of the fundamental properties, composition, motions, and distances of the stars of our galaxy. At the very least, if the 5-year mission is successful, GAIA's results should keep ground-based variable star observers busy on follow-up studies for a generation. The maximum precision of its astrometric measurements should exceed 10 micro-arcsecs and be able to demonstrate proper motion for a great portion of our galaxy and answer many questions about the structure and dynamics of the Milky Way galaxy. Its value for and debt to *EB* studies, at least in its earlier configuration, has been described in Milone (2003).

2.8 Terminology: "Primary Minimum" and "Primary Star"

Quod licet Jovi, non licet bovi   (*Rank has its privileges*)

The deeper minimum in a light curve is called the "primary minimum" by photometrists when the difference in depths of the two minima is clearly discerned, but the designation may be arbitrary in cases where it is not. Photometrists usually compute the decimal fraction of a photometric cycle (the "phase") from the primary minimum. Which component is the "primary star?" An astronomer's background usually dictates the convention: The usage differs among photometrists, spectroscopists, and theoreticians and so is not always consistent. It even differs from one astronomer to another! In the context of photometry, the star being eclipsed at primary minimum is called the "primary star" by convention. Note that this classification is not directly one of size or mass but rather of temperature. For circular orbit binaries, it is rather the star of larger brightness per unit area which is eclipsed at "primary minimum," though this is usually also the more massive component. The main difficulty with this definition is that it leaves in limbo the case of equal minima. The dilemma may be avoided if multiwavelength light curves can resolve the degeneracy, but if all light curves reveal that the surface brightnesses of the components are indeed equal, the choice of primary minimum must be an arbitrary one if based alone on photometry. In spectroscopy, the usage is mixed. In astrophysical investigations of spectral features, the component with the stronger spectral lines, ordinarily the star with the apparently greater luminosity, is usually classified as the "primary star." In radial velocity investigations, the component with the smaller radial velocity amplitude (i.e., the more massive star) is usually designated the "primary star." The definition runs into difficulty in dealing with the case of equal radial velocity amplitudes and must be, again, arbitrary if the masses are measurably equal. Although the more massive star is usually the more luminous and often also the hotter star, there are cases where these associations do not hold.

When theoretical studies are included, the situation becomes even more muddied. In the context of discussing the stellar evolution of a binary in question, the term "primary star" sometimes refers to the *originally* more massive star which can become the lower mass star as a consequence of mass transfer. In celestial mechanics, especially in papers dealing with the restricted three-body problem, both components are called "primaries" in contrast to the massless third body. Finally, again in the general three-body problem, the most massive component is sometimes called the "primary."

Because of these different usages, it is impossible to give a single definitive response to the question of which star should be designated "primary." The only advice we can give is to define the meaning of the term and to assign number 1 or 2 to the components of a system early in any publication.

2.9 Selected Bibliography

- Sterken & Manfroid (1992) for a modern overview of photometry.
- McLean (1997) for a thorough discussion of astronomical imaging.
- Henden & Kaitchuck (1982): A guide to techniques of photometric observing and data reduction.
- Howell (2000): *A Handbook of CCD Astronomy*. One of the best discussions of the hardware, operation, and applications of CCDs in astronomy.
- Wood (1963): *Photoelectric Photometry for Amateurs*. A low-key, nonthreatening introduction.
- Aitken (1964), Binnendijk (1960), and Batten (1973a) – recommended books on double stars. Especially, Binnendijk (1960) is a handy reference with many formulas.
- Hardie (1962) in *Astronomical Techniques* for basic extinction, transformation treatment.
- Young (1974, 1994) for a more sophisticated discussion of extinction and transformation.
- Golay (1974) for definition of color systems and transformations between systems.
- Hall & Genet (1988) provide an overview on many topics of *Photoelectric Photometry of Variable Stars*.
- Tinbergen (1996) for polarimetry.
- Strassmeier (1997) for photometry, spectroscopy, and magnetometry especially of active stars and stellar spots. A book very much recommended from the didactic point of view.
- *Stellar Photometry – Current Techniques and Future Developments* by Butler & Elliot (1993) (proceedings of the IAU Colloquium 136). As the title indicates, a good overview.
- The *Handbook of CCD Astronomy* by Howell (2006) is a new edition of Howell's well-known and very useful practical handbook of CCD astronomy.
- The *Future of Photometric and Polarimetric Standardization* by Sterken (2007) summarizes the developments of absolute photometry up to 2006. Contributions cover the spectrum from the UV to the infrared and discuss the challenges and solutions to the calibrations of modern ground-based investigations and space-based missions.

References

Aitken, R. G.: 1964, *The Binary Stars*, Dover Publications, Philadelphia, PA, 3rd edition
Alcock, C., Allsman, R. A., Alves, D., Axelrod, T. S., Becker, A. C., Bennett, D. P., Cook, K. H., Freeman, K. C., Griest, K., Lacy, C. H. S., Lehner, M. J., Marshall, S. L., Minniti, D., Peterson, B. A., Pratt, M. R., Quinn, P. J., Rodgers, A. W., Stubbs, C. W., Sutherland, W., & Welch, D. L.: 1997, The MACHO Project LMC Variable Star Inventory V: Classification and Orbits of 611 Eclipsing Binary Stars, *AJ* **114**, 326–340
Alcock, C., Alves, D. R., Axelrod, T. S., Becker, A. C., Bennett, D. P., Clement, C. M., Cook, K. H., Drake, A. J., Freeman, K. C., Geha, M., Griest, K., Lehner, M. J., Marshall, S. L., Minniti, D., Muzzin, A., Nelson, C. A., Peterson, B. A., Popowski, P., Quinn, P. J., Rodgers, A. W., Rowe, J. F.,, Sutherland, W., Vandehei, T., & Welch, D. L.: 2004, The MACHO Project Large Magellanic Cloud Variable-Star Inventory. XIII. Fourier Parameters for the First-Overtone RR Lyrae Variables and the LMC Distance, *AJ* **127**, 334–354
Aller, L. H.: 1963, *The Atmospheres of the Sun and Stars*, The Ronald Press, New York, 2nd edition
Andersen, J., Clausen, J. V., Gustafsson, B., Nordström, B., & Vandenberg, D. A.: 1988, Absolute Dimensions of Eclipsing Binaries. XIII. AI Phoenicis: A Case Study in Stellar Evolution, *A&A* **196**, 128–140
Babcock, H. W.: 1958, A Catalogue of Magnetic Stars, *ApJ Suppl.* **3**, 141–210
Babcock, H. W.: 1962, Measurements of Stellar Magnetic Fields, in W. A. Hiltner (ed.), *Astronomical Techniques*, Vol. 2, pp. 107–125, University Chicago Press, Chicago, IL
Barden, S. C.: 1985, A Study of Short-Period RS Canum Venaticorum and W Ursae Majoris Binary Systems – The Global Nature of H-alpha, *ApJ* **295**, 162–170
Bennett, D. P., Becker, A. C., & Tomaney, A.: 2005, Photometric Confirmation of MACHO Large Magellanic Cloud Microlensing Events, *ApJ* **631**, 301–311
Bessell, M. S.: 1979, The Cousins (Cape-Kron) BVRI System, its Temperature and Absolute Flux Calibration, and Relevance for Two-Dimensional Photometry, in A. G. D. Philip (ed.), *Problems of Calibration of Multicolor Photometric Systems*, Vol. 14, pp. 279–296, Dudley Observatory Reports, Schenectady, NY
Biermann, P., Schmidt, G. D., Liebert, J., Stockman, H. S., Tapia, S., Kühr, H., Strittmacher, P. A., & Lamb, D. Q.: 1985, The New Eclipsing Magnetic Binary System E1114+182, *ApJ* **293**, 303–320
Binnendijk, L.: 1960, *Properties of Double Stars*, University of Pennsylvania Press, Philadelphia, PA
Budaj, J. & Richards, M. T.: 2004, A Description of the Shellspec Code, *Contributions of the Astronomical Observatory Skalnate Pleso* **34**, 167–196
Butler, C. J. & Elliot, I. (eds.): 1993, *Stellar Photometry – Current Techniques and Future Developments*, IAU Symposium 136, Cambridge University Press, Cambridge, UK
Butler, R. P., Marcy, G. W., Williams, E., McCarthy, C., Dosanjh, P., & Vogt, S. S.: 1996, Attaining Doppler Precision of 3 m/s, *PASP* **108**, 500–509
Campbell, B., Walker, G. A. H., & Yang, S.: 1988, A Search for Substellar Companions to Solar-Type Stars, *ApJ* **331**, 902–921
Canterna, R.: 1976, Broad-Band Photometry of G and K Stars: The C, M, T_1, T_2 Photometric System, *AJ* **81**, 228–244
Chabrier, G., Segretain, L., & M'era, D.: 1996, Contribution of Brown Dwarfs and White Dwarfs to Recent Microlensing Observations and to the Halo Mass Budget, *ApJ Letters* **468**, L21–L24
Chandrasekhar, S.: 1946a, On the Radiative Equilibrium of a Stellar Atmosphere. X, *ApJ* **103**, 351–370
Chandrasekhar, S.: 1946b, On the Radiative Equilibrium of a Stellar Atmosphere. XI, *ApJ* **104**, 110–132
Cousins, A. W. J.: 1976, VRI Standards in the E Region, *Mem. R. Astron. Soc.* **81**, 25–36
Cousins, A. W. J.: 1978, VRI Photometry of E and F Region Stars, *MNASSA* **37**, 8–10

Deutsch, A. J.: 1958, Harmonic Analysis of the Periodic Spectrum Variables, in B. Lehnert (ed.), *Electromagnetic Phenomena in Cosmical Physics*, IAU Symposium 6, pp. 209–221, Cambridge University Press, Cambridge, UK

Duerbeck, H. W. & Hoffmann, M.: 1994, Principles of Photometry, in G. D. Roth (ed.), *A Compendium of Practical Astronomy*, Vol. I, pp. 319–379, Springer, New York

ESA: 1997, *The Hipparcos and Tycho Catalogues*, Vols. 1–17, Technical Report Sp-1200, ESA, ESA Publications Division, C/O ESTEC, Noordwijk, The Netherlands

Etzel, P. B.: 1988, Spectrophotometry of Eclipsing Binaries, in A. G. D. Philip, D. S. Hayes, & S. J. Adelman (eds.), *New Directions in Spectrophotometry*, pp. 79–89, L. Davis Press, Schenectady, NY

Etzel, P. B.: 1990, Spectrophotometry of Algol-Type Binaries, in I. Ibanoglu (ed.), *Active Close Binaries*, pp. 189–201, Kluwer Academic Publishers, Dordrecht, Holland

Eyer, L. & Cuypers, J.: 2000, Predictions on the Number of Variable Stars for the GAIA Space Mission and for Surveys as the Ground-based International Liquid Mirror Telescope, in L. Szabados & D. W. Kurtz (eds.), *The Impact of Large-Scale Surveys on Pulsating Star Research*, Vol. 203 of *ASP Conference Series*, pp. 71–72, Astronomical Society of the Pacific, San Francisco

Fiorucci, M. & Munari, U.: 2003, The Asiago Database on Photometric Systems (ADPS). II. Band and Reddening Parameters, *A&A* **401**, 781–796

Forbes, J. D.: 1842, The Bakerian Lecture: On the Transparency of the Atmosphere and the Law of Extinction of the Solar Rays in Passing through It, *Philosophical Transactions of the Royal Society, London* **132**, 225–273

Ghedini, S.: 1982, *Software for Photometric Astronomy*, Willmann Bell, Richmond, UK

Gilliland, R. L., Brown, T. M., Kjeldsen, H., McCarthy, J. K., Peri, M. L., Belmonte, J. A., Vidal, I., Cram, L. E., Palmer, J., Frandsen, S., Parthasarathy, M., Petro, L., Schneider, H., Stetson, P. B., & Weiss, W. W.: 1993, A Search for Solar-Like Oscillations in the Stars of M67 with CCD Ensemble Photometry on a Network of 4-m Telescopes, *AJ* **106**, 2441–2476

Golay, M.: 1974, *Introduction to Astronomical Photometry*, Vol. 41 of *Astrophysics and Space Science Library*, D. Reidel, Dordrecht, Holland

Goncharsky, A. V., Ryabchikova, T. A., Stepanov, V. V., Khokhlova, V. L., & Yagola, A. G.: 1983, Mapping of Elements on the Surfaces of Ap Stars. II. Distribution of Eu, Sr, and Si on $\alpha 2$ CVn, χ Ser, and CU Vir., *Sov. Astron.* **27**, 49–53

Goncharsky, A. V., Stepanov, V. V., Khokhlova, V. L., & Yagola, A. G.: 1982, Mapping of Chemical Elements on the Surfaces of Ap Stars. I. Solution of the Inverse Problem of Finding Local Profiles of Spectral Lines, *Sov. Astron.* **26**, 690–696

Gray, D. F. (ed.): 1988, *Lectures on Spectral Analysis: F, G, and K Stars*, "The Publisher", Aylmer, Ontario, Canada

Gray, D. F. (ed.): 1992, *The Observation and Analysis of Stellar Atmospheres*, Cambridge University Press, Cambridge, UK, 2nd edition

Hadrava, P.: 1995, Orbital Elements of Spectroscopic Stars, *A&A Suppl.* **114**, 393–396

Hadrava, P.: 2004, KOREL – User's guide, *Publications of the Astronomical Institute of the Czechoslovak Academy of Sciences* **92**, 15–35

Hale, G. E.: 1908, On the Probable Existence of a Magnetic Field in Sun-Spots, *ApJ* **28**, 315–343

Hall, D. S. & Genet, R. M. (eds.): 1988, *Photoelectric Photometry of Variable Stars*, Willmann Bell, Richmond, UK

Hall, J. S.: 1949, Observations of the Polarized Light from Stars, *Science* **109**, 166–167

Hall, J. S. & Mikesell, A. H.: 1949, Observations of the Polarized Light from Stars, *AJ* **54**, 187–188

Hall, J. S. & Mikesell, A. H.: 1950, Polarization of Starlight in the Galaxy, *Publ., U.S. Naval Obs.* **17**, 275–342

Hardie, R. H.: 1962, Photoelectric Reductions, in W. A. Hiltner (ed.), *Astronomical Techniques*, Vol. 2, pp. 178–208, University of Chicago Press, Chicago, IL

Hauck, B.: 1968, Étude d'une Représentation Tri-dimensionelle des Étoiles de Type Spectral Compris Entre A_0 et G_5, *Publ. de L'Observatoire de Genève* **75**, 133–164 (in Arch. Sci. Genève, Vol. 21)

References

Henden, A. A. & Kaitchuck, R. H.: 1982, *Astronomical Photometry*, Van Nostrand Reinhold Company, New York

Hill, G.: 1982, The Reduction of Spectra-IV: VCROSS, an Interactive Cross-Correlation Velocity Program, *Publ. Dom. Astrophys. Obs.* **16**, 59–66

Hill, G., Fisher, W. A., & Poeckert, R.: 1982, The Reduction of Spectra-III: REDUCE, an Interactive Spectrophotometric Program, *Publ. Dom. Astrophys. Obs.* **16**, 43–58

Hiltner, W. A.: 1949a, On the Presence of Polarization in the Continuous Radiation of Stars. II, *ApJ* **109**, 471–478

Hiltner, W. A.: 1949b, Polarization of Light from Distant Stars by Interstellar Medium, *Science* **109**, 165

Howell, S. B.: 1992, Introduction to Differential Time-Series Astronomical Photometry Using Charged-Coupled Devices, in S. B. Howell (ed.), *Astronomical CCD Observing and Reduction Techniques*, Vol. 23 of *Astronomical Society of the Pacific Conference Series*, pp. 105–128

Howell, S. B.: 2000, *Handbook of CCD Astronomy*, Cambridge Observing Handbooks for Research Astronomers, Cambridge University Press, Cambridge, U.K.

Howell, S. B.: 2006, *Handbook of CCD astronomy*, Cambridge University Press, Cambridge, UK, Handbook of CCD astronomy, 2nd ed., by S.B. Howell. Cambridge observing handbooks for research astronomers, Vol. 5 Cambridge, UK: Cambridge University Press, 2006 ISBN 0521852153

Howell, S. B., Warnock, A. I., & Mitchell, K. J.: 1988, Statistical Error Analysis in CCD Time-resolved Photometry with Applications to Variable Stars and Quasars, *AJ* **95**, 247–256

Howell, S. B., Everett, M. E., Tonry, J. L., Pickles, A., & Dain, C.: 2003, Photometric Observations Using Orthogonal Transfer CCDs, *PASP* **115**, 1340–1350

Howell, S. B., vanoutryve, C., Tonry, J. L., Everett, M. E., and Schneifer, R.: 2005, A search for variable stars and planetary occultations in NGC 2301. II variablility, *PASP* **117**, 1187–1203.

Jacoby, G. H., Hunter, D. A., & Christian, C. A.: 1984, A Library of Stellar Spectra, *ApJ Suppl.* **56**, 257–281

Johnson, H. L.: 1966, Fundamental Stellar Photometry for Standards of Spectral Type on the Revised System of the Yerkes Spectral Atlas, *Ann. Rev. Astron. Astrophys.* **4**, 193–206

Johnson, H. L. & Morgan, W. W.: 1953, Fundamental Stellar Photometry for Standards of Spectral Type on the Revised System of the Yerkes Spectral Atlas, *ApJ* **117**, 313–352

Johnson, J. A., Winn, J. N., Cabrera, N. E., & Carter, J. A.: 2009, A Smaller Radius for the Transiting Exoplanet WASP-10b, *ApJ Letters* **692**, L100–L104

Jordi, C., Carrasco, M., Høg, E., Brown, A. G. A., & Knude, J.: 2004a, *BBP Photometric Systems Evaluation*, Technical Report UB-PWG-028, ESA, ESA Publications Division, C/O ESTEC, Noordwijk, The Netherlands

Jordi, C., Carrasco, M., Høg, E., Brown, A. G. A., & Knude, J.: 2004b, *MBP Photometric Systems Evaluation*, Technical Report UB-PWG-029, ESA, ESA Publications Division, C/O ESTEC, Noordwijk, The Netherlands

Kemp, J. C., Henson, G. D., Barbour, M. S., Kraus, D. J., & Collins, II, G. W.: 1983, Discovery of eclipse polarization in Algol, *ApJ Letters* **273**, L85–L88

Kemp, J. C., Swedlund, J. B., Landstreet, J. D., & Angel, J. R. P.: 1970, Discovery of Circularly Polarized Light from a White Dwarf, *ApJ Letters* **161**, L77–L79

Koen, C. & Eyer, L.: 2002, New Periodic Variables from the Hipparcos Photometry, *MNRAS* **331**, 45–59

Kuerster, M., Hatzes, A. P., Cochran, W. D., Pulliam, C. E., & Dennerl, K.: 1994, A Radial Velocity Search for ExtraSolar Planets Using an Iodine Gas Absorption Cell at the CAT+CES, *The Messenger* **76**, 51–55

Landolt, A. U.: 1983, UBVRI Photometric Standard Stars Around the Celestial Equator, *AJ* **88**, 439–460

Landolt, A. U. & Blondeau, K. L.: 1972, The Calculation of Heliocentric Corrections, *PASP* **84**, 784–809

Marcy, G. W.: 1984, Observations of Magnetic Fields on Solar-Type Stars, *ApJ* **276**, 286–304

McLean, B. J.: 1981, Radial Velocities for Contact Binary Systems – I. W Ursae Majoris and AW Ursae Majoris, *MNRAS* **195**, 931–938

McLean, I. S.: 1997, *Electronic Imaging in Astronomy: Detectors and Instrumentation*, Wiley, Chichester, UK

Mihalas, D.: 1965, Model Atmospheres and Line Profiles for Early-Type Stars, *ApJ Suppl.* **9**, 321–437

Mihalas, D.: 1978, *Stellar Atmospheres*, Freeman, San Francisco, CA, 2nd edition

Miller, B., Budaj, J., Richards, M., Koubský, P., & Peters, G. J.: 2007, Revealing the Nature of Algol Disks through Optical and UV Spectroscopy, Synthetic Spectra, and Tomography of TT Hydrae, *ApJ* **656**, 1075–1091

Milone, E. F.: 1986, The O'Connell Effect Systems CX Canis Majoris, TU Crucis, AQ Monocerotis, and DQ Velorum, *ApJ Suppl.* **61**, 455–464

Milone, E. F. (ed.): 1989, *Infrared Extinction and Standardization*, Springer, Berlin, Germany

Milone, E. F.: 2003, Fundamental Stellar Parameters from Eclipsing Binaries, in U. Munari (ed.), *GAIA Spectroscopy, Science and Technology*, Vol. 298 of *ASP Conference Series*, pp. 303–313, Astronomical Society of the Pacific, San Francisco

Milone, E. F., Chia, T. T., Castle, K. G., Robb, R. M., & Merrill, J. E.: 1980, RW Comae Berenices I. Early Photometry and UBV Light Curves, *ApJ Suppl.* **43**, 339–364

Milone, E. F., Hrivnak, B. J., Hill, B. J., & Fisher, W. A.: 1984, RW Comae Berenices - II. Spectroscopy, *AJ* **90**, 109–114

Milone, E. F., Kallrath, J., Stagg, C. R., & Williams, M. D.: 2004, The Modeling of Binaries in Globular Clusters, *Revista Mexicana AA (SC)* **21**, 109–115

Milone, E. F. & Robb, R. M.: 1983, Photometry with the Rapid Alternate Detection System, *PASP* **95**, 666–673

Milone, E. F., Robb, R. M., Bobott, F. M., & Hansen, C. H.: 1982, Rapid Alternate Detection Systems of the Rothney Astrophysical Observatory, *Appl. Optics* **21**, 2992–2995

Milone, E. F., Stagg, C. R., & Kurucz, R. L.: 1992, The Eclipsing Binary AI Phoenicis: New Results Based on an Improved Light Curve Analysis Program, *ApJ Suppl.* **79**, 123–137

Milone, E. F., Stagg, C. R., & Young, A. T.: 1995, Towards Robotic IR Observatories: Improved IR Passbands, in M. F. Bode (ed.), *Robotic Observatories*, pp. 117–124, Wiley-Praxis, Chichester

Milone, E. F., Williams, M. D., Stagg, C. R., McClure, M. L., Desnoyers Winmil, B., Brown, T., Charbonneau, D., Gilliland, R. L., Henry, G. W., Kallrath, J., Marcy, G. W., Terrel, D., & Van Hamme, W.: 2004, Simulation and Modeling of Transit Eclipses by Planets, in A. J. Penny, P. Artymowicz, A.-M. Lagrange, & S. Russell (eds.), *Planetary Systems in the Universe: Observation, Formation and Evolution*, Vol. 214 of *ASP Conference Series*, pp. 90–92, Astronomical Society of the Pacific, San Francisco

Milone, E. F. & Young, A. T.: 2005, An Improved Infrared Passband System for Ground-based Photometry: Realization, *PASP* **117**, 485–502

Milone, E. F. & Young, A. T.: 2007, Standardization and the Enhancement of Infrared Precision, in C. Sterken (ed.), *The Future of Photometric, Spectrophotometric and Polarimetric Standardization*, Vol. 364 of *Astronomical Society of the Pacific Conference Series*, pp. 387–407

Milone, E. F. & Young, A. T.: 2008, Infrared Passbands for Precise Photometry of Variable Stars by Amateur and Professional Astronomers, *Journal of the American Association of Variable Star Observers (JAAVSO)* **36**, 110–126

Moro, D. & Munari, U.: 2000, The Asiago Database on Photometric Systems (ADPS). I. Census Parameters for 167 Photometric Systems, *A&A Suppl.* **147**, 361–628

Mukherjee, J. D., Peters, G. J., & Wilson, R. E.: 1996, Rotation of Algol Binaries - A Line Profile Model Applied to Observations, *MNRAS* **283**, 613–625

Munari, U.: 2003, *GAIA Spectroscopy: Science and Technology*, Vol. 298 of *Astronomical Society of the Pacific Conference Series*, Astronomical Society of the Pacific, San Francisco

Perryman, M. A. C.: 2002, GAIA: An Astrometric and Photometric Survey of our Galaxy, in V. Vansevicius, A. Kucinskas, & J. Sudzius (eds.), *Census of the Galaxy: Challenges for Photometry and Spectrometry with GAIA*, Vol. 280, pp. 1–10 of *Astrophysics and Space Science*, pp. 1–10, Kluwer Academic Publishers, Dordrecht, Boston, London

References

Perryman, M. A. C., Lindegren, L., Kovalevsky, J., Hoeg, E., Bastian, U., Bernacca, P. L., Creze, M., Donati, F., Grenon, M., van Leeuwen, F., van derMarel, H., Mignard, F., Murray, C. A., Le Poole, R. S., Schrijver, H., Turon, C., Arenou, F., Froeschle, M., & Petersen, C. S.: 1997, The HIPPARCOS Catalogue, *A&A* **323**, L49–L52

Rice, J. B.: 1996, Doppler Imaging of Stellar Surfaces, in K. G. Strassmeier & J. L. Linsky (eds.), *Stellar Surface Structure*, IAU Symposium 176, pp. 19–33, Kluwer Academic Publishers, Dordrecht, Holland

Robb, R. M. & Milone, E. F.: 1982, A Single Night Light Curve of 44i Boo, *Inform. Bull. Variable Stars* **2187**, 1–4

Rucinski, S.: 1999, Determination of Broadening Functions Using the Singular-Value Decomposition (SVD) Technique, in J. B. Hearnshaw & C. D. Scarfe (eds.), *IAU Colloq. 170: Precise Stellar Radial Velocities*, Vol. 185 of *Astronomical Society of the Pacific Conference Series*, pp. 82–90

Rucinski, S. M.: 1992, Spectral-line Broadening Functions of WUMa-type binaries. I - AW UMa, *AJ* **104**, 1968–1981

Rucinski, S. M.: 2002, Radial Velocity Studies of Close Binary Stars. VII. Methods and Uncertainties, *AJ* **124**, 1746–1756

Schiller, S. J. & Milone, E. F.: 1990, Simultaneous Photoelectric and CCD Photometry of the Delta Scuti Star DY Herculis, in A. G. D. Davis, D. S. Hayes, & S. J. Adelman (eds.), *CCDs in Astronomy II. New Methods and Applications of CCD Technology*, pp. 159–165, L. Davis Press, Schenectady, NY

Schlosser, W., Schmidt-Kaler, T., & Milone, E. F.: 1991, *Challenges of Astronomy*, Springer, New York

Simkin, S. M.: 1974, Measurements of Velocity Dispersions and Doppler Shifts from Digitized Optical Spectra, *A&A* **31**, 129–136

Simon, K. P. & Sturm, E.: 1994, Disentangling of Composite Spectra, *A&A* **281**, 286–291

Sterken, C. (ed.): 2007, *Standardization and the Enhancement of Infrared Precision*, Vol. 364 of *Astronomical Society of the Pacific Conference Series*

Sterken, C. and Jaschek, C. (eds.): 1997, *Light Curves of Variable Stars – A Pictorial Atlas*, Cambridge University Press, Cambridge, UK

Sterken, C. & Manfroid, J.: 1992, *Astronomical Photometry – A Guide*, Vol. 175 of Astrophysics and Space Science Library, Kluwer Academic Publishers, Dordrecht, Holland

Stetson, P. B.: 1998, *User's Manual for DAOPHOT II*, Dominion Astrophysical Observatory, Victoria, BC, 1998 edition

Straižys, V.: 1975, The Calibration of the Reddening-Free Parameters of the Vilnius Photometric System in Temperatures, Surface Gravities and Metalicities, in A. G. D. Philip & D. S. Hayes (eds.), *Multicolor Photometry and the Theoretical HR Diagram*, Vol. 9, pp. 65–72, Dudley Observatory Reports, Schenectady, NY

Strassmeier, K. G.: 1994, Rotational-Modulation Mapping of the Active Atmosphere of the RS Canum Venaticorum Binary HD 106225, *A&A* **281**, 395–420

Strassmeier, K. G.: 1997, *Aktive Sterne – Laboratorien der solaren Astrophysik*, Springer, Wien, Austria

Strassmeier, K. G., Boyd, L. J., Epand, D. H., & Granzer, T.: 1997, Wolfgang-Amadeus: The University of Vienna Twin Automatic Photoelectric Telescope, *PASP* **109**, 697–706

Strassmeier, K. G. & Rice, J. B.: 1998, Doppler Imaging of Stellar Surface Structure. VI. HD 129333 = EK Draconis: a Stellar Analog of the Active Young Sun, *A&A* **330**, 685–695

Strömgren, B.: 1966, Spectral Classification Through Photoelectric Narrow-Band Photometry, *Ann. Rev. Astron. Astrophys.* **4**, 433–472

Tinbergen, J.: 1996, *Astronomical Polarimetry*, Cambridge University Press, Cambridge, NY, Melbourne

Torres, S., Camacho, J., Isern, J., & García-Berro, E.: 2008, The Contribution of Red Dwarfs and White Dwarfs to the Halo Dark Matter, *A&A* **486**, 427–435

Vogt, S. S., Penrod, G. D., & Hatzes, A. P.: 1987, Doppler Images of Rotating Stars Using Maximum Entropy Image Reconstruction, *ApJ* **321**, 496–515

Wehlau, W.: 1967, The Spectrum of the Ap Star Chi Serpentis, in R. Cameron (ed.), *The Magnetic and Related Stars*, pp. 441–457, Mono Book Corp., Baltimore, MD

Wilson, R. E.: 1993, From Here to Observables: Beyond Light and Velocity Curves, in E. F. Milone (ed.), *Light Curve Modeling of Eclipsing Binary Stars*, pp. 7–25, Springer, New York

Wilson, W. J. F., Milone, E. F., & Fry, D. J. I. F.: 1993, Studies of Large-Amplitude Delta Scuti Variables. I. A Case Study of EH Librae, *PASP* **105**, 809–820

Wood, F. B.: 1963, *Photoelectric Photometry for Amateurs*, Collier-Macmillan, London

Young, A. T.: 1974, Photomultipliers: Their Cause and Cure, in N. Carleton (ed.), *Methods of Experimental Physics*, Vol. 12a of Part A: *Optical and Infrared*, pp. 1–94, Academic Press, New York

Young, A. T.: 1994, Improvements to Photometry VI. Passbands and Transformations, *A&A* **288**, 683–696

Young, A. T., Genet, R. M., Boyd, L. J., Borucki, W. J., Lockwood, G. W., Henry, G. W., Hall, D. S., Smith, D. P., Baliunas, S. L., Donahue, R., & Epand, D. H.: 1991, Precise Automatic Differential Stellar Photometry Analyzing Eclipsing Binaries, *PASP* **103**, 221–242

Young, A. T., Milone, E. F., & Stagg, C. R.: 1994, On Improving IR Photometric Passbands, *A&A Suppl.* **105**, 259–279

Part II
Modeling and Analysis

Chapter 3
A General Approach to Modeling Eclipsing Binaries

This chapter provides the basis to compute observables for a given set of *EB* parameters and a given set of times or phases. Typical observables are light curves, radial velocity curves, polarization curves, and line profiles. In this chapter the focus is on general considerations; no details of implementation are given. Those are found in Chap. 6 for several light curve models.[1] *EB* data analysis leads to a nonlinear least-squares problem in which observed curves are compared with model curves. The presentation is greatly simplified if we take the following formal approach: We formally define an *eclipsing binary observable curve*, $\mathcal{O}$, as a mathematical object

$$\mathcal{O} := \{(t_k, o_k) \mid 1 \leq k \leq n\},$$

i.e., as a set of n elements in which each element is a pair, (t, o), where t represents an independent, time-related quantity and o is the corresponding observable. The quantity t used as the independent quantity may either represent *time* or the *photometric phase* Φ defined in formula (2.1.1).

In the past, at least, in light curve analysis the phase Φ is used. In more recent years, this has changed. If the period and epoch are to be determined, or if apsidal motion effects are considered, or polarization data and pulse arrival times are included in the analysis, as demonstrated in Sect. 7.3.1, it is necessary to use the time instead of phase, or in addition to phase; Section 3.8 provides an example of how this is done.

The term *light* usually is used in the abstract sense in this book and may represent not only the photometric brightness (i.e., the observed radiant power or flux in a particular passband) but any observable,[2] such as

- the light at a given wavelength;
- the radial velocity;

---

[1] Sometimes we use the expression *light curve model* in a general and abstract sense meaning "a model for computing eclipsing binary observables."

[2] This terminology (without the formal mathematical approach) has appeared already in Wilson (1994).

- polarization;
- photospheric spectral line profile;
- spectral distributions due to circumstellar flows; and
- any other quantity associated with the phase, but also other quantities independent of phase which we call *systemic observables*. Given, say, good Hipparcos data, or if the binary happens to be a member of an assemblage (star cluster, galaxy,...) with known distance D, the parallax π is available and can be considered an additional observable.

It may represent

- a measured value of an observable; and
- a value derived from a *light curve program* (more generally, an "*observables generating program*" as per Wilson (1994) based on a *light curve model* ("model," for short).

Thus an "observable curve" ("observable" for short), may be, for instance

- an observed light curve $\mathcal{O}^{obs}$;
- a calculated light curve $\mathcal{O}^{cal}$;
- a wavelength-dependent light curve $\mathcal{O}^{\lambda}$;
- a radial velocity curve $\mathcal{O}^{vel}$;
- a polarization curve $\mathcal{O}^{pol}$;
- a set of pulse arrival times, $\mathcal{O}^{pul}$.

Before the 1970s an observed light curve $\mathcal{O}^{obs}$ of an *EB* was analyzed following rectification procedures which trace back to the early 1900s. However, the underlying physical models were relatively simple and neglected effects which later turned out to be relevant. Photometric and spectroscopic data were analyzed separately and with different methods.

Today's methods permit analysis of photometric, spectroscopic, and other data simultaneously. If the vector **x** represents all relevant *EB* parameters, for each phase Φ the corresponding observable $o^{cal}(\Phi, \mathbf{x})$, or several observables, $o_c^{cal}(\Phi, \mathbf{x})$, of type c can be computed with a light curve model. For a given set of phases, a whole observable curve $\mathcal{O}^{cal}(\mathbf{x})$ or a set of several curves, $\mathcal{O}_c^{cal}(\mathbf{x}^*)$, can be computed; this problem is denoted as the *direct problem*. The *inverse problem* is to determine a set of parameters $\mathbf{x}^*$ from a set of *EB* observations by the condition that a set of curves[3] $\mathcal{O}_c^{cal}(\mathbf{x}^*)$ best fits a set of observed curves $\mathcal{O}_c^{obs}$. The system parameters **x** are modified according to an iterative procedure until the deviation between the observed curves $\mathcal{O}_c^{obs}$ and the calculated curves $\mathcal{O}_c^{cal}(\mathbf{x}^*)$ becomes minimal in a well-defined sense. The system parameters $\mathbf{x}^*$, corresponding to the observed curves $\mathcal{O}_c^{obs}$, are ordinarily regarded as the solution of a least-squares problem.

In Chap. 4, the inverse problem is discussed. Obviously, in order to tackle the inverse problem we need to be able to solve the direct problem, i.e., the mapping

---

[3] We show that it is advantageous to fit several light curves or even different types of eclipsing binary observations simultaneously.

3.1 System Geometry and Dynamics 77

$\mathbf{x} \to \mathcal{O}^{\text{cal}}(\mathbf{x})$, which is the subject of the present chapter. Each realistic model for computing the observable $\mathcal{O}^{\text{cal}}(\mathbf{x})$ for a given set of parameters consists of three major parts:

1. The physics and geometry of orbits and components.
2. Computation of local radiative surface intensity as a function of local gravity, temperature, chemical composition, and direction. The proper formulation of the radiative physics requires the use of accurate model atmospheres.
3. Computation of the integrated flux in the direction of the observer. This computation must take eclipses into account. The inclusion of other effects such as circumstellar matter, i.e., gas streams, disks, attenuating clouds, etc., may be desirable.

3.1 System Geometry and Dynamics

Orbis scientiarum   *(The circle of the sciences)*

The shapes of the stellar surfaces are either explicitly specified a priori (as, e.g., by spheres and ellipsoids) or, in more sophisticated treatments, determined implicitly by a physical model.[4] The theoretical bases for the modeling of stellar shape distortions are varied. Particular light curve models emphasize one of the following: They adopt Chandrasekhar's (1933a, b) results on the theory of polytropic gas spheres and centrifugal- and tidal-force perturbations (Wood 1971) or the Roche model (see Sect. 3.1.5 for references). If the underlying forces can be determined completely by a potential function, the stellar photospheres are assumed to be equipotential surfaces. Surfaces of constant density then coincide with surfaces on which the potential energy per unit mass is constant and the local gravity and surface orientation are given by the gradient of the potential. This approach is generally applicable if the stars move in circular orbits. Under some limited assumptions it is also a good approximation for eccentric orbits (see comments on page 102).

3.1.1 Coordinates and Basic Geometrical Quantities

Figures 3.1 and 3.2 illustrate the geometry of the coordinate system used in most of the models presented in this book. We introduce for present and future purposes a generalized right-handed Cartesian coordinate system (x, y, z) with origin in the center of mass of a star. The x-axis points to the center of mass of the other star, the z-axis is normal to the orbital plane,[5] and the y-axis is fixed by the "right-handed"

---

[4] We refer to light curve models in which the geometry of components is fixed a priori as "geometric models," and to those based on equipotential surfaces as "physical models."

[5] The rotation axes for orbital and proper rotation of the stars are assumed to be parallel to the normal of the orbital plane.

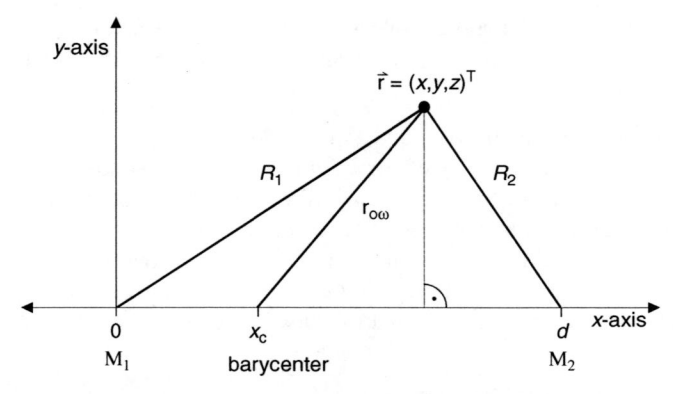

Fig. 3.1 Definition of a right-handed Cartesian coordinate system. The origin is in the center of mass of one of the stars. The x-axis points to the center of mass of the other star, the z-axis is normal to the orbital plane, and the y-axis is fixed by the "right-handed" stipulation

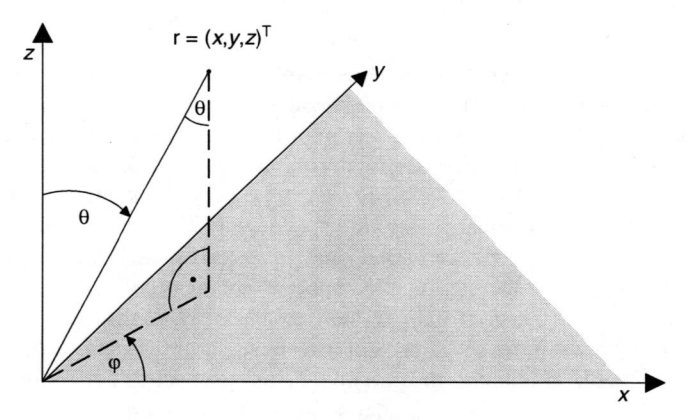

Fig. 3.2 Definition of spherical polar coordinates. The angles Φ and θ denote longitude (zero in the direction toward the companion star, increasing counterclockwise) and colatitude (zero at the "North" pole), respectively

stipulation. This coordinate system is called $\mathcal{C}_1$. Additionally, spherical coordinates (r, θ, Φ) are used, where the unit of r is the relative orbital semi-major axis, a. The radius vector **r** is represented as

$$\mathbf{r} = \begin{pmatrix} x \\ y \\ z \end{pmatrix} = r\mathbf{e}_r, \quad \mathbf{e}_r = \begin{pmatrix} \lambda \\ \mu \\ \nu \end{pmatrix} = \begin{pmatrix} \cos\Phi \sin\theta \\ \sin\Phi \sin\theta \\ \cos\theta \end{pmatrix}, \quad (3.1.1)$$

where r is the modulus of the vector **r**, $\mathbf{e}_r$ is the unit vector, and λ, μ, and ν are the direction cosines. The angles Φ and θ denote longitude (zero in the direction toward the companion star, with Φ increasing counterclockwise) and colatitude (zero at the

3.1 System Geometry and Dynamics

"North" pole), respectively. Next, we introduce the direction cosines (n_x, n_y, n_z) of the surface normal vector

$$\mathbf{n} = (n_x, n_y, n_z)^{\mathrm{T}}. \tag{3.1.2}$$

The formulas to compute $\mathbf{n}$ are different for various classes of surfaces, such as spheres, ellipsoids, Roche equipotentials, and are provided in the appropriate sections. Once $\mathbf{n}$ is known we can compute the angle β between the radius vector $\mathbf{r}$ and the surface normal $\mathbf{n}$ as shown in Fig. 3.3 and get

$$\cos\beta = \mathbf{e}_r \cdot \mathbf{n} = \frac{\mathbf{r}}{r} \cdot \mathbf{n} = \lambda n_x + \mu n_y + \nu n_z, \quad r = |\mathbf{r}|. \tag{3.1.3}$$

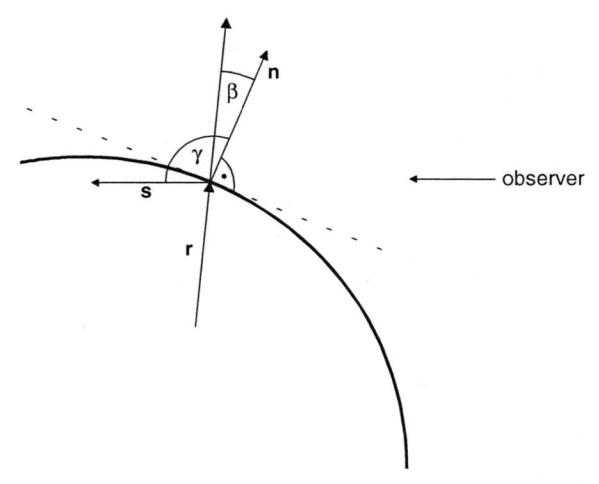

Fig. 3.3 Surface normal and line-of-sight. This figure shows the radius vector $\mathbf{r}$, the normal vector $\mathbf{n}$, the line-of-sight vector $\mathbf{s}$, and the angles β and γ

The distance r from a surface point to the center is a function $r = r(\theta, \Phi; \mathbf{p})$ of angular position, (θ, Φ), and the parameters $\mathbf{p}$ defining the shape of the surface. In these spherical coordinates, as shown in Appendix C.2, the differential volume element $\mathrm{d}V$ is given by

$$\mathrm{d}V = r^2 \sin\theta \, \mathrm{d}\theta \, \mathrm{d}\Phi \, \mathrm{d}r, \tag{3.1.4}$$

and the differential surface element by

$$\mathrm{d}\sigma = \frac{1}{\cos\beta} r^2 \sin\theta \, \mathrm{d}\theta \, \mathrm{d}\Phi. \tag{3.1.5}$$

For discussing eclipse effects it is useful to introduce the *plane-of-sky coordinates* (x^s, y^s, z^s). The origin of this right-handed coordinate system, $\mathcal{P}_1$, is the center of component 1. The traditional sense is such that the x^s-axis is positive away from the observer and coincides with his line-of-sight. As shown in Fig. 3.4, the z^s-axis is up when the y^s-axis is directed to the left. If we want to model polarimetry as in Wilson & Liou (1993, p. 672) and want to keep right-handed coordinate systems, it is necessary to have the positive x^s-axis pointing toward the observer. To derive appropriate formulas let us consider the transformation in detail. In the traditional sense $\mathcal{P}_1$ is related to $\mathcal{C}_1$ as follows. At first we rotate $\mathcal{C}_1$ counterclockwise around its z-axis by $180° - \Phi$, getting an intermediate coordinate system (x', y', z'). This system is rotated counterclockwise around its y'-axis by an angle of $90° - i$ (see Fig. 3.5). Therefore, according to the rotation matrices described in Appendix C.1 we can relate the coordinates by

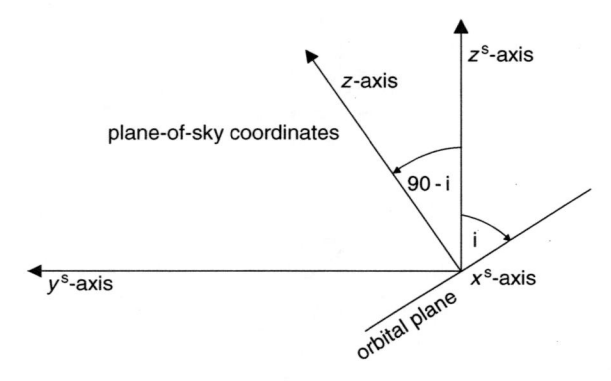

Fig. 3.4 Plane-of-sky coordinates I. The figure shows the plane-of-sky and illustrates the orientation of the orbital plane w.r.t. the plane-of-sky

$$\left(x^s, y^s, z^s\right)^T = R_{y'}(i) R_z(180° - \Phi)(x, y, z)^T \quad (3.1.6)$$

which, with[6] $\sin(180° - \Phi) = \sin \Phi$ and $\cos(180° - \Phi) = -\cos \Phi$, and $\cos(90° - i) = \sin i$ and $\sin(90° - i) = \cos i$ leads to

$$\begin{pmatrix} x^s \\ y^s \\ z^s \end{pmatrix} = \begin{pmatrix} \sin i & 0 & \cos i \\ 0 & 1 & 0 \\ -\cos i & 0 & \sin i \end{pmatrix} \begin{pmatrix} \cos \Phi & \sin \Phi & 0 \\ -\sin \Phi & \cos \Phi & 0 \\ 0 & 0 & 1 \end{pmatrix} \begin{pmatrix} x \\ y \\ z \end{pmatrix} \quad (3.1.7)$$

---

[6] If the photometric phase, Φ, appears in an additive term involving an angle or as the argument of a trigonometric function, e.g., $\sin \Phi$, the term has to be interpreted as $\sin \Phi = \sin \theta(\Phi)$, where the geometric phase or true phase angle $\theta(\Phi)$ is evaluated according to (3.1.19) in the circular, or according to (3.1.37) in the eccentric orbit case.

3.1 System Geometry and Dynamics

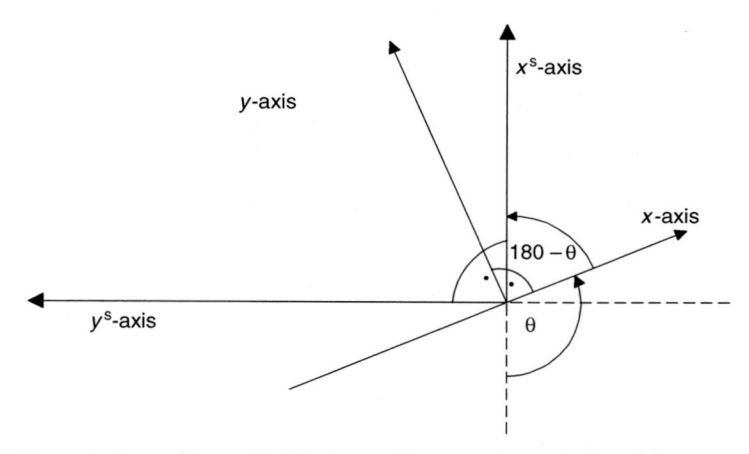

Fig. 3.5 Plane-of-sky coordinates II. This figure relates the plane orthogonal to the plane-of-sky, and the orbital plane

and finally gives

$$\begin{pmatrix} x^s \\ y^s \\ z^s \end{pmatrix} = \begin{pmatrix} x \sin i \cos \Phi + y \sin i \sin \Phi + z \cos i \\ -x \sin \Phi + y \cos \Phi \\ -x \cos i \cos \Phi - y \cos i \sin \Phi + z \sin i \end{pmatrix}. \quad (3.1.8)$$

As a special case we compute the plane-of-sky distance δ between the component centers as a function of phase. If d is the distance between the centers at phase Φ, the plane-of-sky distance δ follows by setting $x = d$, $y = z = 0$ as

$$\delta^2 = (y^s)^2 + (z^s)^2 = d^2 \left(\sin^2 \Phi + \cos^2 i \cos^2 \Phi \right), \quad (3.1.9)$$

or equivalently[7]

$$\delta^2 = d^2 \left(\cos^2 i + \sin^2 i \sin^2 \Phi \right). \quad (3.1.10)$$

The plane-of-sky coordinates just introduced are also useful to represent the line-of-sight vector **S** pointing from the observer to the plane-of-sky. According to our definition of the plane-of-sky coordinate system the observer is located at

$$(x_0^s, y_0^s, z_0^s)^T = (-\infty, 0, 0)^T, \quad (3.1.11)$$

and thus in this coordinate system $\mathbf{s}^s$ is given as

$$\mathbf{s}^s = \frac{\mathbf{S}}{S} = (+1, 0, 0)^T, \quad S = |\mathbf{S}|. \quad (3.1.12)$$

---

[7] Replace $\cos^2 \Phi = 1 - \sin^2 \Phi$, simplify, and replace again $\cos^2 i = 1 - \sin^2 i$.

The inverse transformation associated with (3.1.7) is

$$\begin{pmatrix} x \\ y \\ z \end{pmatrix} = \begin{pmatrix} \cos\Phi & -\sin\Phi & 0 \\ \sin\Phi & \cos\Phi & 0 \\ 0 & 0 & 1 \end{pmatrix} \begin{pmatrix} \sin i & 0 & -\cos i \\ 0 & 1 & 0 \\ \cos i & 0 & \sin i \end{pmatrix} \begin{pmatrix} x^s \\ y^s \\ z^s \end{pmatrix}, \qquad (3.1.13)$$

or

$$\begin{pmatrix} x \\ y \\ z \end{pmatrix} = \begin{pmatrix} \cos\Phi \sin i & -\sin\Phi & -\cos\Phi \cos i \\ \sin\Phi \sin i & \cos\Phi & -\sin\Phi \cos i \\ \cos i & 0 & \sin i \end{pmatrix} \begin{pmatrix} x^s \\ y^s \\ z^s \end{pmatrix}, \qquad (3.1.14)$$

and thus, in the coordinate system $\mathcal{C}_1$, $\mathbf{s} = (s_x, s_y, s_z)^T$ takes the form

$$\mathbf{s} = \begin{pmatrix} \cos\Phi \sin i & -\sin\Phi & -\cos\Phi \cos i \\ \sin\Phi \sin i & \cos\Phi & -\sin\Phi \cos i \\ \cos i & 0 & \sin i \end{pmatrix} \mathbf{s}^s = \begin{pmatrix} \cos\Phi \sin i \\ \sin\Phi \sin i \\ \cos i \end{pmatrix}, \qquad (3.1.15)$$

with direction cosines (s_x, s_y, s_z). The angle γ between the line-of-sight $\mathbf{s}$ and $\mathbf{n}$ follows as (see Fig. 3.3)

$$\cos\gamma := \mathbf{s} \cdot \mathbf{n} = s_x n_x + s_y n_y + s_z n_z. \qquad (3.1.16)$$

3.1.2 Dynamics and Orbits

Points of the stellar surface are considered to belong to an equipotential surface. The mathematics of such level surfaces is similar to that of the zero velocity curves in the *restricted three-body problem* [cf. Szebehely (1967)], in which a particle of negligible mass is subject to gravitational forces of two massive orbiting bodies. Within that framework two cases are distinguished: circular orbits and elliptic or eccentric[8] orbits. We treat them separately because the circular and the eccentric cases require different techniques. More importantly, there are eccentric effects on the components and on the light curves beyond those of the circular case.

We distinguish between absolute and relative orbits. Orbits with absolute orbital semi-major axes a_1 and a_2 have the origin of coordinates at the system barycenter, whereas orbits with the relative orbital semi-major axis a describe the motion with respect to the center of mass of its companion star. Absolute and relative orbits are coupled by

$$a = a_1 + a_2 \qquad (3.1.17)$$

---

[8] We use the terms *eccentric* and *elliptic* orbits synonymously throughout this book.

3.1 System Geometry and Dynamics

and the moment equation

$$a_1 M_1 = a_2 M_2. \quad (3.1.18)$$

In the next two subsections we use the following symbols appropriate for the general case of an elliptic orbit with the relative orbital semi-major axis a and the eccentricity e, $0 < e < 1$; we have the orbital quantities ν, υ, and θ, and the orbital elements ω and i:

- ν, (true) longitude in orbit, measured from some reference point to the star's position in the orbit; $0° \leq \nu < 360°$.
- υ, true anomaly, measured from periastron to the star's position in the orbit; $0° \leq \upsilon < 360°$.
- θ, (true) phase angle or "geometrical phase," i.e., the angle in the orbital plane measured from conjunction in the direction of motion; $0° \leq \theta < 360°$.
- ω, the argument of periastron, i.e., the angle from the ascending node to periastron in the orbital plane (see Fig. 3.6); $0° \leq \omega < 360°$.
- i, orbital inclination, i.e., the angle by which the plane of the true orbit plane tilts out of the plane-of-sky[9] (Fig. 3.6). Note that an edge-on orbit has $i = 90°$. The

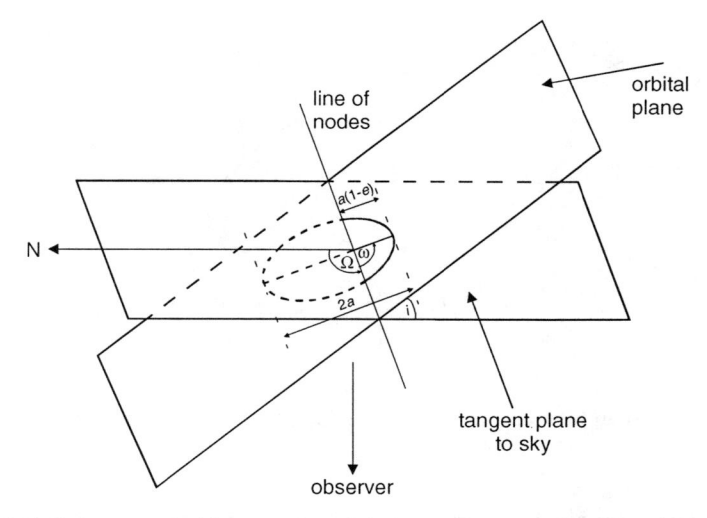

Fig. 3.6 Orbital elements of a binary system. Ω is the position angle (measured in the plane-of-sky) of the ascending node or the position angle of the line of nodes, respectively. ω is the angular distance in the orbital plane between the line of nodes and the periastron in the direction of the motion of the component. N is used to orientate the plane-of-sky and points to North

---

[9] The correct definition of the inclination is an intricate matter related to the orientation of the coordinate system discussed in Sect. 3.1.1. Although $i = 85°$ and $i = 95°$ lead to the same situations concerning light and radial velocity curves, differences appear for modeling polarimetry and interferometry.

inclinations $i = 85°$ and $i = 95°$ can be distinguished by whether the motion as projected onto the plane-of-sky is counter-clockwise or clockwise.

The symbols υ and θ should not be confused with the same symbols used to establish spherical coordinates on the component surfaces introduced in Sect. 3.1.1.

Finally, we use the symbol Φ for the orbital (sometimes also called photometric) phase measured from primary conjunction; $0 \leq \Phi \leq 1$. For circular orbits primary minimum ordinarily, or by convention, coincides with superior conjunction of the primary component, so that $\Phi = 0$ at $\theta = 0$. Whereas in the circular case, photometric phase and true phase angle are simply connected by

$$\theta = \theta(\Phi) = 360° \Phi, \qquad (3.1.19)$$

the geometrical phase in the eccentric case is related to the photometric phase by

$$\theta = \theta(\Phi) = \upsilon + \omega - 90° \qquad (3.1.20)$$

as shown in Fig. 3.7. Note that $\upsilon + \omega$ is the angle from the node to the star.

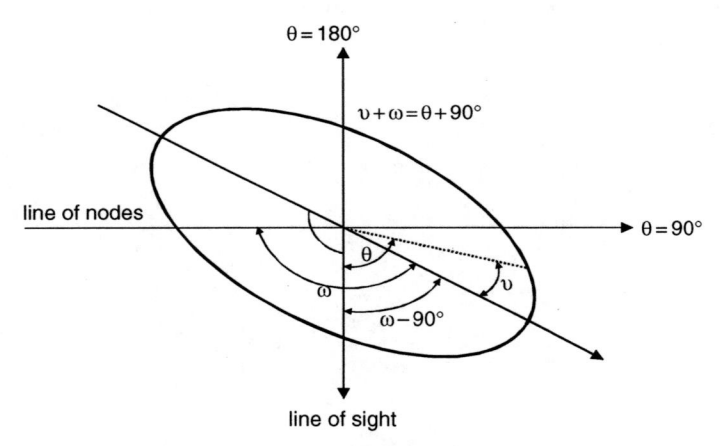

Fig. 3.7 Relationships between phase and orbital quantities. This figure shows the relationships among phase θ, true anomaly υ, and argument/longitude of periastron ω

If the binary's orbit changes[10] in time it might be possible to derive the change, $\dot{P}$, of the orbital period, and the apsidal motion parameter, $\dot{\omega}$, if observation times (rather than phase) are available. If the argument of the periastron, ω_0, is known for a

---

[10] The physical cause can be apsidal motion, orbit around a third body, mass loss and mass transfer, and solar-type magnetic cycles (Hall 1990). *Algol* itself is a good example. It undergoes a 1.783 year cycle as it revolves around *Algol C* and it also has a 32-year magnetic cycle (Søderhjelm 1980). For more details on apsidal motion see page 132.

3.1 System Geometry and Dynamics

reference time, T_0, the instantaneous argument of periastron, $\omega = \omega(t)$, in first-order approximation is given by

$$\omega = \omega_0 + \dot{\omega}(t - T_0). \tag{3.1.21}$$

3.1.2.1 Circular Orbits

Many *EB* systems have circular orbits due to the accumulated effects of tidal forces. The tidal evolution [cf. Hut (1981)] will continually change the orbital and rotational parameters. Ultimately, either an equilibrium state will be reached asymptotically or the two stars will spiral in toward each other at an increasing rate leading to a collision. An equilibrium state is characterized by coplanarity (the equatorial planes of the two stars coincide with the orbital planes), circularity of the orbit, and corotation (the rotation axes and periods of the components equal those of the orbital motion).

In the circular orbit case, the position of each star is a simple function of the phase Φ. Photometric and geometrical phase angles are connected by (3.1.19). The linear distance d between components is then independent of Φ and commonly normalized to $d \equiv 1$. The light curve minima occur at phases $\Phi = 0$ and $\Phi = 0.5$, or $\theta = 0°$ and $\theta = 180°$, respectively.

3.1.2.2 Eccentric Orbits

Although the orbits of many *EB*s are circular (Lucy & Sweeney 1971), some have elliptic orbits and sometimes even high eccentricities [e.g., *HR 6469* with $e = 0.672$ in Scarfe et al. (1994), see Fig. 3.8, showing the orbit]. This is not a great surprise because circularization is a relatively slow process as shown by Hut (1981).

Eccentric orbits have several light curve effects. The eclipse occurring nearer to apastron has the longer duration. In addition, the minima are in general not arranged symmetrically. If t_I and t_{II} denote the times of successive primary and secondary minima, respectively, $t_{II} > t_I$, we have

$$t_{II} - t_I \neq (t_I + P) - t_{II}. \tag{3.1.22}$$

Only when the line of apsides coincides with the line-of-sight is eclipse symmetry reestablished. Primary conjunction occurs at a phase Φ_1 which can be much different from zero. If the orbit is circular the plane-of-sky distance δ between centers takes its minimum at Φ_1. This statement is approximately true in the eccentric orbit case.

The *phase shift* or *displacement of the minima* depends on e and ω and is approximately [cf. Binnendijk (1960, Eq. 384) or Tsesevich (1973)] given by

$$(t_{II} - t_I) - \frac{P}{2} = \frac{P}{\pi} e \cos \omega \left(1 + \csc^2 i\right) \tag{3.1.23}$$

and describes how much the time interval between primary and secondary conjunction differs from a half-period. For $e = 0$ the relation $t_{II} - t_I = P/2$ is reproduced.

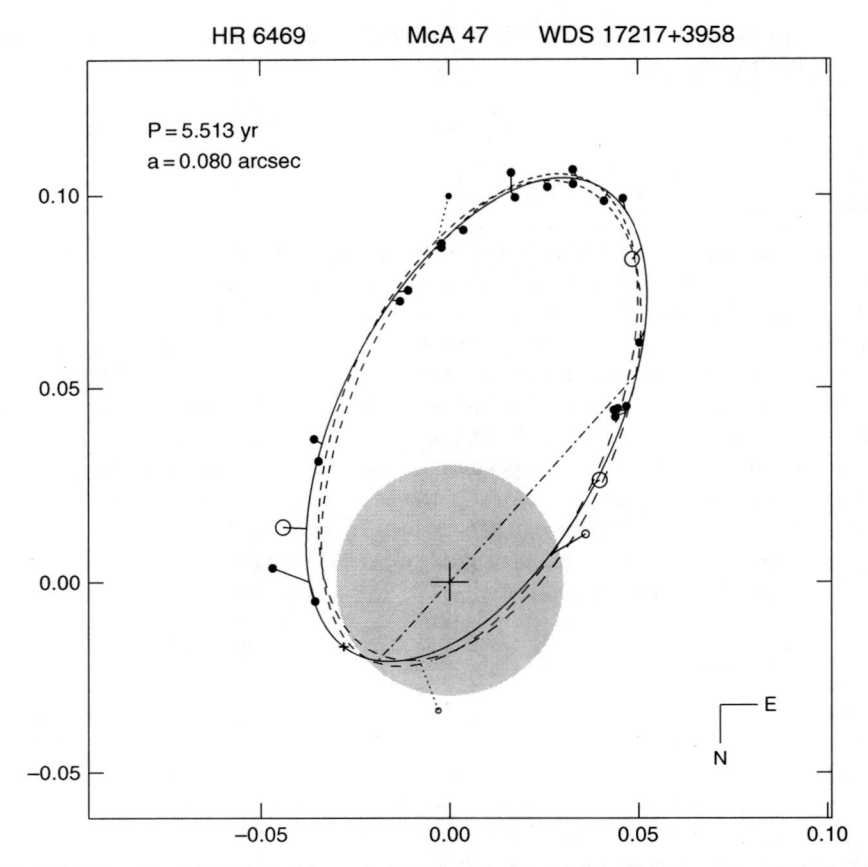

Fig. 3.8 The orbit of *HR 6469* with $e = 0.672$. Fig. 4 in Scarfe et al. (1994), courtesy C. D. Scarfe

A useful relation can be derived from (3.1.23) if i is close or equal to $90°$

$$e \cos \omega = \frac{\pi}{2P} \left(t_{II} - t_I - \frac{P}{2} \right). \tag{3.1.24}$$

Because all quantities on the right-hand side of (3.1.24) can be determined with high accuracy, (3.1.24) can be used to derive a lower limit for e

$$\frac{\pi}{2P} \left(t_{II} - t_I - \frac{P}{2} \right) = e \cos \omega \leq e. \tag{3.1.25}$$

Another useful approximation connects $e \sin \omega$ to the durations, Θ_a and Θ_p, of eclipses at apastron and periastron [cf. Binnendijk (1960, Eq. 385)]:

$$e \sin \omega \approx \frac{\Theta_a - \Theta_p}{\Theta_a + \Theta_p}. \tag{3.1.26}$$

3.1 System Geometry and Dynamics

Note that (3.1.25) and (3.1.26) allow us to derive approximations for e and ω separately because Θ_a and Θ_p can also be measured directly. Modeling of the surface configurations of eccentric *EB*s involves astrophysical considerations and computations beyond the orbital calculations. The shapes and surface gravities of the components are phase dependent, whereas stellar volume and bolometric luminosities are nearly independent of phase.[11] Strictly speaking, the resulting forces for eccentric orbits cannot be described by a potential because the force field is time dependent and therefore nonconservative. If, nevertheless, models do make use of the potential formalism for eccentric orbits (Wilson 1979), it is under this assumption: If a binary can adjust to equilibrium on a timescale short compared to that on which forces vary, an effective potential [Avni (1976), Wilson (1979)] can be defined locally at each point of the orbit without significant inconsistency. The timescale for re-adjustment is that for free nonradial oscillations, which is normally much shorter than an orbital period.

The purely orbital calculations are connected with the Keplerian problem that considers two point masses moving on ellipses around their center of mass. In addition to the orbital elements we need the true anomaly υ, measured from periastron to the star's position in the orbit. The true anomaly υ and the eccentric anomaly E are illustrated in Fig. 3.9 and are related by

$$\tan \frac{\upsilon}{2} = \sqrt{\frac{1+e}{1-e}} \tan \frac{E}{2}. \qquad (3.1.27)$$

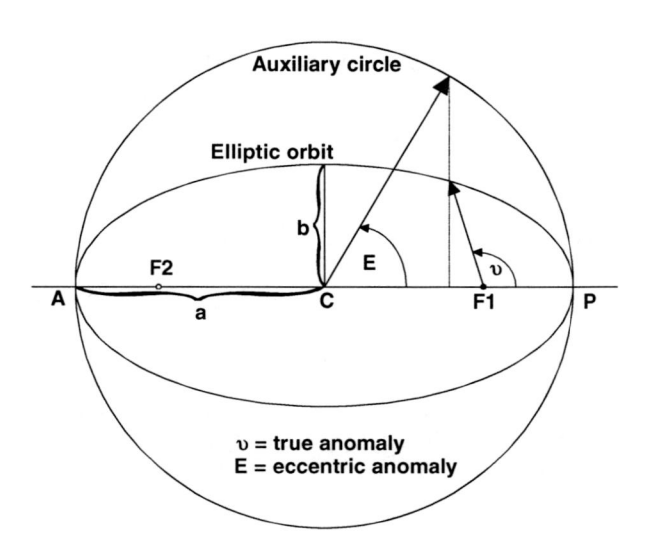

Fig. 3.9 True anomaly and eccentric anomaly. The figure shows the relative orbit of a body around another one located in the focus F1 of the ellipse. The eccentric anomaly E is computed by solving Kepler's equation. Once E is available the true anomaly υ can be computed

---

[11] See comments on page 102.

The eccentric anomaly E is related to the mean anomaly M through Kepler's equation

$$E - e \sin E = M \tag{3.1.28}$$

and thus also to time, t, because M depends on time. Let us now see how M is related to time or photometric phase. Per definition the mean anomaly, M, is the difference between a given orbital phase, Φ, and periastron phase, Φ_{per},

$$M \equiv 360° \Phi_s, \quad \Phi_s = \Phi - \Phi_{per}. \tag{3.1.29}$$

In eccentric orbit scenarios, some attention has to be paid to model the star correctly with regard to apsidal motion and to the computation of M for a given phase Φ. A consistent way[12] is to start with the true anomaly, v_c, of conjunction measured from the ascending node

$$v_c = 90° - \omega. \tag{3.1.30}$$

Computing the eccentric anomaly, E_c, by (3.1.27) and applying (3.1.28) to compute the mean anomaly, M_c, allows us to compute the phase Φ_{per} of periastron passage relative to conjunction according to (3.1.29),

$$\Phi_{per} = 1 - \frac{M_c}{360°}. \tag{3.1.31}$$

From that we derive the phase, Φ_c, of conjunction relative to the adopted zero point of phase, Φ_s:

$$\Phi_c = \frac{M_c + \omega - 90°}{360°} + \Phi_s. \tag{3.1.32}$$

Note that Φ_s is constant and does not depend on ω. Φ_c again gives us the phase, Φ_{per}^0,

$$\Phi_{per}^0 = \Phi_{per} + \Phi_c = \frac{\omega}{360°} + 0.75 + \Phi_s \tag{3.1.33}$$

of periastron passage relative to the adopted zero point of phase; the 0.75 term accounts for ω being measured from the ascending node (270° from conjunction). Now, eliminating Φ by (2.1.1) we are in a position to compute the mean anomaly for a given phase, Φ,

$$M = 360° \left(\Phi - \Phi_{per}^0 \right) = 360° \left(\frac{t - E_0}{P} \right) - \omega - 270°. \tag{3.1.34}$$

---

[12] The Wilson–Devinney program (Wilson, 1979) uses this approach.

3.1 System Geometry and Dynamics

Once M is known, Kepler's equation (3.1.28) is solved (see Appendix C.4) for E, from which υ is derived according to (3.1.27). With known true anomaly υ, the star positions in inertial rectangular barycentric coordinates (within the orbital plane) are given by

$$\xi_1 = -\frac{q}{1+q}d\cos\upsilon, \quad \eta_1 = -\frac{q}{1+q}d\sin\upsilon,$$
$$\xi_2 = +\frac{1}{1+q}d\cos\upsilon, \quad \eta_2 = +\frac{1}{1+q}d\sin\upsilon, \tag{3.1.35}$$

where $q = M_2/M_1$ is the mass ratio of the binary, and the radius vector, d, within the elliptic orbit, is given by

$$d = d(\Phi) = \frac{1-e^2}{1-e\sin(\theta-\omega)} = \frac{1-e^2}{1+e\cos\upsilon} = 1 - e\cos E. \tag{3.1.36}$$

Finally, we can relate υ to the (true) longitude in orbit and also to the (true) phase angle

$$v = \upsilon - \omega, \quad \theta = \theta(\Phi) = \upsilon + \omega - 90°. \tag{3.1.37}$$

So, finally, we coupled the orbital phase Φ and the geometrical phase, θ, through the mean anomaly, M, and Kepler's equation. In the circular case we just had the simple relation (3.1.19).

3.1.3 Spherical Models

De mortuis nihil nisi bonum   *(Of the dead, say nothing but good)*
Diogenes Laertios, I.3n.2, 70

Binaries with two slowly rotating stars sufficiently detached from their limiting lobes are accurately represented by spheres. Stars with radii of the order of 10–15% of their separation as an upper limit fall into this category, and main sequence examples are reasonably common.

The model described here is closely related to the Russell–Merrill model (see Sect. 6.2.1 and Appendix D.1) and its more modern counterpart, the Nelson–Davis–Etzel (*NDE*) model by Nelson & Davis (1972). It involves two spherical stars that move on ellipses around the center of mass. In a binary system with spherical components moving on circular orbits we may encounter a situation as illustrated in Fig. 3.10 and in Fig. 4.8 on page 200. The normal vector, **n**, and β are simply given by

$$\mathbf{n} = \mathbf{e}_r, \quad \beta = 0. \tag{3.1.38}$$

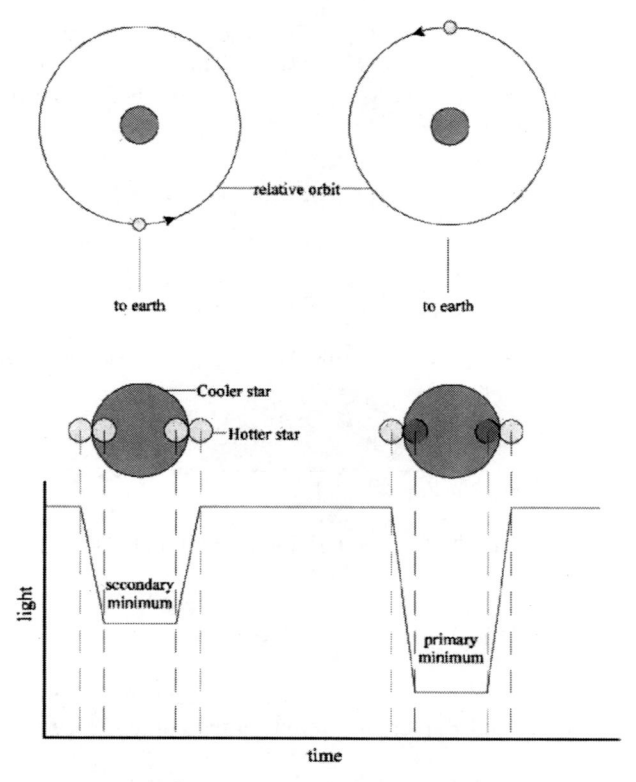

Fig. 3.10 Schematic light curve in the spherical model. The figure shows the light curve of a binary system with spheroidal components moving on circular orbits. The relative orbit of star 2 around star 1 is shown. The inclination is 90°. Note that the light curve has no curvature; however, it would if the figure would be more than schematic. Outside eclipses and during totality the light is constant. Reproduced from Fig. 2 in McVean (1994, p. 7)

In the framework of spherical models, the component eclipsed at the deeper minimum is traditionally called the *primary component*[13] and is labeled with subscript p. In most cases the primary is the one with higher *surface brightness*[14] (note that for $e \neq 0$ this is not necessarily true; however, exceptions are rare). The secondary star is labeled with subscript s.

In the spherical model, the light curve of an *EB* depends on

1. the relative radius r_p of the primary measured in units of the semi-major axis a of the orbit;

---

[13] Note that in most parts of the book we adopt the Wilson–Devinney convention that star 1 is the one eclipsed near phase zero.

[14] *Surface brightness* has the physical dimension of energy/time/solid angle/wavelength unit/unit area. Surface brightness is intensity as "seen" by the observer as he/she looks at the surface of the object.

3.1 System Geometry and Dynamics

2. the ratio $k = r_s/r_p$ of the radii;
3. the fractional luminosity, $L_p/(L_p + L_s)$, of the primary;
4. the inclination i;
5. the center-to-limb variation[15] of surface brightness (limb-darkening coefficients x_p and x_s as in the Russell–Merrill model);
6. the eccentricity e of the orbit and the argument of periastron, ω; and
7. third light, ℓ_3 (extra light of an optical or physical component).

The distinction between third light, ℓ_3, and third luminosity, L_3, is commonly ignored in the spherical and ellipsoidal models, leading to some inconsistencies. Whereas the luminosities L_p and L_s of the components are independent of phase, we really want to compare the total phase-dependent flux ℓ with observed light curves. Thus, although ℓ_3 is independent of phase, it has to be defined consistently with the phase-dependent fluxes ℓ_p and ℓ_s and it has to be added to these quantities as is done in (3.2.50).

Thus, the usual convention in spherical and ellipsoidal models, which normalizes luminosity by

$$L_p + L_s + L_3 = 1, \qquad (3.1.39)$$

has to be carefully checked to keep track of the proper physics. If (3.1.39) is used to normalize luminosity, then L_s need not be specified. Alternatively to L_p, we could also use the mean surface brightness, J_s, of the secondary while fixing $J_p \equiv 1$. Besides numerical reasons related to the modeling of limb-darkening effects, this approach has the following advantage: The ratio of mean surface brightnesses is approximately the ratio of the eclipse depths for stars moving on circular orbits. Using (3.2.31) we get the following expression for the unnormalized luminosity:

$$L_p = 4\pi \left(1 - \frac{x_p}{3}\right) J_p r_p^2, \quad L_s = 4\pi \left(1 - \frac{x_s}{3}\right) J_s r_s^2, \qquad (3.1.40)$$

which shows that the luminosity ratio and the surface brightness relation are connected by

$$\frac{L_s}{L_p} = k^2 \frac{J_s}{J_p} \frac{1 - x_s/3}{1 - x_p/3} = k^2 J_s \frac{1 - x_s/3}{1 - x_p/3} . \qquad (3.1.41)$$

In spherical models, the computation of light works as follows: For a given phase, Φ, the distance, d, between the centers of the stars is computed according to (3.1.36). Next, the projected distance, δ, follows from (3.1.10). If the eclipse condition (3.3.7) is violated, total light is equal to third light plus the flux received from both components. If it is fulfilled, we have to subtract the amount of light lost due

---

[15] Limb darkening is a physical phenomenon in which the intensity is progressively dimmer toward the limb (edge of the visible disk) of a star. The discussion of limb darkening is postponed to Sect. 3.2.4 but, here, we already use some formulas describing this phenomenon.

to the eclipse. If we neglect limb darkening for a moment and consider only stellar disks of uniform surface brightness, the light loss during eclipse is the product of the surface brightness of the eclipsed star and its eclipsed surface area. The orbital computation allows us to decide which component is eclipsed and which is in front. Once we answer this question the problem is reduced to calculating the area of a segment of a circle, i.e., the area between an arc of a circle and its subtending chord. Analytical formulas for this task are found in Nelson & Davis (1972, pp. 618–619).

If we want to treat limb-darkened stars the light loss during eclipse is the flux integral of the surface brightness of the eclipsed star over its eclipsed surface area. We can follow Nelson and Davis's approach evaluating the stellar luminosities by integrating over concentric limb-darkened rings projected onto the stellar disk. Further details about the NDE model and its associated program EBOP are given in Sect. 6.2.2.

3.1.4 Ellipsoidal Models

Autre temps, autres mœurs   *(Other times, other customs)*

The models by Wood (1971, 1972) assume that the forms of the components can be described by triaxial ellipsoids with *semi-axes* a_j, b_j, and c_j, with the major axes along the line of centers at periastron. The orbit is allowed to be eccentric. Usually it is assumed that tidal forces in close binaries require the orbital angular momentum vector and the rotation axes of the stars to be parallel. Furthermore, axial and mean orbital rotation are usually synchronized.

The orbital parameters are the same as in the spherical model with addition of eccentricity e and argument of the periastron ω. In addition, we have six geometric parameters, the *semi-axes* a_j, b_j, and c_j of the ellipsoids. Instead of these parameters the Wood model alternatively also uses the following six dimensionless parameters: $a = a_1/A$, $k = a_2/a_1$, the *ellipticities* $\varepsilon_j = b_j/a_j$ in the orbital plane, and *relative deviations*

$$\varsigma_j = \frac{c_j/b_j}{\varepsilon_j} - 1 \qquad (3.1.42)$$

perpendicular to the orbital plane and normalized w.r.t. ε_j. As outlined in Sect. 2.8, the component eclipsed at the deeper minimum is defined as *star* 1. Thus the ratio $k = a_2/a_1$ can be larger than 1 if *star* 2 is the larger one. For $k \geq 1$ the primary minimum is an *occultation*;[16] for $k < 1$ it is a *transit*.[17]

For triaxial ellipsoids with semi-axes a, b, and c the direction cosines (n_x, n_y, n_z), of the surface normal, **n**, are given by

---

[16] *Occultation* is an eclipse of the smaller star by the larger one.

[17] *Transit* is the passage of the smaller star in front of the larger star.

3.1 System Geometry and Dynamics

$$\mathbf{n} = \begin{pmatrix} n_x \\ n_y \\ n_z \end{pmatrix} = \frac{1}{D} \begin{pmatrix} \lambda/a \\ \mu/b \\ \nu/c \end{pmatrix}, \quad D := \sqrt{\left(\frac{\lambda}{a}\right)^2 + \left(\frac{\mu}{b}\right)^2 + \left(\frac{\nu}{c}\right)^2}. \tag{3.1.43}$$

The distance r from a surface point, $\mathbf{r}$,

$$\mathbf{r} = (\lambda a, \mu b, \nu c)^T, \tag{3.1.44}$$

to the center is given by

$$r = \sqrt{(\lambda a)^2 + (\mu b)^2 + (\nu c)^2}. \tag{3.1.45}$$

Finally, by exploiting

$$\lambda^2 + \mu^2 + \nu^2 = 1, \tag{3.1.46}$$

the differential surface element (3.1.5) takes the form

$$d\sigma = \frac{1}{\cos\beta} r^2 \sin\theta \, d\theta \, d\Phi = Dr^3 \sin\theta \, d\theta \, d\Phi. \tag{3.1.47}$$

The basis for Wood's ellipsoidal model is provided by Chandrasekhar's (1933a) investigations of equilibrium figures in close binary systems, where the companion is assumed to be a point source, and the star itself is described by a polytropic stellar model. In such models (Chandrasekhar, 1939, p. 43) the density ρ varies with the radial coordinate θ according to $\rho \sim \theta^n$ where n is the *polytropic index*. Chandrasekhar analyzed the distortion of such polytropes under the influence of rotation and tides. This leads to an expansion of the potential in spherical harmonics up to order 4, or equivalently, terms of $O(r_0^6)$ are neglected. The radius vector $\mathbf{r}$ from the center of mass to a surface point in direction (λ, μ, ν) is

$$r = r_0 \left[1 + \sum_{k=2}^{4} w_k P_k(\lambda) - \frac{1}{3} v_2 P_2(\nu) \right], \tag{3.1.48}$$

where r_0 is the (dimensionless) radius of a spherical star of identical volume, and the quantities w_k and v_2 are defined by

$$w_k = \Delta_k q \left(\frac{r_0}{d}\right)^6, \quad v_2 = \Delta_2 (1+q) F^2 r_0^3, \tag{3.1.49}$$

and $P_k(\lambda)$ are the Legendre polynomials of degree k:

$$P_2(\lambda) = \frac{3\lambda^2 - 1}{2}, \quad P_3(\lambda) = \frac{5\lambda^3 - 3\lambda}{2}, \quad P_4(\lambda) = \frac{35\lambda^4 - 30\lambda^2 + 3}{8}. \tag{3.1.50}$$

Similarly, as shown by Chandrasekhar (1933b), the gravitational acceleration associated with the deformation of components can be expressed as

$$g = g_0 \left[1 - \sum_{k=2}^{4} (k+2) w_k P_k(\lambda) + \tfrac{4}{3} v_2 P_2(v) \right], \qquad (3.1.51)$$

where g_0 is the acceleration for a spherical star of the same volume. The coefficients of $P_k(\lambda)$ describe the tidal deformations and the contributions to the equatorial ellipticity caused by the first three partial tides. In elliptic orbits, they vary with the size $d = 1 - e \cos E$ of the radius vector.

The coefficient of $P_2(v)$ describes the oblateness caused by the rotation. Δ_2 is a function that depends weakly on the polytropic index n; $\Delta_2(n) \approx 1$. In the limit $n \to 5$ (Roche model), $\Delta_2 \to 1$. Equations (3.1.48) and (3.1.51) contain the expansion factors

$$\Phi(q, r_0, n) := 1 + \tfrac{1}{3}(1+q) r_0^3 \eta(n), \quad \Phi'(q, r_0, n) := 1 - \tfrac{4}{3}(1+q) r_0^3 \eta'(n), \quad (3.1.52)$$

for the radius and the acceleration. $\eta(n)$ and $\eta'(n)$ are functions that approach 1 in the limiting case of the Roche model and are given in Chandrasekhar (1933a, Eq. 44) and Chandrasekhar (1933b, Eq. 101):

	$n = 3$	$n = 4$	$n = 5$
η	0.736	0.898	1
η'	0.811	0.938	1

The factors $\Phi(q, r_0, n)$ and $\Phi'(q, r_0, n)$ express the fact that a star in a binary system has a volume slightly larger than that of a single star of identical mass. Intuitively this is clear because the gravitational acceleration of a single star is reduced by the presence of the companion and the orbital rotation, and thus the density decreases. This systematic deviation needs to be considered when comparing stellar radii derived from EBs with those predicted by models describing stellar structure. For close systems with $r_0 \approx 0.3$ this causes deviations in radius up to 1%.

Expansion (3.1.48) shows that up to $O(r_0^6)$ a rotating, tidally deformed polytrope can be approximated by a triaxial ellipsoid if rotation is sufficiently slow and tides are sufficiently small. This result provides the foundation of ellipsoidal models. In the case of synchronous rotation ($F = 1$) and almost circular orbits, the axes (a, b, c) of the ellipsoid only depend on r_0, mass ratio q, and polytropic index n

$$\begin{pmatrix} a \\ b \\ c \end{pmatrix} = r_0 \begin{pmatrix} 1 + \tfrac{1}{6}(1 + 7q) \Delta_2 r_0^3 \\ 1 + \tfrac{1}{6}(1 - 2q) \Delta_2 r_0^3 \\ 1 - \tfrac{1}{6}(2 + 5q) \Delta_2 r_0^3 \end{pmatrix}. \qquad (3.1.53)$$

In orbits with significant eccentricity, the separation d between the components varies significantly, leading to variable deformation; in the case of asynchronous

3.1 System Geometry and Dynamics

rotation expressed by F (for definition of F, see page 100) the axes depend on $r_d := r_0/d$ according to

$$\begin{pmatrix} a \\ b \\ c \end{pmatrix} = r_0 \begin{pmatrix} 1 - \frac{1}{6}(2 - 4q)\Delta_2 r_d^3 + \frac{1}{2}(1+q)F^2 \Delta_2 r_0^3 \\ 1 - \frac{1}{6}(2 + 5q)\Delta_2 r_d^3 + \frac{1}{2}(1+q)F^2 \Delta_2 r_0^3 \\ 1 - \frac{1}{6}(2 + 5q)\Delta_2 r_d^3 \end{pmatrix}. \qquad (3.1.54)$$

The error in the length of the semi-major axes in this ellipsoid approximation is of the order of qr_d^4. Thus, for extremely close stars with $r_0 \approx \frac{1}{2}d$ the error is about 6% for $q = 1$. For $r_0 \leq \frac{2}{5}d$ the error is smaller than 3%. The approximations up to $O(r_0^6)$ are not valid for the modeling of close (especially contact and over-contact) binary systems. The surface shapes are not correctly represented by triaxial ellipsoids in such cases, and the use of ellipsoidal models to derive photometric mass ratios is inappropriate.

Let us summarize: The Wood model is most useful for sufficiently detached systems, for which the surface distortions are adequately approximated by triaxial ellipsoids. It is certainly better for the analysis of these systems than any model based on spherical stars or rectification. However, for an adequate treatment of severely distorted components, only a model based on equipotential surfaces will suffice.

3.1.5 Roche Geometry and Equipotential Surfaces

Auspicium melioris aeui   *(An omen of a better age)*

The Roche model is based on the following assumptions about mass distribution and orbits.

First, both components are assumed to act gravitationally as point masses (surrounded by essentially massless envelopes). This allows a relatively simple analytical representation of the potential. Theories of stellar structure show that in most cases the approximation of the potential as of two point sources plus a centrifugal potential is sufficient.

Second, it is implicitly assumed that periods of free nonradial oscillations are negligible when compared with the orbital period P, so that the shape of the components is determined by the instantaneous force field. This fact becomes very important for modeling eccentric orbit binaries. The timescale of these oscillations is of the order of the hydrostatic timescale, which for solar type stars is about 15 min. Surfaces of constant potential are assumed to be surfaces of constant density. In particular, this is true for the stellar surface, viz., the visible photosphere. For fixed mass ratio, rotation rates, etc., the stellar surface is parametrized by only one quantity: The potential energy of that surface.

As the basic assumption of the applicability of the Roche model is that the stars must be in hydrostatic equilibrium, strictly speaking, Roche potentials are only valid for components moving in circular orbits and rotating synchronously. The solution of the nonsynchronous problem was first presented by Plavec (1958) and, in an apparently independent work, by Limber (1963). A generalization of Roche potentials to treat eccentric orbits was investigated by Avni (1976). The asynchronous and eccentric solutions were first properly combined by Wilson (1979). In the following subsections, the equations for circular and eccentric orbits are presented separately.

3.1.5.1 Circular Orbits and Synchronous Rotation

Consider point masses moving in circular orbits around their center of mass. Assume an orthogonal right-handed coordinate frame (see Fig. 3.1) with origin at point 1, corotating with the system so that component 2 lies always on the (positive) x-axis and has the vector coordinates $\mathbf{r}_2 = (1, 0, 0)^T$. The z-axis is parallel to the normal vector of the orbital plane. Component masses are labeled M_1 and M_2, and S denotes the center of mass. In this environment, a test particle of unit mass in the atmosphere of component 1 experiences a gravitational plus a centrifugal force. The total force $\mathbf{F}$ acting on the test particle is given by

$$\mathbf{F} = -G \frac{M_1}{R_1^3} \mathbf{r} - G \frac{M_2}{R_2^3} (\mathbf{r} - \mathbf{r}_2) + \omega^2 \mathbf{r}_{0\omega}, \qquad (3.1.55)$$

where $G = 6.673 \cdot 10^{-11}$ m^{3}kg^{-1}s^{-2} is the gravity constant, and R_j denotes the distance of the point $\mathbf{r} = (x, y, z)^T$ from the center of component j. $\mathbf{r}_{0\omega}$ is the vector

$$\mathbf{r}_{0\omega} = \mathbf{M}\mathbf{r} - (x_c, 0, 0)^T, \quad \mathbf{M} := \mathrm{diag}(1, 1, 0), \qquad (3.1.56)$$

originating in $(x_c, 0, 0)^T$ and pointing to $(x, y, 0)^T$, x_c is the position of the center of mass on the x-axis, viz.,

$$x_c = \frac{M_2}{M_1 + M_2} d = \frac{q}{q+1} d, \quad q := \frac{M_2}{M_1}, \qquad (3.1.57)$$

where d is the separation of the components centers, and q denotes the mass ratio.

The force $\mathbf{F}$ per unit mass (this is the surface gravity acceleration $\mathbf{g}$) can be computed as the gradient

$$\mathbf{F} = \mathbf{g} = -\nabla \Psi \qquad (3.1.58)$$

of the potential (Kopal 1959)

$$-\Psi(x, y, z) = G \frac{M_1}{R_1} + G \frac{M_2}{R_2} + \frac{\omega^2}{2} r_{0\omega}^2, \qquad (3.1.59)$$

where

$$r_{0\omega}^2 = (x - x_c)^2 + y^2 = \left(x^2 + y^2\right) - 2xx_c + x_c \qquad (3.1.60)$$

3.1 System Geometry and Dynamics

is the perpendicular distance of the particle from the orbital rotation axis which is parallel to the vector $(x_c, 0, 1)^T$. Whereas the first and second right-hand side terms of (3.1.59) are the gravitational potentials of M_1 and M_2, the third term is the centrifugal potential due to the rotation of the frame of reference.

The relation (3.1.58) expresses that we have a conservative force field, energy is conserved, and the integral

$$\oint \mathbf{F} \cdot d\mathbf{s} = 0 \qquad (3.1.61)$$

along any closed path vanishes.

Let us now transform the potential into a more convenient form. Under the assumptions that the stars revolve in circular orbits and the axial rotation is synchronized with the orbital revolution, the angular[18] velocity ω can be replaced according to Kepler's law by

$$\omega^2 = \frac{4\pi^2}{P^2} = G \frac{M_1 + M_2}{d^3}. \qquad (3.1.62)$$

Substituting (3.1.62) in (3.1.59) and using spherical polar coordinates (3.1.1)

$$x^2 + y^2 = r^2(1 - \nu^2), \quad x = \lambda r, \qquad (3.1.63)$$

leads to a replacement of the physical potential U by the normalized or modified Roche potential Ω

$$\Omega := -\frac{\Psi d}{GM_1} - \Omega_q, \quad \Omega_q := \frac{1}{2} \frac{q^2}{q+1}, \quad d \equiv 1, \qquad (3.1.64)$$

taking the form[19]

$$\Omega(\mathbf{r}; q) = \frac{1}{r} + q \left[\frac{1}{\sqrt{1 - 2\lambda r + r^2}} - \lambda r \right] + \frac{1}{2}(q+1)r^2(1 - \nu^2). \qquad (3.1.65)$$

Note that the constant term Ω_q has been subtracted, as in Kopal's convention (Kopal 1959). Whereas this convention due to (3.1.58) does not change the force field derived from the potential, it destroys the symmetry that otherwise would be conserved between the two component potentials. For sufficiently small values of r we note the asymptotic behavior

---

[18] Note that later, when we also treat eccentric orbits, ω as defined in (3.1.62) will represent the mean orbital angular velocity.

[19] As briefly mentioned in Appendix C.3, this form of the potential can also be used to establish the Roche coordinates, a system of partly orthogonal coordinates (u, v, w). However, these coordinates are not of much practical use.

$$r \ll 1 \quad \Rightarrow \quad \Omega(\mathbf{r}; q) \approx \frac{1}{r}. \tag{3.1.66}$$

Note that r is a dimensionless quantity. This follows from the definition $d \equiv 1$. If d is known in physical units, then r scales accordingly.

If we want to compute the force in physical units, it follows from (3.1.64)

$$\mathbf{F} = \frac{GM_1}{d} \nabla \Omega. \tag{3.1.67}$$

Note that once we know the potential and its gradient we can compute the normal vector, $\mathbf{n}$, at each surface point by

$$\mathbf{n} = \mathbf{n}(\mathbf{r}) = -\frac{\nabla \Omega}{|\nabla \Omega|}, \quad \nabla \Omega = \left(\frac{\partial \Omega}{\partial x}, \frac{\partial \Omega}{\partial y}, \frac{\partial \Omega}{\partial z} \right). \tag{3.1.68}$$

The negative sign in (3.1.68) ensures that the normal vector points inward. The explicit formulas to compute $\nabla \Omega$ in the most general case are provided on page 101.

3.1.5.2 Circular Orbits and Asynchronous Rotation

Already by the early 1950s there was well-established evidence for asynchronous rotation in many close binaries; cf. Struve (1950). Some of the more interesting Algols have rapidly rotating primaries [Van Hamme & Wilson (1990), Wilson (1994)]. Fast rotation strongly deforms a star as is demonstrated for *RZ Scuti* and *RW Persei* in the *Pictorial Atlas* (Terrell et al. 1992). Wilson (1994) also discusses slow or subsynchronous rotation which is pertinent to the study of common envelope evolution.

To model fast and slow rotation binaries, it is necessary to extend the concept of Roche surfaces to asynchronous rotation. It is assumed that the stars rotate uniformly,[20] so that star 1 rotates with angular velocity vector ω_1. We further simplify the dynamics by neglecting minor rotation-induced changes in the mass distribution. We use ω to refer to the angular velocity vector of orbital rotation. The acceleration of a mass element in the rotating frame with center in that star (Fig. 3.11) was derived by Limber (1963) and has the form

$$\mathbf{r}_1'' := \frac{d^2 \mathbf{r}_1}{dt^2} = -\frac{1}{\rho} \nabla p - \nabla \left(G \frac{M_1}{r_1} + G \frac{M_2}{r_2} \right)$$
$$+ \omega \times (\omega \times \mathbf{r}_{01}) + \omega_1 \times (\omega_1 \times \mathbf{r}_1) - 2\omega_1 \times \left(\frac{d\mathbf{r}_1}{dt} \right), \tag{3.1.69}$$

---

[20] An asynchronous theory by Peraiah (1969, 1970) includes even nonuniform rotation. But it seems that it has not been applied to real observations or incorporated into a general light curve program.

3.1 System Geometry and Dynamics

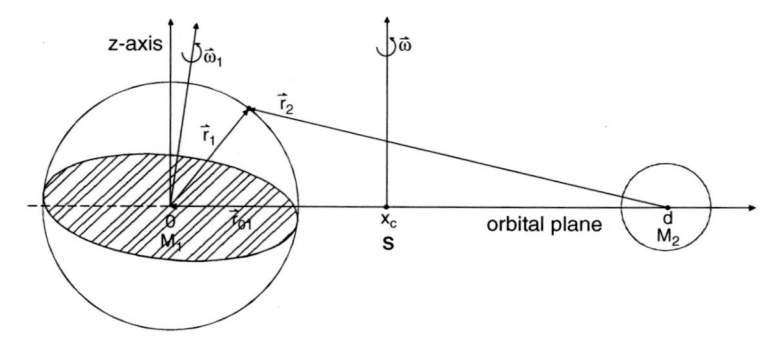

Fig. 3.11 Angular momentum vectors of orbital and stellar rotation

where $\mathbf{r}_1$ is the radius vector from the center of star 1 to the point of interest, $\mathbf{r}_{01}$ is the vector pointing from the center of mass to the center of star 1, ρ is the stellar density, and p is the gas pressure. Note the different sign convention for the potentials. The first term is the force due to pressure gradients in the stars, the second term represents gravity, the last term is the Coriolis force, and the other terms are centrifugal force and an offset from the center of mass.

If (3.1.69) is transformed to the corotating frame of the orbit with center in star 1 defined in Sect. 3.1.5.1, following Limber (1963), all other terms can be expressed by means of an effective potential

$$\Psi_{\text{eff}} := G\frac{M_1}{r_1} + G\frac{M_2}{r_2} + \omega^2 r_{01} x_1 + \tfrac{1}{2}\omega_1^2 r_{\omega_1}^2, \qquad (3.1.70)$$

where r_{ω_1} denotes the distance between the point of interest and the rotation axis of star 1. A special case arises when ω and ω_1 are parallel to each other, i.e., $\omega \times \omega_1 = 0$. In this case, the effective potential takes the form

$$\Psi_{\text{eff}} = G\frac{M_1}{r_1} + G\frac{M_2}{r_2} + \tfrac{1}{2}\omega^2 r_{0\omega}^2 + \tfrac{1}{2}\tilde{\omega}^2 r_{\omega_1}^2 + \omega\tilde{\omega} r_{\omega_1}^2, \quad \tilde{\omega} := \omega_1 - \omega. \qquad (3.1.71)$$

Note that in the limit $\tilde{\omega} = 0$, the effective potential Ψ_{eff} is identical to the potential Ψ in (3.1.59) describing the synchronous case. Here we will consider only the case that ω and ω_1 are parallel. For that case, Fig. 3.12 shows the x, y-plane and the quantities $r_{0\omega}^2$, r_{01}, r_{ω_1}, and x.

If, following Limber (1963), we now assume that the mass motions in star 1 with respect to the rotating frame are negligible, i.e., $\mathbf{r}_1''$, $\mathbf{r}_1'$, and as a consequence the Coriolis forces are small compared to all other terms in (3.1.69), we end up with

$$\nabla p = -\rho \nabla \Psi_{\text{eff}}. \qquad (3.1.72)$$

Thus, under this assumption, according to (3.1.72), surfaces of constant pressure and constant density are identical with the equipotential surfaces of Ψ_{eff}. Thus, from now

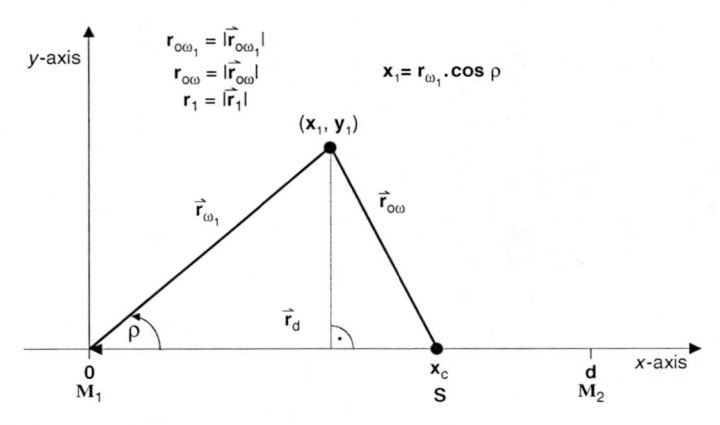

Fig. 3.12 Nonsynchronous rotation. Definition of geometrical quantities in the orbital plane

on, it is sufficient to concentrate on the effective potential Ψ_{eff}. With the geometrical relations

$$r_{\omega_1}^2 = x^2 + y^2, \quad r_{0\omega}^2 = (x - x_c)^2 + y^2 = (x^2 + y^2) - 2xx_c + x_c, \quad (3.1.73)$$

and the definition of the rotational parameter F (the ratio of angular rotation rate to the mean orbital revolution rate ω)

$$F := \frac{\omega_1}{\omega}, \quad (3.1.74)$$

the term involving the centrifugal potential takes the form

$$\begin{aligned}\tfrac{1}{2}\omega^2 r_{0\omega}^2 + \tfrac{1}{2}\tilde{\omega}^2 r_{\omega_1}^2 + \omega\tilde{\omega} r_{\omega_1}^2 &= \tfrac{1}{2}\omega^2 \left[r_{0\omega}^2 + (F-1)^2 r_{\omega_1}^2 + 2(F-1) r_{\omega_1}^2 \right] \\ &= \tfrac{1}{2}\omega^2 \left[r_{0\omega}^2 + (F^2 - 1) r_{\omega_1}^2 \right] \\ &= \tfrac{1}{2}\omega^2 \left[F^2 (x^2 + y^2) - 2xx_c + x_c \right]. \end{aligned} \quad (3.1.75)$$

Combining (3.1.71) and (3.1.75) and proceeding as in Sect. 3.1.5.1 yields

$$\Omega(\mathbf{r}; q) = \frac{1}{r} + q \left[\frac{1}{\sqrt{1 - 2\lambda r + r^2}} - \lambda r \right] + \tfrac{1}{2} F^2 (q+1) r^2 \left(1 - \nu^2\right). \quad (3.1.76)$$

So, if we neglect the Coriolis term, the only difference between the potential including uniform asynchronous rotation and the original one in (3.1.65) is that the centrifugal term is multiplied by F^2. But note that for asynchronous rotation there are separate potential systems for the two stars. The dynamical extension of the Roche models including asynchronous rotation not only complies with more realistic physical conditions but also allows us to model spectral line broadening, as discussed in Sect. 3.6.

3.1 System Geometry and Dynamics

Common envelope evolution is thought to lead to cataclysmic variables [see, for instance, Warner (1995)] which contain white dwarf stars and erupt as classical novae, recurrent novae, dwarf novae, and the novae-like variables (sometimes called *UX UMa* stars). If one component of a binary undergoes evolutionary expansion the binary's outer envelope may begin to engulf the companion. If synchronism cannot be maintained, the orbit decays in a tight spiral as the orbital motion becomes faster than the rotation, which cannot keep up through the usual tidal locking mechanism [see the review articles by Webbink (1992, 2008), Taam & Bodenheimer (1992), Iben & Livio (1994) for references to original work in this field].

3.1.5.3 Eccentric Orbits and Asynchronous Rotation

At each phase Φ in the eccentric two-body problem the position and separation $d = d(\Phi)$ of the components follow from Kepler's equation (3.1.28). The force field on any third object is time dependent and therefore nonconservative. This precludes the existence of a static potential field and a relation such as (3.1.58). If, however, a binary component can readjust to equilibrium on a timescale short compared to that on which forces vary (orbital period P), Wilson (1979) has shown that it is possible to define the effective potential

$$\Omega(\mathbf{r}; q, d) = \frac{1}{r} + q \left[\frac{1}{\sqrt{d^2 - 2\lambda dr + r^2}} - \frac{\lambda r}{d^2} \right] + \frac{1}{2} F^2 (q+1) r^2 (1 - \nu^2). \quad (3.1.77)$$

This potential may be used without significant inconsistencies, *if* the timescale for nonradial oscillations is much smaller than the orbital period P. In the eccentric orbit case d depends on phase Φ instead of $\Omega(\mathbf{r}; q, d)$ so we also use the notation $\Omega(\mathbf{r}; q, \Phi)$ in the context of eccentric orbits to indicate that the potential and stellar surface depend on phase. We also need the gradient $\nabla \Omega$, i.e., the partial derivatives

$$\frac{\partial \Omega}{\partial x} = -\frac{x}{r^3} + \frac{q(d-x)}{\tilde{r}^3} + F^2(q+1)x - \frac{q}{d^2}, \quad (3.1.78)$$

$$\frac{\partial \Omega}{\partial y} = -y \left[\frac{1}{r^3} + \frac{q}{\tilde{r}^3} - F^2(q+1) \right], \quad (3.1.79)$$

$$\frac{\partial \Omega}{\partial z} = -z \left[\frac{1}{r^3} + \frac{q}{\tilde{r}^3} \right], \quad \begin{matrix} r^2 = x^2 + y^2 + z^2 \\ \tilde{r}^2 = (d-x)^2 + y^2 + z^2 \end{matrix} \quad (3.1.80)$$

and for the secondary component in the same coordinate system [see Wilson (1979), Eq. 6)]

$$\frac{\partial \Omega}{\partial x} = \frac{q(d-x)}{\tilde{r}^3} - \frac{x}{r^3} - F^2(q+1)(1-x) - \frac{1}{d^2}. \quad (3.1.81)$$

Note that in the circular-synchronous case, $d = 1$ and $F = 1$, (3.1.78) and (3.1.81) give the same expression. The partial derivatives w.r.t. y and z are the same, anyway.

The gradient $\nabla\Omega$ is needed to compute the normal vector, **n**, according to (3.1.68) and the local gravity as described in Sect. 3.2.1.

Because the potential (3.1.77) is phase dependent, binary stars moving in elliptic orbits will change their shapes accordingly. The potential formalism is, of course, only an approximation. What is really needed is an analysis of the response of the stellar surface to varying tidal forces, including nonradial oscillations. Although, a rigorous analysis and computation of the instantaneous volume taking into account that stellar matter is compressible, has not yet been worked out, eccentric binary modeling is often based on the following assumption: The shape of a star varies along the orbit, but it is expected that its volume V remains essentially constant [Avni (1976), Wilson (1979)]. For polytropic stars Hadrava (1986) has formally proven that the contact of the stellar surface with its Roche lobe can occur only at periastron. Therefore the stellar surface may be parametrized by the periastron potential Ω_p which then yields the periastron volume V_p assumed constant over phase. The phase-dependent potential can then be found from V_p. In what follows we pick up Wilson's (1979) argument. A star's critical lobe size sets an upper limit for its size. In the circular, synchronous case, the maximum size is the Roche lobe (however, if this size is exceeded, we still can have an over-contact binary). In the eccentric case the effective critical lobe size is the one which causes the star to fill its critical lobe exactly at periastron. Hut (1981) shows that rotation will tend to synchronize to the periastron angular rate because of the strong dependence of the tidal force on distance. The periastron-synchronized F is given by Hut (1981, Eq. 44)

$$F^2 = \frac{(1+e)^4}{(1-e^2)^3} = \frac{1+e}{(1-e)^3}. \qquad (3.1.82)$$

Analogous to the Lagrangian point L_1^p in the circular case with synchronous rotation, the equilibrium point $x_{L_1^p}$ of vanishing effective gravity is, for a given F and periastron separation $d_p = 1 - e$, the solution of the equation

$$\frac{\partial\Omega}{\partial x}(x_{L_1^p}, y=0, z=0) = 0. \qquad (3.1.83)$$

The solution of this equation is further discussed in Appendix E.12. The potential Ω_p corresponding to $(x_{L_1^p}, 0, 0)$ yields V_p of the star by a volume integration. Whereas $\Omega = \Omega(\Phi)$ varies along the orbit, the volume V of the star is kept constant, $V = V_p$.

It should be noted that for $e \neq 0$ or $F \neq 1$ no over-contact configuration can be stable but now a new configuration enters the stage: double-contact. For a further discussion of binary morphologies we refer to Sect. 3.1.6.

As we have seen, the classical Roche model allows only for gravitational and centrifugal forces. The modifications for eccentric orbits and asynchronous uniform rotation make it possible to analyze a much larger group of binaries. The extended Roche model provides a physically reasonable basis for the description of the geometrical structure and, as we will see in Sect. 3.1.6, evolutionary processes of most systems of intermediate to late spectral type which are not too strongly magnetized. However, in very early spectral-type binaries the interaction between radiation and matter may become important because the radiation pressure increases with the fourth power of the effective temperature.

3.1.5.4 Approaches Including Radiation Pressure

As seen in the literature references below there have been many efforts to extend the Roche model and to include the radiation pressure expected in hot stars. Although these efforts have not yet led to a consistent and commonly accepted model, due to several deficiencies, it seems appropriate to discuss them briefly, to comment on their deficiencies, to point the reader to the problems involved in including radiation pressure, and hopefully, to raise further interest in the subject.

In very hot stars the radiation pressure is due to the interaction between electromagnetic radiation and matter and can be important. The radiation pressure decreases the effect of gravity, depends on the momentum transfer associated with absorbed or scattered photons, and is a complicated function of the local conditions. Because a large fraction of the momentum transfer is due to absorption in prominent ultraviolet resonance lines, the problem is related to the radiative acceleration of stellar winds [see, for instance, Castor et al. (1975) or Hearn (1987)].

Dynamically, radiation pressure leads to complicated situations. Stars with radiative envelopes have solutions at depth that are insensitive to surface boundary conditions. Thus, controlled by the optical depths, not too far below the surface, the state variables including the total radiation pressure [cf. Mihalas (1978, Eq. 1–46, p. 17)]

$$P^R = \tfrac{1}{3}aT^4, \quad a = 4\frac{\sigma}{c} = 7.5647 \cdot 10^{-15} \text{ erg} \cdot \text{cm}^{-3}\text{K}^{-4} \qquad (3.1.84)$$

will be constant on the standard Roche equipotential surfaces. Accordingly, to get the shape of the photosphere, we integrate the structure equations inward along normals to these potential surfaces and determines the starting height by requiring asymptotically the constancy of state variables on equipotential surfaces. The solution thus obtained necessarily has horizontal pressure gradients in the surface layers of nonspherical stars (Kippenhahn and Weigert, 1989), but they become vanishingly small in deep layers. These gradients will give rise to "geostrophic winds" analogous to the Earth's jet stream. Because the depth of the photospheres of hot stars on the main sequence is about 1% of the radius, this is the order of magnitude of the deviations from Roche geometry that we could expect in best cases (Lucy 1997). For very hot stars and certainly for WR components, the photosphere is formed in the star's radiatively driven wind and large deviations from Roche geometry will arise as the problem becomes nonstatic.

In a binary system the radiation pressure influences not only the shape of the surface by the gravitational force field but also deforms the companion's surface directly (this might be called the *outer* radiation pressure effect).

Despite the physical effects and complexity mentioned above some early and simple attempts to include radiation pressure have been made by Schuerman (1972), Kondo & McCluskey (1976), Vanbeveren (1977, 1978), Djurasevic (1986), and Zhou and Leung (1987). These approaches have in common that they use a modified force field, and they consider only the *inner* radiation pressure effect due to the radiation of the star itself. They replaced the potential GM_1/r_1 of the hotter component (and if necessary also that of the secondary accordingly) by

$$G(1-\delta)\frac{M_1}{r}, \quad \delta := \frac{F^R}{F^G}, \qquad (3.1.85)$$

where G is the gravitational constant, and F^G is the force due to gravity

$$F^G = G\frac{M_1}{r^2}. \qquad (3.1.86)$$

Assuming that δ is constant, the potential (3.1.59) in the binary system is now supposed to be

$$-\Psi_{\text{rad}}(x,y,z) = G(1-\delta_1)\frac{M_1}{R_1} + G(1-\delta_2)\frac{M_2}{R_2} + \frac{\omega^2}{2}r_{0\omega}^2. \qquad (3.1.87)$$

Note that F^R accounts only for the interaction of stellar matter with the star's own radiation field and is derived as follows. At first, the radiation pressure, P^R, acting on a unit surface element is given by

$$P^R = \frac{1}{c}\int_0^\infty \int_\omega I_\nu \cos^2\gamma \, d\omega d\nu, \qquad (3.1.88)$$

where γ is the angle between the surface normal and the incident radiation, $d\omega$ is the solid angle element, and I_ν is the monochromatic intensity in the frequency interval $d\nu$ around ν. If ρ denotes the mass density, force and radiation pressure are coupled by

$$\mathbf{F}^R = -\frac{1}{\rho}\nabla P^R. \qquad (3.1.89)$$

Equation (3.1.89) is true if P^R includes all radiation pressure contribution from both stars. However, in the papers above, using the monochromatic average opacity κ_ν of the envelope and absorption coefficient $k_\nu = \kappa_\nu/\rho$ per unit mass, a plane parallel radiative or spherically symmetric transfer equation is assumed, and P^R is replaced by the *inner* radiation pressure. Then ∇P^R is replaced by the radial derivative of P^R and the radiation force per unit mass follows as

$$F^R = -\frac{1}{\rho}\frac{\partial P^R}{\partial r} = \frac{1}{c\rho}\int_0^\infty \int_\omega \kappa_\nu I_\nu \cos\gamma \, d\omega d\nu = \frac{1}{4\pi r^2 c}\int_0^\infty k_\nu I_\nu d\nu, \qquad (3.1.90)$$

3.1 System Geometry and Dynamics

where $c = 2.9979 \cdot 10^8$ m/s is the speed of light, and k_ν is an average absorption coefficient of the envelope per unit mass. Thus, using (3.1.86), we finally get a constant expression for δ

$$\delta = \frac{F^R}{F^G} = \frac{1}{4\pi c G M_1} \int_0^\infty k_\nu I_\nu d\nu. \quad (3.1.91)$$

The assumption $\nabla P^R = \partial P^R / \partial r$ is true only for spherical stars. For nonspherical stars, gravitation and flux-proportional *inner* radiation pressure do not vary with the inverse square of the distance, r. Nevertheless, based on this approach, the shape of equipotentials under the influence of the inner radiation pressure has been investigated by several authors: Djurasevic (1986), Zhou & Leung (1987), Drechsel et al. (1995), and Niedsielska (1997). Figure 3.13 (courtesy Drechsel) shows the meridional intersections of equipotential surfaces of a binary system with mass ratio, $q = 1$, for different δ_1 values. The top part shows the shrinking of a fixed equipotential surface ($\Omega_1 = 3.75$) with increasing δ_1; the bottom part demonstrates the influence of increasing inner radiation pressure on the extent of the Roche lobe of the primary.

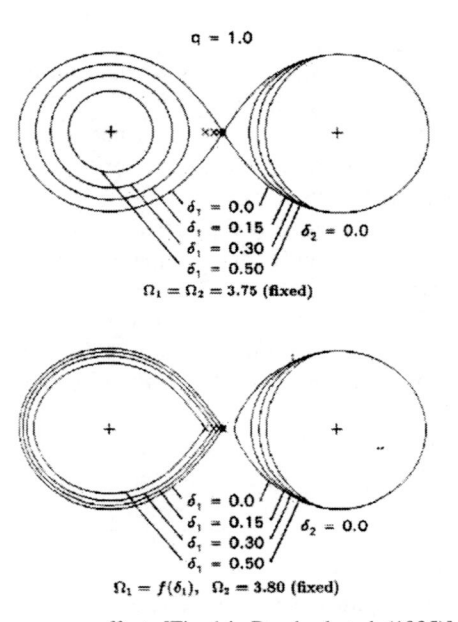

Fig. 3.13 *Inner* radiation pressure effects [Fig. 1 in Drechsel et al. (1995)]. Courtesy H. Drechsel

This approach, although used by many authors, has not been without criticism. Howarth (1997) shows that the inner radiation pressure does not change the stellar figure at all. His arguments are based on radiative equilibrium and von Zeipel's law (see page 117). For a lobe-filling star the gravity at the inner Lagrangian point,

L_1^p, is zero and thus according to von Zeipel's law, the temperature (and hence the inner radiation pressure) is also zero, and thus cannot change the location of L_1^p. His mathematical argumentation is: According to von Zeipel's law the flux vector **F** is proportional to the gradient of the gravitational potential [see Eq. (3.2.10)]. The radiative and gravitational acceleration of star 1 are antiparallel and coupled by

$$\mathbf{a}_{\text{rad}} = -\delta \mathbf{g}. \tag{3.1.92}$$

This leads to the effective surface gravity acceleration

$$\mathbf{g}_{\text{eff}} = \mathbf{g} + \mathbf{a}_{\text{rad}} = (1 - \delta)\mathbf{g}. \tag{3.1.93}$$

According to (3.1.58), **g** is the (negative) gradient of the potential Ψ, so it is also possible to represent $\mathbf{g}_{\text{eff}}$ as the gradient of the effective potential

$$\Psi_{\text{eff}} = (1 - \delta)\Psi. \tag{3.1.94}$$

The potential Ψ_{eff} differs from the modified potential Ψ_{rad}; Ψ_{eff} follows from Ψ by simple scaling. On page 97 we derived the dimensionless potential Ω from Ψ by dividing it by GM_1. Note that if we divide Ψ_{eff} by $(1-\delta)GM_1$ we get the same dimensionless potential Ω. That clearly tells us that the inner radiation pressure does not change the shape of the components.

The description of the radiation pressure also needs to consider the incoming radiation of the companion (*outer* radiation pressure effect). Even under mild conditions it is no longer possible to derive an analytical expression describing the equipotential surface. Drechsel et al. (1995) treat the photosphere as a deformable membrane subject to the radiation of the companion and compute iteratively its shape. However, if inward integrations were made, enormous unbalanced pressure gradients would be found in deep layers. Nevertheless, because it is the first time that the inner radiation effect and the radiation pressure of the companion are considered separately, we briefly sketch their approach coded into a light curve program.[21]

To account for the consequences of irradiation of the companion, Drechsel et al. (1995) introduced two functions $\delta_j^* = \delta_j^*(\theta, \varphi, r)$,

$$\delta_j^*(\theta, \varphi, r) := \frac{F^R(\theta, \varphi, r)}{F^G(\theta, \varphi, r)}, \tag{3.1.95}$$

depending on the local coordinates of a surface point on component j. These functions vary according to

$$0 \leq \delta_j^*(\theta, \varphi, r) \leq \delta_j \tag{3.1.96}$$

---

[21] They used a circular orbit version of the Wilson–Devinney program. It is also the first time that the outer radiation pressure has been coded into a light curve program.

3.1 System Geometry and Dynamics

and take their maximum values at the intersection points of the lines connecting both mass centers with the stellar surfaces ($\vartheta = 0°$) and the minimum values at the stellar horizons ($\vartheta = 90°$).

For a given surface, the integration of the incident flux, $\mathcal{F}(\theta, \varphi, r)$, on a unit area element located at (θ, φ, r), is very similar to that used to compute the reflection effect. The computation of $\delta_j^*(\theta, \varphi, r)$ is performed by the formula

$$\delta_j^*(\theta, \varphi, r) = \delta \frac{F^R(\theta, \varphi, r)}{F_0^R(\theta, \varphi, r)} = \delta \frac{\mathcal{F}(\theta, \varphi, r)}{\mathcal{F}_0(\theta, \varphi, r)}, \qquad (3.1.97)$$

where $F^R(\theta, \varphi, r)$ and $F_0^R(\theta, \varphi, r)$ are the radiation forces associated with $\mathcal{F}(\theta, \varphi, r)$ and $\mathcal{F}_0(\theta, \varphi, r)$, and $\mathcal{F}(\theta, \varphi, r)$ is a reference flux incident on a unit area element located at (θ, φ, r) perpendicular to the direction of $\mathcal{F}_0(\theta, \varphi, r)$. The reference flux enables us to couple $\delta_j^*(\theta, \varphi, r)$ to the inner radiation pressure parameter δ. Whereas the computation of $\mathcal{F}(\theta, \varphi, r)$ involves all eclipse effects, $\mathcal{F}_0(\theta, \varphi, r)$ does not. The reference radiation force is computed as

$$F_0^R(\theta, \varphi, r) = \delta F^G(\theta, \varphi, r) \qquad (3.1.98)$$

which, using (3.1.95), leads to the first part of equation (3.1.97). Although formula (3.1.97) looks simple, the computations are complicated by the fact that the reference flux can be computed only for a predefined orientation of the irradiated surface. The surface in turn adjusts itself according to incident flux. Thus an iterative procedure is necessary which assumes first a surface normal pointing to the mass center of the irradiating star. The surface normal is then improved until convergence.

Once the functions $\delta_j^*(\theta, \varphi, r)$ are known, the modified Roche potentials

$$\Omega_1^{\text{rad}}(\mathbf{r}_1; q) = \frac{1-\delta_1}{r_1} + q\left[\frac{1 - \delta_2^*(\theta_1, \varphi_1, r_1)}{\sqrt{1 - 2\lambda r_1 + r_1^2}} - \lambda r_1\right] + \tfrac{1}{2}(q+1)r_1^2\left(1 - \nu^2\right) \qquad (3.1.99)$$

for and in the coordinate frame of component 1 and

$$\Omega_2^{\text{rad}}(\mathbf{r}_2; q) = q\frac{1-\delta_2}{r_2} + \frac{1 - \delta_1^*(\theta_2, \varphi_2, r_2)}{\sqrt{1 - 2\lambda r_2 + r_2^2}} - \lambda r_2 + \tfrac{1}{2}(q+1)r_2^2\left(1 - \nu^2\right) + \frac{1-q}{2} \qquad (3.1.100)$$

for and in the coordinate frame of component 2 are used to compute the stellar surfaces and surface normal vectors. That in turn leads to new values of $\delta_j^*(\theta, \varphi, r)$ and so on.

As computed by Drechsel et al. (1995), in extreme cases such as the one shown in Fig. 3.14 increasing radiation pressure can force the secondary to switch from *inner* to *outer* contact configuration. So besides changing the stellar shapes the system configuration can be changed completely due to the shift of the positions of the Lagrangian points and the altered shapes and extents of the Roche lobes. In

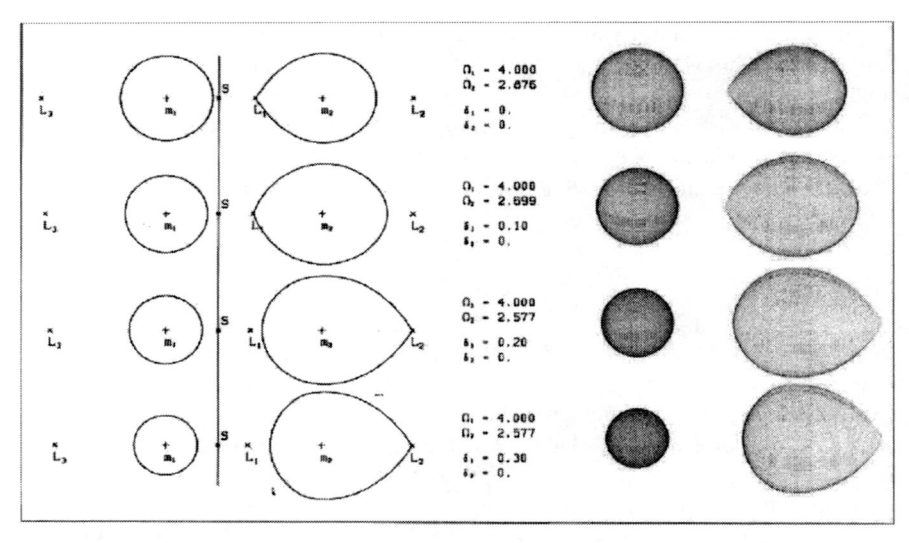

Fig. 3.14 Effects of full radiation pressure [Fig. 3 in Drechsel et al. (1995)]. Courtesy H. Drechsel

scenarios with δ_1 and δ_2 of a few percent Drechsel (1997, private communication) reports that the radii of the stars change only by a few percent as well.

Although the modeling of the inner and outer radiation pressure effects of the previous paragraph are based on doubtful assumptions, the computations indicate that radiation pressure can have drastic effects in binary systems and thus require us again to be careful regarding the physical assumptions. Let us therefore summarize the assumptions and their deficiencies and point the reader to the physics to be considered.

Even the modeling of the inner radiation pressure needs to incorporate the gradient of the radiation pressure as done by Howarth (1997), not the radial derivative, because the local physics involves the entire force field. The reason is that a local point on the surface sees only the entire force field, not the separate gravities of the two stars and not the centrifugal force. The outer radiation pressure will lead to horizontal pressure gradients in the surface layers causing instabilities and fluctuations on the surfaces of the stars. Thus the problem is not static. The "potential functions" $\Omega^{\rm rad}$ in the Drechsel et al. membrane formalism are not potential functions in the strict sense because their gradient does not generate the net force field. At best, we can hope that if radiation pressure is sufficiently small the potential is only slightly perturbed and that $\nabla \Omega^{\rm rad}$ approximates the force.

So despite many efforts there is no consistent model for Roche geometry including radiation pressure. If the radiation pressure is negligible as in most stars there is no need to consider it. If it becomes relevant (e.g., in Wolf–Rayet binaries or X-ray binaries) the stars are so hot that the radiation pressure effects become very important and require a dynamical treatment. In these cases, there are additional

3.1 System Geometry and Dynamics

effects such as radiation-driven colliding stellar winds, as discussed in Sect. 3.4.4.4, that require further modifications of our binary model.

3.1.6 Binary Star Morphology

Whereas the original classification of *EB*s was phenomenological (for types EA, EB, EW see Sect. 1.2.2), based on observed light curves, morphological classification based on equipotentials provide more physical insight. Associated with the concept of equipotentials are "limiting surfaces" or "limiting lobes." A limiting lobe is the volume enclosed by a limiting surface. The usefulness of morphological classifications is that each of the stable configurations is generated by a structural–evolutionary process.

Let us start with the circular orbit and synchronous case. The equipotentials of (3.1.65) are identical with the surfaces of zero relative velocity in the *restricted three-body problem* [Szebehely (1967), Kopal (1978)]. There exist five Lagrangian points $L_i^p, i = 1, \ldots, 5$, characterized by the requirement,[22] $\nabla \Omega = 0$. The Lagrangian point, L_1^p, is also called the *inner* Lagrangian point and is of particular relevance for *EB* stars because it is critical to the concepts of detached, semi-detached, and over-contact binaries. L_1^p lies between the two stars (see Fig. 3.15), and at that point surfaces of equal potential coalesce in such a manner that the

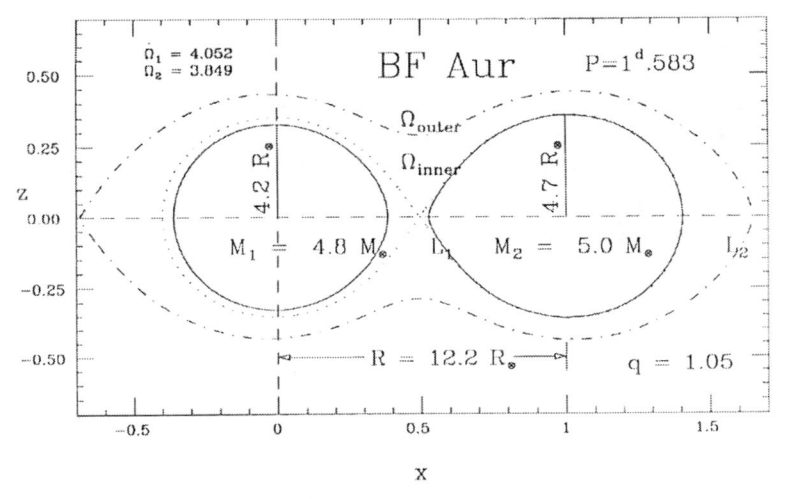

Fig. 3.15 Lagrangian points L_1^p and L_2^p in the *BF Aurigae* system. This is Fig. 3 in Kallrath & Kämper (1992)

---

[22] In the more general cases of eccentric orbits or asynchronous rotation we will use the term *equilibrium points* rather than Lagrangian points.

surfaces passing through L_1^p are the largest closed equipotentials enveloping the two stars separately. L_1^p marks the *inner Lagrangian surface* and the *Roche lobes* of the components; the relative sizes of the Roche lobes depend directly on the mass ratio such that the star with greater mass has the larger lobe. If one of the stars fills its Roche lobe (*semi-detached binary*), it may overflow the critical surface, transferring mass to its companion through L_1^p. If both stars satisfy this condition, we call the system a *contact binary*.[23] The modified potential at the inner Lagrangian surface is called Ω^I and that at the outer, Ω^O. Note that these quantities depend only on q. The latter potential marks the effective limit of the binary; matter beyond this surface is lost from the binary system through the outer Lagrangian point, L_2^p. When a particle leaves the binary through L_2^p its energy is too small to escape to infinity. However, it is then no longer forced to corotate with the binary and, for most mass ratios, acquires enough energy by gravitational interaction with the binary to spiral to infinity. If components are in contact, i.e., $\Omega^I \geq \Omega \geq \Omega^O$, then Ω describes the surface of the common envelope. Such a system is an *over-contact binary*.

We are now in a position to connect the notions of lobe-filling stars and the values of Roche potential values. If only one component accurately fills its Roche lobe the system is *semi-detached*. If neither fills its Roche lobe, it is *detached*. The computation of L_1^p and of the critical potentials Ω^I and Ω^O is explained in Appendix E.12. The degree of contact is measured by the contact parameter, f, sometimes called the *fill-out factor* or *parameter*:

$$f = \frac{\Omega^I - \Omega}{\Omega^I - \Omega^O}, \quad \Omega \leq \Omega^I. \tag{3.1.101}$$

Note that $f = 0$ when the component fills its lobe, i.e., $\Omega = \Omega^I$; and $f = 1$ when $\Omega = \Omega^O$, but when one of the components is within its Roche lobe, the meaning of the contact parameter can be extended: $f < 0$ for that component.

The fill-out factors need to be computed for each component separately in each component's own reference frame. They are only reasonably defined for circular and synchronous orbits ($d = 1$ and $F_1 = F_2 = 1$). Thus we have

$$f_1 := f(\Omega_1, q), \quad f(\Omega, q) := \frac{\Omega^I(\Omega, q) - \Omega}{\Omega^I(\Omega, q) - \Omega^O(\Omega, q)}, \quad \Omega \leq \Omega^I(\Omega, q), \tag{3.1.102}$$

where the functions $\Omega^I(\Omega, q)$ and $\Omega^O(\Omega, q)$ are evaluated as described in Appendix E.12. To compute the fill-out factor of component 2, it is necessary to transform Ω_2 into the coordinate system of component 2:

$$\Omega_2' = q'\Omega_2 + \tfrac{1}{2}(1 - q'), \quad q' = \frac{1}{q}. \tag{3.1.103}$$

---

[23] This configuration is a special case ($e = 0$, $F = 1$) of what on page 113 is called a *double-contact binary*. As is discussed later, a true *contact binary* is not likely to exist. Nevertheless, it is interesting from a mathematical point of view.

3.1 System Geometry and Dynamics

The inverse transformation to (3.1.103) is

$$\Omega_2 = q\Omega_2' + \tfrac{1}{2}(1-q), \qquad (3.1.104)$$

which enables us to compute f_2

$$f_2 := f(\Omega_2', q'). \qquad (3.1.105)$$

The fill-out factor should not be confused with a similar term, f^R, also described as the *fill-out parameter*, introduced by Rucinski (1973):

$$f^R = \begin{cases} \Omega^I/\Omega, & \text{if } \Omega > \Omega^I, \\ \dfrac{\Omega^I - \Omega}{\Omega^I - \Omega^O} + 1, & \text{if } \Omega < \Omega^I. \end{cases} \qquad (3.1.106)$$

Detached systems have $\Omega > \Omega^I$ and $f^R < 1$, lobe-filling components have $f^R = 1$, and over-contact systems are described by $1 \leq f^R \leq 2$. Similar to f_1 and f_2, for computing f_1^R and f_2^R we have to apply the coordinate transformations described above.

Another definition being used in Binary Maker 3.0 is

$$f^{BM} = \begin{cases} \Omega^I/\Omega - 1, & \Omega > \Omega^I \\ f, & \Omega \leq \Omega^I \end{cases}, \qquad (3.1.107)$$

which is properly normalized for detached systems between $-1 < f^{BM} \leq 0$ as well.

For circular orbits and synchronous rotation the Roche potential approach led to the morphological types of detached, semi-detached, and over-contact binaries. The names and the full set of categories were used first by Kopal (1954) but the term "over-contact" dates back to Kuiper (1941) who already understood the relevant principles. Unfortunate or not, many authors have used the adjective *contact*, rather than over-contact, for binaries with common envelopes. However, in a more general context (namely eccentric orbits, asynchronous rotation) a consistent approach is possible only if we have a concept in which the word *contact* has a meaning in the sense "in contact with a critical surface." To stress again, the word *contact* refers to contact with a limiting surface (not necessarily with the other component). In that sense, the term "semi-contact" would be a more accurate usage than "semi-detached." For the case of circular orbits and synchronous rotation, the "degree of contact" can be quantitatively described by the term *contact parameter*, or sometimes, *fill-out factor*, f, defined according to (3.1.101). It measures the degree to which a component fills its Roche lobe: 0 if the potential matches that of the inner, and 1 if it matches the outer Lagrangian surface.

For circular orbits and synchronous rotation the limiting surfaces are the inner Lagrangian surface (the Roche lobes) and the outer Lagrangian surface. These simple scenarios already explain many observed configurations and enable us to link them to evolutionary states taking into account evolutionary expansion, gravitational radiation, mass loss and exchange, and magnetic braking. The first three morphological types are the following:

1. **Detached** binary systems, where both components are within their lobes. The fill-out factor of each component is negative. If the components are small compared with their Roche lobes, their shapes closely approximate spheres. Whereas morphology and evolutionary state are related for semi-detached and over-contact binaries, such a connection does not exist in detached systems.
2. **Semi-detached** systems, with one component within its critical lobe, whereas the other exactly fills its lobe ($f = 0$ for this component). This morphological type includes Algols, cataclysmic variables, and some X-ray binaries,[24] in which one component is highly evolved and in which mass transfer occurs. *Algol itself* may serve as an example of the Algol type. In general, the lobe-filling star can lose matter through the inner Lagrangian point.
3. **Over-contact** systems or common envelope binaries, where each component has a surface larger than its Roche lobe. Mechanical equilibrium requires that the surfaces match in potential. That is, the common surface must coincide with a single equipotential above the Roche lobes ($0 < f_{1,2} \leq 1$, and $f_1 = f_2$). Configurations are limited by the outer Lagrangian surface. This morphological type explains W UMa stars very well. Although in all over-contact binaries the more massive star is larger than its companion, Binnendijk (1965, 1970) defined two subclasses of W UMas observationally: *W-type* and *A-type* W UMa stars on the basis of the larger star being cooler or hotter than the other, respectively; i.e., the primary minimum being an occultation or a transit. So we have *A-type* contact systems, in which the more massive star has the greater surface brightness and the *W-type* systems, in which the more massive and larger star has less surface brightness. Our present interpretation [Lucy (1976), Flannery (1976), Robertson & Eggleton (1977), Wilson (1978), and Lucy & Wilson (1979)] is that W-types are formed by slightly over-contact binaries with moderate mass ratio such as 0.4–0.6 and with components close to the zero age main sequence (ZAMS), and that A-types are somewhat evolved – on the main sequence but not on the ZAMS. Configurations with one component larger than its critical lobe while the other is not do not have closed surface equipotentials and are not expected to exist for more than a few orbits. However, the early (extremely brief) stages of common envelope evolution specifically involve exactly this configuration [cf. Webbink (1992), Taam & Bodenheimer (1992), or Iben and Livio (1994)]. Binaries in which both components exactly fill their Roche lobe ($f_1 = f_2 = 0$; the true *contact* system as we might use the term) could in principle exist. But no mechanism is known by which they could come into existence and they are not expected to be stable against small perturbations. Small effects caused by evolutionary changes lead to either the semi-detached or over-contact scenario.

---

[24] The basic model of X-ray binaries is a close binary system with a "normal" star (main sequence or giant, in exceptional cases also a degenerate star) filling its Roche lobe and transferring matter to the compact object, a neutron star, or a black hole (Krautter 1997).

3.1 System Geometry and Dynamics

4. **Double-contact** system (Wilson, 1979), where each component fills its lobe (see Fig. 3.16) exactly, and at least one rotates supersynchronously. For asynchronous rotation ($F \neq 1$) or eccentric orbits ($e \neq 0$), over-contact binaries can no longer exist. The extreme case is a centrifugally limited binary or a double-contact system, where two components fill their limiting lobes but do not touch each other. β *Lyrae* and *V356 Sagittarii* are likely candidates. What is the astrophysical meaning of double-contact binaries? It is observed that some Algol primaries (among those with primaries well within their Roche lobes) rotate much faster than synchronously, some even close to or approximately at the centrifugal limit. An underlying physical process to account for that fast rotation is spin-up by the accretion process. As described in Wilson (1994), gas transferred from the contact component arrives with considerable angular momentum and converts orbital to rotational angular momentum. The outer envelope of the primary component now spins-up. Rather than the star expanding to reach the lobe, the limiting lobe contracts to meet the star. The secondary component rotates synchronously and already fills its limiting lobe (the ordinary Roche lobe).

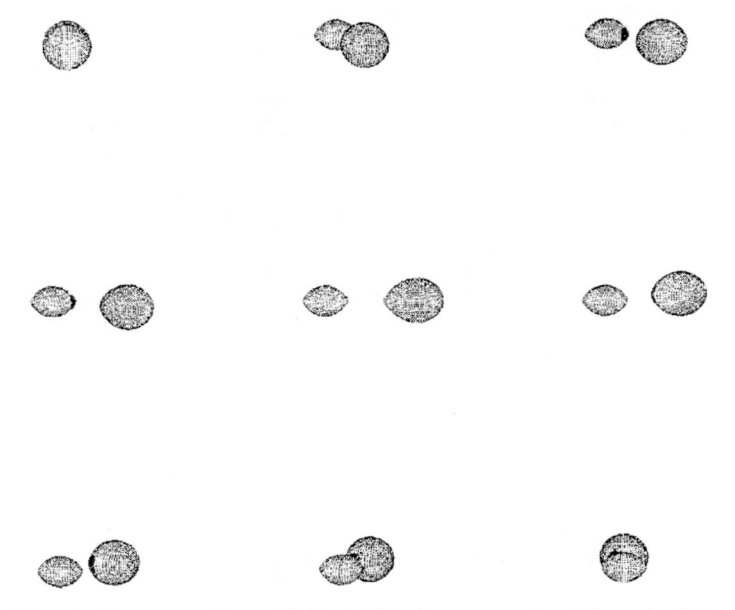

Fig. 3.16 The double-contact binary *RZ Scuti*. This figure, reproduced from the *Pictorial Atlas* (Terrell et al. 1992, p. 342) and provided by Dirk Terrell, shows the shape of *RZ Scuti* at phases 0, 0.05, ..., 0.5. Courtesy D. Terrell

In summary, the following stable configurations can occur:

- detached: both components are smaller than the critical or limiting lobe;
- semi-detached: one component is smaller than the critical lobe, while the other fills its critical lobe at periastron;

- double-contact: each component exactly fills its critical lobe (again, at periastron); and
- over-contact for $F = 1$ and $e = 0$ only: common envelope binary.

In Chap. 4, we show and discuss how a priori knowledge about the configuration of a binary system can be used as a constraint. The Wilson–Devinney program, for example, implements such explicit constraints by different modes of operation.

3.2 Modeling Stellar Radiative Properties

Ignorantia legis neminem excusat (*Ignorance of the law excuses none*)

The computation of the flux emitted from the binary components requires the integration of local quantities over the surfaces. In the Roche potential models, especially in the circular orbit and synchronous rotation case, the stellar surface S' is defined as the set of all points $\mathbf{r}_s$ on the equipotential surface (3.1.65) specified by Ω_0,

$$S' = S'(q, \Omega_0) := \{\mathbf{r}_s | \Omega(\mathbf{r}_s, q) = \Omega_0\}. \tag{3.2.1}$$

For each star, in the circular orbit and synchronous rotation case, the surface is parametrized by only two quantities: q and Ω_0. For fixed q, the larger Ω_0 the smaller the star, and vice versa. The surface defined by the set S' of vectors or points $\mathbf{r}_s$ has the surface area $S = \int_{S'} d\sigma$. Scaling with R^2 gives the real surface measure. The differential surface element $d\sigma$ in spherical coordinates was given in (3.1.5), repeated here for convenience,

$$d\sigma = \frac{1}{\cos \beta} r^2 \sin \theta d\theta d\varphi. \tag{3.2.2}$$

Corresponding to the equipotential condition

$$\Omega(\mathbf{r}_s, q) = \Omega_0, \quad \mathbf{r}_s = (r_s, \theta, \varphi), \tag{3.2.3}$$

it is possible to define the function[25]

$$r_s : [-\pi, \pi] \times [0, 2\pi] \to \mathbb{R}^+, \quad (\theta, \varphi) \to r_s(\theta, \varphi), \tag{3.2.4}$$

which gives the distance of a surface point $\mathbf{r}_s$ to the center of the star. The computation of the surface area follows as

$$S = S(\Omega_0) = \int_{S'} d\sigma = \int_0^{2\pi} \int_0^{\pi} \frac{1}{\cos \beta} r_s^2(\theta, \varphi) \sin \theta d\theta d\varphi \tag{3.2.5}$$

and the volume is

---

[25] In order to define a function we explicitly define the domain of its argument, here θ and φ.

3.2 Modeling Stellar Radiative Properties

$$V = V(\Omega_0) = \int_0^{2\pi} \int_0^{\pi} \int_0^{r_s(\theta,\varphi)} r_s^2(\theta, \varphi) \sin\theta \, dr \, d\theta \, d\varphi. \quad (3.2.6)$$

In the eccentric orbit case, there is no potential in the strict sense as discussed in Sect. 3.1.5.3. However, as discussed on page 102, there is good physical reasoning that the volume remainsconstant along the orbit. Thus, the shapes of the stars vary with phase but we require that the volume remains constant. At first, $V = V(\Omega_p)$ is computed according to (3.2.6) where Ω_p is the potential at periastron. Note that Ω_p plays the rôle Ω_0 played in the circular orbit case. Then, for a given phase Φ, the corresponding Roche potential Ω_Φ is computed. Ω_Φ is derived from the requirement that it yields a stellar surface which leads to the correct volume, i.e., $V(\Omega_\Phi) = V$. Therefore, to compute Ω_Φ the following iterative procedure based on Wilson (1979):

$$\Omega_\Phi^{(k)} \rightarrow \left(\Omega(\mathbf{r}_s; q, \Phi) = \Omega_\Phi^{(k)}\right) \rightarrow r_s^{(k)}(\theta, \varphi) \rightarrow V\left(\Omega_\Phi^{(k)}\right) \quad (3.2.7)$$

is applied until, after a number of iterations, $k = 0, \ldots, K$,

$$\left| V\left(\Omega_\Phi^{(K)}\right) - V \right| \leq \varepsilon \quad (3.2.8)$$

a predefined tolerance is achieved. The result is again a radial function $r_s(\theta, \varphi)$ defining the stellar photosphere.

The computation of the flux emitted by the stellar photosphere is based on several assumptions about the underlying photospheric physics. These include the choice of a model atmosphere and several physical effects:

- gravity brightening;
- limb darkening;
- reflection effect; and
- blackbody radiation, gray atmosphere, or a model atmosphere.

In addition, special physical effects such as dark or bright spots on the star surfaces might be included.

3.2.1 Gravity Brightening

Hydrostatic equilibrium is equivalent to constant density and pressure on equipotential surfaces. If we assume that density ρ, temperature [26] T, and pressure p are related to each other by an equation of state, e.g., the ideal gas law,

---

[26] Note that this temperature is the local thermodynamic temperature which differs conceptually from the effective temperature defined as a function of bolometric flux.

$$p = R\rho T, \quad R = 8.31451 \text{ J} \cdot \text{mol}^{-1}\text{K}^{-1}, \tag{3.2.9}$$

with the universal gas constant R, then on equipotentials we see constant temperature as well. So, the star in hydrostatic equilibrium is homogeneous on each equipotential.

A rotating star differs from a nonrotating star in shape, local surface gravity acceleration, and surface brightness. It develops oblateness and a pole-to-equator variation in surface brightness. This variation is called gravity brightening (sometimes, especially in the older literature, called *gravity darkening*). This phenomenon, for radiative envelopes, has its origin in the temperature gradient in and near the surface. The atmosphere is assumed to be locally plane-parallel (that implies we need to consider only one geometrical dimension), but irradiated from below by a radiative flux varying across the stellar surface. According to von Zeipel's theorem [27] this flux and, due to the Stefan–Boltzmann law (3.2.15), the temperature are also determined by the gradient (w.r.t. the optical depth τ) of the source function in the subphotospheric layers. The result is that this flux is proportional to the effective surface gravity acceleration g at the given point of the surface. A modern derivation of this result is found in Kippenhahn & Weigert (1989, Eq. 42.6, p. 436) and reads

$$\mathbf{F} = \frac{4ac}{3\kappa\rho}T^3\frac{dT}{dU}\mathbf{g} = -k(U)\mathbf{g}, \quad \mathbf{g} = -\nabla U, \tag{3.2.10}$$

where $\mathbf{F}$ is the vector of radiative energy flux and $\mathbf{g}$ is the effective gravitational acceleration consisting of gravitational and centrifugal acceleration. The proportionality factor $k(U)$ describes the conduction of this radiative transport and depends only on the potential U because the temperature, $T = T(U)$, and opacity $\kappa(\rho, T) = \kappa(U)$ depend only on potential (Kippenhahn & Weigert, 1989, p. 436). If we want to compute the radiative flux on a given equipotential it varies only with $\mathbf{g}$ and is antiparallel to $\mathbf{g}$.

Gravity brightening is thus described by a relation between the local effective temperature T_l (or the local bolometric flux) and the *local surface gravity acceleration*. For our purpose is sufficient to consider only the modulus of $\mathbf{g}$, i.e., according to (3.1.58) we get

$$g = |\mathbf{g}| = |\nabla U| = g_0 |\nabla \Omega|, \tag{3.2.11}$$

where g_0 is a proportionality factor. Thus, for each surface point $\mathbf{r}_s \in S'$ we express the local surface gravity acceleration in terms of potential gradient components of the gradient of the Roche potential,

---

[27] There are three papers by von Zeipel (1924a, b, c) related to the radiative equilibrium of distorted stars. Relevant to our problem is Eq. (36) in the first paper, and Eqs. (90) and (91) in the third paper.

3.2 Modeling Stellar Radiative Properties

$$g_l = g(\mathbf{r}_s) := g_0|\nabla \Omega(\mathbf{r}_s)| = g_0\sqrt{\left(\frac{\partial \Omega}{\partial x}\right)^2 + \left(\frac{\partial \Omega}{\partial y}\right)^2 + \left(\frac{\partial \Omega}{\partial z}\right)^2}. \quad (3.2.12)$$

Further in the book the subscript l will indicate "local." Once g_l is known, the *local bolometric flux* $F_l = F(\mathbf{r}_s)$ is computed according to

$$F_l = F_p \left(\frac{g_l}{g_p}\right)^g, \quad g = \begin{cases} 1.00, & \text{von Zeipel theorem,} \\ \approx 0.32, & \text{Lucy's law,} \end{cases} \quad (3.2.13)$$

where the index p refers to the pole of a star, and the exponent g should not be confused with the modulus of **g**. The upper part of (3.2.13) summarizes the *von Zeipel theorem* by von Zeipel (1924a) for stars in radiative equilibrium. Lucy's law established by Lucy (1967) gives the relation for stellar envelopes in convective equilibrium. The exponent 0.32 is an estimate derived numerically from tables of convective stellar envelopes.

Let us make a few more remarks on the von Zeipel theorem. The radiative equilibrium in regions with different **g** (and consequently also the optical depths of particular equipotentials) gives rise to temperature gradients along the equipotential surface. We should expect that this temperature gradient leads to a mass flow (meridional[28] circulation) parallel to the surface which tends to homogenize the physical conditions in the layer. Thus, strictly speaking, it is not possible for a rotating and tidally distorted star to be in hydrostatic and radiative equilibrium simultaneously. The complicated dependence of the temperature on optical depth in the photosphere in radiative equilibrium immediately violates the underlying assumptions of the homogeneity on equipotentials. If we give up the radiative equilibrium assumption and assume that the horizontal homogenization would be effective in the photosphere and lead to weaker temperature variations on equipotentials, then we can use the result due to Hadrava (1987, 1988), who showed that the flux would vary, under these circumstances, such that

$$F \sim g^{0.56}, \quad (3.2.14)$$

and that limb darkening would also depend on g. To close the discussion on von Zeipel's theorem we conclude that a rigid analysis of the problem should include hydrodynamic calculations of meridional circulations in the atmosphere and also the computation of three-dimensional radiative transfer (giving up the assumption of a local plane parallel atmosphere), keeping in mind that the theorem is an approximation which in many binaries seems to represent the situation appropriately.

As is obvious from (3.2.13), gravity brightening is strong in distorted stars hot enough to have radiative envelopes. Early-spectral-type close binaries such as *TU Muscae* [cf. Andersen & Grønbech (1975)] are the best examples. In these systems,

---

[28] Meridional circulation is for instance described by Tassoul (1978, Chap. 8) or Kippenhahn and Weigert (1989, pp. 437–443).

the bolometric flux is directly proportional to local gravity. Because bolometric flux and local effective temperature are coupled by the Stefan–Boltzmann law

$$F_l \sim T_l^4, \quad (3.2.15)$$

from (3.2.13) with $\beta := g/4$ we can derive a similar relation for the local *effective temperature*

$$T_l = T_p \left(\frac{g_l}{g_p}\right)^\beta, \quad \beta = \frac{g}{4} = \begin{cases} 0.25, & \text{von Zeipel-theorem,} \\ \approx 0.08, & \text{Lucy's law,} \end{cases} \quad (3.2.16)$$

where T_p is the polar effective temperature. The polar temperature is the highest temperature on the star's surface and therefore higher than the "spectroscopically observed" effective temperature T_{eff}, which is some kind of average (weighted with aspect effects). If the stars move in elliptic orbits, in contrast to T_{eff} the polar effective temperature, T_p, varies with phase. Therefore, Wilson (1979) recommends use of the mean surface effective temperature T_{eff} as input parameter. For a star with surface S, T_{eff} can be defined through the bolometric *luminosity* $\mathcal{L}$ and the Stefan–Boltzmann law,

$$\mathcal{L} = \sigma S T_{\text{eff}}^4, \quad \sigma = 5.6705 \cdot 10^{-8} \text{ J} \cdot \text{m}^{-2}\text{s}^{-1}\text{K}^{-4}. \quad (3.2.17)$$

On the other hand, T_{eff} may be computed from the average flux over the surface

$$T_{\text{eff}}^4 = \frac{1}{S} \int_{S'} F_l \mathrm{d}S' = \frac{1}{S} \int_{S'} F_p \left(\frac{g_l}{g_p}\right)^g \mathrm{d}S' \quad (3.2.18)$$
$$= \frac{1}{S} \int_{S'} T_p^4 \left(\frac{g_l}{g_p}\right)^g \mathrm{d}S',$$

whence Wilson (1979) derives the reference temperature at the pole

$$T_p = T_{\text{eff}} \left[S \bigg/ \int_{S'} \left(\frac{g_l}{g_p}\right)^g \mathrm{d}S' \right]^{0.25}. \quad (3.2.19)$$

Note that T_p and T_{eff} are equal for spherical stars. Now that we have the local temperature, T_l, and local surface gravityacceleration, g_l, we are able to compute the monochromatic or bolometric fluxes and intensities from stellar atmosphere models, as follows. From the gravity brightening law we compute the local bolometric flux. The local bolometric flux plus a stellar atmosphere model enable us to compute the bolometric and monochromatic intensities.

3.2 Modeling Stellar Radiative Properties

3.2.2 Stellar Atmosphere Models

For a given chemical composition, stellar atmosphere models allow us to compute the monochromatic intensity $I_\lambda(T_{\text{eff}}, \log g, \gamma)$ as a function of the effective temperature, T_{eff}, the logarithm, $\log g$, of the surface gravity, and the aspect angle, γ. They allow integration over photometric passbands for the computation of bolometric intensities. Frequently used stellar atmosphere models are those by Mihalas (1965), the Uppsala model atmospheres by Gustafsson et al. (1975), and the Kurucz (1979, 1993) stellar atmosphere models. Some light curve programs (e.g., the Wilson–Devinney program) have included stellar atmosphere corrections. Others have empirical or semi-empirical corrections [Hill & Rucinski (1993); Linnell (1991)]. Milone et al. (1992) and Van Hamme & Wilson (2003) apply Kurucz's stellar atmospheres to the Wilson–Devinney light curve model. The requirement to use an accurate model for the radiation physics becomes crucial when the binary components have very different temperatures. The use of stellar atmospheres is most valuable to analyze light curves simultaneously at two or more wavelengths. In order to have a consistent model it is important to incorporate $\log g$ correctly. Otherwise there would be only one radiative parameter, namely T_{eff}. The consequence would be that a computed eclipse may be too deep in one passband and not deep enough in another.

3.2.3 Analytic Approximations for Computing Intensities

Light curves directly show relative radiative power or observed radiative flux because the telescope collects integrated light from the stellar system and does not resolve the details of the surface. However, it is useful to think of the process of emission at the stars' surfaces.

The computation of intensities by means of stellar atmosphere models is very time consuming, so usually a simple analytical approximation is used to calculate the specific intensity [cf. Mihalas (1978, p. 2)] at the surface of the stars, namely, the *local monochromatic intensity* $I_l(\lambda)$ which has units of energy/unit surface area/time/solid angle/wavelength. In the simplest case the computation starts with blackbody radiation,[29] i.e., $I_l(T_l, \lambda) = B_\lambda(T_l)$ with $B_\lambda(T_l)$ being the *Planck function*

$$B_\lambda(T) := \frac{2hc^2}{\lambda^5} \wp(\lambda, T), \quad \wp(\lambda, T) := \frac{1}{e^{hc/k\lambda T} - 1}, \quad (3.2.20)$$

---

[29] A *blackbody* is a (hypothetical) perfect radiator of light that absorbs and reemits all radiation incident upon it.

where Planck's constant, $h = 6.62608 \cdot 10^{-34}$ Js, Boltzmann's constant $k = 1.3807 \cdot 10^{-23}$ J/K, and the speed of light, $c = 2.9979 \cdot 10^8$ m/s. Alternatively, we also write the Planck function in the form

$$B_\nu(T) := \frac{2\pi \nu^3}{c^2} \frac{1}{e^{h\nu/kT} - 1}. \tag{3.2.21}$$

The blackbody assumption holds strictly only where there is no net flux of radiation (and thus in no real star; after all, we *see* the radiation that emerges from its surface). It is a useful starting point, and sometimes not a bad approximation to real surfaces, but it is only a very rough approximation for most real stars. Although an ideal radiator has no limb darkening, we can regard (3.2.20) as representing the emergent intensity normal to the surface and introduce a limb-darkening factor, as in Sect. 3.2.4.

3.2.4 Center-to-Limb Variation

Neglecting the center-to-limb variation (CLV) is equivalent to assuming that the stellar disks have uniform brightness. There is ample evidence from the Sun and other stars, however, that surface brightness varies from mid-disk to the limb. Stellar surface imaging by *microlensing*[30] [cf. Sasselov (Sasselov 1998a, b)] is used to measure stellar CLV. CLV, as the term is used in this section, is the dependence of intensity on angular distance from the surface normal (see Fig. 3.17). It arises because temperature increases with depth in stellar atmospheres, and the line-of-sight at the limb does not penetrate into high-temperature regions as does the line-of-sight through the disk center. Therefore, in order to compute the intensity at an arbitrary point, a factor $D(\mu)$ needs to be computed. Let γ denote the angle (sometimes called the aspect angle) between the surface normal $\mathbf{n}(\mathbf{r}_s)$ in point $\mathbf{r}_s$ and the arbitrary direction $\mathbf{e}$, in which radiation is emitted, so that

$$\mu := \cos\gamma = \cos\gamma(\mathbf{r}_s) = \mathbf{n}(\mathbf{r}_s) \cdot \mathbf{e}(\mathbf{r}_s, \Phi), \quad 0 \le \mu \le 1. \tag{3.2.22}$$

Limb brightening is important only for chromospheric and coronal emission and far-ultraviolet light curves, so we will concentrate in this section only on *limb darkening*, which affects the visible radiation from a stellar photosphere. The simple and traditional monochromatic limb-darkening law is

$$D_\lambda(\mu) = 1 - x_\lambda(1 - \mu) = 1 - x_\lambda + x_\lambda \cos\gamma \tag{3.2.23}$$

with a limb-darkening coefficient x_λ. Note the wavelength dependence of x indicated by the subscript λ. As discussed at the end of this section, similar coefficients

---

[30] Microlensing occurs if the light of a star is refracted and amplified by the gravity field of another star just moving through the line-of-sight.

3.2 Modeling Stellar Radiative Properties

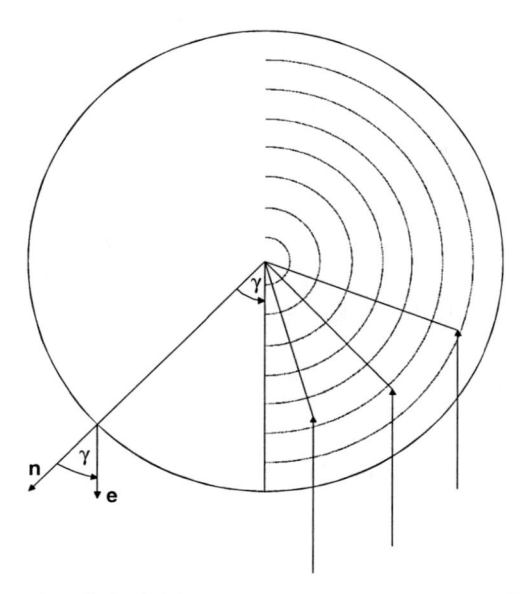

Fig. 3.17 Center-to-limb variation. This figure shows the aspect angle γ (angle between normal vector n and radiation emission direction e) appearing in the mathematical formulation of the limb darkening. The right part of the figure illustrates that the depth of the atmosphere region (and thus temperature) accessible to an observer varies with the aspect angle γ

and laws can be established for the bolometric case. For notational convenience we drop the wavelength dependence in the rest of this section. The associated (monochromatic or bolometric) flux $\mathcal{F}$ received from the (spherical) stellar disk with radius R that is displayed in Fig. E.8 can be computed as

$$\mathcal{F} = \int_0^{2\pi} \int_0^R \mathrm{d}F(u,\varphi). \tag{3.2.24}$$

The contribution of a differential surface element to the flux is

$$\mathrm{d}F(u) = I(u)u\mathrm{d}u\mathrm{d}\varphi \tag{3.2.25}$$

with the intensity distribution (here, for the linear limb-darkening law)

$$I(u) = D(\mu)I_0 = (1 - x + x\cos\gamma)I_0, \quad I_0 = 1, \tag{3.2.26}$$

where I_0 is the normal emergent intensity; for spherical stars this is the normal emergent intensity at disk center. Observing the (spherical star) relation

$$\sin\gamma = \frac{u}{R}, \tag{3.2.27}$$

the differential du and r can be eliminated using

$$du = R\cos\gamma\, d\gamma. \tag{3.2.28}$$

It then follows that

$$\mathcal{F} = \int_0^{2\pi}\int_0^{\pi/2} d\mathcal{F}(\gamma,\varphi), \quad d\mathcal{F}(\gamma,\varphi) := D_\lambda \cos\gamma\, I_0 R^2 \sin\gamma\cos\gamma\, d\gamma\, d\varphi, \tag{3.2.29}$$

with the effective limb-darkening factor, $\mathcal{D}$,

$$\mathcal{D} = \frac{\mathcal{F}}{I_0 R^2} = \int_0^{2\pi}\int_0^{\pi/2}\left[(1-x)\sin\gamma\cos\gamma\, d\gamma\, d\varphi + x\sin\gamma\cos^2\gamma\right]d\gamma\, d\varphi, \tag{3.2.30}$$

we eventually get (assuming a unit disk, i.e., $R \equiv 1$)

$$\mathcal{F} = \mathcal{D}I_0, \quad \mathcal{D} = \pi\left(1 - \frac{x}{3}\right). \tag{3.2.31}$$

The linear limb-darkening law is a one-parameter law. It is only a very rough representation of the actual emergent intensity. Accuracy is increased if we consider two-parameter, nonlinear limb-darkening laws. These laws and their coefficients are derived from stellar atmosphere models [see Van Hamme (1993) and references therein], e.g., by least-squares fitting of the chosen expression to the normalized intensities of the atmosphere model, tabulated as a function of μ. Some approaches impose the condition of conservation of total emergent flux and some do not.

The most simple class of nonlinear relations involves polynomials, such as

$$D(\mu) = 1 - x(1-\mu) - y(1-\mu)^p. \tag{3.2.32}$$

For (3.2.32) the associated flux $\mathcal{F}^{(P)}$ over the entire disk is

$$\mathcal{F}^{(P)} = \pi\left(1 - \frac{x}{3} - \frac{y}{\frac{1}{2}p^2 + \frac{3}{2}p + 1}\right)I_0, \tag{3.2.33}$$

where I_0 is again the normal emergent intensity.

In particular, Linnell (1984) has used the quadratic limb-darkening law ($p = 2$)

$$D(\mu) = 1 - x(1-\mu) - y(1-\mu)^2 =: 1 - u_1 - u_2 + u_1\cos\gamma + u_2\cos^2\gamma \tag{3.2.34}$$

with limb-darkening coefficients $u_1 = x + 2y$ and $u_2 = -y$. In that case the monochromatic flux $\mathcal{F}^{(2)}$ received from the stellar disk displayed in Fig. 3.17 is given by

$$\mathcal{F}^{(2)} = \pi\left(1 - \tfrac{1}{3}x - \tfrac{1}{6}y\right)I_0 = \pi\left(1 - \tfrac{1}{3}u_1 - \tfrac{1}{2}u_2\right)I_0. \tag{3.2.35}$$

3.2 Modeling Stellar Radiative Properties

The Wilson–Devinney program includes the logarithmic law

$$D(\mu) = 1 - x(1-\mu) - y\mu \ln \mu, \quad \mathcal{F}^{\mathrm{LOG}} = \pi \left(1 - \tfrac{1}{3}x + \tfrac{2}{9}y\right) I_0, \quad (3.2.36)$$

proposed by Klinglesmith & Sobieski (1970). The square-root law

$$D(\mu) = 1 - x(1-\mu) - y\left(1 - \sqrt{\mu}\right), \quad \mathcal{F}^{\mathrm{SRL}} = \pi \left(1 - \tfrac{1}{3}x - \tfrac{1}{5}y\right) I_0, \quad (3.2.37)$$

has been investigated by Díaz-Cordovés & Giménez (1992) and is now also included as an option in the WD-program.

Note that the nonlinear limb-darkening laws reduce to the linear law (3.2.23) if $y = 0$. Unlike x in the linear law, the coefficients y are not restricted to nonnegative values. In least-squares analyses we should check how strongly x and y are correlated; we should not adjust both. Currently in WD, only x is adjustable.

Whatever limb-darkening law is used, the local intensity I follows:

$$I = I(\cos \gamma; g, T, \lambda) = D_\lambda(\mu) I_{\mathrm{N}}(\cos \gamma = 1; g, T, \lambda), \quad (3.2.38)$$

where $I_{\mathrm{N}}(\cos \gamma = 1; g, T, \lambda)$ is the local normal monochromatic intensity, and γ, g, and T are also local quantities. The most simple case is to assume I_{N} to be equal to the blackbody radiation defined in (3.2.20). More accurate modeling requires that I be computed from a model atmosphere, with such local effects as spots, prominences, faculae, and gas streams included.

More complicated limb-darkening laws have been proposed for the Sun, and the form of the limb darkening varies with wavelength, especially when the radiation comes predominantly from regions other than the visible photosphere. Thus the center-to-limb variation for the Sun from 200 to 300 nm may be fitted with logarithmic among other limb-darkening laws; cf. Kjeldseth-Moe & Milone (1978). In the far-ultraviolet, below ~ 160 nm, limb brightening occurs because ultraviolet arises primarily from the chromosphere where temperature *increases* with height.

Bolometric limb-darkening coefficients can be obtained by numerically integrating the model monochromatic intensities over all wavelengths. Bolometric coefficients can then be derived similar to the monochromatic coefficients. Van Hamme (1993a) lists bolometric limb-darkening coefficients as well as monochromatic coefficients derived from Kurucz's model atmospheres.

3.2.5 Reflection Effect

In a binary system the presence of a companion star leads to an increased radiative brightness on the side that faces toward the companion. The cause is heating by the radiant energy of the companion. That in turn leads to an increase of the temperature calculated according to (3.2.15). Because heating by mutual irradiation is the physical cause, it is somewhat misleading to use the expression *reflection effect*. However, in very hot binaries a considerable fraction of the incident radiation is simply scattered by free electrons, so in that case the term *reflection effect* is reasonably appropriate.

As illustrated in Fig. 3.19, one effect of reflection on binary star light curves is to raise the light around the secondary eclipse relative to that near the primary eclipse. Another is to produce a concave-upward curvature between eclipses (curvature opposite to that from ellipsoidal variation). The reflection effect is usually modeled by mean global parameters such as the *bolometric albedo*, without consideration of the microphysics. For binaries whose components have similar temperatures and are close to but not actually over-contact, it may be necessary to consider multiple reflection [see, for instance, Kitamura and Yamasaki (1984) or Wilson (1990)]. *BF Aurigae* [cf. Kallrath & Kämper (1992), Van Hamme (1993b), Kallrath & Strassmeier (2000)] is an example for such a binary. The first star heats the second star, and the (now warmer) second star then heats the first star more than otherwise expected because of its own raised temperature. This process is iterative, leading to higher temperatures on the facing hemispheres. Tassoul & Tassoul (1983) investigated gradient-induced diffusion as another potentially important effect, at least close to the terminator regions of the reflection-illuminated hemispheres.

Heating caused by reflection can be a strong effect. The current champion is *HZ Herculis* with a $1.^{m}5$ reflection amplitude (cf. Lyutyi et al. 1973). Figure 3.18 shows the light variation of the close binary *V664 Cassiopeiae*, the nucleus of the planetary nebulae HFG 1. According to Grauer et al. (1987), Bond et al. (1989), and Bond & Livio (1990), an extremely hot primary heats one hemisphere of a larger and cooler main sequence companion in this noneclipsing binary classified as a reflection variable according to the *General Catalog of Variable Stars* [cf. Sterken & Jaschek (1997)]. The separation is sufficiently small so that reflection produces significant variability of the total light.

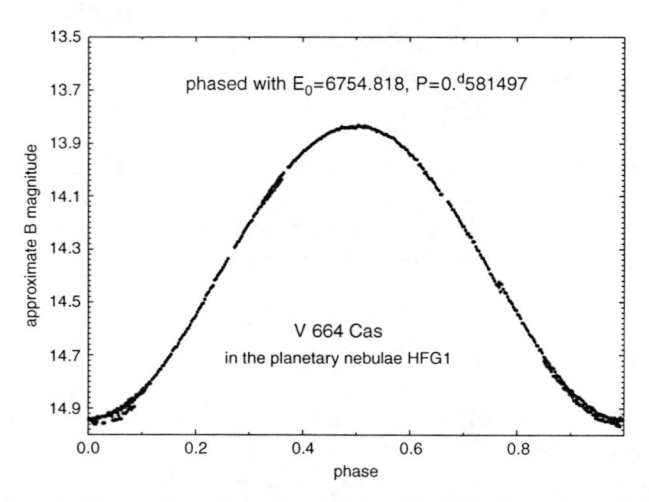

Fig. 3.18 Light variation caused purely by the reflection effect. The figure shows the B light curve of *V664 Cassiopeiae*, the close binary nucleus located in the planetary nebulae HFG 1. Data, courtesy Howard E. Bond

3.2 Modeling Stellar Radiative Properties

In a quantitative picture, let subscript t refer to that star for which the reflection heating is to be investigated (the *target* or *irradiated* star). The index s refers to the irradiating *source* star. In an Algol-type system the irradiating star would be the smaller, hotter, "primary" component; this hotter component is our initial source star s. In the same Algol system, the irradiated star would be a yellow subgiant or giant. Here it is our initial target star, t. In Wilson–Devinney argot, these are stars 1 and 2, respectively.

Let T_l denote the local temperature at a surface point, $\mathbf{r}_l$, of the target star, computed according to (3.2.13). A conventional approach in light curve reflection modeling is to compute, for each surface element, the ratio of the integrated bolometric irradiance flux F_s (i.e., incident flux) (coming from our source star s) to the local "undisturbed" bolometric flux, F_t, and to derive a *modified effective temperature* T_l' according to

$$T_l' = \sqrt[4]{R_t}\, T_l, \qquad R_t := 1 + A_t \frac{F_s}{F_t}. \tag{3.2.39}$$

Here A_t is the bolometric albedo of the target star specifying the local ratio of the reradiated to the incident energy over all wavelengths. The local reflection factor, R, is the ratio of the total radiated flux (including the fraction due to reflection) and the internal flux according to the gravity brightening law, so that $R \geq 1$. For atmospheres in radiative equilibrium, and therefore for local energy conservation, $A = 1$. For stars in convective equilibrium, the albedo may be lower ($0 \leq A \leq 1$) which follows from the thermodynamical requirement that the entropy in deeper convection regions is the same on both irradiated and nonirradiated hemispheres. It is reasonable to follow Rucinski (1969) and set $A = 0.5$ for convective atmospheres. This value has been derived from computational experiments. We can interpret $A = 0.5$ as follows: A star with a convective envelope locally reradiates about half of the external heating energy, while the rest emerges from the entire surface.

In order to compute R_t we need to compute the ratio of F_s/F_t of bolometric fluxes. Let T_l' be the increased temperature at $\mathbf{r}_l$ due to the absorbed and reprocessed flux from the other component. Further, let ρ be the distance between the point, $\mathbf{r}_l$, and the center of the irradiating source star; $F_t(T_l)$, the local bolometric flux at $\mathbf{r}_l$; F_s, the bolometric irradiance flux from star s received at $\mathbf{r}_l$; and A_t, the bolometric albedo (the fraction of F_s which is "reflected"). Once T_l' is known the monochromatic flux follows from Planck's law or from a model atmosphere.

The computation of T_l' is based on the following assumption: The temperature, T_l', and the sum of the effective irradiance and internal flux, $A_t F_s + F_t$, are coupled by Stefan–Boltzmann's law according to

$$\frac{T_l'}{T_l} = \sqrt[4]{\frac{A_t F_s + F_t}{F_t}} = \sqrt[4]{1 + \frac{A_t F_s}{F_t}}. \tag{3.2.40}$$

F_t is computed similarly to (3.2.13). Flux conservation guarantees that

$$\int_{S'} F_t d\sigma' = L_t^{\text{bol}}, \tag{3.2.41}$$

i.e.,

$$F_t = F_t(\mathbf{r}_l) = L_t^{\text{bol}} \frac{|\nabla\Omega|^g}{\int_{S'} |\nabla\Omega|^g d\sigma'}. \tag{3.2.42}$$

The accuracy of reflection modeling depends on how the incident flux, F_s, is computed. First consider a simple inverse square law treatment, and some corrections for penumbral and ellipsoidal effects. If it is assumed that bolometric flux decreases with the square of the distance ρ, the incident flux F_s is given by[31]

$$F_s = F_s(\mathbf{r}_l) = L_s^{\text{bol}} \frac{\cos\varepsilon}{4\pi\rho^2}, \tag{3.2.43}$$

where ε is the angle between the direction toward $\mathbf{r}_l$ and the normal vector $\mathbf{n}(\mathbf{r}_l)$. The ratio F_s/F_t is

$$\frac{F_s}{F_t} = \frac{L_s^{\text{bol}}}{L_t^{\text{bol}}} \frac{\cos\varepsilon}{4\pi\rho^2} \frac{\int_{S'} |\nabla\Omega|^g d\sigma'}{|\nabla\Omega|^g}. \tag{3.2.44}$$

In Wilson et al. (1972) a physically more realistic model is presented in which ellipsoidal geometry is assumed for the irradiating star. It is explicitly considered that the irradiating star might only be partially above the local horizon. Relation (3.2.44) for F_s/F_t is therefore modified by two factors E (for ellipsoidal correction) and P (for penumbra correction) which describe a more detailed geometry. In Appendix E.29 a detailed derivation of the relation for E and P is given, which leads to

$$\frac{F_s}{F_t} = \frac{L_s^{\text{bol}}}{L_t^{\text{bol}}} \frac{\cos\overline{\varepsilon}}{4\pi\rho^2} \frac{\int_{S'} |\nabla\Omega|^g d\sigma'}{|\nabla\Omega|^g} EP, \tag{3.2.45}$$

where $\overline{\varepsilon}$ is also defined in Appendix E.29. The computation of the ratio of bolometric luminosities is further discussed in Appendix E.5. So, finally, we can compute the reflection factor

$$R_t = 1 + A_t \frac{F_s}{F_t}. \tag{3.2.46}$$

Eventually, in Kitamura & Yamasaki (1984) and Wilson (1990), we find an accurate computation of the incident flux integrated over the visible surface of the irradiating

---

[31] This is the case, for instance, if the irradiating star is spherical.

3.2 Modeling Stellar Radiative Properties

star. In that case (compare Appendix E.25 where ε is replaced by γ_A) we have an expression of the form

$$F_s = F_s(\mathbf{r}_l) = \int_{S'''} I(\cos\gamma; g, T, \lambda) \frac{\cos\gamma \cos\varepsilon}{4\pi\rho^2} d\sigma, \qquad (3.2.47)$$

where S''' indicates that we integrate over that part of the irradiating star's surface that is visible from $\mathbf{r}_l$. Note that all quantities in the integrand of (3.2.47) depend on source star properties.

Figure 3.19 illustrates the reflection effect and how light curves change when the albedo, A_1, of the hotter star is varied from 0 to 1.

The impinging radiation is not merely reflected from the receiving star but heats up the impacted surface, which then reradiates[32] toward the irradiating star (source star). Multiple reflections are thus needed to treat the effect properly, and the flux at each point of both components must be integrated to consider each subsequent reflection. The process is, of course, iterative because each reflection produces higher temperatures on both initial target and source stars. Iterations are stopped when the multiple reflection computations come to a constant distribution of surface effective temperature, or we might simply ask for a certain number of iterations.

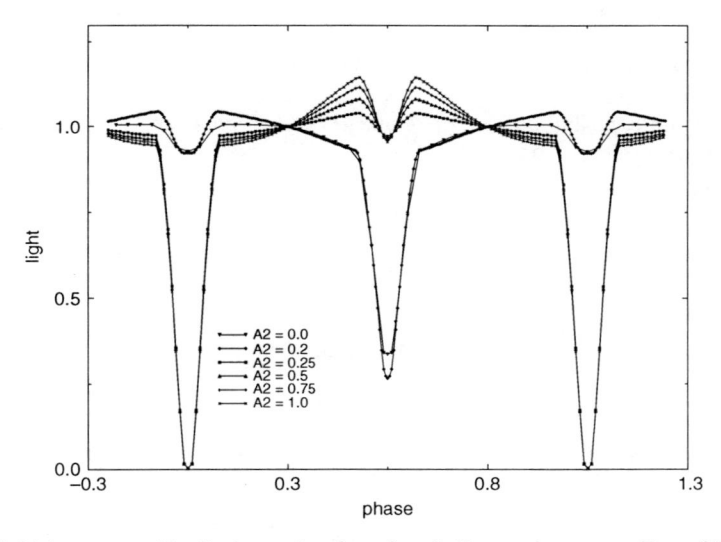

Fig. 3.19 Light curves with albedo varying from 0 to 1. Temperatures were $T_1 = 20,000$ and $T_2 = 3,600$ K. Light curves were produced with Binary Maker 2.0 (Bradstreet 1993) with the albedo A_2 varying between 0 and 1

---

[32] The multiple reflection effect is part of the Wilson–Devinney model of the early 1990s (Wilson 1990). Multiple reflection seems to be significant in only a small fraction of binaries.

The multiple reflection effect involves many iterative computations and thus much computing time, it is advised to pay close attention to the structure and logic of the computations (see Appendix E.25).

3.2.6 Integrated Monochromatic Flux

Eventually, the *monochromatic flux* or *light* from component j is

$$\ell_j(\Phi) = \int_{S'} \chi(\mathbf{r}_s) I(\cos\gamma; g, T, \lambda) \cos\gamma \, d\sigma, \tag{3.2.48}$$

where χ_s is the characteristic function defined in (3.3.2) ensuring that we consider only those points $\mathbf{r}_s$, or equivalently, we integrate only over those parts of the stellar surface which are visible to the observer on Earth. In spherical polar coordinates (3.2.48) takes the form

$$\ell_j(\Phi) = \int_0^\pi \int_0^{2\pi} \chi(\mathbf{r}_s) I(\cos\gamma; g, T, \lambda) \frac{\cos\gamma}{\cos\beta} r^2 \sin\theta \, d\varphi \, d\theta. \tag{3.2.49}$$

In most light curve programs, for reasons of efficiency, the integrand is evaluated only if $\chi_s(\Phi) = 1$.

Adding all contributions of the binary system, an observer at unit distance receives the total flux $\ell(\Phi)$,

$$\ell(\Phi) = \ell_1(\Phi) + \ell_2(\Phi) + \ell_3, \tag{3.2.50}$$

while an observer at distance D receives the flux

$$\ell_D(\Phi) = \frac{1}{D^2} \ell(\Phi). \tag{3.2.51}$$

The light ℓ_3 of a third source, usually a third star far away from the binary system, is ordinarily assumed to be independent of time or phase Φ. Astrophysical interpretations of third light are discussed in Sect. 3.4.1.

3.3 Modeling Aspect and Eclipses

Exitus acta probat   *(The end justifies the means)*

The computation of $\ell_j(\Phi)$ in (3.2.49) requires not only the stellar surface S' defined in (3.2.1), but in addition, for considering eclipses, the *visible stellar surface S''*:

$$S'' = S''(q, \Omega_0, F, e; \Phi) := \{\mathbf{r}_s \mid \mathbf{r}_s \in S' \text{ and } r_s \text{ visible}\}. \tag{3.3.1}$$

3.3 Modeling Aspect and Eclipses

The attribute "visible" means $\mathbf{r}_s \in S'$ so that the observer receives flux emitted from a point $\mathbf{r}_s$. In particular, this requires that $\mathbf{r}_s$ be on the side of the star facing the observer and that the other component does not eclipse the point $\mathbf{r}_s$. The characteristic function, $\chi(\mathbf{r}_s)$, summarizes this condition:

$$\chi(\mathbf{r}_s) := \begin{cases} 1, & \mathbf{r}_s \text{ visible,} \\ 0, & \mathbf{r}_s \text{ not visible.} \end{cases} \quad (3.3.2)$$

This function is the product of two characteristic functions, $\chi^A(\mathbf{r}_s)$ and $\chi^B(\mathbf{r}_s)$. The first one is the *horizon* and *self-eclipse function*

$$\chi^A(\mathbf{r}_s) := \begin{cases} 1, & \mathbf{r}_s \text{ is not beyond the component's own horizon,} \\ 0, & \mathbf{r}_s \text{ is beyond the horizon.} \end{cases} \quad (3.3.3)$$

The second one is the *companion eclipse function*

$$\chi^B(\mathbf{r}_s) := \begin{cases} 1, & \mathbf{r}_s \text{ is not eclipsed by the companion,} \\ 0, & \mathbf{r}_s \text{ is eclipsed by the other component.} \end{cases} \quad (3.3.4)$$

According to the orientation of the normal vector, $\mathbf{n}$, and the line-of-sight vector, $\mathbf{s}$, introduced in Sect. 3.1.1, and as illustrated in Fig. 3.20, for convex surfaces, i.e., positive curvature on the whole surface, $\chi^A(\mathbf{r}_s)$ can be expressed by the aspect angle γ and the relation

$$\chi^A(\mathbf{r}_s) = \begin{cases} 1, & \text{if } \cos \gamma < 0, \\ 0, & \text{else.} \end{cases} \quad (3.3.5)$$

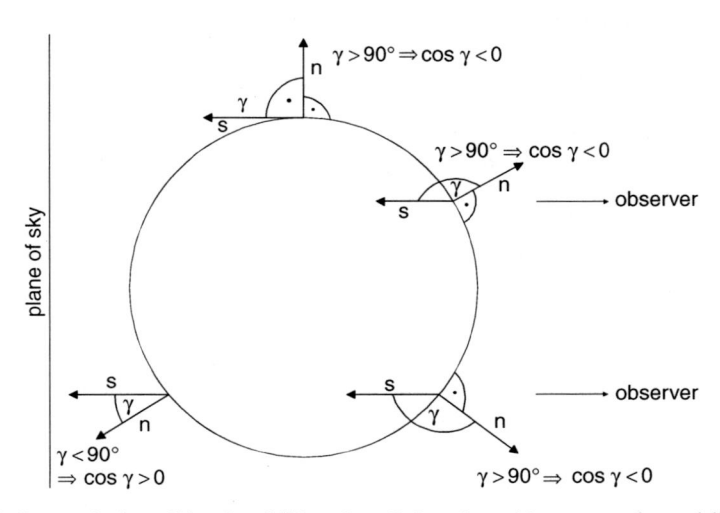

Fig. 3.20 Geometrical condition for visible points. Only points with $\cos \gamma < 0$ are visible to the observer

The situation is more complicated for surfaces that have local negative curvatures such as over-contact equipotentials. In that case, self-eclipses are possible. We have outlined in a formal way how to restrict the range of integration over the visible part, but we also need to explain how $\chi(\mathbf{r}_s)$ is computed in real light curve programs. Thus we need a procedure that indicates whether or not a point on a star is visible from Earth. A necessary condition for a surface point $\mathbf{r}_s$ to be visible is that its associated angle $\gamma = \gamma(\Phi)$ between $\mathbf{s}$ and $\mathbf{n}(\mathbf{r}_s)$ computed according to (3.1.16) fulfills the condition

$$\cos \gamma < 0. \tag{3.3.6}$$

Thus, in principle all those points fulfilling (3.3.6) on both stars can be identified and establish our function $\chi^A(\mathbf{r}_s)$. Depending on phase an eclipse might occur or not. In the trigonometric relations involving the phase, Φ, we rather need the true phase angle, θ, defined in Sect. 3.1.2. Thus, whenever a term such as $\sin \Phi$ or $\cos \Phi$ occurs, it means rather $\sin \theta$ or $\cos \theta$, where θ is computed according to (3.1.19) in the circular case and according to (3.1.37) in the eccentric case. Figure 3.21 shows the eclipse geometry and units. Although treating the eclipse geometry requires many subtle details the basic ideas are simple. The first step is a sort of global checking whether at phase Φ an eclipse is possible at all or not. For spherical stars with relative radii r_1 and r_2, an eclipse occurs only if δ, the *projected (plane-of-sky) distance* between the centers of the components, fulfills the relation

$$\delta \leq \frac{r_1 + r_2}{d}, \quad r_j := \frac{\mathcal{R}_j}{a}, \quad j = 1, 2, \tag{3.3.7}$$

where δ is computed according to (3.1.9) in Sect. 3.1.1, i.e.,

$$\delta^2 = \left(y^s\right)^2 + \left(z^s\right)^2 = d^2(\Phi)\left(\sin^2 \Phi + \sin^2 i \cos^2 \Phi\right). \tag{3.3.8}$$

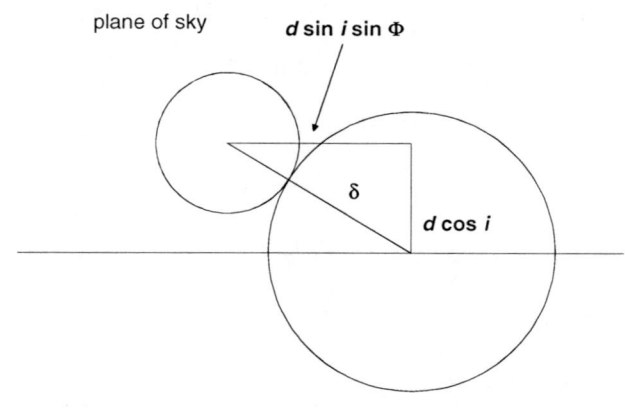

Fig. 3.21 Projected plane-of-sky distance. The projected plane-of-sky distance can be used to test whether an eclipse can occur

Within the Roche model the largest value for $r_j(\theta, \varphi)$ might be used as the radius r_j and then (3.3.7) might be applied.

If an eclipse is not excluded, the next step is to identify which star is in front (this star cannot be eclipsed at this phase). In Appendix E.21 we describe how this test is performed in the WD program. The next step is to compute and represent the horizon of the front star in the plane-of-sky coordinates; this step is full of tricky details related to numerical accuracy and varies among different light curve programs. Once this representation is available, the grid points representing the surface of the distant star can be tested w.r.t. eclipse by comparing its plane-of-sky coordinates with those of the horizon. This procedure establishes the function $\chi^B(\mathbf{r}_s)$.

3.4 Sources and Treatment of Perturbations

In omnia paratus   (*Ready for all things*)

3.4.1 Third Light

The light l_3 of a third source, usually a third star far away from the binary system,[33] is assumed to be independent of time or phase Φ. Third light has become more interesting now that companions can be discovered with the new generation of interferometers. The presence of the light of a third star decreases the depths of both eclipses because addition of a constant to a positive function diminishes its "fractional" or "percent" variation. As third light decreases the depth of eclipses it roughly simulates a system with lower inclination. A hot companion may make a greater relative contribution to system flux in the ultraviolet, whereas a cooler companion will be a stronger contributor in the infrared. In the intermediate or far infrared, emission from circumstellar dust can contribute to the background. If the third body has a spectral type different from the close binary components, the added flux will be different in different passbands, and the third light may be modeled to determine the nature and apparent brightness of the third star.

In *44i Bootis*, third light is contributed by the primary component of a visual binary star system in which the secondary is the *EB*. The angular semi-major axis is only $\sim$3.8 arc-sec (Linshan et al. 1985), and the separation over the next half-century will not exceed $\sim$2.5 arc-sec. The third component is brighter than the hotter, more luminous component of the over-contact binary (Hill et al. 1989) by at least a magnitude in V and so contributes a significant amount of "third light," unless the component somehow can be excluded from the measurement. It is also difficult to exclude its scattered light from spectra.

---

[33] The binary system and the third star may establish a gravitationally bound triple system. By "far away," however, we mean compared to the separation of the eclipsing binary components. Typically, the third component may be seen as a very close visual binary with the combined light of the eclipsing system, or may not even be resolved optically.

Sometimes, as is the case for *VV Orionis*, there is spectroscopic evidence for a third body in the form of disturbances of the radial velocity curve [cf. Scarfe et al. (1994)]. In that particular case, however, Van Hamme & Wilson (2007) have shown that the light and radial velocity curves of *VV Orionis* can be fitted without third light. It may be possible to demonstrate the existence of a third body by analysis of times of minima. A linear relation between *Observed–Computed* times of minimum plotted versus time implies a simple correction to the period; a parabola implies a constant rate of period change; and a sinusoid implies *apsidal motion*[34] or variation in arrival time (*light-time effect*) due to orbital motion of the close binary around the system center of mass (binary plus third body). Apsidal motion can be due to gravitational perturbations (e.g., by a third body), finite nonspherical mass distribution of the stars (not point masses), and general relativistic effects [cf. Quataert et al. (1996)].

In light curve modeling, third light ℓ_3 is usually added to the computed light as a constant

$$\ell(\Phi) = \ell_1(\Phi) + \ell_2(\Phi) + \ell_3. \tag{3.4.1}$$

The partial derivative $\partial \ell / \partial \ell_3$ is therefore constant, namely

$$\frac{\partial \ell}{\partial \ell_3}(\Phi) = 1, \tag{3.4.2}$$

which might be exploited in derivative-based least-squares analysis.

3.4.2 Star Spots and Other Phenomena of Active Regions

As is observed on our own Sun (Fig. 3.22), stars can have spots. Stellar surface imaging by *microlensing* [cf. Sasselov (1998a, b)] shows directly that spots are present on other stars as well. A star spot is a region with higher or lower temperature than the surrounding photosphere, and thus it modifies the local flux. By way of physical analogy to the Sun, we should expect magnetic spots to result from convection in the outer envelope and differential rotation. Accompanying phenomena include small "pores,"[35] umbral and penumbral dark regions, and spot groups. Additional phenomena are bright *flocculi* (Latin for tufts of wool) or plages (French for the white sand on a beach) – most easily seen in the lines of Ca II H&K and Hα – and faculae ("little torches"), sometimes called "white-light plages." Finally, we mention

---

[34] The term *apsidal motion* refers to the rotation of lines of apsides of an eccentric binary orbit, or the rotation of the periastron. Apsidal motion is caused by perturbations to the $1/r$ gravitational potential and indicates deviations from Keplerian elliptic motion.

[35] *Pores* are small sunspots. This is a term in solar research. Pores are usually at the resolution limit on visual images of the solar disk. We would need an enormous number of these to make a difference to hemispherical flux values, of course.

3.4 Sources and Treatment of Perturbations   133

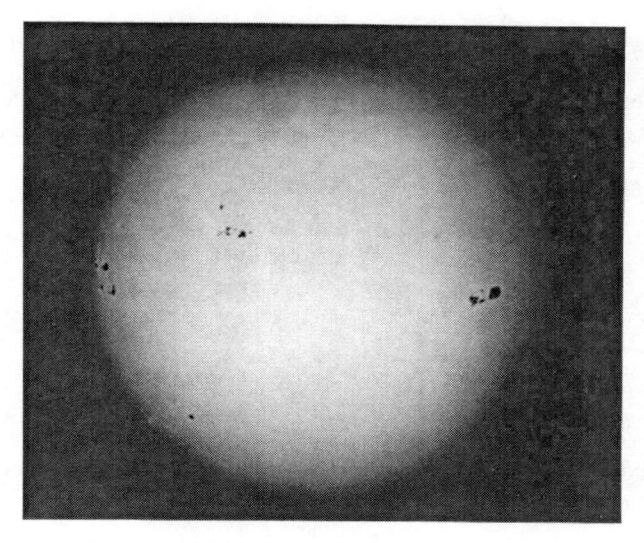

Fig. 3.22 Sun spots. This white light image of the Sun was taken in Calgary on September 3, 1988, in the early afternoon (20:05 UT) by Fred M. Babott. It shows spot groups, spot umbrae, penumbrae, pores, and toward the limb, faculae. Courtesy F. M. Babott

solar prominences, best seen in Hα, but also seen in "white light," which are bright off the limb but may appear as dark filaments when projected onto the photospheric disk.

In the light curve context, many researchers have included star spots in light curve models. Poe & Eaton (1985) give most of the historical and technical background for the analysis of spotted stars. A more recent review is by Linnell (1993). In this book, we characterize spots, as in the Wilson–Devinney model, by four parameters: latitude θ, longitude Φ, angular radius ρ, and temperature factor t_f. Some authors use the colatitude $\theta^c := 90° - \theta$ instead of θ. The spots subtend circular solid angles at the star centers and, in their ideal form (infinitely fine grid of the surface), are essentially circular areas on the surface, except for the effect of the star's asphericity, and the ellipticity of features, such as round spots, produced by foreshortening, most noticeable at the limbs. Φ is the longitude of the spot-center. The reference direction is the line of centers, and Φ increases as in a right-handed system (set up separately for each component). The spot-center colatitude θ^c is zero at the North ($+z$) pole and increases to 180° at the other pole. As seen from the North pole, the binary orbit is described counterclockwise. The spot angular radius ρ is half the angle subtended by the spot at the center of the star. The spot temperature factor t_f is the ratio between the local surface temperature and the local undisturbed temperature. Due to reflection and gravity effects, the surface temperature across a spot may not be constant. Temperature factors less than and greater than unity correspond to cool spots and hot spots, respectively. As noted above, such a characterization of spot regions can at best be only a rough approximation of the true physical situation.

Below, we give a mathematical representation of that simplified approach, as it exists in the Wilson–Devinney model.

Let the angles θ^c and φ represent a point on the surface of the star. In addition θ_s and φ_s refer to the coordinates of a particular surface spot with radius ρ_s. From the cosine law of spherical trigonometry, the angular distance Δ_s of the point (θ, φ) to the center of the spot s follows as

$$\cos \Delta_s = \cos \theta^c \cos \theta_s^c + \sin \theta^c \sin \theta_s^c \cos(\varphi - \varphi_s), \qquad (3.4.3)$$

where θ^c and θ_s^c denote the colatitudes, and $\cos(\varphi - \varphi_s)$ can be computed as

$$\cos(\varphi - \varphi_s) = \cos \varphi \cos \varphi_s + \sin \varphi \sin \varphi_s. \qquad (3.4.4)$$

From (3.4.3) it is easy to check whether the point (θ, φ) "lies" within the spot. If so, the spot-free local temperature T_l in (θ, φ) is modified by the temperature factor t_f. This is summarized in the formula

$$T_l^{SC} = T_l \begin{cases} t_f, & \text{if } \Delta_s \leq \rho_s, \\ 1, & \text{if } \Delta_s > \rho_s, \end{cases} \qquad (3.4.5)$$

where T_l^{SC} denotes the local temperature after the spot correction has been applied.

A slightly more realistic model is that of Hill & Rucinski (1993), which allows for elliptical spot regions, with the major axis as an optionally adjustable parameter. However, elliptical spots may cause problems because the least-squares problem may easily become overparametrized.

A further complication arises from the dynamics of a binary system: Except in the synchronously rotating, circular orbit case, the physical surface of a star moves w.r.t. (model) grid elements, so that the spot longitude can be a function of time. Magnetic spots, as observed on our Sun, take part in the motion of the surface. In contrast, an accretion hot spot on an asynchronously rotating star (in circular orbit) could remain fixed with the grid. In an eccentric orbit the grid rotates at a nonuniform rate so a time-dependent spot longitude transformation[36]

$$\varphi(\Phi) = 360° F \Phi - \frac{180°}{\pi} (\upsilon - \upsilon_0) + \varphi_0 \qquad (3.4.6)$$

needs to be applied, where F is the rotation parameter (3.1.74), Φ is the orbital phase, and υ is the true anomaly; subscript 0 refers to conjunction. Note that Φ and υ need not be in the ranges 0 and 1 and 0 to 2π, respectively, so that the effects of longitude drift for spots can be followed over many orbital cycles.

---

[36] Some kinds of spots (such as accretion hot spots) do not rotate with a star. For such cases, the transformation is not applied; therefore, $\varphi(\Phi) = $ const.

At least some of the difficulties occurring in the analysis of *EB* light curves which show unequally high consecutive light curve maxima[37] might be overcome by including star spots in the model [Yamasaki (1982), Milone et al. (1987), Hill et al. (1989, 1990)], although there are probably several physical effects that produce unequal maxima (Davidge & Milone 1984).

As other light curve parameters, spots and their parameters $(\theta, \varphi, \rho, t_f)$ can be estimated by least-squares methods. Budding & Zeilik (1987) and Zeilik et al. (1988) use adjustable spot parameters to represent the light curves of short-period systems with *RS CVn*-like phenomena. Historically, a Wilson–Devinney spot parameter (t_f) was first optimized by Milone et al. (1987), in the light curve analysis of the contact system *RW Comae Berenices*; by repeated trials with slightly readjusted values, a parabola of Σwr^2 versus t_f was constructed and solved for the minimum. This procedure was extended to all spot parameters by Milone et al. (1991). Spots were automatically adjusted within the Wilson–Devinney program for the first time by Kang & Wilson (1989). All of this work was done with the Wilson–Devinney program of the early 1980s.

Spot parameters differ greatly from other light curve parameters, such as the mass ratio or inclination, in that they can vary significantly on a relatively short timescale (say days).[38] A steady change in spot longitude with time, permits, in principle, the determination of spot migration periods. However, the determination of spot parameters requires abundant, accurate, synoptic data and careful analysis. Complete light curves must be obtained before the spot or spot groups achieve perceptible motion in longitude. Spot fitting ideally should be subject to determinacy tests (Banks & Budding 1990).

Must spots be used to model light curve perturbations? Even though most light curve analysts would argue that their presence is likely, and in the interest of achieving a more physically realistic picture of the binary system, they should be modeled, there are some doubts. As we have noted, the O'Connell effect may have more than one origin, and without further substantial evidence, such as molecular absorption features characteristic of M-stars in stars of otherwise higher temperatures [cf. Vogt (1979), or Ramsey & Nations (1980)], or Doppler imaging from line-profile analysis, the assumption of a spot cause usually is not justified. Milone et al. (1987) demonstrated that an analysis following rectification of the light curve produced no significant differences in parameters from those modeled with dark spots placed on either component, except for the parameters T_2 and i, where the differences were

---

[37] Sometimes called the *O'Connell effect* (Davidge & Milone 1984), named after D. J. K. O'Connell (1951), who demonstrated that unequal maxima in light curves was *not* a "periastron effect" because it is found predominantly in systems with circular orbits. Wesselink suggested the new usage, which has become widely accepted. The effect has also been called the "Kwee effect." Unfortunately, it has become a practice to use these names as catch-all terms for a variety of physical effects. As defined, it is purely phenomenological.

[38] Active prominences vary on a scale of hours and flares even over minutes. Modeling such effects would involve some stochasticity and is therefore not covered here.

nevertheless small. Some effects spots have on light curves could also be produced by circumstellar matter clouds (see Sect. 3.4.4.5).

Nevertheless, in some cases stellar surface imaging by microlensing shows spots. In many other cases the following indications in light and color curves support the existence of spots:

- phases of minima agree in, e.g., V and in V-I but not in U-B;
- color amplitudes increase when going to $V \rightarrow R \rightarrow I$ as expected for cool spots; and
- V-I, V-K, and U-B excess.

As spots have relatively short life times and change in size quickly, one would like to fit curve-dependent spot parameters, and be able to make allowance for differential stellar rotation and latitude migration of spot groups. A disadvantage is the increased number of parameters leading to uniqueness problems.

3.4.3 Atmospheric Eclipses

The term "atmospheric eclipse" refers to an eclipse of a star by one with an extended atmosphere. The classical example is that of ζ Aurigae, in which the width of the eclipse is greater in the ultraviolet than in longer wavelengths [see Fig. 1 in Wilson (1960, p. 441), based in turn on Roach & Wood (1952)]. We extend this idea to include eclipses of underlying radiation by overlying material of a different nature than chromospheric and coronal layers in their average, "quiet" state. Readers interested in extended atmospheres themselves are referred to Wehrse (1987) or Wolf (1987).

Atmospheric eclipses occur in binary systems in which at least one component has an extended atmosphere. The *EB V444 Cygni* (WN5+O6) is such a system, with partial eclipses at both primary (at $\lambda = 424.4$ nm the depth is 0.225 in light units normalized to 1) and secondary minimum (depth 0.141) and with insignificant reflection and ellipticity effects. Here the WN5 component has an extended atmosphere and is in front at primary minimum, which is an atmospheric eclipse. For simplicity the stars are considered as spheres and the orbit as circular. Light curve modeling of early-type stars with essentially spherical geometry has been carried out by a number of investigators, and Roche geometry modeling of eclipses by translucent plasma clouds in the atmospheres of early-type systems was developed by Kallrath and Milone at the University of Calgary (Milone 1993), and is now in the 2007 WD version. Such clouds might be produced by stellar wind interactions.

The mathematical formalism (see Sect. 6.4) usually includes a term such as

$$I^A(\xi) = I^0 \left[1 - e^{-\tau(\xi)}\right], \tag{3.4.7}$$

which represents the amount of radiation absorbed by the extended atmosphere of the WN5 component. Here I^0 denotes the brightness at the center of the disk of the normal O6 star, and $\tau(\xi)$ is the optical depth along the line-of-sight intersecting the

3.4 Sources and Treatment of Perturbations

WR component at a distance ξ from its center when the disk of the WR component is viewed by *transillumination*. Transillumination means, in addition to the light reflecting and originating from the surface of the WR disk, we see light from the O6 star passing through the extended atmosphere.

In the past light curves of early-type systems such as *DQ Cep* and *V444 Cyg* were analyzed by treating atmospheric eclipses as perturbations to photospheric eclipses but now it is possible, and perhaps more appropriate, to use a consistent model including all physical effects in a binary system simultaneously.

3.4.4 Circumstellar Matter in Binaries

Circumstellar matter in binaries may occur in gas streams, rings and disks, clouds, and boundary regions of colliding stellar winds. Distinction among these categories depends on one's focus. Gas streams may establish rings and disks, stellar winds might be considered as special kinds of gas streams, and boundary layers of colliding stellar winds might be considered as clouds with a special geometry. What the categories have in common is that they absorb, reemit, scatter, and thereby redirect the star light.

The existence of circumstellar material was detected first by Wyse (1934) who found Balmer emission in several Algol-type binaries. That this material is often manifested as an accretion disk has been known since the pioneering observations of *RW Tauri* by Joy (1942) in the early 1940s. Spectra showed double-peaked emission line features that could be explained if they arose from a ring of material circling the hotter and more massive star. Struve painstakingly observed Algols for many years [see Struve (1944), for example], and a large number of high-quality spectra demonstrated that these features were characteristic of Algols, not anomalies in a handful of systems. About the same time, Kuiper (1941), in a paper on β Lyrae, pointed out the importance of the inner Lagrangian point L_1^p in understanding gas flows in interacting binaries.

Spectroscopic and photometric observations showed another peculiarity of Algols. The hotter, more massive primaries were clearly main sequence stars, but the less massive secondaries had radii much too large for the main sequence (i.e., they were evolved subgiants or giants). The so-called *Algol paradox* begs the question: How does a binary evolve to the point where the low mass star is evolved, but the high mass star is still on the main sequence?

Roche geometry provides the key that solves the puzzle. See any of the sources Crawford (1955), Hoyle (1955, pp. 197–200), or Pustylnik (2005) for a wonderful explanation of the modern resolution of the Algol paradox. Algol-type systems are stable because the less massive star is filling its Roche lobe. As matter is transferred, the physical size of the lobe increases because of the increasing separation of the stars, which slightly detaches the star from the lobe. The star is expanding on a nuclear timescale, so mass transfer events tend to be sporadic and of small scale.

Hoyle (1955) and Crawford (1955) proposed that Algols were systems that had experienced large-scale mass transfer and that the less massive secondary had originally been the more massive star. It evolved first, overflowed its Roche lobe, and transferred enough mass to the other star to reverse the mass ratio. Morton (1960) showed that a binary system with the more massive star filling its Roche lobe is unstable to mass transfer. When the more massive star loses mass through the L_1^p point, its Roche lobe shrinks because its mass becomes smaller and the separation of the stars decreases. Although Morton failed to take the latter fact into account, the decreasing separation only increases the rate of mass transfer and creates a classic instability. The mass transfer is very rapid in the early stages and becomes continually slower toward the end of the process. A further important point is that the equilibrium radius of the mass-losing star can change as a result of mass loss. Whether it increases or decreases depends on the star's evolutionary state. Highly evolved stars (that still retain their envelopes) tend to increase their radii upon losing mass. Detailed reviews of the evolution of Algols were published by Plavec (1968) and Paczynski (1971). Batten (1973b, 1976) discusses and reviews observations of the flow of matter within binary systems. Gas streams can become evident in spectroscopic data and may influence radial velocities, or produce peculiar light curve disturbances. The situation in the detached system *VV Orionis* is discussed in some detail in Sect. 3.4.4.2. In addition to Batten's reviews, the whole proceedings of the IAU Symposium 73, edited by Batten (1973a), is a useful source and covers many aspects of the topic, e.g., Plavec's (1973) review on the evolutionary aspects of circumstellar matter in binary systems. Finally, we mention the proceedings of the IAU symposium, edited by Appenzeller & Jordan (1987), which cover many aspects of circumstellar matter around single stars as well.

Spectroscopic observations made with the International Ultraviolet Explorer led Plavec (1980) to identify a group of systems (the *W Serpentis stars*) that appeared to be in the rapid phase of mass transfer. In some of these, such as β Lyrae [Wilson (1974); Wilson & Terrell (1992)], the gainer is completely engulfed by the circumstellar material. Wilson (1981) worked out structural models of these thick disks and proposed that the gainer has been spun up by the accreting material to the centrifugal limit, thus preventing the material from settling onto the star. However, viscous and tidal interactions will eventually decrease the angular momentum of the stars, allowing the disk material to be accreted. This leaves the system in a state where both stars fill their limiting lobes. For the less massive secondary, the limiting lobe is the classical Roche lobe. For the rapidly rotating primary, the limiting lobe is bounded by the equipotential that has the equatorial material rotating at the centrifugal limit. Systems in such a configuration, the so-called double-contact binaries, were predicted by Wilson (1979) and analysis of observations shows that these systems do indeed exist (Wilson et al. 1985). Over time, tidal forces will slow the rotation of the primary to synchronism, and it then will become an Algol-type system.

The later stage of mass transfer in an Algol-type system is small scale and sporadic and is understood[39] as follows. A key point is that the stars move farther apart

---

[39] A more complete discussion of this problem would also involve the equilibrium radius of the mass-losing star; cf. Plavec (1968).

to conserve angular momentum when flow is from the less to the more massive star. The subgiant star expands on a nuclear timescale (i.e., slowly). Each small burst of transferred matter produces relatively large increases in the critical lobe size as the binary conserves orbital angular momentum, i.e., the separation of the stars increases, and the star slightly detaches from the lobe. Thus the transfer process tends to turn itself off, and proceeds in small episodes as the star undergoes its evolutionary expansion. This is the slow phase of mass transfer. Early in the mass transfer process the situation is quite different. The mass-losing star then is the more massive one, and as mass is transferred the stars come together to conserve angular momentum, and the lobe size decreases. Thus, mass transfer tends to run away and is limited only by the thermal timescale of the envelope (rapid phase of mass transfer). Mezzetti et al. (1980) provide statistics of 55 Algol-type *EB*s related to mass transfer and mass loss.

3.4.4.1 Gas Streams

Whereas light curve models imply static or quasi-static physics, circumstellar gas streams, if modeled correctly, require a fully dynamical treatment based on the equations of radiation hydrodynamics. A full treatment may be beyond present-day computing power, but the rapid increase in computing power over the last few decades has made it possible to make the models more and more realistic and to model the gas flow in Algol-type systems, as well as the radiation that arises from the gas.

When insufficient computer power was available, gas streams, and also rings and disks, were often modeled with multiple particle trajectories; see, e.g., Gould's (1959) particle path model (for details see Sect. 3.4.4.2), Kruszewski's (1967) analysis on exchange of matter in close binary systems and ring formation, or Smak's (1978) analysis of the escape of particles from disks in close binary systems. For a review of problems of gaseous motions within binary stars we refer to Huang (1973).

Prendergast (1960) showed that the mean free path of gas particles was much smaller than the separation of the two stars, indicating that a hydrodynamical treatment of the problem was necessary. Due to a lack of computing power at the time, Prendergast's solutions were limited by simplifying assumptions, such as ignoring the pressure gradient terms in Euler's equation and assuming hydrostatic support perpendicular to the orbital plane. An improved treatment was given by Prendergast & Taam (1974), who simulated solutions of the Boltzmann equation rather than solve a set of difference equations. Unfortunately, their scheme had an inherent artificial viscosity which was coupled to the grid resolution, but the results of their application to a system similar to *U Cephei* were intriguing, indicating, among other things, that mass transfer was nonconservative (i.e., some mass was ejected from the binary).

Another important contribution to understanding mass transfer in Algols was by Lubow & Shu (1975). Using matched asymptotic expansions, they developed a semi-analytical model of the gas flow. Exploiting the existence of a parameter labeled ε

$$\varepsilon = \frac{a}{\omega d}, \quad a^2 = \frac{kT}{m}, \tag{3.4.8}$$

with a being the isothermal sound speed, ω the orbital rotation rate, and d the binary separation, they reduced the parameter space of the equations to that of one parameter: the binary mass ratio. They also treated gas flow near the Lagrangian point, L_1^p.

Lin & Pringle (1976) outlined a two-dimensional, many-body approach to the problem where the gainer is of negligible size compared to the stars' separation, such as in cataclysmic variables and X-ray binaries. Their's was apparently the first fully Lagrangian method for treating gas flows in close binaries. The scheme included viscosity, but did not treat pressure gradients. Based on their simulations, Lin and Pringle concluded that disks can be well defined and comparable in size to the Roche lobe.

Whitehurst (1988b) extended the Lin and Pringle model to include pressure gradients as outlined by Larson (1978). Larson's approach, assuming that the particles are extended, deformable gas clouds, is somewhat simplistic, although not too confining because the disk is dominated by angular momentum transport. Whitehurst's major simplifying assumption was that the energy dissipated in particle interactions was instantaneously radiated away through the disk surface. This makes the pressure calculations somewhat crude, but it was obviously the next step to take in the development of models of mass transfer. Whitehurst (1988a) applied his model to the SU UMa star *Z Chamaeleontis* and achieved impressive agreement with observed light curves, containing superhumps.[40]

With the advent of 8-m class telescopes and more sensitive detectors, the next few years promise great progress in the study of circumstellar material in Algols. Newly developed models and observational techniques (especially polarimetry) should greatly improve our understanding of mass transfer in binaries, and therefore binary star evolution.

3.4.4.2 Gas Stream in the *VV Orionis System*

Based on a particle path model, Duerbeck (1975) has studied gas streams in *VV Orionis*. He discusses the consequences of circumstellar matter for the light curve, equivalent widths of the hydrogen lines, and the Hβ-index. His analysis gives us an illustrative example of how special features can be added to an otherwise standard *EB* analysis.

The disturbances or irregular features in the light curve during phases 0.6 and 0.7 are interpreted as light loss caused by scattering through particles of a stream. The basic parameter describing this gas stream is its particle density. The geometry

---

[40] Superhumps are periodic increases in brightness of 20–30% that occur during superoutbursts (outbursts lasting 10–14 days as opposed to normal outbursts lasting 2–3 days). The superhump period is somewhat longer than the orbital period of the system and is explained by the precession of the (elliptical) accretion disk.

3.4 Sources and Treatment of Perturbations

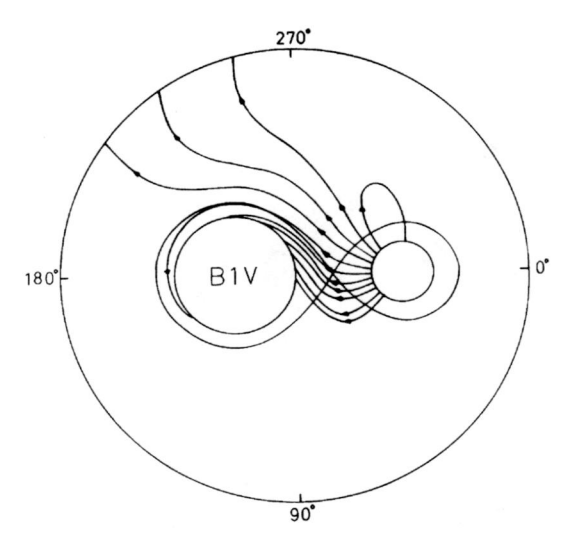

Fig. 3.23 Trajectories in a binary system. This figure, Fig. 9 in Duerbeck (1975), illustrates the first hypothesis. There is one gaseous stream present; the trajectories are computed according to Gould (1959). Courtesy H. W. Duerbeck

and its flow pattern are taken from Gould's (1959) particle path model. The model computes trajectories within the framework of the restricted three-body problem and uses an ejection velocity of 400 km/s. Figure 3.23 shows typical trajectories or orbits. There exist escape orbits, connecting trajectories, and loops originating and ending on the same component. As shown in Fig. 3.24, neglecting the pressure gradient, the trajectories form a gas stream which can absorb light.

Under reasonable assumptions it is shown that the gas stream is almost completely ionized, which allows concentration on electron scattering. In that case, neglecting multiple light scattering, the transmitted intensity I_T after passing the gas stream is related to the incident light intensity I_0 by

$$I_T = I_0 e^{-kx}, \quad k = 6.655 \cdot 10^{-25} n_e \text{ cm}^2, \tag{3.4.9}$$

where x is the path length and n_e is the electron number density in cm^{-3}.

Duerbeck's approach was to derive n_e directly from comparing the observed light curve with the light intensity I_0 derived from the Russell–Merrill model, and independently from the equivalent widths of lines. Two hypotheses were checked against observations by applying (3.4.9) to the "unperturbed" light computation:

1. One gas stream ejected is from that part of the secondary's surface, which is heated by the hotter component. From the observed light curve, i.e., transmitted light I_T, Duerbeck estimated that about 10% of the light of the secondary is lost by scattering. This yields $x = 10^{-12}$ cm and $n_e = 1.5 \cdot 10^{11}$ cm^{-3} and agrees well with the values $10^{11} \leq n_e \cdot \text{cm}^3 \leq 2.6 \cdot 10^{11}$ derived from the equivalent widths of hydrogen lines.

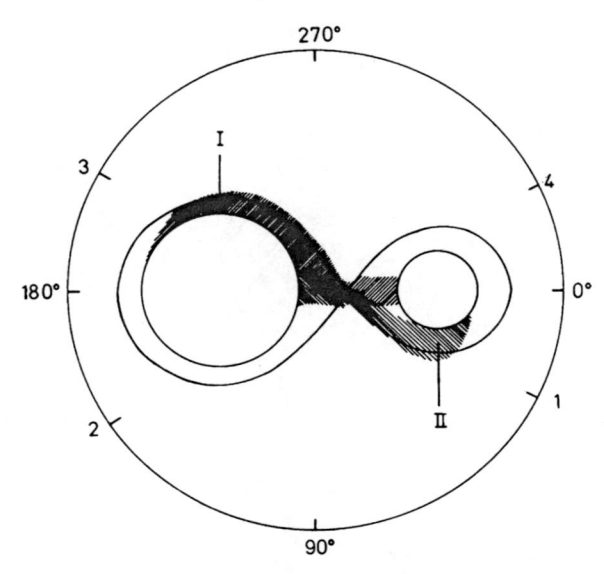

Fig. 3.24 Gas streams in *VV Orionis*. This figure, Fig. 10 in Duerbeck (1975), shows the two gas streams involved in the second hypothesis. Courtesy H. W. Duerbeck

2. Two gas streams (indicated as I and II in Fig. 3.24) exist, i.e., in addition to the hypothesis above, another gas stream flow from the primary to the secondary should be present. This explains both the disturbance at phase 0.35 and the asymmetry of the primary minimum.

In both cases, the discussion involves only the geometry of the gas stream projected onto the orbital plane. Although the model produces plausible results and qualitatively explains features in the observed light curve, it might be worthwhile to reanalyze the data with a modern light curve program based on Roche geometry including some special features for streams or clouds.

3.4.4.3 Disks and Rings

Algols and other binaries with slow or intermediate mass transfer develop thin disks. To consider the disks and rings in the model requires the computation of the gas flow, radiative properties (in particular, emission line strengths and profiles), and the calculation of the spectral energy distribution by spatial integration taking into account Doppler shifts and eclipses. In some cases, disks can be approximated as non-self-gravitating and governed by celestial mechanics in the first approximation, but with nonnegligible viscous and pressure interactions.

Różyczka & Schwarzenberg-Czerny (1987) present and solve two-dimensional hydrodynamical models for the stream–disk interaction in cataclysmic binaries, focusing on the collision region. As an example of a fully three-dimensional stream–accretion disk computation we refer to Dgani et al. (1989) who computed the

3.4 Sources and Treatment of Perturbations 143

time-dependent interaction between the stream from the inner Lagrangian point and the accretion disk, and the response of the disk to an increase in the mass transfer rate.

Terrell (1994) applied the method of *Smoothed Particle Hydrodynamics* (SPH) to the problem and also made a detailed computation of the spectral energy distribution of the radiation emitted by the gas. Pressure gradients were calculated in the usual manner for SPH [see Monaghan (1992) and references therein], but the viscosity was computed with a newly developed algorithm [Terrell & Wilson (1993); Terrell (1994)]. Viscosity was modeled by allowing particles within a specified distance of one another to exchange momentum, with close encounters being stronger than more distant ones. This scheme avoids problems inherent in earlier models such as artificial acceptance/rejection of interactions based on a Cartesian grid, as in the Lin and Pringle scheme. Terrell also computed Hα line profiles by coupling the radiative transfer code of Drake & Ulrich (1980) to his hydrodynamics code and found

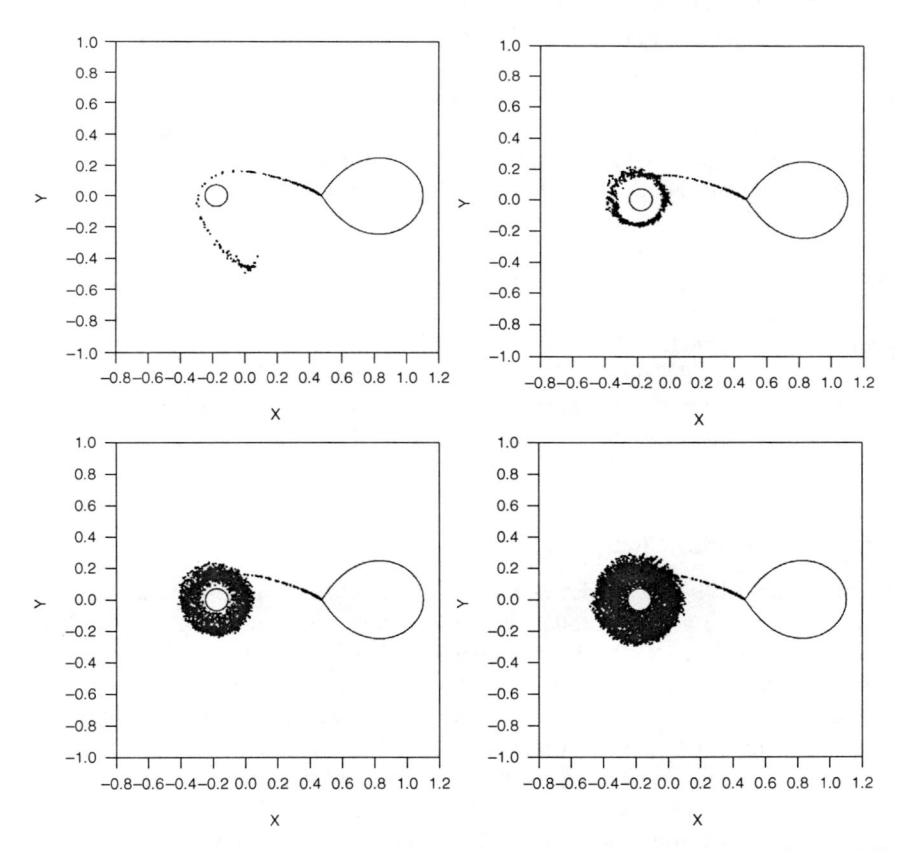

Fig. 3.25 Disk formation in *SX Cassiopeiae*. This plot, part of Fig. 1 in Terrell and Wilson (1993, p. 32), shows the disk formation at 0.32, 1.11, 3.18, and 7.96 orbital revolutions caused by an episodic mass flow event. Courtesy D. Terrell

reasonably good matches with the observed profiles of several Algol-type binaries. He also used the more extensive radiative transfer code of Ko & Kallman (1994) which treats several atomic species, but further work must be done to improve the efficiency of the calculations. Terrell & Wilson (1993) computed the disk matter distribution and motion. Figure 3.25 shows their result for *SX Cassiopeiae* disk images at 0.32, 1.11, 3.18, and 7.96 orbital revolutions. Based on these calculations, their objective is to derive observable quantities such as spectralline profiles or polarization properties of photospheric radiation scattered by the disk.

3.4.4.4 Stellar Winds

In addition to the radiation pressure effects discussed in Sect. 3.1.5.4, additional complicated physics is required in hot binary systems. If the components of a binary are hot and produce a sufficiently large radiation pressure as in some Wolf–Rayet stars [cf. Pollock (1987) and White & Long (1989)], they establish radiation-driven hypersonic counterflowing stellar winds [see, for instance, Castor et al. (1975) or Hearn (1987)]. An X-radiation emitting sheet (Prilutskii & Usov 1976) between both stars can block out light emitted toward the observer by one of the stars and significantly modify the light curve and other binary observables. Interested readers should consult Siscoe & Heinemann 1974 and Campbell (1997).

Kallrath (1991) and Stevens et al. (1992) computed the hydrodynamical properties of such colliding winds and the properties of the contact discontinuity established in binary systems. Although Neutsch et al. (1981) and Neutsch & Schmidt (1985) use only simple models for the interface in the context of binary star analysis and compute the effect of this boundary layer on line profiles in *HD 152270*, their approach is nevertheless instructive.

In hot binaries the opacity increases at the wind interface so as to affect the radiative flux between components. The interface might be considered an attenuating region and could be incorporated into the model as described in Sect. 3.4.4.5.

3.4.4.5 Attenuating Clouds

As in Wilson (1998, 1999) let us use the term "cloud" (more precisely defined, below) to refer to a circumstellar light-attenuating gas or dust region. The attenuation might be due to Thomson scattering, to scattering with (arbitrary) power law wavelength dependence (such as Rayleigh scattering), to continuum opacities, and true absorption, e.g., discrete wavelength absorption features as ultraviolet Balmer lines. Circumstellar matter follows dynamical trajectories, so we might expect there to be little effect from attenuating regions that are fixed in a coordinate frame that rotates with the binary, but such is not entirely the case. A few binaries [e.g., *RZ Sct*, *AX Mon* (Elias et al. 1997)] have approximately stationary loci of circumstellar gas that extinct light and distort the light curves. A boundary layer produced by colliding stellar winds in hot binary systems is another example. In general, the loci may be stream–stream, stream–disk, or stream–wind interaction regions. Efforts to represent such light curve distortions via bright or dark star spots sometimes rule out a spot explanation and point to essentially fixed attenuation regions (hereafter, "clouds,"for brevity). The Wilson–Devinney model includes n_c spherical semi-transparent clouds specified by their locations (x, y, z) (in a rectangular frame that corotates with the stars; coordinate frame C_1 defined in Sect. 3.1.1), cloud radius r, density ρ, electron density n_e, and mean molecular weight per free electron μ_e. The part of the line-of-sight that passes through the various clouds is computed individually for the lines-of-sight to all surface points and individually for all clouds. Regions of variable density can be made by nesting individual clouds. Regions of nonspherical shape can be approximated by overlapping spherical clouds. Each cloud is allowed its own attenuation law, [41] whose general form is

$$\frac{d\tau}{ds} = \sigma_e n_e + (\kappa_\lambda + \kappa_{sb})\rho, \qquad (3.4.10)$$

where τ is the optical thickness, σ_e is the Thomson scattering cross-section per electron, s is the distance along the line-of-sight (in cm), κ_λ is a wavelength-dependent opacity, and κ_{sb} is an additional opacity for a specific passband (κ in cm^2/g). The κ_{sb} term might represent, for example, opacity due to absorption lines averaged over a particular passband. The κ_λ term is

$$\kappa_\lambda = \kappa_0 \lambda^\alpha, \qquad (3.4.11)$$

where κ_0 and α are input quantities. Each cloud has its individual κ_0, α, and κ_{sb}. However, to make it easy to change the κ_{sb} of all clouds together, the κ_{sb}'s are not entered directly as individual members. Instead we enter an overall κ_{sb} and the fractions f_c that applies for each cloud. Thus

$$\kappa_{sbc} = f_c \kappa_{sb}, \qquad (3.4.12)$$

where all the f_c can be unity if κ_{sbc} is to be the same for all clouds c, or non-unity if κ_{sbc} is to differ from cloud to cloud. The model computes absolute lengths from the system geometry, including the orbital semi-major axis. A first application is to *AX Monocerotis* (Elias et al. 1997).

3.5 Modeling Radial Velocity Curves

Motu proprio (By one's own motion)

Radial velocities usually are extracted from spectra taken at modest spectral dispersions (1–3 nm/mm of reciprocal linear dispersion) through the averaged measure-

---

[41] At present the clouds only attenuate starlight that pass through them, but they may be made to scatter starlight toward the observer in a future program version.

ments of absorption line shifts on CCD images. Since they first came into use, the physical size of CCD chips has restricted their wavelength registration range, and as a consequence, velocities have been based on fewer lines than was the case with photographic or even Reticon detectors of the more recent past.

A powerful means of acquiring good spectral resolution over a large spectral range is through Echelle spectra. These spectra have a large number of orders, each of which has a short spectral range. A cross-disperser is needed to separate the orders. Both background subtraction and ghost images require extra attention in processing, but the advantages of this two-dimensional spectroscopic technique outweigh the disadvantages if enough flux is available. A frequently used method is cross-correlation (see Sect. 2.2.1) of the program star spectrum against that of a velocity standard, with due allowance taken for the reduction of velocities to the Sun. The resulting radial velocities indicate the Doppler shifts of one or both components. Whether Doppler shifts of both components are visible depends on the relative brightness of the less luminous component.

Proximity effects in close binary systems distort not only light curves but also radial velocity curves. The curves are affected by a star's nonsphericity, surface brightness distribution, line strength variation over the surface, aspect dependence of spectral line strength, and eclipses. In binaries with strong tidal distortions or reflection effect heating, these effects have to be accounted for when estimating masses or other quantities derived from velocities.

Even in the case of rotating spheres in a binary the eclipse of part of one of the components results in a phase-dependent shift of the estimated velocity and reflects the dominance of one of the eclipsed star's limbs. In particular, the rotation of the partially eclipsed component distorts the velocity curve. Schlesinger (1909, p. 134) gives already a clear explanation of this *"rotation effect"*: "The rotation of the bright star has another consequence in certain parts of the orbit. In general we obtain light from the whole disk and the observed velocity is equal to that of the center of the star. Just before and just after light minimum, however, this is not the case; before the minimum the bright star is moving away from us and part of its disk is hidden by the dark star. The part that remains visible has on the whole an additional motion away from us on account of the rotation; the observed velocity will therefore be greater than the orbital. On the other hand, just after minimum the circumstances are reversed so that the observed velocity is less than the orbital." Now this effect is called the *Rossiter effect*[42] after Rossiter (1924).

Radial velocity measurements differ from those expected for point sources in other ways as well. That is illustrated in Fig. 3.26 which shows the radial velocity

---

[42] Rossiter also uses the term "rotational effect." For β Lyrae he measured an amplitude of 13 ± 2 km/s and in his paper he wrote: "This is called *rotational effect*..... This is, I believe, the first time that this rotational effect has been isolated and measured and eliminated from the least-squares adjustment of the elements. Professor Schlesinger has suspected it in δ Librae (he referred directly to the page 134 of Schlesinger's paper) and ...."

3.5 Modeling Radial Velocity Curves

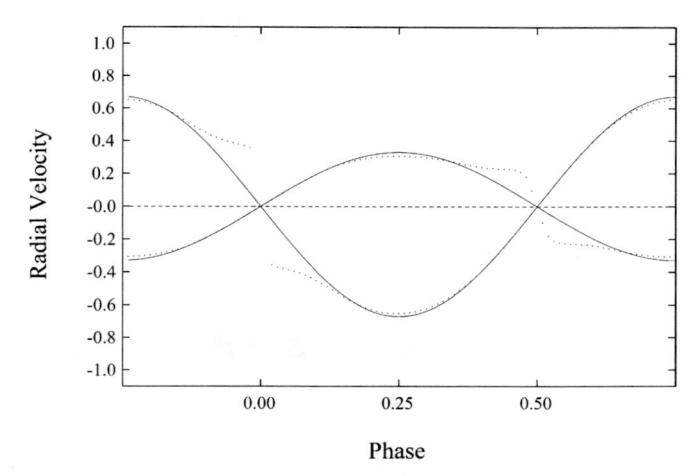

Fig. 3.26 Modeling of the Rossiter effect in *AB Andromedae*. This plot, produced with Binary Maker 2.0 using the parameter file *aband* in the examples collection, shows the radial velocity curves for a point source (*solid line*) and a distorted binary (*dotted line*)

curves of the W UMa-type[43] EB AB Andromedae. The solid line shows the radial velocity curve we would expect if both stars were point sources. The dotted line is the one including proximity effects. Note the great difference near eclipse phases.

Any program capable of computing photometric light curves should be able to compute radial velocity curves, with only slight extra programming. Intuitively, we expect that measured radial velocities do not correspond to those associated with point masses moving on Keplerian ellipses but rather to the "centers-of-light." The effective radial velocities are the average radial velocities weighted with the local intensities over the stellar surface.

In order to derive results which are independent of the period P and semi-major axis a, the radial velocities may be treated as follows. For radial velocity curves produced by a pair of point masses (see Fig. 3.27) it has been common practice to consider

$$\omega a = \frac{K_1 + K_2}{\sin i}, \quad q_{sp} = \frac{M_2}{M_1} = \frac{a_1}{a_2} = \frac{K_1}{K_2}, \tag{3.5.1}$$

where the undistorted radial velocity amplitudes K_1 and K_2 directly give the spectroscopic mass ratio q_{sp}. The quantity ω (not to be confused with the argument of periastron) is the time-averaged angular velocity of the orbital motion. If we consider deformed stellar surfaces and compute the radial velocity contribution from each surface element, it is not appropriate to keep K_1 and K_2 in the analysis because they are not uniquely defined in the presence of proximity effects. The

---

[43] The Rossiter effect is very significant in strongly distorted eclipsing binaries with short periods as is the case in W UMa-type stars.

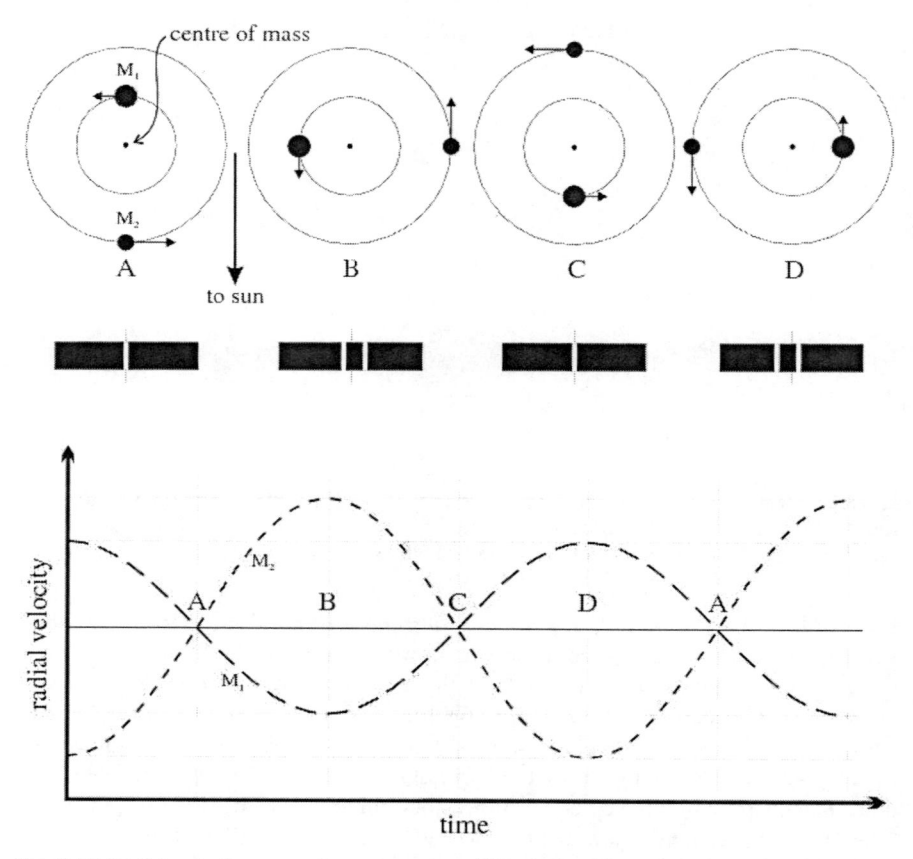

Fig. 3.27 Radial velocity curves for point masses. This figure (Courtesy J. D. Mukherjee) shows the radial velocities for a binary system with two point masses

natural parameters to replace K_1 and K_2 are the directly physical parameters a_1 and a_2. An arbitrary surface element $d\sigma$ at position $\mathbf{r} = \mathbf{r}(r; \lambda, \mu, \nu)$ on the surface of the primary component, seen from the direction $\mathbf{s} = (s_x, s_y, s_z)$, produces the contribution $V(d\sigma)$ to the radial velocity curve

$$\frac{1}{\omega a} V(d\sigma) = v_c + v, \quad v_c = -\frac{q}{1+q} s_y = -\frac{a_1}{a} s_y, \quad (3.5.2)$$

where v_c is the (dimensionless) constant radial velocity of the center-of-mass of the primary, and v is the local radial velocity of $d\sigma$ in a corotating coordinate system with origin in the center-of-mass of the primary component. Note that velocities are given in units of $a\omega$. In those units, in a rotating coordinate frame centered at component 1, v is given by [compare Wilson & Sofia (1976, pp. 183–184)]

3.5 Modeling Radial Velocity Curves

$$v = \frac{r}{a}(\lambda s_y - \mu s_x), \qquad (3.5.3)$$

where r is the distance from the origin, and λ, μ, s_x, and s_y are the direction cosines defined in (3.1.1) and (3.1.15). The effective radial velocity is now computed by averaging over the surface, weighted by the intensity I_λ of $d\sigma$ in the direction s to the observer. A finite grid-point value is computed for each phase Φ:

$$\Delta V_\lambda = \frac{\int I_\lambda v d\sigma}{\int I_\lambda d\sigma}, \qquad (3.5.4)$$

where $\int I_\lambda d\sigma$ is the integrated flux $\ell(\Phi)$ at phase Φ. This yields the radial velocity curve $V(\Phi)$:

$$\frac{1}{\omega a} V(\Phi) = V_c(\Phi) + \Delta V_\lambda(\Phi) \qquad (3.5.5)$$

of the primary, with an analogous expression for the secondary. In absolute units we eventually get

$$V_j(\Phi) = -a_j \omega \sin\Phi \sin i + \omega a \Delta V_{\lambda j}(\Phi) + \gamma, \quad j = 1, 2. \qquad (3.5.6)$$

Here a constant velocity, γ, is added to account for the velocity of the center-of-mass of the binary system.

More realistic modeling of the proximity effects and their influences on the radial velocity curves might consider the variation of line strength. To measure line strength the line equivalent width is used. On stars with negligible winds, it depends on effective temperature, surface gravity, and aspect angle alone. The formalism discussed above computes the global mean radial velocities based on flux-weighted local velocities, but ideally they should also be weighted by equivalent width. Van Paradijs et al. (1977) seem to be the first to have done so. Van Hamme & Wilson (1994) explored the effects of line strength weighting for a small number of binaries and found that the difference between flux-only and [flux/equivalent width] weighted proximity effects were, at most, a few percent of the velocity amplitude in those examples. Van Hamme & Wilson (1997) extended the approach to eccentric orbits and asynchronous rotation. Again the result is confirmed that for high mass X-ray binaries or binaries with extreme mass ratios the effect of the variation of line-strength is of the order of a few percent. Nevertheless, observers are encouraged to list the lines used to measure the radial velocities in their papers. In order to exploit the line-weighting formalism correctly it is recommended to publish the radial velocities for individual lines when line-strength effects are expected to be significant.

3.6 Modeling Line Profiles

Divide et impera *(Divide and conquer)*
after Philip of Macedon (359–336 B.C.)

Accurate modeling of line profiles enables us to estimate stellar rotation rates as has been discussed in Sect. 2.2.3. In the context of light curve modeling it is natural to use a theory of stellar line broadening for the local profile and a binary star model for the rotational theory (Mukherjee et al. 1996). That strategy avoids the overhead produced by stellar atmosphere theory. The relevant input quantities are the effective temperature, T, the damping constant, Γ, including both natural and collisional damping, the number of absorbers, N_f, the microturbulent velocity, v^{tur}, the ratio of continuum scattering opacity to total continuum opacity, ρ, the monochromatic ratio of line opacity to continuum opacity, β_ν, at frequency ν, and the fraction, $1-\varepsilon$, of absorbed photons which are scattered. This leads to the auxiliary quantity, λ_ν, defined as

$$\lambda_\nu := \frac{(1-\rho) + \varepsilon \beta_\nu}{1 + \beta_\nu}. \tag{3.6.1}$$

Mukherjee et al. (1996, Sect. 3) obtain the emergent intensity, $I_\nu(0, \mu)$, for given μ defined in equation (3.2.22):

$$I_\nu(0, \mu) = (a + p_\nu \mu) + \frac{\left(p - \sqrt{3}a\right)(1 - \lambda_\nu)}{\sqrt{3}\left(1 + \sqrt{\lambda_\nu}\right)\left(1 + \sqrt{3\lambda_\nu}\mu\right)}. \tag{3.6.2}$$

The expressions for evaluating a, b, and the probability, p_ν, that a photon is thermalized in an interaction with an atom or ion are given in Mihalas (1978, pp. 312–313).

For the continuum intensity, $I_c(0, \mu)$, Mukherjee et al. obtain

$$I_c(0, \mu) = (a + b\mu) + \frac{\left(b - \sqrt{3}a\right)\rho}{\sqrt{3}\left(1 + \sqrt{1-\rho}\right)\left(1 + \sqrt{3(1-\rho)}\mu\right)}. \tag{3.6.3}$$

Finally, they obtain the residual intensity, relative to the continuum:

$$r_\nu(\mu) = \frac{I_\nu(0, \mu)}{I_c(0, \mu)}, \tag{3.6.4}$$

which is then integrated by their light curve program to yield the residual flux. The velocity needed for the calculation of a line profile is ΔV_λ as computed in (3.5.4).

Line profiles are generated for each surface element. Figure 3.28 shows how the intrinsic line profile changes as T, Γ, N_f, and v^{tur} are changed. To get the line profile for the entire star the local line profiles are weighted according to the flux from each of these areas. Figure 3.29 shows an example of profiles computed at

3.6 Modeling Line Profiles

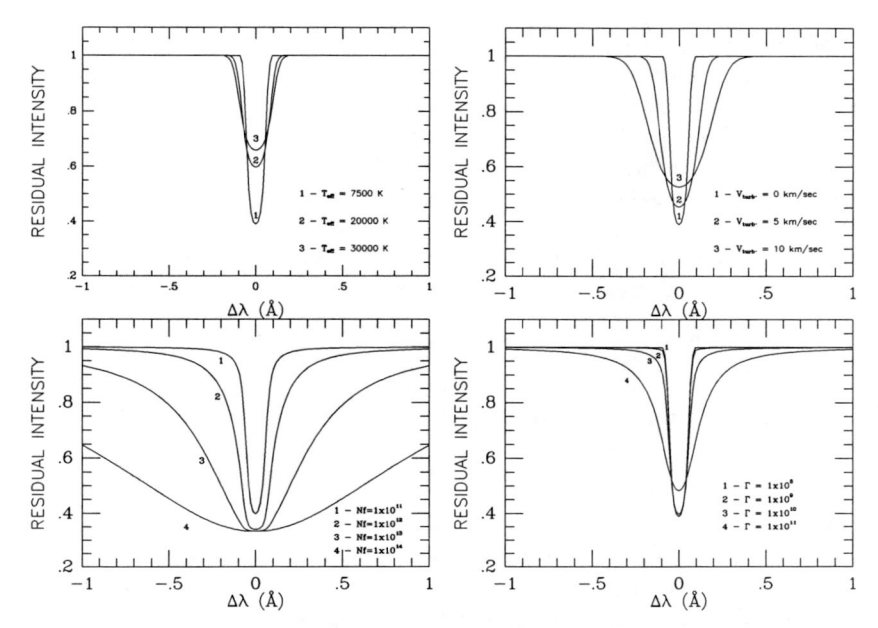

Fig. 3.28 Variation in the intrinsic line profile. This plot, Fig. 1 in Mukherjee et al. (1996), shows the influence of several parameters on the intrinsic line profile ($\mu = 1$): clockwise from lower left, the number of absorbers, effective temperature, microturbulent velocity, and the damping constant. Courtesy J. D. Mukherjee

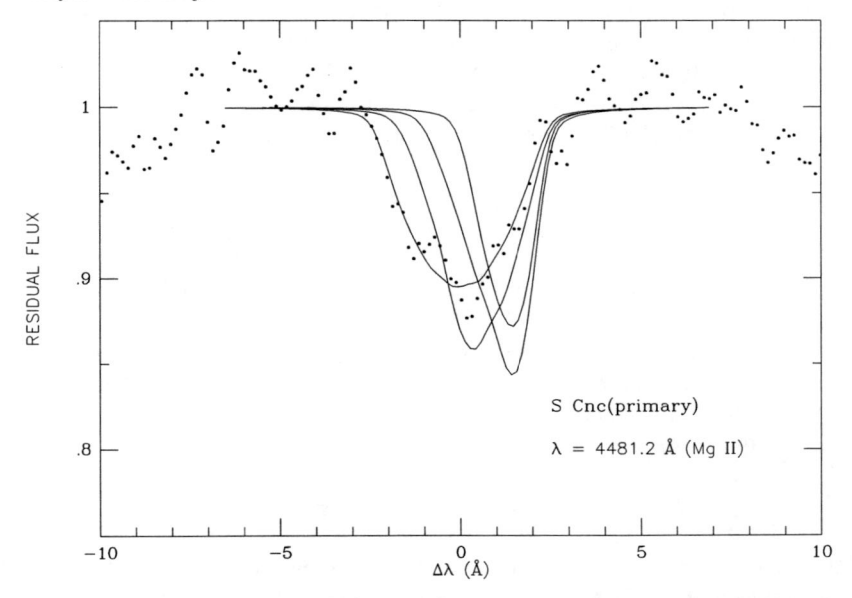

Fig. 3.29 Line profiles and Rossiter effect. This plot, Fig. 4 in Mukherjee et al. (1996), shows the Rossiter effect on the line profiles in *S Cancri*. The phase of observation is 0.184. The orbital shifts corresponding to phases 0.05 and 0.184 differ by too small an amount to show significantly. Courtesy J. D. Mukherjee

several phase shifts (0.01, 0.02, 0.03, 0.05) for *S Cnc*. The phase of the observations is 0.184. The orbital shifts corresponding to phases 0.05 and 0.184 differ by too small an amount to show up significantly. In order to compare the computed line profiles with observed ones they are convolved with the instrumental profile. The instrumental profile is usually represented by a Gaussian given in terms of full width at half maximum. The last step in getting the final line profile is to take phase smearing into account by averaging computed line profiles from the beginning and end of the observation interval. In case the interval is sufficiently small the profile can be calculated at the phase corresponding to the middle of the interval.

3.7 Modeling Polarization Curves

Embarras de choix   (*Embarrassment of choice*)

As an example of how additional observables may be incorporated into a light curve model we can consider circumstellar polarization based on scattering in optically thin stellar envelopes.

Chandrasekhar (1946a, b) stimulated interest in polarized radiation from binary stars when he showed that eclipses would break the disk symmetry and allow limb polarization to be observed. In a binary system with a nonspherically symmetric circumstellar envelope, various processes (such as scattering, reflection, and Zeeman splitting of spectral lines in the presence of a magnetic field) can produce a small contribution to the intrinsic linear and/or circular polarization in the total light from the system. The type and degree of polarization will vary over the binary period and will depend on the polarizing mechanism (Thomson scattering, Rayleigh scattering, and others), the distribution of scattering material, and the aspect of the system as seen by an observer.

In principle, polarization data provide a chance to derive the inclination i and thus masses of binary stars in the case of *noneclipsing* spectroscopic binaries because in this case i is not available from either light curves or radial velocity curves. However, the main polarization mechanism(s) need to be identified and the model must be rather good.

In a binary star system, two main sources of polarization have to be considered: *Photospheric polarization*, namely the limb polarization effect predicted by Chandrasekhar (1946a, b) and *circumstellar polarization*. The additional observables are the Stokes quantities

$$Q = P\cos 2\theta_p, \quad U = P\sin 2\theta_p, \qquad (3.7.1)$$

where θ_p is the position angle (measured conventionally counterclockwise from North) of maximum signal as an analyzer is rotated, and

3.7 Modeling Polarization Curves

$$P = \frac{F_{max} - F_{min}}{F_{max} + F_{min}}. \qquad (3.7.2)$$

Net photospheric polarization should be zero for a centrally symmetric star face even if there is significant polarization at the limb, as expected when electron scattering is important. Eclipses break the symmetry and can lead to net observable polarization, called the limb eclipse effect. Although many attempts have been made to detect the limb eclipse effect, its unambiguous detection remains elusive. Kemp et al. (1983) claimed to have detected it in *Algol*, but this claim is not universally accepted. Wilson & Liou (1993) presented an analysis of the Kemp et al. (1983) observations and argued that their model, based on the Wilson–Devinney light curve model, showed that the eclipse effect caused a significant part of the rapid variation of the Stokes quantities during primary eclipse.

Wilson & Liou (1993) used the relations outlined in Brown et al. (1978) to compute polarization arising from circumstellar matter. Later, Terrell and Wilson (unpublished) combined a gas flow program [Terrell (1994), Terrell & Wilson (1993)] with the Wilson–Liou program to compute observable polarization curves. As can be seen in Fig. 3.30, the polarization curves are very sensitive to the location of and physical conditions in the circumstellar gas. In these simulations, the ionization scheme was relatively simple, but the results show that polarimetry can be a very effective tool. The ionization is computed either from the Saha equation (telling us the relative populations of two adjacent stages of ionization)

$$\frac{N_{i+1}}{N_i} = B \frac{(kT)^{3/2}}{n_e} e^{-\chi_i/kT}, \qquad (3.7.3)$$

or the Boltzmann–Saha equation (giving the number of atoms available for a transition and so to produce a given spectral line)

$$\frac{N_{is}}{N} \approx \frac{N_{is}}{N_{i-1} + N_i + N_{i+1}} = \frac{N_{is}}{N_i} \bigg/ \left[\frac{N_{i-1}}{N_i} + 1 + \frac{N_{i+1}}{N_i} \right] \qquad (3.7.4)$$

with

$$\frac{N_{is}}{N_i} = C e^{-\chi_s/kT}. \qquad (3.7.5)$$

The Saha and Boltzmann–Saha equations involve the following symbols: N_{is} the relative number of atoms in any state of excitation, s, of a stage of ionization i; N, the sum populations of all ionization states; n_e, the electron density; χ_i, the ionization potential between adjacent stages of ionization; and some proportionality constants, B and C, which include several atomic constants.

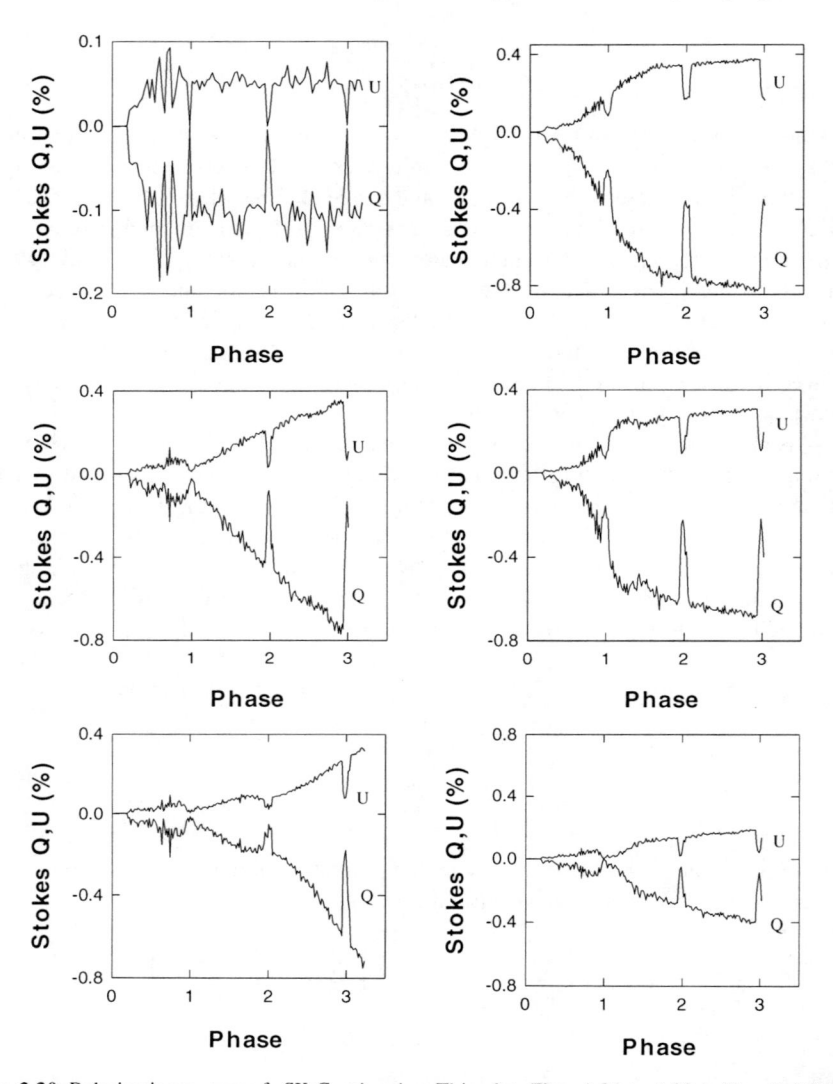

Fig. 3.30 Polarization curves of *SX Cassiopeiae*. This plot, Figs. 4-24 to 4-29 in Terrell (1994), shows mass transfer events of several durations. The upper curves are the *U* curves, the lower curves the *Q* curves. Courtesy D. Terrell

Polarization observations of Algol-type binaries are relatively rare for several reasons. Although bright, Algol has low circumstellar activity. The more active systems are too dim for existing telescope/polarimeter combinations. However, polarimetry has seen increased interest in recent years, and the combination of more sensitive polarimeters and the new 8-m class telescopes will be able to collect and detect enough photons to achieve reasonable signal-to-noise ratios. Although there are very few observations to compare to the model at present, the availability of larger telescopes equipped with efficient polarimeters promises to make polarimetry a very powerful probe of interacting binaries in the next few years.

A desirable polarization model would be based on the solution of the hydrodynamic flow equations, including viscosity and pressure, providing the electron densities of circumstellar gas elements for many points in space. Today there are no well-developed programs to do the full dynamical circumstellar computations.[44] In order to compute Q and U, limb and circumstellar polarization are combined as in Wilson & Liou (1993). The limb polarization is computed from the Wilson–Devinney light curve model, and the circumstellar polarization is computed by the relations given in Brown et al. (1978) assuming optically thin pure electron scattering.

3.8 Modeling Pulse Arrival Times

Binaries that contain an X-ray pulsar are very rich data sources. Besides radial velocity and light curves they also provide pulse arrival times. As in the previous sections, we need to be able to compare observed and theoretical quantities, in this case pulse arrival times in addition to other observables. Therefore, following Wilson & Terrell (1998), we write the pulse arrival time in Heliocentric Julian Day Number, τ, as

$$\tau = \tau_{\text{ref}} + S(n - n_{\text{ref}})P_p + \Delta t - \Delta t_{\text{ref}}, \tag{3.8.1}$$

with τ_{ref} being the arrival time of a reference pulse (which defines pulse phase zero) and Δt and Δt_{ref} the light time delays due to orbit crossing for a given pulse and the reference pulse, respectively. S is the number of days in a second of time (1/86400), P_p is the pulse period in seconds, and n is an integer assigned to an observed pulse.

Equation (3.8.1) establishes our model for computing the pulse arrival times. The delays Δt depend implicitly on the common light curve parameters q, i, a, and e according to

$$\Delta t = R_\odot \frac{Sadq \sin i}{c(1 + q)} \cos(360°\theta), \tag{3.8.2}$$

where $R_\odot$ is the radius of the Sun in kilometers, d is the instantaneous separation of the two stars in units of a, θ is the geometrical phase defined in (3.1.20), and c is the speed of light in kilometers per second.

---

[44] For some specific applications, it certainly could be done – probably within a year or less of development. The real problem is that there are almost no published observations to test the idea (or sufficiently accurate and numerous ones, covering at least several consecutive orbits). So there is little motivation at present to do the calculations.

Wilson & Terrell (1998) proceed as follows to compute Δt. The pulse ephemeris is used to obtain time. The first two terms on the right-hand side of (3.8.1) are interpreted as time, τ_p, kept by the pulsar clock and measured in Heliocentric Julian Day Number, i.e.,

$$\tau_p = \tau_{\text{ref}} + S(n - n_{\text{ref}})P_p. \quad (3.8.3)$$

This time, t_p, is coupled to the orbital phase, Φ, according to (2.1.1):

$$\Phi = \Phi_s + frac\left\{\frac{\tau_p - T_0}{P}\right\}, \quad (3.8.4)$$

where T_0 denotes the reference epoch, P is the orbital period, and Φ_s is a constant offset which in most cases is simply zero. The next step is to compute the mean anomaly M according to (3.1.34). This gives the true anomaly, υ, the geometrical phase, θ, and, finally, the separation, d.

So, the pulse arrival model involves the following set of adjustable parameters:

$$\tau_{\text{ref}}, q, i, a, e, \omega, E_0, P_p, P.$$

As noted by Wilson & Terrell (1998), the parameters can be adjusted by differential corrections and analytic derivatives exist for nine parameters.

An analysis by Wilson & Terrell (1994, 1998) of data from a binary containing an X-ray pulsar is briefly outlined in Sect. 7.3.1.

3.9 Self-Consistent Treatment of Parallaxes

As pointed out in Chap. 1 (page 22) the distance D or parallax π of a binary can (in favorable cases) be derived if both light and radial velocity curves are available. If parallax data are available for the binary, for instance, from the Hipparcos mission [cf. Rucinski & Duerbeck (1997)], or if it is a member of an ensemble (e.g., star cluster, galaxy) with known distance, $n_\pi \geq 1$ measured values π_k of the parallax might be available. Thus, on the one hand, the parallax is a systemic observable and, on the other hand, it is a model parameter to be estimated by the least-squares analysis.

Let us now consider the parallax π as a parameter in the EB model, and let D denote the distance of the binary connected to π by

$$\pi = \frac{k_\pi}{D}. \quad (3.9.1)$$

If the distance is measured in parsecs and the parallax is measured in arc-seconds, the constant is $k_\pi = 1$. To couple the parallax to the binary model it is more convenient to measure the distance in units of the semi-major axis a.

First, we include the parallax both as an observable and also as an adjustable parameter. Second, instead of the normalized light or flux $\ell(\Phi)$ usually used in light curve analysis, the flux $\ell_D(\Phi)$ in absolute physical dimensions [energy/time/wavelength/unit receiver area] must be used in the least-squares analysis. Note that this requires absolute calibration of the photometric systems as discussed in Sect. 5.1.2.3.

The addition of parallaxes to light curve analysis slightly extends the least-squares function (see Chap. 4). The contribution of parallax as an observable is

$$\sum_{k=1}^{n_\pi} w_k \left(o_{\pi k}^{\text{obs}} - o_\pi^{\text{cal}}\right)^2 = \sum_{k=1}^{n_\pi} w_k \left(\pi_k^{\text{obs}} - \pi^{\text{cal}}\right)^2, \quad \pi^{\text{cal}} = \pi, \tag{3.9.2}$$

where the weights are derived from the standard deviations of the parallaxes according to (A.3.4). With a derivative-based least-squares algorithm the analytic derivatives are

$$\frac{\partial o_\pi^{\text{cal}}}{\partial \pi} = 1, \quad \frac{\partial o_{\text{rv}}^{\text{cal}}}{\partial \pi}(\Phi) = 0, \quad \frac{\partial \ell_D}{\partial \pi}(\Phi) = -2\frac{\pi}{k_\pi}\ell_D(\Phi), \tag{3.9.3}$$

which follows directly from (3.2.51) and (3.2.1). As the computation of radial velocities does not explicitly depend on the parallax, the radial velocities' partial derivatives are zero. An obvious point is that, in the absence of radial velocity curves (semi-major axis a is not known in that case) π cannot be determined because $a/D = a\pi/k_\pi = \text{const}$ and a is unconstrained.

3.10 Chromospheric and Coronal Modeling

The extension of *observables* to include spectrometric data and very narrow line profile information, the availability of X-ray and ultraviolet data from space platforms, infrared and even radio data, make it possible to model the details of stellar chromospheres and coronae with improved accuracy.

It is well known that strong spectral lines, such as Hα, or Ca II H&K, originate much higher in the photosphere than does the continuum. From rocket and space platforms, such as NRLs stigmatic solar spectrograph data on Skylab, the far ultraviolet emission regions have been mapped in great detail. Of particular interest are the He II spectroheliographic images at 30.4 nm which map coronal holes. Figure 3.31 above shows the Sun in the far-ultraviolet region.

Ability to model these features requires capability in a light curve code that does not exist. The disk of the Sun, for example, is silhouetted by the active regions and chromospheric network behind the limb; it is dark in the far-ultraviolet and X-ray regions. The emission in many passbands comes exclusively from active regions, and the optical depth may be so low outside these regions that only patches of the Sun may be visible in an otherwise dark field. The emission may arise

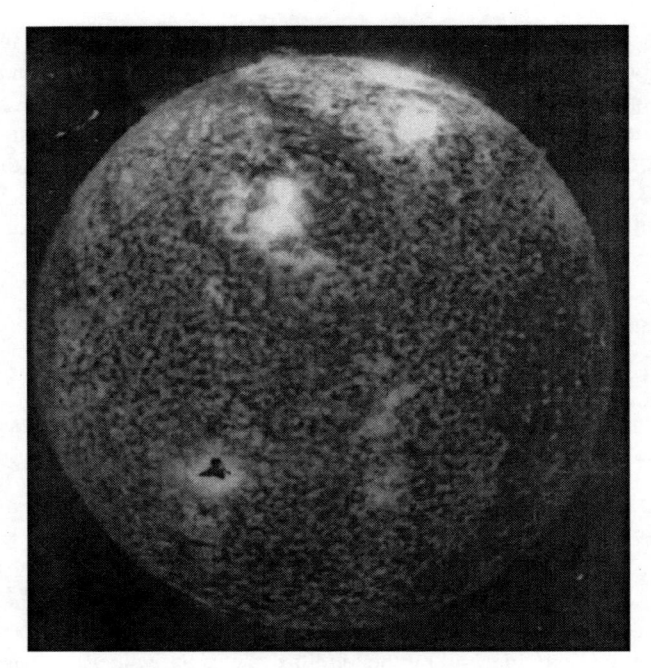

Fig. 3.31 He II spectroheliographic image of the Sun. This Skylab photograph (Experiment S082A) shows the Sun at 30.4 nm. Courtesy R. Tousey, US Naval Research Laboratory (NRL)

solely from an annulus around the dark disk. There is every reason to suspect that a similar situation is to be found in other stars with convective envelopes, at least of solar type. As described in Strassmeier (1997, Chap. 9), many observational data, e.g., the Ca II H&K lines, are also available for active stars and active chromospheres.

3.11 Spectral Energy Distribution

Bona diagnosis, bona curatio   (*Good diagnosis, good cure*)

Plotting flux (essentially brightness) versus frequency or wavelength of light of an astronomical object gives its spectral energy distribution (SED). Studies of individual *EB*s (cf. Siviero et al. (2004) and Marrese et al. (2005) for example) have shown that including flattened SEDs may be used as an external check of the model solution, where individual spectral lines of echelle spectra are compared with Kurucz (1993) model atmospheres. Flattened spectra are used instead of flux-calibrated spectra because *observed* echelle spectra are very challenging to flux-calibrate. There are good ways to flux-calibrate spectra but echelle spectra are used mainly for RV studies. Hence flattened spectra. Despite the fact that some information is lost, fortunately there are many spectral lines and their profiles and equivalent widths are strongly dependent on T_{eff}, $\log(g/g0)$, v_{rot}, and [M/H].

SED data are useful in solving the inverse *EB* problem as discussed by Prša & Zwitter (2005b). Their program PHOEBE described in Sect. 8.2 already takes a step in that direction by using a synthetic spectra database to test whether flattened, wavelength-calibrated spectra match synthetic spectra within a given level of significance. As the spectra depend on $[T_{\text{eff}}, \log(g/g0), v_{\text{rot}}]_{1,2}$ and metallicity, they can in favorite cases provide valuable insight to break the often experienced problem of degeneracy among light curve parameters (often, Roche potentials Ω_1, Ω_2, and inclination i), or to support (for well determined radii) the determination of the yielded synchronicity parameters F_1 and F_2, because the only way to compensate for the change in rotational velocities for any predetermined radii is to break the corotation presumption. This may be especially important in analysis of well-detached systems, as demonstrated by Siviero et al. (2004).

3.12 Interstellar Extinction

Το αγκάθι από μικρό αγκυλώνει (A thorn stings even if it's small)

In the field of *EB* analysis interstellar extinction and reddening usually have not been treated as part of *EB* models. A binary star appears fainter if its light passes through regions of the interstellar medium filled with dust and gas particles causing absorption and scattering. If scattering by dust or grain solids is the main cause the process is roughly described by Mie scattering. As in Mie scattering the amount of scattered light in optical wavelengths decreases with wavelength, more blue light is removed, i.e., the apparent *B* brightness decreases (percentwise) more than the apparent *V* brightness, and objects appear reddened. Because the difference $B - V$ increases with extinction, the color excess

$$E(B - V) = (B - V) - (B - V)_0 \quad (3.12.1)$$

is a useful measure of interstellar extinction. The quantity $(B - V)_0$ is the intrinsic color index of the object. The $(U - B)$ color excess is defined similarly. The effect of interstellar extinction on the *V* band is described by the attenuation, A_V, expressed in magnitudes. As the ratio $R = A_V/E(B - V) \approx 3.1$ is a good approximation to most directions across the sky, an estimated value of A_V is derived directly from the observed color excess by

$$A_V = 3.1 E(B - V). \quad (3.12.2)$$

Regardless of how A_V (or other passband attenuations) has been estimated, in *EB* analysis it is traditionally subtracted uniformly for all phases from photometric magnitudes, i.e.,

$$(V - A_V) - M_V = 5\log D - 5, \qquad (3.12.3)$$

where M_V is the absolute magnitude in the V band and D is distance. Prša & Zwitter (2005a, b) questioned whether this type of correction is adequate, especially if interstellar extinction and the color difference between the binary components are large.

Instead, Prša & Zwitter (2005a) treated reddening systematically in the context of data fitting. They determined $E(B-V)$ from multi-color *EB* light curves by comparing several color indices in-and-out of eclipse and demonstrated that estimation of $E(B-V)$ from least-squares analysis requires light curves in three or more bands. As interstellar extinction and reddening depend on wavelength, one has to integrate over the wavelength of a passband instead of using a simple effective wavelength, λ_{eff}, in the calculations.

Wilson (2008) remarks in his development of the direct distance estimation scheme that, although interstellar extinction increases distance estimates, its associated reddening decreases temperature estimates. Reduced theoretical temperatures reduce predicted absolute fluxes and so decrease distance estimates. Thus, in regard to distance determined from light curve analyses, extinction and reddening partly offset one another and, accordingly, the overall effect of extinction on distance determination is less than one might suppose. Wilson (2008, Sect. 7) also investigated the possibility of determining the attenuation A through the least-squares analysis. Note that attenuations in different passbands are connected through the *Jason Cardelli & Mathis* approximation functions and thus only one attenuation A is a free parameter. Although this is indeed possible given accurate absolute light curves in three passbands, it is not very practical as the sensitivity with respect to the calibration of the light curves is too strong. Small deviations in the calibration lead to significantly wrong values of A. The situation might be improved if the three bands are widely separated in wavelength.

3.13 Selected Bibliography

This section is intended to guide the reader to recommended books or articles on the physics involved in modeling *EB*s, or binaries in general.

- The review article by Wilson (1994) gives an excellent overview of *Light Curve Models*. It provides a historical view and discusses the embedded astrophysics.
- The Proceedings of IAU Symposium 51 provides many useful contributions on *gas streams* (Batten, 1973b).
- Readers interested in the *Structure and Evolution of Close Binary Systems* are pointed to the Proceedings of IAU Symposium 73 (Eggleton et al. 1976).
- On the topic of stellar atmospheres, Alter's (1963) *Astrophysics: The Atmospheres of the Sun and Stars* has excellent physical insights. *Theoretical Astrophysics* by Ambartsumyan (1958) is a classic work; an excellent book. *Stellar Atmospheres* by Mihalas (1978) is the most detailed reference on this topic.

- A compact source on the *Theory of Rotating Stars* is provided by Tassoul's (1978) book.
- Warner (1995) provides in his book *Cataclysmic Variable Stars* useful material on mass transfer and accretion disks. This book also contains background on other astrophysical topics relevant to *EB* modeling.
- The *Tidal Evolution in Close Binary Systems* is discussed and quantitatively investigated in the excellent paper by Hut (1981).
- The book *Magnetohydrodynamics in Binary Stars* by Campbell (1997) gives an outline of early work in binary stars and introduces the fundamentals of magnetohydrodynamics and binary star theory. It also covers X-ray binary pulsars, accretion disk magnetism, and stellar and disk winds.
- A brief review on *X-ray binaries*, their classification, and observational facts is given by Krautter (1997).
- A useful introduction into the field of active stars, stellar spots, active chromospheres, and stellar magnetic fields is provided in the book *Aktive Sterne* by Strassmeier (1997).
- *New Techniques and Limitations of Light Curve Analysis* by Hadrava (2005) provides an overview on light curve modeling and analysis with a historical introduction very recommended to the reader.

References

Aller, L. H.: 1963, *Astrophysics: The Atmospheres of the Sun and Stars*, The Ronald Press, New York, 2nd edition

Ambartsumyan, V. A.: 1958, *Theoretical Astrophysics*, Pergamon Press, New York

Andersen, J. & Grønbech, B.: 1975, The Close O-type Eclipsing Binary TU Muscae, *A&A* **45**, 107–115

Appenzeller, I. & Jordan, C. (eds.): 1987, *Circumstellar Matter*, IAU Symposium 122, Kluwer Academic Publishers, Dordrecht, Holland

Avni, Y.: 1976, The Eclipse Duration of the X-Ray Pulsar 3U 0900-40, *ApJ* **209**, 574–577

Banks, T. & Budding, E.: 1990, Information Limit Optimization Technique Applied to AB Doradus, *Ap. Sp. Sci.* **167**, 221–234

Batten, A. H.: 1973a, Discussion of Observations of the Flow of Matter Within Binary Systems, in A. H. Batten (ed.), *Extended Atmospheres and Circumstellar Matter in Spectroscopic Binary Systems*, IAU Symposium 51, pp. 1–21, D. Reidel, Dordrecht, Holland

Batten, A. H. (ed.): 1973b, *Extended Atmospheres and Circumstellar Matter in Spectroscopic Binary Systems*, IAU Symposium 51, D. Reidel, Dordrecht, Holland

Batten, A. H.: 1976, Notes on the Interpretation of Observations of Circumstellar Matter in Binary Systems, in P. Eggleton, S. Mitton, & J. Whelan (eds.), *Structure and Evolution of Close Binary Systems*, IAU Symposium 73, pp. 303–310, D. Reidel, Dordrecht, Holland

Binnendijk, L.: 1960, *Properties of Double Stars*, University of Pennsylvannia Press, Philadelphia, PA

Binnendijk, L.: 1965, The W Ursae Majoris Systems, *Kleine Veröffentlichungen Bamberg* **4**, 36–51

Binnendijk, L.: 1970, The Orbital Elements of W Ursae Majoris Systems, *Vistas* **12**, 217–256

Bond, H. E., Ciardullo, R., Fleming, T. A., & Grauer, A. D.: 1989, HFG1: A Planetary Nebula with a Close-Binary Nucleus, in S. Torres-Peimbert (ed.), *Planetary Nebulae*, IAU Symposium 131, p. 310, Kluwer Academic Publishers, Dordrecht, Holland

Bond, H. E. & Livio, M.: 1990, Morphologies of Planetary Nebulae Ejected by Close-Binary Nuclei, *ApJ* **355**, 568–576

Bradstreet, D. H.: 1993, Binary Maker 2.0 – An Interactive Graphical Tool for Preliminary Light Curve Analysis, in E. F. Milone (ed.), *Light Curve Modeling of Eclipsing Binary Stars*, pp. 151–166, Springer, New York

Brown, J. C., McLean, I. S., & Emslie, A. G.: 1978, Polarisation by Thomson Scattering in Optically Thin Stellar Envelopes. II. Binary and Multiple Star Envelopes and the Determination of Binary Inclinations, *A&A* **68**, 415–427

Budding, E. & Zeilik, M.: 1987, An Analysis of the Light Curves of Short-Period RS CVn stars: Starspots and Fundamental Properties, *ApJ* **319**, 827–835

Campbell, C. G.: 1997, *Magnetohydrodynamics in Binary Stars*, No. 216 in Astrophysics and Space Science Library, Kluwer Academic Publishers, Dordrecht, Holland

Castor, J. I., Abbott, D. C., & Klein, R. I.: 1975, Radiation-Driven Winds in Of Stars, *ApJ* **195**, 157–174

Chandrasekhar, S.: 1933a, The Equilibrium of Distorted Polytropes. III. The Double Star Problem, *MNRAS* **93**, 462–471

Chandrasekhar, S.: 1933b, The Equilibrium of Distorted Polytropes. IV. The Rotational and the Tidal Distortions as Functions of the Density Distribution, *MNRAS* **93**, 538–574

Chandrasekhar, S.: 1939, *An Introduction to the Study of Stellar Structure*, Dover Publications, New York

Chandrasekhar, S.: 1946a, On the Radiative Equilibrium of a Stellar Atmosphere. X, *ApJ* **103**, 351–370

Chandrasekhar, S.: 1946b, On the Radiative Equilibrium of a Stellar Atmosphere. XI, *ApJ* **104**, 110–132

Crawford, J. A.: 1955, On the Subgiant Components of Eclipsing Binary Systems, *ApJ* **121**, 71–76

Davidge, T. J. & Milone, E. F.: 1984, A Study of the O'Connell Effect in the Light Curves of Eclipsing Binaries, *ApJ Suppl.* **55**, 571–584

Dgani, R., Livio, M., & Soker, N.: 1989, On the Stream-Accretion Disk Interaction: Response to Increased Mass Transfer Rate, *ApJ* **336**, 350–359

Díaz-Cordovés, J. & Giménez, A.: 1992, A New Nonlinear Approximation to the Limb-Darkening of Hot Stars, *A&A* **259**, 227–231

Djurasevic, G.: 1986, Critical Equipotential Surfaces in Close Binary Systems, *Ap. Sp. Sci.* **124**, 5–25

Drake, S. A. & Ulrich, R. K.: 1980, The Emission-Line Spectrum from a Slab of Hydrogen at Moderate to High Densities, *ApJ Suppl.* **42**, 351–383

Drechsel, H., Haas, S., Lorenz, R., & Gayler, S.: 1995, Radiation Pressure Effects in Early-Type Close Binaries and Implications for the Solution of Eclipse Light Curves, *A&A* **294**, 723–743

Duerbeck, H. W.: 1975, The Eclipsing Binary System VV Orionis, *A&A Suppl.* **22**, 19–47

Eggleton, P., Mitton, S., & Whelan, J. (eds.): 1976, *Structure and Evolution of Close Binary Systems*, IAU Symposium 73, D. Reidel, Dordrecht, Holland

Elias, N. M., Wilson, R. E., Olson, E. C., Aufdenberg, J. P., Guinan, E. F., Gudel, M., Van Hamme, W., & Stevens, H. L.: 1997, New Perspectives on AX Monocerotis, *ApJ* **484**, 394–411

Flannery, B. P.: 1976, A Cyclic Thermal Instability in Contact Binary Stars, in P. Eggleton, S. Mitton, & J. Whelan (eds.), *Structure and Evolution of Close Binary Systems*, IAU Symposium 73, p. 331, D. Reidel, Dordrecht, Holland

Gould, N. L.: 1959, Particle Trajectories Around Close Binary Systems, *AJ* **64**, 136–139

Grauer, A. D., Bond, H. E., Ciardullo, R., & Fleming, T. A.: 1987, The Close-Binary Nucleus of the Planetary Nebula HFG 1, *BAAS* **1**, 643

Gustafsson, B., Bell, R. A., Eriksson, K., & Nordlund, A.: 1975, A Grid of Model Atmospheres for Metal-Deficient Giant Stars I., *A&A* **42**, 407–432

Hadrava, P.: 1986, Roche Lobe in Eccentric Orbits, *Hvar. Obs. Bull.* **10**, 1–10

Hadrava, P.: 1987, Model Atmospheres of Binary Components, *Publ. Astron. Inst. CS Acad. Sci.* **70**, 263–266

Hadrava, P.: 1988, Models of Atmospheres of Contact Binary-Components: Incompatibility of Hydrostatic and Radiative Equilibria, *Publ. Astrophys. Obs. zu Potsdam* **114**, 11–13

Hadrava, P.: 2005, New Techniques and Limitations of Light Curve Analysis, *Ap. Sp. Sci.* **296**, 239–249

Hall, D. S.: 1990, Period Changes and Magnetic Cycles, in I. Ibanoglu (ed.), *Active Close Binaries*, pp. 95–119, Kluwer Academic Publishers, Dordrecht, Holland

Hearn, A. G.: 1987, Theory of Winds from Hot Stars, in I. Appenzeller & C. Jordan (eds.), *Circumstellar Matter*, IAU Symposium 122, pp. 395–408, Kluwer Academic Publishers, Dordrecht, Holland

Hill, G., Fisher, W. A., & Holmgren, D.: 1989, Studies of the Late-Type Binaries. I. The Physical Parameters of 44i Bootis ABC, *A&A* **211**, 81–98

Hill, G., Fisher, W. A., & Holmgren, D.: 1990, Studies of Late-Type Binaries. IV. The Physical Parameters of ER Vulpeculae, *A&A* **238**, 145–159

Hill, G. & Rucinski, S. M.: 1993, LIGHT2: A light-curve modeling program, in E. F. Milone (ed.), *Light Curve Modeling of Eclipsing Binary Stars*, pp. 135–150, Springer, New York

Howarth, I. D.: 1997, The Effect of Radiation Pressure on Equipotential Surfaces in Binary Systems, *Observatory* **117**, 335–338

Hoyle, F.: 1955, *Frontiers of Astronomy*, Harper & Brothers, New York

Huang, S.-S.: 1973, Problems of Gaseous Motion Around Stars, in A. H. Batten (ed.), *Extended Atmospheres and Circumstellar Matter in Spectroscopic Binary Systems*, IAU Symposium 51, pp. 22–47, D. Reidel, Dordrecht, Holland

Hut, P.: 1981, Tidal Evolution in Close Binary Systems, *A&A* **99**, 126–140

Iben, I. & Livio, M.: 1994, Common Envelopes in Binary Star Evolution, *PASP* **105**, 1373–1406

Joy, A. H.: 1942, Observations of RW Tauri at Minimum Light, *PASP* **54**, 35–37

Kallrath, J.: 1991, Numerical Hydrodynamics of Counter Flowing Binary Stellar Winds, *A&A* **247**, 434–446

Kallrath, J. & Kämper, B.-C.: 1992, Another Look at the Early-Type Eclipsing Binary BF Aurigae, *A&A* **265**, 613–625

Kallrath, J. & Strassmeier, K.: 2000, The BF Aurigae system - A Close Binary at the Onset of Mass Transfer, *Astronomy and Astrophysics* **362**, 673–682

Kang, Y. W. & Wilson, R. E.: 1989, Least-Squares Adjustment of Spot Parameters for Three RS CVn Binaries, *AJ* **97**, 848–865

Kemp, J. C., Henson, G. D., Barbour, M. S., Kraus, D. J., & Collins, G. W.: 1983, Discovery of Eclipse Polarization in Algol, *ApJ Lett.* **273**, L85–L88

Kippenhahn, R. & Weigert, A.: 1989, *Stellar Structure and Evolution*, Springer, Berlin, Germany

Kitamura, M. & Yamasaki, A.: 1984, A Detailed Reflection Model for Close Binary Systems with Equal Components, *Ann. Tokyo Astr. Obs.* **20**, 51–74

Kjeldseth-Moe, O. & Milone, E. F.: 1978, Limb Darkening 1945-3245 Å for the Quiet Sun from Skylab Data, *ApJ* **226**, 301–314

Klinglesmith, D. A. & Sobieski, S.: 1970, Nonlinear Limb Darkening for Early Type Stars, *AJ* **75**, 175–182

Ko, Y. & Kallman, T. R.: 1994, Emission Lines from X-Ray-Heated Accretion Disks in Low-Mass X-Ray Binaries, *ApJ* **431**, 273–301

Kondo, Y. & McCluskey, G. E.: 1976, Mass Flow in Close Binary Systems, in P. Eggleton, S. Mitton, & J. Whelan (eds.), *Structure and Evolution of Close Binary Systems*, IAU Symposium 73, pp. 277–282, D. Reidel, Dordrecht, Holland

Kopal, Z.: 1954, A Study of the Roche Model, *Jodrell Bank Ann.* **1**, 37–57

Kopal, Z.: 1959, *Close Binary Systems*, Chapman & Hall, London

Kopal, Z.: 1978, *Dynamics of Close Binary Systems*, D. Reidel, Dordrecht, Holland

Krautter, J.: 1997, X-Ray Binaries, in C. Sterken & C. Jaschek (eds.), *Light Curves of Variable Stars – A Pictorial Atlas*, Cambridge University Press, Cambridge, UK

Kruszewski, A.: 1967, Exchange of Matter in Close Binary Systems. IV. Ring Formation, *Acta Astron.* **17**, 297–310

Kuiper, G. P.: 1941, On the Interpretation of Beta Lyrae and Other Close Binaries, *ApJ* **93**, 133–177
Kurucz, R. L.: 1979, Model Atmospheres for G, F, A, B, and O Stars, *ApJ Suppl.* **40**, 1–340
Kurucz, R. L.: 1993, New Atmospheres for Modelling Binaries and Disks, in E. F. Milone (ed.), *Light Curve Modeling of Eclipsing Binary Stars*, pp. 93–102, Springer, New York
Larson, R. B.: 1978, A Finite-Particle Scheme for Three-Dimensional Gas Dynamics, *J. Comp. Phys.* **27**, 397–409
Limber, D. N.: 1963, Surface Forms and Mass Loss for the Components of Close Binaries – General Case of Non-Synchronous Rotation, *ApJ* **138**, 1112–1132
Lin, D. N. C. & Pringle, J. E.: 1976, Numerical Simulation of Mass Transfer and Accretion Disc Flow in Binary Systems, in P. Eggleton, S. Mitton, & J. Whelan (eds.), *Structure and Evolution of Close Binary Systems*, IAU Symposium 73, pp. 237–252, D. Reidel, Dordrecht, Holland
Linnell, A. P.: 1984, A Light Synthesis Program for Binary Stars, *ApJ* Suppl. **54**, 17–31
Linnell, A. P.: 1991, Does SV Centauri Harbour an Accretion Disk? *ApJ* **379**, 721–728
Linnell, A. P.: 1993, Light Synthesis Modeling of Close Binary Stars, in E. F. Milone (ed.), *Light Curve Modeling of Eclipsing Binary Stars*, pp. 103–111, Springer, New York
Linshan, Y., Zongyuan, C., & Dosa, P.: 1985, *General Catalogue of Ephemerides and Apparent Orbits of 736 Visual Binary Stars*, Science and Technology, Shanghai, China
Lubow, S. H. & Shu, F. H.: 1975, Gas Dynamics of Semidetached Binaries, *ApJ* **198**, 383–405
Lucy, L. B.: 1967, Gravity-Darkening for Stars with Convective Envelopes, *Zeitschr. f. Astrophys.* **65**, 89–92
Lucy, L. B.: 1976, W Ursae Majoris Systems with Marginal Contact, *ApJ* **205**, 208–216
Lucy, L. B.: 1997, *Comments on Radiation Pressure in Binary Systems*, private communication
Lucy, L. B. & Sweeney, M. A.: 1971, Spectroscopy Binaries with Circular Orbits, *AJ* **76**, 544–556
Lucy, L. B. & Wilson, R. E.: 1979, Observational Tests of Theories of Contact Binaries, *ApJ* **231**, 502–513
Lyutyi, V. M., Syunyaev, R. A., & Cherepashuk, A. M.: 1973, Nature of the Optical Variability of HZ Herculis (Her X-1) and BD +34°3815 (Cyg X-1), *Sov. Astron.* **17**, 1–6
Marrese, P. M., Milone, E. F., Sordo, R., & Williams, M. D.: 2005, Gaia and the Fundamental Stellar Parameters from Double-Lined Eclipsing Binaries, in C. Turon, K. S. O'Flaherty, & M. A. C. Perryman (eds.), *The Three-Dimensional Universe with Gaia*, Vol. 576 of *ESA Special Publication*, pp. 599–*
McVean, J. R.: 1994, Analysis of Eclipsing Binaries in the Globular Cluster M71, MSc Thesis, Department of Physics and Astronomy, University of Calgary, Calgary, AB
Mezzetti, M., Giuricin, G., & Mardirossian, F.: 1980, Mass Loss and Mass Transfer in Algols: A Check on Some Current Theoretical Views, *A&A* **83**, 217–225
Mihalas, D.: 1965, Model Atmospheres and Line Profiles for Early-Type Stars, *ApJ Suppl.* **9**, 321–437
Mihalas, D.: 1978, *Stellar Atmospheres*, Freeman, San Francisco, CA, 2nd edition
Milone, E. F. (ed.): 1993, *Light Curve Modeling of Eclipsing Binary Stars*, Springer, New York
Milone, E. F., Groisman, G., Fry, D. J. I. F., & Bradstreet, H.: 1991, Analysis and Solution of the Light and Radial Velocity Curves of the Contact Binary TY Bootis, *ApJ* **370**, 677–692
Milone, E. F., Stagg, C. R., & Kurucz, R. L.: 1992, The Eclipsing Binary AI Phoenicis: New Results Based on an Improved Light Curve Analysis Program, *ApJ Suppl.* **79**, 123–137
Milone, E. F., Wilson, R. E., & Hrivnak, B. J.: 1987, RW Comae Berencis. III. Light Curve Solution and Absolute Parameters, *ApJ* **319**, 325–333
Monaghan, J. J.: 1992, Smoothed Particle Hydrodynamics, *Ann. Rev. Astron. Astrophys.* **30**, 543–574
Morton, D. C.: 1960, Evolutionary Mass Exchange in Close Binary Systems, *ApJ* **132**, 146–161
Mukherjee, J. D., Peters, G. J., & Wilson, R. E.: 1996, Rotation of Algol Binaries – A Line Profile Model Applied to Observations, *MNRAS* **283**, 613–625
Nelson, B. & Davis, W. D.: 1972, Eclipsing-Binary Solutions by Sequential Optimization of the Parameters, *ApJ* **174**, 617–628

References

Neutsch, W. & Schmidt, H.: 1985, Expanding Envelopes of Binary Stars, *Ap. Sp. Sci.* **109**, 249–257

Neutsch, W., Schmidt, H., & Seggewiss, W.: 1981, A Model for the Expanding CIII Envelope of the Wolf–Rayet Spectroscopic Binary HDL52270, *Acta Astron.* **31**, 197–205

Niedsielska, Z.: 1997, Periodic Orbits and Accretion Disks, *CMDA* **67**, 205–213

O'Connell, D. J. K.: 1951, The So-Called Periastron Effect in Close Eclipsing Binaries, *Riverview College Observatory Publ.* **II(6)**, 85–99

Paczynski, B.: 1971, Evolutionary Processes in Close Binary Stars, *Ann. Rev. Astron. Astrophys.* **9**, 183–208

Peraiah, A.: 1969, Gravity Darkening in the Components of Close Binary Systems, *A&A* **3**, 163–168

Peraiah, A.: 1970, Theoretical Light Changes in Close Binaries, *A&A* **7**, 473–480

Plavec, M. J.: 1958, Dynamical Instability of the Components of Close Binary Systems, *Mem. Soc. Roy. Sci. Liege* **20**, 411–420

Plavec, M. J.: 1968, Mass Exchange and Evolution of Close Binaries, *Adv. in Astron. Astrophys.* **6**, 201–278

Plavec, M. J.: 1973, Evolutionary Aspects of Circumstellar Matter in Binary Systems, in A. H. Batten (ed.), *Extended Atmospheres and Circumstellar Matter in Spectroscopic Binary Systems*, IAU Symposium 51, pp. 216–259, D. Reidel, Dordrecht, Holland

Plavec, M. J.: 1980, IUE Observations of Long Period Eclipsing Binaries: A Study of Accretion onto Non-Degenerate Stars, in M. J. Plavec, D. M. Popper, & R. K. Ulrich (eds.), *Close Binary Stars: Observations and Interpretation*, pp. 251–261, D. Reidel, Dordrecht, Holland

Poe, C. H. & Eaton, J. A.: 1985, Star Spot Areas and Temperatures in Nine Binary Systems with Late-Type Components, *ApJ* **289**, 644–659

Pollock, A. M. T.: 1987, The Einstein View of the Wolf–Rayet Stars, *ApJ* **320**, 283–295

Prendergast, K. H.: 1960, The Motion of Gas Streams in Close Binary Systems, *ApJ* **132**, 162–174

Prendergast, K. H. & Taam, R. E.: 1974, Numerical Simulation of the Gas Flow in Close Binary Systems, *ApJ* **189**, 125–136

Prilutskii, O. F. & Usov, V. V.: 1976, X-Rays from Wolf–Rayet Stars, *Sov. Astron.* **20**, 2–4

Prša, A. & Zwitter, T.: 2005a, Influence of Interstellar and Atmospheric Extinction on Light Curves of Eclipsing Binaries, *Ap. Sp. Sci.* **296**, 315–320

Prša, A. & Zwitter, T.: 2005b, A Computational Guide to Physics of Eclipsing Binaries. I. Demonstrations and Perspectives, *ApJ* **628**, 426–438

Pustylnik, I. B.: 2005, Resolving the Algol Paradox and Kopal's Classification of Close Binaries with Evolutionary Implications, *Ap. Sp. Sci.* **296**, 69–78

Quataert, E. J., Kumar, P., & Ao, C. O.: 1996, On the Validity of the Classical Apsidal Motion Formula for Tidal Distortion, *ApJ* **463**, 284–296

Ramsey, L. W. & Nations, H. L.: 1980, HR 1099 and the Starspot Hypothesis for RS Canum Venaticorum Binaries, *ApJ* **239**, L121–L124

Roach, F. E. & Wood, F. B.: 1952, An Interpretation of the Photometric Observations of Zeta Aurigae, *Annales d'astrophysique* **15**, 21–53

Robertson, J. A. & Eggleton, P. P.: 1977, The Evolution of W Ursae Majoris Systems, *MNRAS* **179**, 359–375

Rossiter, R. A.: 1924, On the Detection of an Effect of Rotation During Eclipse in the Velocity of the Brighter Component of β Lyrae, and on Constancy of the Velocity of this System, *ApJ* **60**, 15–21

Różyczka, M. & Schwarzenberg-Czerny, A.: 1987, 2-D Hydrodynamical Models of the Stream–Disk Interaction in Cataclysmic Binaries, *Acta Astron.* **37**, 141–162

Rucinski, S. M.: 1969, The Photometric Proximity Effects in Close Binary Systems. I. The Distortion of the Components and the Related Effects in Early Type Binaries, *Acta Astron.* **19**, 125–153

Rucinski, S. M.: 1973, The W UMa-Type Systems as Contact Binaries. I. Two Methods of Geometrical Elements Determination. Degree of Contact, *Acta Astron.* **23**, 79–120

Rucinski, S. M. & Duerbeck, H. W.: 1997, Absolute-Magnitude Calibration for the W UMa-Type Systems Based on HIPPARCOS Data, *PASP* **109**, 1340–1350

Sasselov, D. D.: 1998a, Surface Imaging by Microlensing, in R. Donahue & J. Bookbinder (eds.), *Cool Stars, Stellar Systems, and the Sun*, No. 154 in 10th Cambridge Workshop, ASP Conference Series, pp. 383–391, Astronomical Society of the Pacific, San Francisco, CA

Sasselov, D. D.: 1998b, The Chromaticity of Microlensing, in R. Ferlet & J. P. Maillard (eds.), *Variable Stars and the Astrophysical Return of Microlensing Surveys*, pp. 141–146, Editions Frontiers, Paris, France

Scarfe, C. D., Barlow, D. J., Fekel, F. C., Rees, R. F., Lyons, R. W., Bolton, C. T., McAlister, H. A., & Hartkopf, W. I.: 1994, The Spectroscopic Triple System HR 6469, *AJ* **107**, 1529–1541

Schlesinger, F.: 1909, The Algol Variable δ Librae, *Publ. Allegheny Obs. Univ. of Pittsburgh* **1**, 123–134

Schuerman, D. W.: 1972, Roche Potentials Including Radiation Pressure, *Ap. Sp. Sci.* **19**, 351–358

Siscoe, G. L. & Heinemann, M. A.: 1974, Binary Stellar Winds, *Ap. Sp. Sci.* **31**, 361–374

Siviero, A., Munari, U., Sordo, R., Dallaporta, S., Marrese, P. M., Zwitter, T., & Milone, E. F.: 2004, Asiago eclipsing binaries program. I. V432 Aurigae, *A&A* **417**, 1083–1092

Smak, J.: 1978, The Escape of Particles from Disks in Close Binaries, in A. N. Zytkow (ed.), *Nonstationary Evolution of Close Binaries*, No. 2 in Symposium of the Problem Commission "Physics and Evolution of Stars", pp. 111–116, PWN – Polish Scientific Publishers, Warsaw, Poland

Søderhjelm, S.: 1980, Geometry and Dynamics of the Algol System, *A&A* **89**, 100–112

Sterken, C. and Jaschek, C. (eds.): 1997, *Light Curves of Variable Stars – A Pictorial Atlas*, Cambridge University Press, Cambridge, UK

Stevens, I. R., Blondin, J. M., & Pollock, A. M. T.: 1992, Colliding Winds from Early-Type Stars in Binary Systems, *ApJ* **386**, 265–287

Strassmeier, K. G.: 1997, *Aktive Sterne – Laboratorien der solaren Astrophysik*, Springer, Wien, Austria

Struve, O.: 1944, The Spectroheliographic Problem of U Cephei, *ApJ* **99**, 222–238

Struve, O.: 1950, *Stellar Evolution*, Princeton University Press, Princeton, NJ

Szebehely, V.: 1967, *Theory of Orbits: The Restricted Problem of Three Bodies*, Academic Press, London

Taam, R. E. & Bodenheimer, P.: 1992, The Common Envelope Evolution of Massive Stars, in E. P. J. V. den Heuvel & S. A. Rappaport (eds.), *X-Ray Binaries and Recycled Pulsars*, Vol. 377 of *NATO ASI Science Series C: Mathematical and Physical Sciences*, pp. 281–291, Kluwer Academic Publishers, Dordrecht, Holland

Tassoul, J.-L.: 1978, *Theory of Rotating Stars*, Princeton Series in Astrophysics, Princeton University Press, Princeton, NJ

Tassoul, J.-L. & Tassoul, M.: 1983, Meridional Circulation in Rotating Stars. IV – The Approach to the Mean Steady State in Early Type Stars, *ApJ Suppl.* **264**, 298–301

Terrell, D.: 1994, Circumstellar Hydrodynamics and Spectral Radiation in Algols, PhD thesis, Department of Astronomy, University of Florida, Gainesville, FL

Terrell, D., Mukherjee, J. D., & Wilson, R. E.: 1992, *Binary Stars – A Pictorial Atlas*, Krieger Publishing Company, Malabar, FL

Terrell, D. & Wilson, R. E.: 1993, Spectral Energy Distributions of Circumstellar Gas in Algols, in E. F. Milone (ed.), *Light Curve Modeling of Eclipsing Binary Stars*, pp. 27–37, Springer, New York

Tsesevich, V. P. (ed.): 1973, *Eclipsing Variable Stars*, A Halsted Press Book, Wiley, New York

Van Hamme, W.: 1993a, New Limb-Darkening Coefficients for Modeling Binary Star Light Curves, *AJ* **106**, 2096–2117

Van Hamme, W.: 1993b, The New Wilson Reflection Treatment and the Nature of BF Aurigae, in E. F. Milone (ed.), *Light Curve Modeling of Eclipsing Binary Stars*, pp. 53–68, Springer, New York

Van Hamme, W. & Wilson, R. E.: 1990, Rotation Statistics of Algol-Type Binaries and Results on RY Geminorum, RW Monocerotis, and RW Tauri, *AJ* **100**, 1981–1993

Van Hamme, W. & Wilson, R. E.: 1994, Binary Star Radial Velocities Weighted by Line Strength, *Mem. Astron. Soc. Ital.* **65**, 89–92

References

Van Hamme, W. & Wilson, R. E.: 1997, Radial Velocity Proximity Effects for Selected Examples, in E. F. Milone (ed.), *Proceedings of the AAS 1997 Meeting – Binary Section*, pp. 1–10, University of Calgary, Calgary, AB

Van Hamme, W. & Wilson, R. E.: 2003, Stellar Atmospheres in Eclipsing Binary Models, in U. Munari (ed.), *GAIA Spectroscopy: Science and Technology*, Vol. 298 of *Astronomical Society of the Pacific Conference Series*, pp. 323–328, San Francisco

Van Hamme, W. & Wilson, R. E.: 2007, Third-Body Parameters from Whole Light and Velocity Curves, *ApJ* **661**, 1129–1151

Van Paradijs, J., Takens, R. J., & Zuiderwijk, E. J.: 1977, Systematic Distortions of the Radial Velocity Curve of HD 77581 (Vela X-1) Due to Tidal Deformation, *A&A* **57**, 221–227

Vanbeveren, D.: 1977, The Influence on the Critical Surface of Radiation Pressure, X-Rays and Asynchronisation of Both Components in a Binary System, *A&A* **54**, 877–882

Vanbeveren, D.: 1978, The Influence of the Radiation Pressure Force on Possible Critical Surfaces, *Ap. Sp. Sci.* **57**, 41–51

Vogt, S. S.: 1979, A Spectroscopic and Photometric Study of the Star Spot on HD 224085, *PASP* **91**, 616

von Zeipel, H.: 1924a, Radiative Equilibrium of a Double-Star System with Nearly Spherical Components, *MNRAS* **84**, 702–719

von Zeipel, H.: 1924b, The Radiative Equilibrium of a Rotating System of Gaseous Masses, *MNRAS* **84**, 665–683

von Zeipel, H.: 1924c, The Radiative Equilibrium of Slightly Oblate Rotating Stars, *MNRAS* **84**, 684–701

Warner, B. (ed.): 1995, *Cataclysmic Variable Stars*, Cambridge University Press, Cambridge, UK

Webbink, R. F.: 1992, Common Envelope Evolution and Formation of Cataclysmic Variables and Low-Mass X-Ray Binaries, in E. P. J. V. den Heuvel & S. A. Rappaport (eds.), *X-Ray Binaries and Recycled Pulsars*, Vol. 377 of *NATO ASI Science Series C: Mathematical and Physical Sciences*, pp. 269–280, Kluwer Academic Publishers, Dordrecht, Holland

Webbink, R. F.: 2008, Common Envelope Evolution Redux, in E. F. Milone, D. A. Leahy, & D. W. Hobill (eds.), *Astrophysics and Space Science Library*, Vol. 352 of *Astrophysics and Space Science Library*, pp. 233–257

Wehrse, R.: 1987, Theory of Circumstellar Envelopes, in I. Appenzeller & C. Jordan (eds.), *Circumstellar Matter*, IAU Symposium 122, pp. 255–266, Kluwer Academic Publishers, Dordrecht, Holland

White, R. L. & Long, K. S.: 1989, X-Ray Emission from Wolf–Rayet Stars, *ApJ* **310**, 832–837

Whitehurst, R.: 1988a, Numerical Simulations of Accretion Disks. I. Superhumps – A Tidal Phenomenon of Accretion Disks, *MNRAS* **232**, 35–51

Whitehurst, R.: 1988b, Numerical Simulations of Accretion Disks. II. Design and Implementation of a New Numerical Method, *MNRAS* **233**, 529–551

Wilson, O. C.: 1960, Eclipses by extended atmospheres, in J. L. Greenstein (ed.), *Stellar Atmospheres*, Vol. VI of *Stars and Stellar Systems*, pp. 436–465, University of Chicago Press, Chicago, IL

Wilson, R. E.: 1974, Binary Stars – A Look at Some Interesting Developments, *Mercury*, pp. 4–12

Wilson, R. E.: 1978, On the A-Type W Ursae Majoris Systems, *ApJ* **224**, 885–891

Wilson, R. E.: 1979, Eccentric Orbit Generalization and Simulaneous Solution of Binary Star Light and Velocity Curves, *ApJ* **234**, 1054–1066

Wilson, R. E.: 1981, A Generalization of the Henyey and Integration Methods for Computing Stellar Evolution, *A&A* **99**, 43–47

Wilson, R. E.: 1990, Accuracy and Efficiency in the Binary Star Reflection Effect, *ApJ* **356**, 613–622

Wilson, R. E.: 1994, Binary-Star Light-Curve Models, *PASP* **106**, 921–941

Wilson, R. E.: 1998, *Computing Binary Star Observables (Reference Manual to the Wilson–Devinney Program*, Department of Astronomy, University of Florida, Gainesville, FL, 1998 edition

Wilson, R. E.: 1999, A Fluorescence and Scattering Model for Binaries, in R. Dvorak (ed.), *Modern Astrometry and Astrodynamics*, Österreichische Akademie der Wissenschaften, Vienna, Austria

Wilson, R. E.: 2008, Eclipsing Binary Solutions in Physical Units and Direct Distance Estimation, *ApJ* **672**, 575–589

Wilson, R. E., DeLuccia, M. R., Johnston, K., & Mango, S. A.: 1972, Photometry and Differential Correction Analysis of Algol, *ApJ* **177**, 191–208

Wilson, R. E. & Liou, J.-C.: 1993, Quantitative Modeling and Impersonal Fitting of Algol Polarization Curves, *ApJ* **413**, 670–679

Wilson, R. E. & Sofia, S.: 1976, Effects of Tidal Distortion on Binary-Star Velocity Curves and Ellipsoidal Variation, *ApJ* **203**, 182–186

Wilson, R. E. & Terrell, D.: 1992, Learning About Algol Disks – Learning from Algol Disks, in S. F. Dermott, J. H. Hunter, & R. E. Wilson (eds.), *Astrophysical Disks*, Annals of the New York Academy of Sciences, pp. 65–74, New York Academy of Sciences

Wilson, R. E. & Terrell, D.: 1998, X-Ray Binary Unified Analysis: Pulse/RV Application to Vela X1/GP Velorum, *MNRAS* **296**, 33–43

Wilson, R. E., Van Hamme, W., & Pettera, L. E.: 1985, RZ Scuti as a Double Contact Binary, *ApJ* **289**, 748–755

Wolf, B.: 1987, Some Observations Relevant to the Theory of Expending Envelopes, in I. Appenzeller & C. Jordan (eds.), *Circumstellar Matter*, IAU Symposium 122, pp. 409–425, Kluwer Academic Publishers, Dordrecht, Holland

Wood, D. B.: 1971, An Analytic Model of Eclipsing Binary Star Systems, *AJ* **76**, 701–710

Wood, D. B.: 1972, *A Computer Program for Modeling NonSpherical Eclipsing Binary Star Systems*, Technical Report X-110-72-473, GSFC, Greenbelt, MD

Wyse, A. B.: 1934, A Study of the Spectra Of Eclipsing Binaries, *Lick. Obs. Bull.* **464**, 37–51

Yamasaki, A.: 1982, A Spot Model for VW Cephei, *Ap. Sp. Sci.* **85**, 43–48

Zeilik, M., deBlasi, C., Rhodes, M., & Budding, E.: 1988, A Half-Century of Starspot Activity on SV Camelopardalis, *ApJ* **332**, 293–298

Zhou, H. N. & Leung, K.-C.: 1987, The Influence of Radiation Pressure on Equipotential Surfaces in High-Temperature Binary Systems, *Ap. Sp. Sci.* **141**, 257–270

Chapter 4
Determination of Eclipsing Binary Parameters

The determination or estimation of physical parameters from *EB* light curves and/or radial velocity curves is an inverse problem and can be formulated as a nonlinear least-squares problem. It is solved by optimizing the agreement between the calculated light and the observed light curve. The parameter vector **x** corresponding to minimum deviation is the system solution, and the calculated light curve produced from it is said to be the *best fit* to the data. A measure of the deviation is the weighted sum of the squared residuals.

In this chapter, we formulate the inverse problem and describe numerical methods used in the *EB* community to solve the least-squares problem. Intentionally, we give a rather formal approach closer to today's education of physicists and mathematicians. In addition, Appendix A provides a more general background on least-squares problems. Regarding the analysis of light curve data, we suggest some rules for estimating initial parameters and for interpreting and checking the consistency of light curve solutions; In addition, we discuss the interpretation of errors associated with the derived parameters. Of all these issues, the interpretation of the solution is of greatest importance because it transforms the mathematical results into useful physical information.

4.1 Mathematical Formulation of the Inverse Problem

Per aspera ad astra (*Through arduous labors, to the stars*)
after Seneca (4 B.C.–A.D. 65)

For a set $\{t_{ck} \mid 1 \leq k \leq n_c\}$ of given timelike quantities and a given m-dimensional parameter vector $\mathbf{x} \in \mathbb{R}^m$, an observable $\mathcal{O}_c$ of curve-type c, $1 \leq c \leq C$

$$\mathcal{O}_c^{\text{cal}}(\mathbf{x}) := \left\{ (t_{ck}, o_{ck}^{\text{cal}}) \mid o_{ck}^{\text{cal}} := o_c^{\text{cal}}(t_{ck}), 1 \leq k \leq n_c \right\}, \qquad (4.1.1)$$

may be calculated at n_c points t_{ck} in time based on a light curve model, in the way described in Chap. 3. This problem, the mapping

$$\mathbf{x} \rightarrow \left\{ \mathcal{O}_1^{\text{cal}}(\mathbf{x}), \ldots, \mathcal{O}_c^{\text{cal}}(\mathbf{x}), \ldots, \mathcal{O}_C^{\text{cal}}(\mathbf{x}) \right\} \qquad (4.1.2)$$

is called the *direct problem*. Note that we consider C observables simultaneously. For present purposes, it is not essential that different observables in simultaneous light curve fitting are measured at the same time; in general we have $t_{c_1 k} \neq t_{c_2 k}$.

The problem of finding the parameter vector **x** which best fits a set of C given observed light curves $\{\mathcal{O}_1^{obs}, \ldots, \mathcal{O}_c^{obs}, \ldots, \mathcal{O}_C^{obs}\}$

$$\mathcal{O}_c^{obs} := \left\{ (t_{ck}, o_{ck}^{obs}) \mid 1 \leq k \leq n_c \right\}, \quad 1 \leq c \leq C, \tag{4.1.3}$$

with each of them consisting of n_c data points is called the *inverse problem*. The inverse problem yields the mapping $\{\mathcal{O}_1^{obs}, \ldots, \mathcal{O}_c^{obs}, \ldots, \mathcal{O}_C^{obs}\} \to $ **x**, of course, only in the sense of a statistical estimation.

In order to formulate the inverse problem, we introduce the column vectors **o** and **c** containing the observables o_{ck} of the sets $\mathcal{O}_c^{obs}$ and $\mathcal{O}_c^{cal}(\mathbf{x})$:

$$\mathbf{o} := (o_{11}^{obs}, o_{12}^{obs}, \ldots, o_{1n_1}^{obs}, \ldots, o_{c1}^{obs}, o_{c2}^{obs}, \ldots, o_{cn_c}^{obs}, \ldots, o_{C1}^{obs}, o_{C2}^{obs}, \ldots, o_{Cn_C}^{obs})^T, \tag{4.1.4}$$

$$\mathbf{c} := (o_{11}^{cal}, o_{12}^{cal}, \ldots, o_{1n_1}^{cal}, \ldots, o_{c1}^{cal}, o_{c2}^{cal}, \ldots, o_{cn_c}^{cal}, \ldots, o_{C1}^{cal}, o_{C2}^{cal}, \ldots, o_{Cn_C}^{cal})^T, \tag{4.1.5}$$

and the vector $\mathbf{d}(\mathbf{x})$ of light residuals (observed minus calculated):[1]

$$\mathbf{d}(\mathbf{x}) := \mathbf{o} - \mathbf{c}(\mathbf{x}), \quad d_{ck}(\mathbf{x}) := o_{ck}^{obs} - o_{ck}^{cal}. \tag{4.1.6}$$

Note that the total number of data points involved in the problem is

$$n := n_1 + \cdots + n_c + \cdots + n_C. \tag{4.1.7}$$

In addition, let $\mathbf{w} := (w_1, \ldots, w_n)^T \neq \mathbf{0}$ be a column vector, the components of which represent the nonnegative weights of the individual observations. Furthermore, it is convenient to define a matrix **W** which, in the case of uncorrelated data,[2] is a diagonal matrix

$$\mathbf{W} := \text{diag}(w_{11}, w_{12}, \ldots, w_{1n_1}, \ldots, w_{c1}, w_{c2}, \ldots, w_{cn_c}, \ldots, w_{C1}, w_{C2}, \ldots, w_{Cn_C}). \tag{4.1.8}$$

In addition, we define the weighted residual vector $\mathbf{R}(\mathbf{x})$

$$\mathbf{R}(\mathbf{x}) = \sqrt{\mathbf{W}} \cdot \mathbf{d}(\mathbf{x}), \tag{4.1.9}$$

$$R_{ck}(\mathbf{x}) = \sqrt{w_{ck}} \left(o_{ck}^{obs} - o_{ck}^{cal} \right). \tag{4.1.10}$$

Weighting is an important part of the data analysis, so in Sect. 4.1.1.5 we comment further on it.

---

[1] The overdetermined system $\mathbf{d}(\mathbf{x}) = \mathbf{0}$ is also called *equations of conditions*.

[2] Unlike the analysis of photographic plates in astrometry where errors in x and y are strongly correlated, eclipsing binary light curve and radial velocity curve data are not correlated with each other.

4.1 Mathematical Formulation of the Inverse Problem

To solve the inverse problem and to derive the unknown parameter vector $\mathbf{x}$ describing the physics of the binary from the observed curves $\mathcal{O}_c^{\text{obs}}$, it appears reasonable to vary $\mathbf{x}$ until the deviation between the calculated observable $\mathcal{O}_c^{\text{cal}}(\mathbf{x})$ and $\mathcal{O}_c^{\text{obs}}$ attains a minimum in a well-defined sense. In data analysis several criteria are used. Mathematically they can be expressed as the problem to minimize the norm $\|\mathbf{R}\|_p$ with $p \in \{1, 2, \infty\}$. In this book at most places, we will use $p = 2$. For the case $p = 1$, we refer the reader to Branham (1990, Chap. 6). An alternative approach to data analysis is *robust estimation* [cf. Press et al. (1992, pp. 694–698)].

Since the days of Gauß (see Sect. 4.1.4), it has been customary[3] to use the weighted sum of squared residuals ($p = 2$)

$$f(\mathbf{x}) := \mathbf{R}^T \mathbf{R} = \sum_{\nu=1}^{n} R_\nu^2(\mathbf{x}) = \mathbf{d}^T \mathbf{W} \mathbf{d} = \sum_{\nu=1}^{n} w_\nu \left(o_\nu^{\text{obs}} - o_\nu^{\text{cal}}\right)^2 \quad (4.1.11)$$

as a measure of this deviation if we are convinced that the distribution of the errors can be described by a normal distribution. A parameter vector $\mathbf{x}$ in a light curve model is called the *light curve solution* or *system solution*, $\mathbf{x}_*$, if

$$f_* \equiv f(\mathbf{x}_*) = \min\{f(\mathbf{x}) \mid \mathbf{x} \in \mathcal{S} \subset \mathbb{R}^m\}. \quad (4.1.12)$$

The subset $\mathcal{S} \subset \mathbb{R}^m$, called the feasible region, is implicitly determined by a set of constraints as in (A.2.2), i.e., functional relations (often equations or inequalities) applied to components of the parameter vector $\mathbf{x}$. Whereas f_* is unique, $\mathbf{x}_*$ is very often not. In nonlinear models, this might be caused by the structure of the model; in linear models, correlations between parameters undermine the uniqueness of $\mathbf{x}_*$. Concerning the minimization of the quadratic form or function $f(\mathbf{x})$, it is equivalent to minimize the corresponding function $f(\mathbf{x})$ defined as the standard deviation σ^{fit}

$$f(\mathbf{x}) = \sigma^{\text{fit}} = \sqrt{\frac{n}{n-m-1} \bigg/ \sum_{\nu=1}^{n} w_\nu} \sqrt{\mathbf{R}^T \mathbf{R}}. \quad (4.1.13)$$

This gives a measure independent of the number m of free parameters and the number of data points, n. Thus, the inverse problem of light curve analysis is reduced to the following problem: Given a light curve model and a parameter vector $\mathbf{x} = (x_1, x_2, \ldots, x_m)^T$, we seek a solution in multidimensional parameter space minimizing the quadratic form $\mathbf{R}^T \mathbf{R}$. Equivalently, as a measure for the quality of the fit, we minimize σ^{fit}, which is normalized by the number of observed data points and free parameters.

---

[3] Gauß showed that the weighted sum of squares gives the most probably correct results. In modern terminology, the least-squares procedures, under the assumptions described in Appendix A.3, provide a *maximum likelihood estimator* (Brandt, 1976, Chap. 7).

Light curve analysis leads to a typical situation for fitting model functions (here, the light curve model) to data. In this case, it is a nonlinear parameter estimation, which in this form is a special case of unconstrained optimization. However, strictly speaking, we are confronted with a nonlinear *constrained* optimization problem. The optimization is subject to bounds

$$l_i \leq x_i \leq u_i. \tag{4.1.14}$$

An example is $l_i = 0$ and $u_i = 1$ for the eccentricity e. Another constraint, seldomly mentioned explicitly, is a nonlinear and implicit relation among the mass ratio q, the Roche potentials Ω_1 and Ω_2, and the rotation factors F_1 and F_2. This relation (note that f_1 and f_2 denote the filling (or "fill-out") factors; they should not be confused with $f(\mathbf{x})$)

$$\text{"}f_{1,2} < 0\text{"} \vee \text{"}0 < f_1 = f_2 < 1\text{"} \vee \text{"sign}(f_1) + \text{sign}(f_2) = -1\text{"} \tag{4.1.15}$$

guarantees that a feasible binary configuration is produced. These configurations are

- detached: two stars inside their critical lobes ($f_1 < 0$, $f_2 < 0 \Leftrightarrow \text{sign}(f_1) + \text{sign}(f_2) = -2$); or
- semi-detached: one star filling its critical lobe whereas the other one is well inside its critical lobe ($f_1 < 0$, $f_2 = 0$ or $f_1 = 0$, $f_2 < 0 \Leftrightarrow \text{sign}(f_1) + \text{sign}(f_2) = -1$); or
- over-contact: both stars establish an over-contact binary in the synchronous case ($0 < f_1 = f_2 < 1$); or finally,
- double-contact: both stars filling their critical lobes in the asynchronous case ($f_1 = f_2 = 1$).

The necessary and sufficient conditions for the existence of a minimal point $\mathbf{x}_*$ of (4.1.12) can be expressed in terms of the first and second derivatives and some Lagrange multipliers (see Appendix A.2). The nonlinear structure of the problem requires an iterative solution algorithm, and it forces us to distinguish between global and local minima. $\mathbf{x}_*$ is called *global minimum* if for all $\mathbf{x} \in S$ the relation

$$f(\mathbf{x}_*) \leq f(\mathbf{x}) \tag{4.1.16}$$

holds. In contrast, $\mathbf{x}_*$ is called local minimum, if $\mathbf{x}_*$ satisfies (4.1.16) only for a local sphere $B_\varepsilon(\mathbf{x}_*)$

$$B_\varepsilon(\mathbf{x}_*) := \{\mathbf{x} \in S \mid \|\mathbf{x} - \mathbf{x}_*\|_2 < \varepsilon\} \tag{4.1.17}$$

around $\mathbf{x}_*$ with a suitable $\varepsilon > 0$. Finally, an algorithm for solving our least-squares problem is said to converge globally if from an arbitrary initial point $\mathbf{x}_0 \in S$ it converges to a local (or global) minimum $\mathbf{x}_*$. However, unless $f(\mathbf{x})$ is a convex function, there is no algorithm available which could be proven to converge to the global minimum $\mathbf{x}_*$. So, in solving nonlinear least-squares problems, at best we can

4.1 Mathematical Formulation of the Inverse Problem

prove that we have reached a local minimum. We might have reached the global minimum but we are usually not able to prove it formally.

4.1.1 The Inverse Problem from the Astronomer's Perspective

4.1.1.1 The Input Database

The analysis of a photometric light curve alone cannot provide absolute dimensions of the stars or the orbit. The reason for this is a scaling property: if all geometric properties of a binary system are doubled, original light curves can be reproduced by shifting to a larger distance. The distance and surface brightnesses, which would either individually or in combination determine the scale of the model, are known only under exceptional circumstances, e.g., if the binary is a member of a well-studied star cluster.

Light curves can provide relative quantities (radii in terms of the semi-major axis a, information on temperature, relative luminosities, perhaps the photometric mass ratio, the shapes of the stars) and the orientation and eccentricity of the orbit (inclination i, argument of the periastron, ω). Radial velocity curves can provide $a \sin i$, i.e., the scaling factor a in physical units if i is known. When a and the mass ratio q are known, the masses can be found unambiguously from (4.4.14) and (4.4.16). Note that in order to derive definite masses and orbital dimensions [viz. (4.4.16) and (4.4.17)] from a radial velocity curve, the inclination needs to be known. Thus, the absolute determination of *EB* parameters requires at least one light curve and radial velocity curves for both components. The following tables list combinations of observables needed to derive certain binary parameters in favorable cases:

1 = at least one photometric light curve;
2 = only one radial velocity curve;
3 = both radial velocity curves, but no light curve;
4 = at least one photometric light and one radial velocity curve; and
5 = at least one photometric light curve and both radial velocity curves:

	1	2	3	4	5
$a_1 \sin i$ or $a_2 \sin i$		✓	✓	✓	✓
$a \sin i$, $a_{1,2} \sin i$, $M_{1,2} \sin^3 i$			✓		✓
a, $a_{1,2}$, $M_{1,2}$, $\mathcal{R}_{1,2}$, $\mathcal{L}_{1,2}$, d				(✓)	✓
e, ω, P	✓	✓	✓	✓	✓
γ		✓	✓		✓
q_{sp}			✓		✓
q_{ph}	(✓)			(✓)	(✓)
i, $\mathcal{R}_{1,2}/a$, L_2/L_1, $g_{1,2}$, $A_{1,2}$, $F_{1,2}$, $x_{1,2}$, ℓ_3	✓			✓	✓
T_2	✓	(?)	(?)	✓	✓

The photometric mass ratio q_{ph} indicated by ($\sqrt{}$) can be determined with much confidence only for lobe-filling or over-contact systems. Unfortunately, the incorrect notion that q_{ph} derives mainly from ellipsoidal variation has gained moderately widespread acceptance. The actual situation is that q_{ph} comes from coupling the Roche configuration to the radii of the stars. Thus, the same light curve characteristics that define the radii[4] also define q_{ph}, but only when the radii can be related to the equipotential configuration. There are two distinct cases. In a semi-detached binary the radius of the lobe-filling star fixes the lobe radius, and thus the mass ratio. Best results are obtained for complete eclipses (Terrell & Wilson 2004). In an over-contact binary with complete eclipses, the ratio of the radii is fixed by elementary considerations [see (4.4.1)]. The ratio of radii in turn is essentially a unique function of the mass ratio for over-contact equipotentials, with minor dependence on the degree of over-contact (Wilson 1978).

Essentially all published values of photometric Fs are for rapidly rotating Algols. Two conditions help in determining the values of F. First, eclipse circumstances (shape and depth) are altered by rotationally induced oblateness. Second, and more subtle, the proximity effects due to reflection and ellipsoidal variation of the secondary star are effectively enhanced by the reduced brightness of the fast rotating primary for observers near the orbit plane.

The question mark in parentheses (?) indicates a chance that the temperature(s) can be derived from spectral features. In order to compute T_2 from a light curve solution, T_1 has to be known in advance (e.g., derived from color or spectral type); of course we could also fix T_2 and adjust T_1. The quantity $a_1 \sin i$ can be obtained from a single-lined system, and $a \sin i$ from a double-lined system to give lower limits for the orbital size $a = a_1 + a_2$. In ideal cases, an eclipsing, double-lined system therefore provides everything needed. Whether or not a given system does so in fact is a matter to be determined. The light curve of an *EB* depends nonlinearly on the parameters, so solving the inverse problem, and thus minimizing $F(\mathbf{x})$ or $f(\mathbf{x}) = \sigma^{fit}(\mathbf{x})$, requires one to navigate all the pitfalls of nonlinear multiparameter fitting.

4.1.1.2 General Problems of Nonlinear Parameter Fitting

Binary parameters can become correlated: Changes in one parameter can nearly be compensated by a combination of changes in other parameters. Unfortunately, this problem is usually present in light curve analysis (Wilson 1983), and it is an intrinsic property of the problem. The problem can be imagined geometrically by considering the hyperspace formed by the sum-of-squares of the residuals, plotted as a function of the parameters (see Figs. 4.1, 4.2, 4.3 and 4.4). The valleys in this hypersurface are long and narrow, so that a small error in the direction taken by the solution vector at each step can cause the algorithm to run up and down the sides of the narrow valley instead of along its axis which would bring it much faster to the solution. The

---

[4] In detached systems, correlations of q_{ph} with Ω_1, Ω_2, and other parameters make it almost impossible to derive accurate photometric mass ratios. In lobe-filling or over-contact binaries either of Ω_1 or Ω_2 is eliminated from the adjustable parameter list.

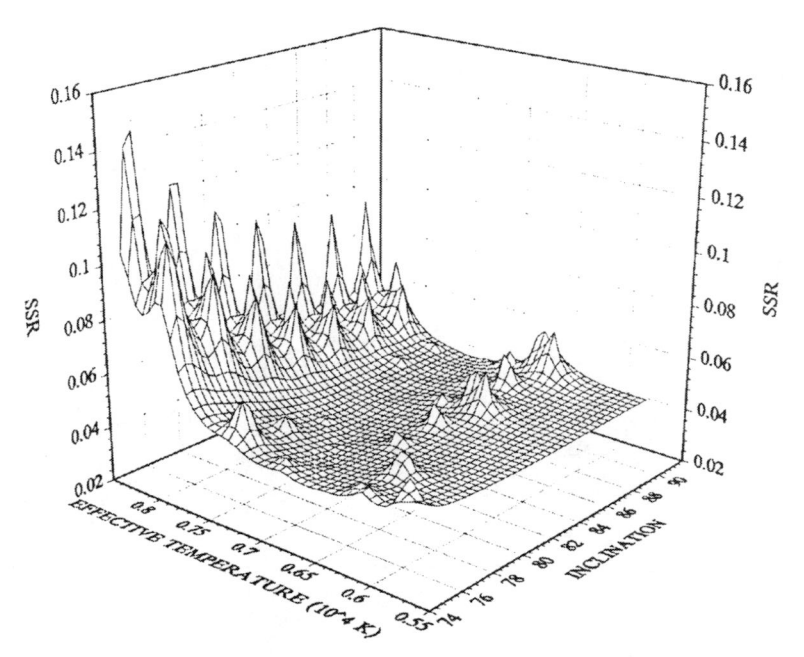

4.1 Mathematical Formulation of the Inverse Problem

Fig. 4.1 Sum of the squared residuals versus parameters. Example of weighted sum of squared residuals $\sum wr^2$ versus T and i. The plot [Fig. 1 in Stagg & Milone (1993, p. 80)] shows data produced during the modeling of the binary *NH30* in the globular cluster *NGC 5466*

first problems of correlations are numerical difficulties in solving the linear algebra problem [the rows of the Jacobian matrix J (see Appendix A.3.3) are nearly linear combinations of each other]. This difficulty may be enhanced if the linearization makes use of numerical partial derivatives (finite-difference approximations). The situation becomes even worse when the linearized problem is solved by using the normal equations (4.3.9). In that case, the condition number of J is squared.

The second problem is the nonlinearity itself, which bends the long, narrow valleys and prevents iterations from taking a large step along the valley axis, before it starts to run up one side of the valley. Consequently, step-size limiting algorithms are commonly used to prevent Differential Corrections from making the solution worse rather than better.

4.1.1.3 Special Problems of Nonlinear Parameter Fitting in Light Curve Analysis

In addition to these frequently encountered problems, a few aspects are especially important in connection with light curve analysis.

First, the influence of parameters on the shape of the light curve is strongly phase dependent. Slight changes of the temperature T_2 of the secondary star may show up only near the eclipses. The albedos A_1 and A_2 have an effect mainly on the shoulders of the minima. Spots affect the light curve only when they appear on a part of the

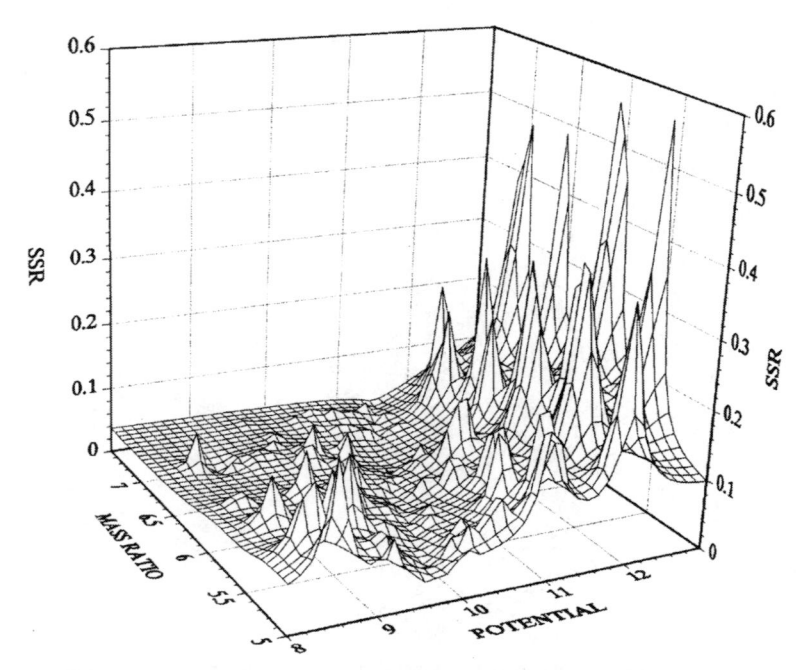

Fig. 4.2 Sum of the squared residuals versus parameters. Example of weighted sum of squared residuals $\sum wr^2$ versus q and Ω. The plot [Fig. 3 in Stagg and Milone (1993, p. 83)] shows data produced for the over-contact binary *H235* in the open cluster *NGC 752*

star visible to the observer. Thus, the derivative, $\partial_{x_p} l_k^{\text{cal}}(\mathbf{x})$ of calculated light $l_k^{\text{cal}}(\mathbf{x})$ with respect to a spot parameter, x_p, say, longitude, is zero over a range of phase values.

The second problem that may occur is the existence of local minima in parameter hyperspace (see Figs. 4.1, 4.2, 4.3 and 4.4). It is very difficult to prove uniqueness in the m-dimensional parameter space S as that requires initial solutions in all the hyperspace valley regions. Such questions become important when analyzing light curves with primary and secondary minima of similar depths, for example. In the case of *BF Aurigae*, two well-separated local minima of σ^{fit} occur (Kallrath & Kämper 1992). One reason is the dual possibility of atransit or occultation eclipse at primary minimum. The two assumptions lead to solutions with σ^{fit} values of comparable size. Such nonuniqueness problems can sometimes (as in the *BF Aurigae* system) be overcome by additional information, such as spectral line ratios (in an SB2). Milone et al. (1991) dealt with such uniqueness questions by perturbing the parameters of their *TY Bootis* solution by about 10%, one at a time, and iterating each time to a new solution. This technique still begs the question about the location of the deepest possible minimum in parameter space because the range of parameters explored is limited in such a procedure. In the case of *TY Bootis*, confidence in the solution was enhanced somewhat by good radial velocity curves. Different models, though, can arise from different assumptions about the assumed

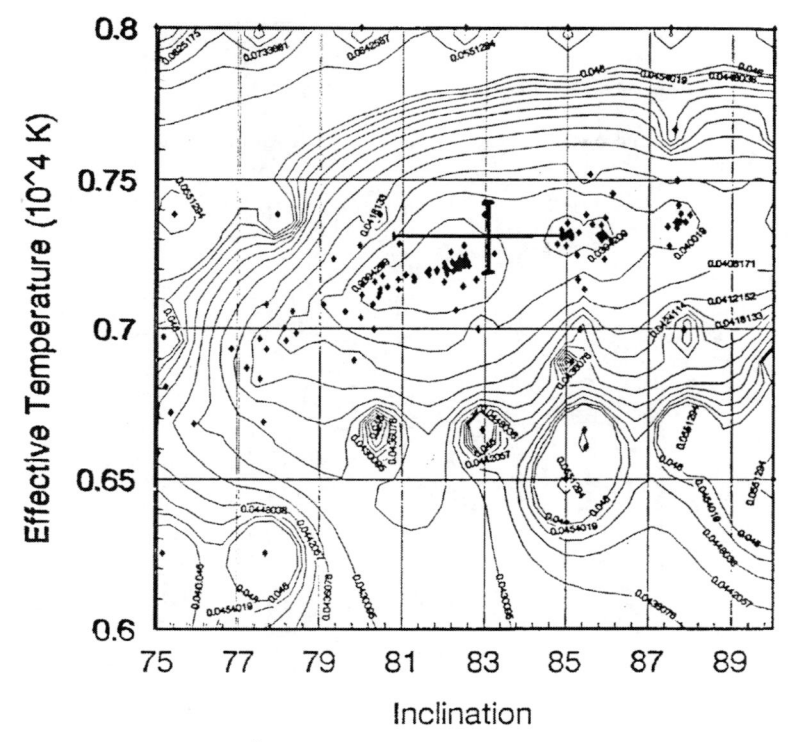

4.1 Mathematical Formulation of the Inverse Problem 177

Fig. 4.3 Contour plot of $\sum wr^2$ versus T and i. The plot [Fig. 2 in Stagg and Milone (1993, p. 82)] shows data produced during the modeling of *NH30*

temperature of one of the components. What temperature scale should we use when deciding the temperature of one of the components? Detailed atmospheres for these kinds of systems are sorely needed before color indices can be de-reddened and used in models.

There are two aspects of analyzing data sets that highlight the problem of strong parameter correlation: The numerical difficulties and the uniqueness problem. The numerical difficulties can be overcome, for instance, by the method of multiple subsets described in Sect. 4.3.2. "Overcoming" in this context means that the algorithm finds a parameter solution with a value of the least-square function close or almost identical to the minimum value. This formulation brings up the second problem: The uniqueness problem. We sometimes face very flat minima with similar values in the least-squares function around the solution. If light is scaled to unity, differences in σ^{fit} smaller than, say, 10^{-3} are usually not statistically significant.

The correlation or uniqueness problem can often be overcome if it is possible to reduce the number of free parameters. For example, limb-darkening coefficients might be taken from a model atmosphere; albedos and gravity brightening parameters might be fixed. Such decisions have to be made with care because they could bias a solution toward an incorrect model. We have to bear in mind that this

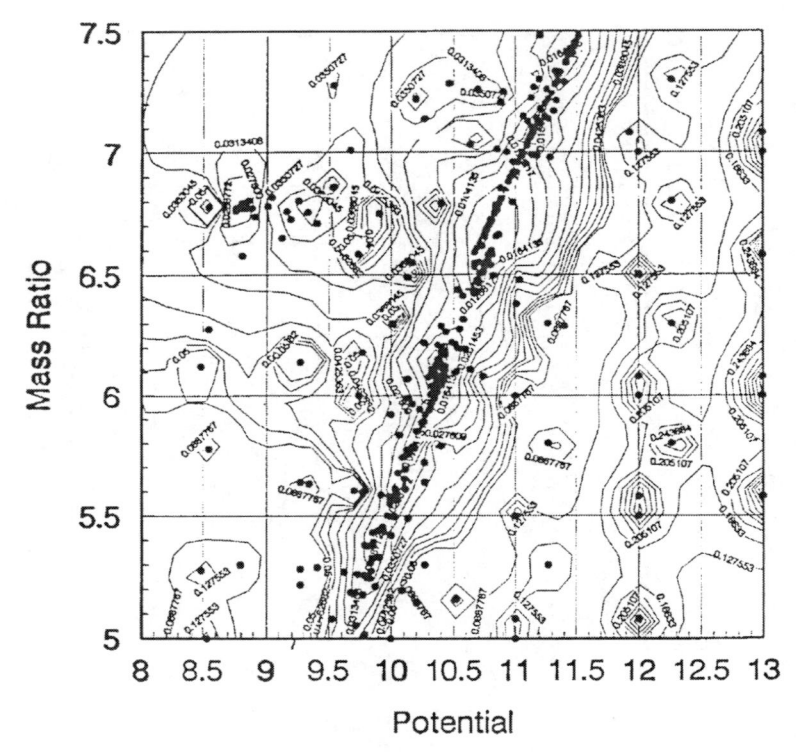

Fig. 4.4 Contour plot of $\sum wr^2$ versus q and Ω. The plot [Fig. 4 in Milone (1993, p. 84)] shows data produced during the modeling of *H235*

procedure can artificially improve determinacy but introduce the wrong physics. Perhaps the most honest approach is to solve the problem, find a decent fit and a corresponding parameter vector, and clearly state the uniqueness problem. In addition, we should try to establish confidence limits for the parameters.

4.1.1.4 On the Use of Constraints

In mathematical optimization problems (Appendix A.2) and constrained least-squares problems (Sects. 4.2.2 and Appendix A.4), constraints are relations among parameters. In optimization theory, constraints are implicit relations connecting several parameters and decreasing the size of the solution space. Sometimes, if constraints are available in explicit form, they can be used to eliminate unknown parameters directly [see, for instance, (4.1.19)]; explicit constraints reduce the dimensionality of the solution space, i.e., the number of adjustable parameters. In some light curve programs, such as the Wilson–Devinney code, explicit constraints are exploited directly in the model. If emission-line activity or secular period change indicates that one component of the binary fills its Roche lobe, the lobe potential

4.1 Mathematical Formulation of the Inverse Problem

can be replaced by the critical value. The eclipse-duration constraint (Wilson, 1979) is another example.

As described in Appendix E.11 for the special case of circular orbits and synchronous rotation, the relation

$$\Theta_e = \Theta_e(i, \Omega, q) \tag{4.1.18}$$

contains the semi-duration Θ_e of the X-ray eclipse and allows us to eliminate the inclination i, the potential Ω, or the mass ratio q. Usually, (4.1.18) is inverted w.r.t. Ω, i.e.,

$$\Omega = \Omega(i, q, \Theta_e), \tag{4.1.19}$$

which expresses the fact that the X-ray eclipse duration puts a limit on the size of the optical star (the X-ray star has negligible dimension).

4.1.1.5 Assignment of Weights

In the context of light curve analysis, a weight assigned to each data point can be regarded as the product of three (hopefully) independent factors

$$w = w^{\text{intr}} w^{\text{flux}} w^c, \tag{4.1.20}$$

with the following meaning as in Wilson (1979):

1. w^{intr} is an intrinsic weight. If normal points are used, w^{intr} is often taken to be the number of data points averaged to produce the "normal" (sometimes called, "binned data") point;
2. w^{flux} is a flux-dependent weight (see below); and
3. w^c is a curve-dependent weight.

We discuss the importance of each factor in turn. For unbinned data, most observers take $w^{\text{intr}} = 1$; however, experience suggests that binning data saves computer time and provides a measure of scatter for each binned point. The calculation of the standard deviation of each binned point obviously requires the use of correct values for the uncertainty in each individual point, if known, or, instead, one can avoid reliance on individual datum weights by applying the factor w^{flux}, as we shall see below. Some practitioners prefer to use merely the number of individual data points in each bin for w^{intr}, irrespective of scatter in those data. If this is done, if the light-dependent weights are selected appropriately, and if the data do indeed scatter accordingly with the flux level, then the effect may be nearly the same as if the weights of the individual data points were calculated from their intrinsic errors and applied directly. But there are no guarantees! Like the issue of dome flats versus sky flats in CCD photometry (see Sect. 2.1.4), the issue of binning or not binning raises the specter of religious warfare among practitioners. If you want to be safe, perform a solution both ways, but compute the weights correctly in each case!

w^{flux} may be set to 1 if and only if the weight factor w^{intr} is computed so as to reflect the actual intrinsic error in each observation. If this is not the case, w^{flux} must be selected carefully. Let $\Delta\ell$ denote the standard deviation of a single observation in units of normalized light ℓ. The precision of a photometric observation usually depends on the star brightness. The dependence of the precision on light level ℓ is governed by the source of the noise. Figure 4.5, adapted from the classical review article by Code & Liller (1962, p. 285), shows the noise contributions to various regimes of star brightness. Note that for the brightest stars, "seeing" or scintillation dominates; for fainter stars, "shot noise" is most important, and for the faintest stars, fluctuations in the sky background contribute significantly to the noise. The sky background and level of seeing vary from site to site and night to night, of course, so the figure should be considered a rough guide only and not a prescription for any specific case. In the context of light curve analysis, Linnell and Proctor (1970) define the relation between $\Delta\ell$ and ℓ by an exponent b, $b \in \{0, 0.5, 1\}$, as follows:

$$\Delta\ell = \beta \cdot \ell^b \qquad (4.1.21)$$

yielding a weight of

$$w^{\text{flux}} = \frac{1}{\ell^{2b}} \qquad (4.1.22)$$

if we chose the constant β as unity. For photoelectric observations of most stars that can be observed with some modicum of precision, in a "photometric sky," shot noise is usually the most important contribution to the noise. In this case, and otherwise

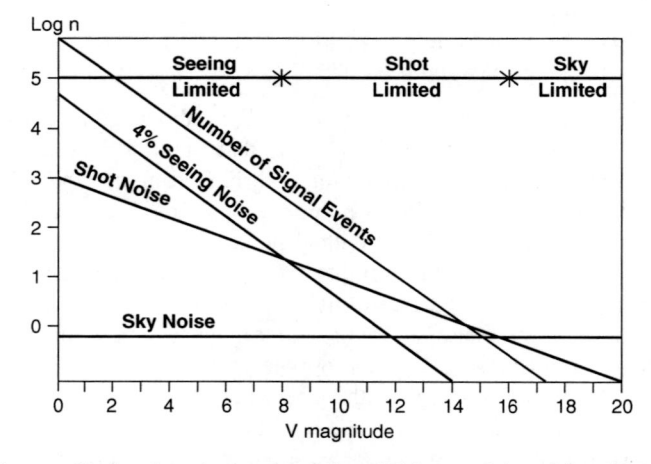

Fig. 4.5 Noise contributions in various regimes of star brightness. Adapted from Fig. 1 in Code and Liller (1962, p. 285). For the brightest stars, "seeing" or scintillation dominates; for fainter stars, "shot noise" is most important; and for the faintest stars, fluctuations in the sky background contribute significantly to the noise. Notice the region where the shot noise due to the sky background ("Sky Limited") becomes dominant over the shot noise from the star

4.1 Mathematical Formulation of the Inverse Problem

ideal circumstances, e.g., full quantum efficiency, the errors are Poisson-distributed and are determined by photon statistics. This means that the expected uncertainty in an observation of n photons is $\sqrt{n}$; therefore, $b = 0.5$. If the errors are constant on the magnitude scale, such as those due to scintillation noise, the major source of noise for bright stars in an otherwise perfect sky, one should use $b = 1$. This arises because of the relation between magnitude differences and light fluctuations: $\Delta m = 1.086 \Delta \ell / \ell$ so that if $\Delta m = $ const we have $\Delta \ell \propto \ell$. Finally, if the major source of the noise is "read noise" from a CCD chip, or detector noise from a photoconductive device, the errors are independent of light level and $b = 0$. See Young et al. (1991) for a detailed assessment of the sources of photometric error.

In many cases, the appropriate choice of flux-dependent weight is not clear because several sources of noise can occur together. Variable sky transparency can be a major source of noise and may contribute in more than one way, especially if the observations are made near a bright city sky, where increased cloudiness means also an increase in the sky brightness and the shot noise associated with it. In such a case, not even clever instruments such as the *Rapid Alternate Detection System* (see Chap. 2) or other two-star photometers may be able to improve much on the error in the data for the following reason: As the star dims, it also undergoes irregular fluctuations as cloud transparency varies, and, on top of this, the fluctuations in sky brightness as well as the shot noise in the sky flux all contribute to the noise level. Clearly the separation of the comparison and program stars on the sky and the spatial and temporal scales of cloud variation play critical roles in determining how well these compensatory photometry systems can overcome the intrinsic difficulties. This having been said, we hasten to note that we would not be without one! Many, many nights of useful data collected on otherwise useless nights attest the value of RAD systems because we can sometimes make up deficits in precision by increases in integration time or through the superposition of data. The main difficulty arises when the data are phase limited to an extent that longer integration or superposition is impractical.

The use of a nonzero value for b assumes that the data have not been binned or otherwise averaged. If they *have* been binned, and the standard deviations of the means are used to establish the weights of each averaged or binned point, the resulting weight should have already contained some light-dependent factor; in such a case, the appropriate value for b is 0. Otherwise, the weighting factors will no longer be independent, and systematic trends in the residuals may result.

The curve-dependent weights w^c are important for the simultaneous analysis of either a set of several light curves or possibly a set of light and radial velocity curves (see Sect. 4.1.1.6). Each light or velocity curve has its own variance which can be used to weight the data in that curve. Usually, but not always, light curves in shorter wavelengths, such as the Johnson U, show larger scatter than those in longer wavelengths, such as V or I. The scatter may be intrinsic, such as active region emission, or arise from variable atmospheric extinction, to which ultraviolet photometry is especially susceptible, or may also come from auroral emission. The latter explanations are less likely if the data are obtained from the variable and comparison stars simultaneously, as in a CCD or multichannel photometer, but spatial variations in auroral emission are possible. In any case, we need to provide a weight or a measure of accuracy for the light curve as a whole. In the Wilson–Devinney program, a variance σ_c is input. Usually, σ_c is the standard deviation of a single observation (rather than of the mean), and thus the associated weight follows (see Appendix A.3 for reasoning) as

$$w^c = \frac{1}{10,000\sigma_c^2}. \qquad (4.1.23)$$

The factor 10,000 ensures that we get weights of convenient size. This error or standard deviation may be computed from differences between consecutive, phased observations within a specified interval, such as 0.005^p (or more in the maximum part of an EA-type light curve). The computed error will then be of the form

$$\sigma = \sqrt{\frac{\sum_{i=1}^{n}(\Delta\ell_i)^2}{p(n-1)}}, \qquad (4.1.24)$$

where p is the number of pairings, probably less than $2n$, where n is the number of independent observations i because not all parts of the light curve may be equally well covered. Alternatively, σ may be computed from the fit error of a truncated Fourier series representation of the light curve outside of the eclipse. This method is not without pitfalls, however. Perturbations may influence the light curve and cause higher-order terms in the Fourier series of the form

$$\ell(\Phi) = \sum_{m=1}^{M} [A_m \sin(m\Phi) + B_m \cos(m\Phi)], \qquad (4.1.25)$$

where M is restricted such that the representation is reasonably realistic ($M = 2$; in over-contact systems, perhaps $M = 4$ or more depending on perturbations). If M is too small, the fit error will be larger, and the weight of the light curve in the modeling process will be lower than that of other light curves. This is not a good idea if the perturbations themselves are to be modeled, say, with star spot simulations, because the leverage for spot parameter determinations will then be shortened. If the modeler is not convinced of the physical existence of spots but nevertheless wishes to model the system as well as possible, the only alternative at present is to rectify the light curve of those perturbations by subtraction or division of the "unexplained" sine term components of the Fourier representation, as prescribed by the Russell–Merrill method in Sect. 6.2.1.

4.1.1.6 Simultaneous Fitting

EBs can be very rich data sources. Often several photometric passbands such as B and V or u, v, b, and y are available. Ideal observation sets include one or two radial velocity curves, in addition. It is important to stress that all these data are included in the analysis to yield consistent results. Simultaneous data fitting does not mean that all observations must have been made at the same time. If polarization data or apsidal motion data are included, then these data were taken at different times, possibly separated by decades. Simultaneous data fitting means that all available observations are used simultaneously in one least-squares fit. To be sure, combining all these data from different sources requires careful weighting (see Sect. 4.1.1.5). But if a consistent least-squares solution in such a situation can be found it deserves respect! The real binary system has a definite inclination i, mass ratio q, and so on, so all the data should yield the nonwavelength-dependent properties of the model identically within the error limits. From the physical point of view, model fits in separate passbands yielding different values for i or q are difficult to interpret. At best they indicate a problem with the model or some of the fixed parameters. From the numerical point of view, the "separate curve" approach involves too many free parameters. Simultaneous fitting improves the safety at which parameters are determined. Of course, with more constraints, or fewer free parameters, the fit of the individual curves may look less attractive. A typical light curve fitting in different passbands (say, u, v, b, and y), done separately, might have the following free parameters:

$$u \text{ curve}: i \; q \; \Omega_1 \; \Omega_2 \; T_2 \; L_1^u$$
$$v \text{ curve}: i \; q \; \Omega_1 \; \Omega_2 \; T_2 \; L_1^v$$
$$b \text{ curve}: i \; q \; \Omega_1 \; \Omega_2 \; T_2 \; L_1^b$$
$$y \text{ curve}: i \; q \; \Omega_1 \; \Omega_2 \; T_2 \; L_1^y.$$

In the separate curve fitting, each curve has six adjustable parameters. The result may be 4 nice looking fits produced by a total of 24 parameters. In a simultaneous fit (representing the correct physics), there would be only $5 + 4 \cdot 1 = 9$ free parameters (five global and four curve-dependent ones). This demonstrates how simultaneous fitting can drastically reduce the number of free parameters.

A remarkable analysis of an eclipsing X-ray binary is described in Sect. 7.3.1. The analysis by Wilson & Terrell (1994) includes B and V light curves, an optical (He I) radial velocity curve, pulse arrival times, and estimations of X-ray eclipse duration. This analysis is remarkable both for the rich physics and the advantages of simultaneous fitting that it illustrates.

For a systematic discussion of simultaneous fitting of curves (or multi-experiment analysis) from a mathematical point of view, the reader is referred to Schlöder & Bock (1983) and Schlöder (1988). These references also contain efficient algorithms to solve such problems.

4.2 A Brief Review of Nonlinear Least-Squares Problems

Ab origine   (*From the beginning*)

The inverse problem formulated in Sect. 4.1 is a typical nonlinear least-squares problem which in turn is an optimization problem with a special structure. More details are given in Appendix A. Here we provide a brief overview of nonlinear unconstrained and constrained least-squares.

4.2.1 Nonlinear Unconstrained Least-Squares Methods

Legendre (1805) first published the method of least-squares, applying it to data from the 1795 French meridian arc survey as an example. In 1809, Gauß in *Theoria Motus Corporum Coelestium* derived the justification for the method in terms of the normal error law, showed how to obtain the errors of the estimated parameters, and also how nonlinear problems could be linearized, so that the method could be applied to the problem of nonlinear parameter estimation. He also claimed he had been using the method since 1795, 10 years before Legendre's work was published. It appears that Gauß had indeed been using the method as he claimed, but had not appreciated its wider importance until Legendre's publication. For a more detailed discussion of this priority conflict, we refer the reader to Stigler (1986) or Schneider (1988).

Since the time of Gauß, numerical methods for solving several types of least-squares problems, e.g., those with probablistic constraints [Eichhorn (1978) and Eichhorn & Standish (1981)] and those involving differential equations [Bock (1981), Kallrath et al. (1993) and Kallrath (1999)] have been developed and improved, and there is still much active research in that area. For a review of the methods of least-squares as known and used in astronomy, we refer to Jefferys (1980, 1981) and Eichhorn (1993).

A popular method, sometimes also called "damped least-squares" or DLS for short, is the Levenberg–Marquardt algorithm proposed independently by Levenberg (1944) and Marquardt (1963). It modifies the eigenvalues of the normal equation matrix and tries to reduce the influence of eigenvectors related to small eigenvalues [cf. Dennis & Schnabel (1983)].

Since its original invention by Levenberg (1944) over half a century ago, there have been numerous comparisons of DLS with other methods for nonlinear least-squares, such as simple step-cutting (solving a line-search problem and reducing the step-size if necessary) and derivative-free methods. Although there was a time when the Broyden–Fletcher–Goldfarb–Shanno secant method was thought to be superior, there now seems to be general agreement that DLS is similarly effective, and much simpler to code. For this reason, it can be found in commercially available software packages. DLS handles problems with exponential or logarithmic nonlinearities nicely and works well on problems that have only power-law nonlinearities. To achieve global convergence (i.e., convergence even for ill-chosen initial points), it is necessary to choose the damping strategy carefully.

Damped (step-size cutting) Gauß–Newton algorithms that control the damping by natural level functions [Deuflhard & Apostolescu (1977, 1980) Bock (1987)] seem to be superior to Levenberg–Marquardt-type schemes and can be more easily extended to nonlinear constrained least-squares problems. To avoid confusion

4.2 A Brief Review of Nonlinear Least-Squares Problems

about the use of the word "damping," we clarify that we use it for both the damped Levenberg–Marquardt algorithm (which modifies the system matrix) and step-cutting algorithms.

4.2.2 Nonlinear Constrained Least-Squares Methods

A common basic feature and limitation of least-squares methods used in astronomy, but seldomly explicitly noted, is that they require some explicit model to be fitted to the data. Most models are based on physical laws or include geometry properties, and very often lead to differential equations which may, however, not be solvable in a closed analytical form. Thus, such models do not lead to explicit functions we want fit to data. We rather need to fit an implicit model (represented by a system of differential equations or another implicit model). Therefore, it seems desirable to develop least-squares algorithms that use the differential equations as constraints or side conditions to determine the solution implicitly. By a multiple shooting approach (Bock 1987), such differential equation-based side conditions can be discretized and represented as equality constraints. The demand for, and the applications of such techniques, are widespread in science, especially in the rapidly increasing fields of nonlinear dynamics in physics and astronomy, nonlinear reaction kinetics in chemistry (Bock 1981), nonlinear models in material sciences (Kallrath et al. 1999) or biology (Baake & Schlöder 1992), and nonlinear systems describing ecosystems in biology or the environmental sciences.

Formally, we want to be able to solve a least-squares problem with n_c constraints as a constrained optimization problem of the type

$$\min_{\mathbf{x}} \left\{ \mathbf{F}_1(\mathbf{x})^2 \mid \mathbf{F}_2(\mathbf{x}) = 0 \text{ or } \geq 0 \in \mathbb{R}^{n_c} \right\}. \qquad (4.2.1)$$

Using the nomenclature of Appendix A.3 and the definition (4.1.9) of the residual vector including weights, $\mathbf{F}_1(\mathbf{x})$ follows as

$$\mathbf{F}_1(\mathbf{x}) = \mathbf{R}(\mathbf{x}), \quad \mathbf{R}(\mathbf{x}) = \mathbf{Y} - \mathbf{F}(\mathbf{x}) = \sqrt{\mathbf{W}} \left[\mathbf{o} - \mathbf{c}(\mathbf{x}) \right]. \qquad (4.2.2)$$

This usually large, constrained, structured nonlinear problem is solved by a damped generalized Gauß–Newton method; cf. Bock (1987). Here we describe only the equality constrained case. Starting with an initial guess $\mathbf{x}_0$, the variables are iterated via

$$\mathbf{x}_{k+1} = \mathbf{x}_k + \alpha_k \Delta \mathbf{x}_k \qquad (4.2.3)$$

with a damping constant α_k restricted to the range $0 < \alpha_{\min} \leq \alpha_k \leq 1$. In order to compute the increment $\Delta \mathbf{x}_k$, we substitute $\mathbf{x}$ in (4.2.1) by $\mathbf{x}_k + \Delta \mathbf{x}_k$ and linearize the problem around $\mathbf{x}_k$. Then $\Delta \mathbf{x}_k$ is the solution of the linear, equality constrained least-squares problem

$$\min_{\Delta \mathbf{x}_k} \left\{ \|J_1(\mathbf{x}_k)\Delta \mathbf{x}_k + \mathbf{F}_1(\mathbf{x}_k)\|_2^2 \mid J_2(\mathbf{x}_k)\Delta \mathbf{x}_k + \mathbf{F}_2(\mathbf{x}_k) = 0 \right\} \quad (4.2.4)$$

with the Jacobian matrices (see Appendix A.1)

$$J_i(\mathbf{x}_k) := \frac{\partial \mathbf{F}_i}{\partial \mathbf{x}}(\mathbf{x}_k). \quad (4.2.5)$$

Under appropriate assumptions about the regularity of the Jacobians J_i, there exists a unique solution $\Delta \mathbf{x}_k$ of the linear problem and a unique linear mapping J_k^+ [called *generalized inverse* as introduced by Bock (1981)] obeying the relations

$$\Delta \mathbf{x}_k = -J_k^+ F_1(\mathbf{x}_k), \quad J_k^+ J_k J_k^+ = J_k^+, \quad J_k^T := [J_1(\mathbf{x}_k)^T, J_2(\mathbf{x}_k)^T]. \quad (4.2.6)$$

The solution $\Delta \mathbf{x}_k$ of the linear problem follows uniquely from the Kuhn–Tucker conditions (Kuhn & Tucker 1951)

$$J_1^T J_1 \Delta \mathbf{x}_k - J_2^T \lambda_c + J_1^T \mathbf{F}_1 = 0, \quad J_2 \Delta \mathbf{x}_k + F_2 = 0, \quad (4.2.7)$$

where $\lambda_c \in \mathbb{R}^{n_c}$ is a vector of Lagrange multipliers.

For the numerical solution $\Delta \mathbf{x}_k$ of the linear constrained problem (4.2.4), several structure-exploiting methods, for example the one by Bock (1981), have been developed which compute special factorizations of J_1 and J_2 and thus implicitly, but not explicitly, the generalized inverse J^+. The availability of the Jacobians J_1 and J_2 allows rank checks in each iteration and automatic detection of violations of the regularity assumptions. In this case, automatic regularization and computation of a relaxed solution is possible (Bock 1981).

The iteration (4.2.3) can be forced to converge globally to a stationary point of the problem if the damping factors α_k are chosen appropriately. In the treatment of a large number of practical problems, strategies based on *natural level functions* have proven to be very successful [cf. Bock (1987), the brief summary in Kallrath et al. (1993), or Appendix A]. In the region of local convergence of the full step method, the algorithm converges linearly to a solution that is stable against statistical variations in the observations. An iterate, $\mathbf{x}_k$, is accepted as solution $\mathbf{x}_*$ of the nonlinear constrained problem if a scaled norm of the increments $\Delta \mathbf{x}_k$ is below a user-specified tolerance. As the Jacobians and their decompositions are available in each iteration, covariance and correlation matrices are easily computable (Bock 1987) for the full variable vector $\mathbf{x}_*$.

4.3 Least-Squares Techniques Used in Eclipsing Binary Data Analysis

Inter nos; inter vivos   (*Between ourselves; between the living*)

This section describes the least-squares methods used and implemented in several light curve programs.

4.3.1 A Classical Approach: Differential Corrections

Differential corrections can be understood as an undamped Gauß–Newton method, or in accordance with what has been said above, as an undamped Levenberg–Marquardt algorithm. Thus, it is a special case of the techniques described in the previous section. However, as the Differential Correction method is widespread within the *EB* community[5] and has achieved venerable status – it was first used to determine the parameters of *EB*s by Wyse (1939) and Kopal (1943) – it deserves detailed discussion.

The nonlinear least-squares problem is linearized in the following way: starting from the necessary condition $\nabla f(\mathbf{x}) = 0$ and an initial solution $\mathbf{x}_k \in \mathbb{R}^n$ present at the beginning of the kth iteration, a correction vector $\Delta \mathbf{x}_k$ is defined such that $\mathbf{x}_k + \Delta \mathbf{x}_k$ obeys the necessary condition. With the unweighted residual vector $\mathbf{d}(\mathbf{x}) \in \mathbb{R}^N$, the least-squares function (4.1.11) leads to the necessary condition

$$\nabla f(\mathbf{x}) = 0 \quad \Leftrightarrow \quad 2A(\mathbf{x})W\mathbf{d}(\mathbf{x}) = 0, \quad A \in M(n, N), \tag{4.3.1}$$

where $M(n, N)$ denotes the sets of matrices with n rows and N columns, and A is given by

$$A_{i\nu}(\mathbf{x}) := \frac{\partial}{\partial x_i} d_\nu(\mathbf{x}) = -\frac{\partial}{\partial x_i} o_\nu^{\mathrm{cal}}(\mathbf{x}), \quad A \in M(n, N). \tag{4.3.2}$$

Replacing the unknown solution $\mathbf{x}$ by $\mathbf{x}_k + \Delta \mathbf{x}_k$ yields the necessary condition

$$A(\mathbf{x}_k + \Delta \mathbf{x}_k) W \mathbf{d}(\mathbf{x}_k + \Delta \mathbf{x}_k) = 0, \tag{4.3.3}$$

or component-wise

$$\sum_{\nu=1}^{N} A_{i\nu}(\mathbf{x}_k + \Delta \mathbf{x}_k) w_\nu d_\nu(\mathbf{x}_k + \Delta \mathbf{x}_k) = 0, \quad i = 1, \ldots, n. \tag{4.3.4}$$

Taylor-series expansion up to first-order derivatives gives

$$\left(A(\mathbf{x}_k) + \Delta \mathbf{x}_k^T G \right) W \left(\mathbf{d}(\mathbf{x}_k) + A^T(\mathbf{x}_k) \cdot \Delta \mathbf{x}_k \right) \doteq 0,$$

or with the Hessian matrix G_{ij}^ν [interpret this as an $M(n, n)$ matrix at each data point indexed by ν]

---

[5] Many photometric light curve data have been analyzed with the Wilson–Devinney program (Wilson & Devinney, 1971) the first to use the method of Differential Corrections with a physical light curve model.

$$G_{ij}^v := \frac{\partial}{\partial x_i}\left(\frac{\partial}{\partial x_j}d_v(\mathbf{x})\right), \quad G \in M(n, n, N). \tag{4.3.5}$$

Multiplication (and neglecting second-order terms in $\Delta \mathbf{x}_k$) finally leads to

$$A(\mathbf{x}_k) \cdot W \cdot \mathbf{d}(\mathbf{x}_k) + \left[\mathbf{d}^T(\mathbf{x}_k) \cdot W \cdot G(\mathbf{x}_k) + A(\mathbf{x}_k) \cdot W \cdot A^T(\mathbf{x}_k)\right]\Delta \mathbf{x}_k = 0, \tag{4.3.6}$$

or component-wise for $i = 1, \ldots, n$ [see, for instance, Powell (1964a)]

$$\sum_{v=1}^{N}\left\{\begin{array}{c} A_{iv}(\mathbf{x}_k)w_v d_v(\mathbf{x}_k) \\ +\sum_{j=1}^{n}\left[G_{ij}^v(\mathbf{x}_k)w_v d_v(\mathbf{x}_k) + A_{iv}(\mathbf{x}_k)A_{jv}(\mathbf{x}_k)\right]\Delta \mathbf{x}_k \end{array}\right\} = 0. \tag{4.3.7}$$

Under the "small residual assumption"

$$\left\|\mathbf{d}^T(\mathbf{x}_k) \cdot W \cdot G(\mathbf{x}_k)\Delta \mathbf{x}_k\right\| \ll \left\|A(\mathbf{x}_k) \cdot W \cdot A^T(\mathbf{x}_k)\Delta \mathbf{x}_k\right\|, \tag{4.3.8}$$

we get the equations

$$[A(\mathbf{x}_k) \cdot W \cdot A^T(\mathbf{x}_k)] \cdot \Delta \mathbf{x}_k = -A(\mathbf{x}_k) \cdot W\mathbf{d}(\mathbf{x}_k), \tag{4.3.9}$$

which is the normal equation of the linear least-squares problem

$$\min_{\mathbf{x}_k}\left\|\sqrt{W}\left(\mathbf{y} - A^T\mathbf{x}_k\right)\right\|_2, \quad \mathbf{y} = -\mathbf{d}(\mathbf{x}_k). \tag{4.3.10}$$

Note the similarity between the method described here and the formal procedure in Appendix A.3.3 with $J = A^T$. Iterations are continued with

$$\mathbf{x}_{k+1} = \mathbf{x}_k + \Delta \mathbf{x}_k \tag{4.3.11}$$

until one of the stopping criteria described in Sect. 4.4.2 terminates the algorithm.

To compute the correction vector $\Delta \mathbf{x}_k$ in the kth iteration

$$\Delta \mathbf{x}_k = -\hat{C} \cdot A(\mathbf{x}_k) \cdot W\mathbf{d}(\mathbf{x}_k), \quad \hat{C} := C^{-1} = (\hat{c}_{ij}), \tag{4.3.12}$$

we need the inverse of the matrix

$$C = (c_{ij}) := A(\mathbf{x}_k) \cdot W \cdot A^T(\mathbf{x}_k). \tag{4.3.13}$$

The inverse of C is just the *covariance matrix* $\hat{C}$. Under the assumptions discussed, for example, by Press et al. (1992, pp. 690–693), the diagonal elements of the covariance matrix $\hat{C}$ provide a measure for the probable error δ (see Sect. 4.4.3).

In summary, the method of Differential Corrections belongs to a class of Newton-type methods without second derivatives. It makes use of the fact that the gradient

4.3 Least-Squares Techniques Used in Eclipsing Binary Data Analysis

and the Hessian matrix of $f(\mathbf{x})$ have a special structure but is based on the premise that eventually first-order terms will dominate second-order terms, an assumption that is not justified when the residuals of the solution are very large because the second-order term (4.3.8) contains the product of second derivatives and residuals. In practice, in light curve analysis, it turns out that this assumption seems to be valid, i.e., the residuals at the solutions are small enough. The bounds (4.1.14) are not taken into account in the method of Differential Corrections. The procedure can be applied successfully if an initial solution $\mathbf{x}_0$ close enough to the solution $\mathbf{x}_*$ is known and if the correlations between the parameters are not too large. However, these requirements are not always met. The following disadvantages are sometimes mentioned in the context of Differential Corrections. Note that some of the following points (1 and 3) apply to any derivative-based least-squares method:

1. The initial solution $\mathbf{x}_0$ might not be well known and not located in the local convergence region. The result can be divergence or oscillation. It is not unusual to observe higher standard deviations σ^{fit} after a few iterations.
2. Most of the partial derivatives $\partial_{x_i} d_\nu(x)$ can be derived only numerically for light curve solutions. This means that to adjust m parameters, $m+1$ or $2m+1$ light curves have to be calculated at each iteration, depending on whether asymmetric [see (4.5.25)] or symmetric [see (4.5.26)] finite differences are used to compute the partial derivatives. Accuracy depends critically on the choice of the increments Δx_i and is difficult to control (see Sect. 4.5.5). As is shown in Sect. 4.5.4, for some parameters, it is possible to calculate derivatives analytically.
3. Convergence problems, due to both nonlinearity and correlations, have long been observed in applications of Differential Corrections (Wilson & Biermann 1976). The linearized normal equations are sometimes ill-conditioned with condition numbers of the order of 10^6, i.e., parameters are strongly correlated. In light curve analysis, the problem is pronounced when the mass ratio is an adjustable parameter. Wilson & Biermann (1976) cope with these problems and solve alternatively and iteratively for subsets of parameters in successive iterations (see Sect. 4.3.2).
4. The need for precision, hence grid density, for the calculation of the theoretical light curve by numerical quadrature of the flux over the star's surface is obviously much greater than for a direct search method. Computing-time costs and/or restricted memory have prevented many light curve analyzers from using all observed data points. Instead, they form *normal points* and analyze them to estimate the parameters. Although we do not recommend this procedure enthusiastically, because there are subjective decisions to be made in the binning/averaging procedure such as the width of the bins in different parts of the light curve, in practice the results are not significantly different in most cases. See Sect. 4.1.1.5 for further discussion of the intrinsic merits of binning and not binning. Machines are now becoming so fast and memory so inexpensive that normal points are no longer needed.

4.3.2 Multiple Subset Method and Interactive Branching

To overcome some of the limitations of Differential Corrections, Wilson & Biermann (1976) introduced the *method of multiple subsets*, by which they could separate the most correlated parameters during the optimization process (which still proceeds by Differential Corrections). If $\mathcal{P}$ denotes the set of free or adjustable parameters, then this method solves for parameter subsets $\mathcal{P}_i \subset \mathcal{P}$ in each iteration. $\mathcal{P}$ is separated into two or (rarely) more, not necessarily disjoint subsets $\mathcal{P}_i$, but, of course, we expect that the union of all, say, N^S subsets

$$\bigcup_{i=1}^{N^S} \mathcal{P}_i = \mathcal{P} \tag{4.3.14}$$

coincides with the original set $\mathcal{P}$. The solution proceeds by alternating subset solutions. As the derivatives were already calculated for the full set $\mathcal{P}$, by interactive branching several combinations of free parameters can be established without much additional computing time. Associated with these subsets are normal equations which have smaller dimension and a smaller condition number. The standard errors of the parameters are derived by a final run on the whole parameter set (ignoring the corrections). The method requires some interactive intervention of the human analyzer but nevertheless many authors proved its efficiency (Wilson 1988). The process of applying the subset with the smallest predicted σ^{fit} has turned out to be an effective way to arrive at the deepest minimum in most cases. Intuitively, this method can be understood as a decomposition technique. In the example of two-dimensional minimization in the x- and y-directions, this decomposition would lead to a minimization in the x- and y-direction separately, i.e., fixing y and minimizing with respect to x, fixing x and minimizing with respect to y, and so on.

The multiple subset method addresses the difficulties caused by correlations between parameters. To overcome problems related to nonlinearity, a damping strategy is required (see Sect. 4.3.3, or Appendix A.1).

4.3.3 Damped Differential Corrections and Levenberg–Marquard Algorithm

A frequently used method to increase the convergence region of nonlinear least-squares problems is the Levenberg–Marquardt algorithm (Levenberg 1944, Marquardt 1963), sometimes also called the Marquardt method. Early users of the Levenberg–Marquardt algorithm in the context of *EB* light curves were Hill (1979) and Djurasevic (1992). Wilson & Terrell (1998) also mentioned the use of this method in the Wilson–Devinney program.

The damped least-squares algorithm implemented in WD95 (Kallrath et al. 1998), a version of the Wilson–Devinney program described in Chap. 7, is based on the normal equation and modification of the system matrix (4.3.13) in Sect. 4.3.1 and

4.3 Least-Squares Techniques Used in Eclipsing Binary Data Analysis

is identical to the Levenberg–Marquardt strategy for damping. In its simplest form, the key idea of the Levenberg–Marquardt algorithm is to add a multiple of the identity matrix $\mathbb{1}$, $\lambda_k \mathbb{1}$, to the Hessian in (4.3.9) with a suitable nonnegative damping factor λ_k. In place of the normal equation (4.3.9), we get a regular system of linear equations of the form

$$[C_k + D_k]\Delta x_k = -A(x_k) \cdot Wd(x_k), \qquad (4.3.15)$$

where D_k is a diagonal matrix usually chosen as $D_k = \lambda_k \mathbb{1}$. In general, the crucial point of the method is to choose the damping properly[6] as discussed by Moré (1978). A heuristic procedure to choose λ_k is given in Kallrath et al. (1998) and looks like this:

1. for given x_k compute $f(x)$ and establish the normal equation, in particular, compute C;
2. for the initial damping constant λ pick a modest value, for example, $\lambda = 10^{-4}$, and set $\lambda_0 = \lambda$;
3. the diagonal elements c_{jj} are replaced by $(1 + \lambda_k)c_{jj}$;
4. solve the modified normal equation (4.3.15) for Δx_k and set the potential iterated vector $x' = x_k + \Delta x_k$;
5. compute $f(x')$;
6. acceptance test: if $f(x') \leq f(x_k) \Rightarrow x_{k+1} = x'$ and $\lambda_{k+1} = 0.1\lambda_k$, goto step 2;
7. $f(x') > f(x_k) \Rightarrow$ goto step 8;
8. if $k = 1$ set $\lambda_1 = 1$, then goto step 2; and
9. if $k \geq 2$ set $\lambda_k \leftarrow 3\lambda_k$, goto step 2.

The question remains to define a condition for terminating the procedure. That topic is discussed in Sect. 4.4.2.

As the Levenberg–Marquardt algorithm is sometimes referred to as a "damped least-squares method" let us consider this point for a moment. A good property of the Levenberg–Marquardt method is its numerical stability caused by the fact that the matrix $R'(x_k)^T R'(x_k) + \lambda_k \mathbb{1}$ is positive definite even for rank-deficient Jacobians $R'(x_k)$ (see Appendix A.3.3 for nomenclature). However, often the "solution" is not the minimum of $f(x)$ but only premature termination. Great care is needed because the damping effect of adding $\lambda \mathbb{1}$ to the Hessian hides to some extent the uniqueness problem associated with rank-deficient Jacobians.

4.3.4 Derivative-Free Methods

Derivative-free optimization is performed with direct search algorithms which were reviewed by Murray (1972) and Lootsma (1972). Among others, this group of

---

[6] In light curve analysis, the method seems not to depend on the choice of λ strongly if λ is chosen between 10^{-4} and 10^{-8}.

algorithms includes the Fibonacci line search, and the Simplex algorithm[7] invented by Spendley et al. (1962). The Simplex algorithm becomes efficient if there are more than two or three parameters to be adjusted. In the context of light curve analysis, this algorithm was first used by Kallrath & Linnell (1987). Since then several authors have successfully applied this method. Experiences with the Simplex algorithm in the context of light curve analysis can be found in Kallrath (1993). More recently, Powell's Direction Set Method has been implemented by Prša & Zwitter's (2005) in the light curve package PHOEBE.

4.3.4.1 The Simplex Algorithm

The Simplex algorithm does not require any derivatives to be computed but compares only the function values $f_i := f(x_i)$ at the $m + 1$ vertices of a simplex in parameter space, S. In the first versions of the method by Spendley et al. (1962), the simplex was moved through S by an operation called *reflection*. The operation bears a close relationship to the *method of steepest descent*. A more efficient version by Nelder & Mead (1965) uses three additional operations: *expansion*, *contraction*, and *shrinkage* which move the simplex, adapt it to the hypersurface, and iterate it with decreasing volume to the solution $\mathbf{x}_*$. The basic idea is to eliminate the worst vertex by one of the four operations and to replace it with a better one. Figure 4.6 demonstrates these geometrical operations. $f(x) = \sigma^{\text{fit}}$ has been minimized w.r.t. inclination i and mass ratio q. In the initial simplex $S^{(0)}$, the vertex with the highest value σ^{fit} has been marked with 0. Via reflection, $S^{(0)}$ is mapped onto $S^{(1)}$. $S^{(2)}$ is the result of contracting $S^{(1)}$ toward the center of the simplex. This is a simulated case, so the real solution $x_* = (i, q)_*$ is known and marked as a circle in the $(i - q)$-plane. The construction of $S^{(0)}$, introduced by Kallrath & Linnell (1987) and used in the light curve packages, deviates from the usual procedure as expressed in (4.3.21) because that appeared to be advantageous in *EB* light curve analysis.

Let us consider an arbitrary real-valued continuous function $f(\mathbf{x})$ depending on a vector $\mathbf{x} \in S \subset \mathbb{R}^m$, and $m+1$ vertices of a simplex $S^{(0)}$ constructed below and embedded in $\mathbb{R}^m$. Let $V_i^{(k)}$, $1 \leq k \leq m + 1$, be the vertices of the simplex $S^{(k)}$ with coordinates $x_i^{(k)}$ after k iterations. For convenience we represent $S^{(k)}$ as an $(m + 1) \times (m + 1)$-dimensional matrix $S_{ij}^{(k)}$

$$S_{ij}^{(k)} := \begin{cases} S_{ij}^{(k)}, & \text{for } 1 \leq j \leq m, \\ f_i := f(\mathbf{x}_i), & \text{for } j = m + 1. \end{cases} \quad (4.3.16)$$

The following definitions hold for each iteration, therefore the superscript k is neglected. In particular, we define the function value f^h with highest value as

$$f^h := \max \{f_i, 1 \leq i \leq m + 1\}, \quad (4.3.17)$$

---

[7] Not to be confused with Dantzig's Simplex algorithm used in Linear Programming. First published by Dantzig (1951), this Simplex algorithm was introduced by Dantzig in 1947.

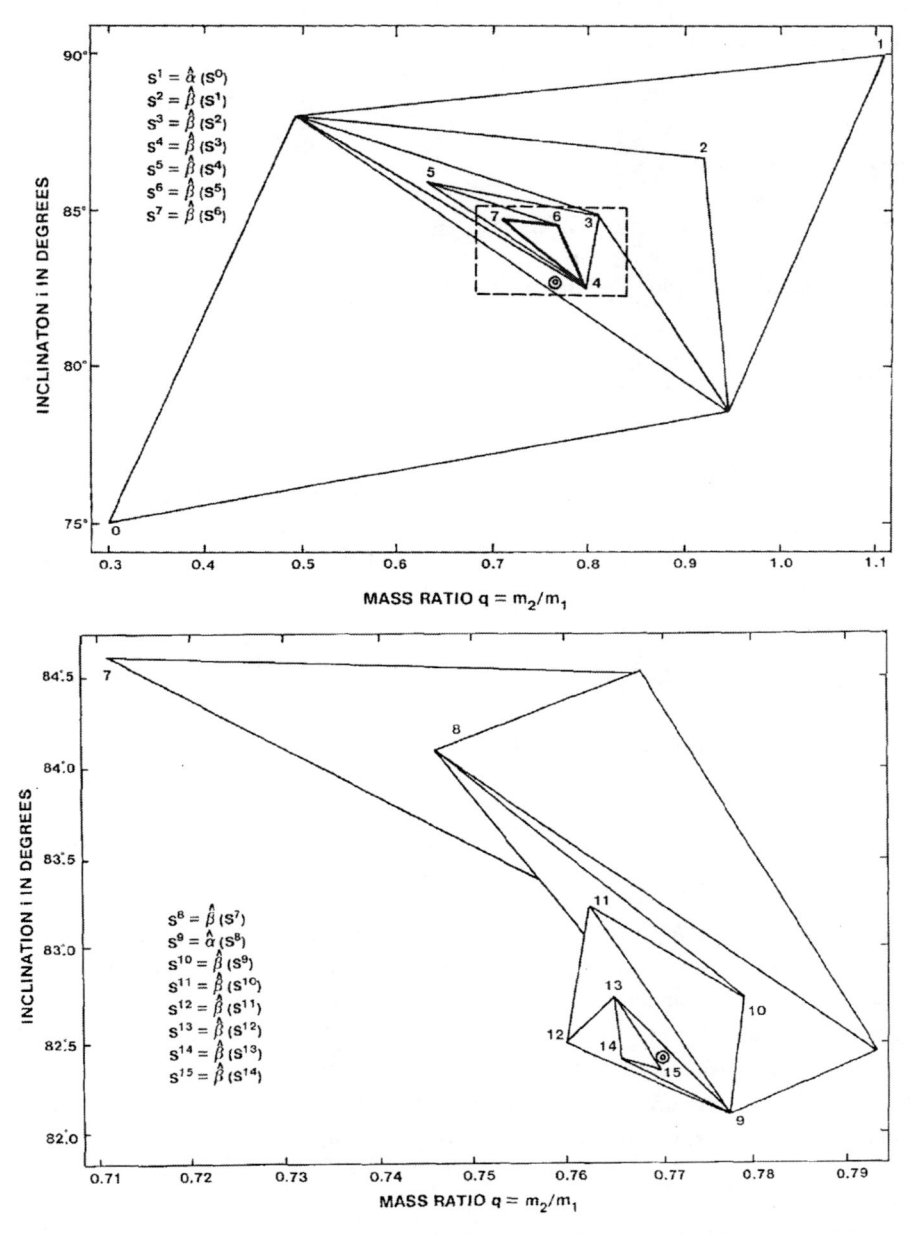

4.3 Least-Squares Techniques Used in Eclipsing Binary Data Analysis

Fig. 4.6 The geometry of the Simplex algorithm. This figure [Figs. 1 and 2 in Kallrath and Linnell (1987, p. 349)] shows the change of simplex shape while moving through a $(i - q)$-parameter plane. The upper part shows the first seven iterations, the lower part the iteration while moving toward the local minimum at $i = 82°\!\!.43$ and $q = 0.770$.

the second highest value as

$$f^s := \max\{f_i, 1 \le i \le m+1, i \ne h\}, \qquad (4.3.18)$$

and the lowest function value as

$$f^l := \min\{f_i, 1 \le i \le m+1\}, \qquad (4.3.19)$$

and eventually the center and the geometrical center of the simplex as

$$\mathbf{x}^{\text{cal}} := \frac{1}{m} \sum_{i=1 | i \ne h}^{m+1} \mathbf{x}_i, \quad \mathbf{x}^g := \frac{1}{m} \sum_{i=1}^{m+1} \mathbf{x}_i. \qquad (4.3.20)$$

Note that for the computation of the center $\mathbf{x}^{\text{cal}}$, the vertex $\mathbf{x}^h$ is omitted.

The initial simplex $S^{(0)}$ depends on the number of parameters m, an initial vector $\mathbf{x}_1^{(0)}$, and a scaling quantity $\mathbf{s} := (s_1, s_2, \ldots, s_m)$ determining the size of $S^{(0)}$. s_j is the size of $S^{(0)}$ projected onto the x_j-axis. Spendley et al. (1962) and Yarbro & Deming (1974), based on statistical considerations, derive the auxiliary quantities

$$(p, q) := \frac{1}{\sqrt{2m}} \left(\sqrt{m+1} + m - 1, \sqrt{m+1} - 1 \right), \qquad (4.3.21)$$

and eventually construct $S^{(0)} = S^{(0)}(m, \mathbf{x}_1^{(0)}, \mathbf{s})$ as

$$x_{ij}^{(0)} := x_{1j}^{(0)} + s_j \begin{cases} 0, i = 1, \\ p, j = i - 1. \\ q, \text{else}. \end{cases} \qquad (4.3.22)$$

According to that definition, we have $x_{ij}^{(0)} \ge x_{1j}^{(0)}$. The geometrical interpretation is that the already known simplex is constructed from one of its vertices. However, if information such as an initial solution $\mathbf{x}_1^{(0)}$ about the minimum of $f(\mathbf{x})$ is available, it seems better to construct the simplex $S^{(0)} = S^{(0)}(m, \mathbf{x}_1^{(0)} - q\mathbf{s}, \mathbf{s})$, whose geometrical center $\mathbf{x}^g$ approximately coincidences with $\mathbf{x}_0$, i.e., $\mathbf{x}^g \approx \mathbf{x}_1^{(0)}$. In order to cover a parameter space defined by simple bounds, the simplex $S^{(0)} = S^{(0)}(m, \mathbf{x}_1^{(0)}, \mathbf{s}/p)$ is appropriate because its vertices lie on the edges of a hypercube.

The iteration process is summarized as a flow chart in Fig. 4.7 using the following definitions:

$$\begin{aligned}
\mathbf{x}^\alpha &= \mathbf{x}^\alpha(\mathbf{x}) := (1+\alpha)\mathbf{x}_c - \alpha\mathbf{x}; & 0 < \alpha, \\
\mathbf{x}^\beta &= \mathbf{x}^\beta(\mathbf{x}) := (1-\beta)\mathbf{x}_c + \beta\mathbf{x}; & 0 < \beta < 1, \\
\mathbf{x}^\gamma &= \mathbf{x}^\gamma(\mathbf{x}) := (1+\gamma)\mathbf{x}_c + \gamma\mathbf{x}; & 0 < \gamma, \\
\mathbf{x}^\delta &= \mathbf{x}^\delta(\mathbf{x}) := \mathbf{x}^l + \delta(\mathbf{x} - \mathbf{x}^l); & 0 < \delta.
\end{aligned} \qquad (4.3.23)$$

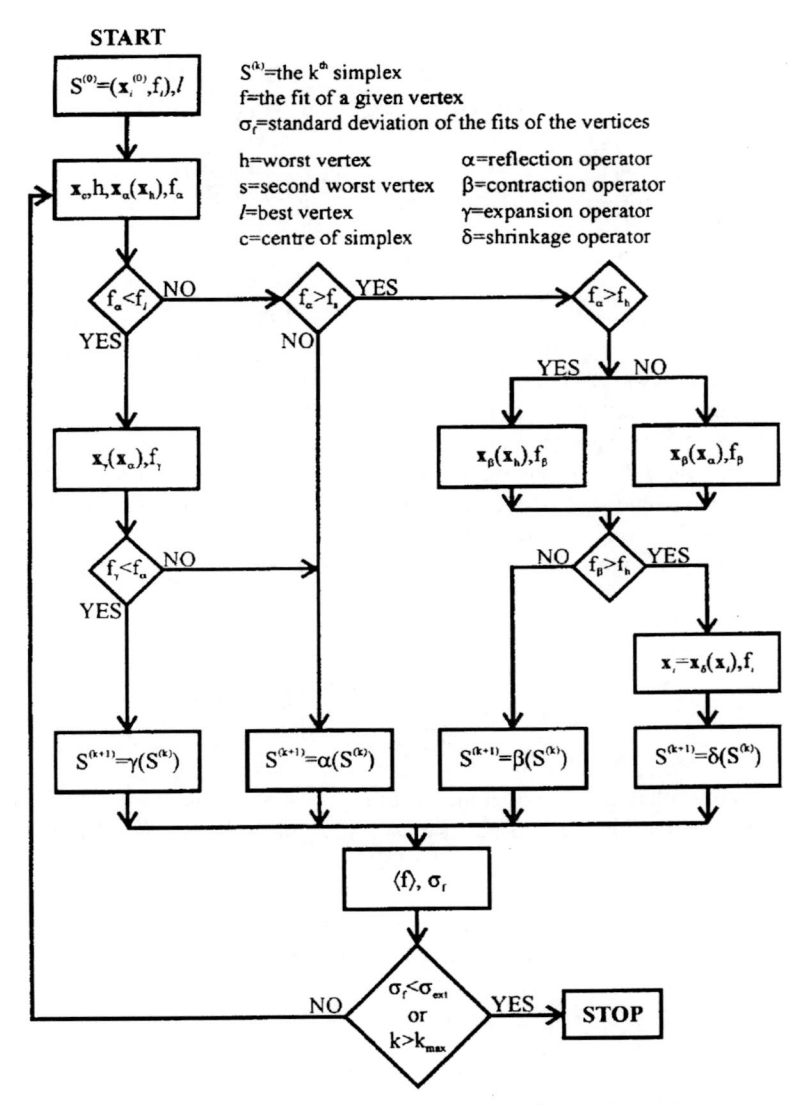

4.3 Least-Squares Techniques Used in Eclipsing Binary Data Analysis

Fig. 4.7 Flow chart of the Simplex algorithm. This figure [Fig. 9 in Kallrath and Linnell (1987, p. 356)] represents the rules embedded in the Simplex algorithm

The four operations $\hat{\alpha}$ (reflection), $\hat{\beta}$ (contraction), $\hat{\gamma}$ (expansion), and $\hat{\delta}$ (shrinkage) are applied according to the following scheme:

$$\begin{array}{lll}
\text{reflection} & S^{(k+1)} = \hat{\alpha}\left(S^{(k)}\right), \mathbf{x}^h \to \mathbf{x}^\alpha(\mathbf{x}^h) & \overset{if}{f^s \geq f^\alpha > f^l,} \\
\text{expansion} & S^{(k+1)} = \hat{\gamma}\left(S^{(k)}\right), \mathbf{x}^h \to \mathbf{x}^\gamma(\mathbf{x}^h) & f^\gamma < f^\alpha < f^l, \\
\text{contraction} & S^{(k+1)} = \hat{\beta}\left(S^{(k)}\right), \mathbf{x}^h \to \begin{array}{l}\mathbf{x}^\beta(\mathbf{x}^h) \\ \mathbf{x}^\beta\left(\mathbf{x}^\alpha(\mathbf{x}^h)\right)\end{array} & \begin{array}{l}f^h < f^\alpha, \\ f^\alpha \leq f^h,\end{array} \\
\text{shrinkage} & S^{(k+1)} = \hat{\delta}\left(S^{(k)}\right), \mathbf{x}^h \to \mathbf{x}^\delta(\mathbf{x}^i), i \neq l & f^\beta \geq f^h.
\end{array} \quad (4.3.24)$$

By careful selection of the parameters α, β, γ, and δ, the four operations can be controlled. The values $\alpha = 1.0$, $\beta = 0.35$, $\gamma = 2.0$, and $\delta = 0.5$ are recommended by Parkinson & Hutchinson (1972). For most light curve analyses, this set of parameters is used, but sometimes $\beta = 0.5$ is also substituted.

From the definitions of the operations, it follows that the volume of the simplex and the function value at that vertex with minimal function value are monotonically decreasing functions of k unless the operation *expansion* occurs. The iterations are halted by one of three stopping criteria. The first is the number $k_{\max}$ of iterations. The second stops iterations when all mean errors, σ_j, of the column average, $\bar{S}_j$,

$$\sigma_j^2 = \frac{1}{m}\sum_{i=1}^{m+1}\left(S_{ij} - \bar{S}_j\right)^2, \quad \bar{S}_j = \frac{1}{m+1}\sum_{i=1}^{m+1} S_{ij}, \quad (4.3.25)$$

become smaller than a given tolerance ε_j, i.e., $\sigma_j \leq \varepsilon_j$ for all j. Third, iterations may be stopped if f^l becomes smaller than a threshold value f^t. If $f(\mathbf{x})$ represents a least-squares function, e.g., the standard deviation of the fit, σ^{fit}, then f^t may be set to the inner noise, σ^{data}, of the data.

As the Simplex algorithm is a direct search method, it cannot give standard errors for the parameters. It is also difficult to add additional parameters during the course of iterations.

4.3.4.2 Powell's Direction Method

Powell's Direction Set method (Powell 1964a,b) is an efficient method for finding the unconstrained minimum of a function of n variables without calculating derivatives. The method has a quadratic termination property, i.e., quadratic functions are minimized in a predetermined number of operations, performing $n^2 + O(n)$ line search steps along conjugate directions in parameter space. In Appendix A.1 we briefly mentioned variable metric and conjugate direction methods. The method was first applied to light curve analysis by Prša & Zwitter's (2005b). A more recent write-up of Powell's method by Vassiliadis & Conejeros (2008) provides the computational details (line search, choice of direction), and a discussion about the termination criterion.

4.3.4.3 Simulated Annealing

Simulated annealing (SA) dates back to Metropolis et al. (1953) and is a function-evaluation-based improvement or stochastic search method often used to solve combinatorial optimization problems; see Pardalos & Mavridou (2008) for a recent description. It is capable to escape local minima and can, in principle, find the global optimium, however, without any guarantee. During its iterations, SA accepts not only better than previous solutions, but also worse quality solutions controlled probabilistically through a control parameter T. The basis requirements are an initial starting point, a neighborhood concept and a cooling scheme involving a few tuning parameters. For continuous optimization problems, the neighborhood concept leads to parameter variations of the form:

$$x_i \to x'_i = x_i + r \cdot v_i,$$

where x_i is a parameter, r is a random number in the range $[-1, +1]$, and v_i is the step length for this parameter. The function to be minimized, $f(x)$, is then computed for the new values of the parameter, to give the value $f'(x')$. In our program, this function is the sum of squares of the weighted residuals. If $f' < f$, the function is said to have moved downhill, and the values x' are taken as the new optimum values. All downhill steps are accepted and the process repeats from this new point. An uphill step may be accepted, allowing the parameter values to escape from local minima. Uphill movements (for example, a value f' such that $f' > f$) are accepted with probability $p = e^{(f-f')/T}$, where f is the previously found minimum value of the function to be minimized, f' is the most recently computed value of the function, and T is the current value of the "temperature," the thermodynamic analogue, not the light curve parameter. If p is larger than a random number between 0 and 1, the new point is accepted (the "Metropolis criterion"). Both the values of v_i and T are adjusted. The step length is changed such that a fixed percentage – this is one of the tuning parameters mentioned above – of the new evaluations is accepted. When more points are accepted, v is increased, and more points may be "out of bounds," that is, beyond the parameter limits that are established in a constraints file. The temperature is decreased after a specified number of iterations, after which the temperature is changed to $T' = r_T T$, where $0 \leq r_T \leq 1$. In the course of iterations, the step length, r_T, decreases and the algorithm terminates ideally close to the global optimum. The computational cost to reach this state can be high, as temperature annealing has to be sufficiently slow. A pseudocode of SA and detailed discussion of the parameters and temperature adjustment are presented by Alizamir et al. (2008).

SA has been applied to *EB* research by Milone & Kallrath (2008), in a slightly modified form named *Adapative Simulated Annealing* by Prša (2005). SA is also part of `NightFall`, a freely available amateur code for modeling eclipsing binary stars described in Sect. 8.3.

4.3.5 Other Approaches

In *EB* light curve analysis, other optimization techniques have been used in addition to the methods described in the previous subsections. Napier (1981) used a sequential (*creeping random*) search technique for fitting the parameters in his tri-axial ellipsoid geometry light curve model. Because this light curve model did not require a large amount of computer time, Napier was able to find a solution parameter set within minutes. A similar approach, namely a "controlled random search optimization procedure" based on the *Price algorithm* (Price 1976) has been applied to the Wilson–Devinney model by Barone et al. (1988). In this case, the efficiency was low. Barone et al. (1998) reported solutions which required about 30,000 iterations and 11 days of CPU time on a VAX 8600. Finally, we mention Metcalfe (1999) who used genetic algorithms to analyze *EB*s.

Various semi-empirical modifications of Differential Corrections were outlined by Khaliullina & Khaliullin (1984) and implemented in a computer program capable of automatic iteration (which, however, is designed to analyze the light curves of *EB*s with spherical components only). Hill (1979), in his program LIGHT2, which is based on Roche geometry, may have been the first to change his light curve solver from Differential Corrections to the Marquardt (1963) algorithm, which has much improved convergence properties and has become a standard of nonlinear least-squares routines.

Recently, the WD program has been enhanced by a damped least-squares solver based on the idea of Levenberg (1944). Further material on the Levenberg–Marquardt & Marquardt algorithm can be found in Moré (1978) and provides the basis for light curve package WD95 by Kallrath et al. (1998) and later versions, that uses the *Simplex algorithm* for initial search and global investigation, and switches to a damped least-squares solver when approaching a local minimum. The Levenberg–Marquardt procedure efficiently combines a steepest descent method with a step-size controlled Gauß–Newton algorithm. A similar step in this direction has already been performed by Plewa (1988) with the program MINW. MINW combines the Simplex algorithm, a variable metric method [Fletcher (1970), or Appendix A1], and the computation of the Hessian matrix for statistical purposes and error estimation.

4.4 A Priori and A Posteriori Steps in Light Curve Analysis

Ex necessitate rei   (*From the necessity of the case*)

4.4.1 Estimating Initial Parameters

The heuristic rules summarized in this section may be useful for estimating some of the initial parameters and also for checking the consistency of light curve solutions as well as interpreting their physical meanings. It may happen that the least-squares

4.4 A Priori and A Posteriori Steps in Light Curve Analysis

adjustment, for some reason, produces a seriously inadequate solution. A formal criterion to detect an inadequate solution is to compare the standard deviation of the fit, σ^{fit}, and the internal noise, σ^{data}, of the data. The solution is acceptable only if $\sigma^{\text{fit}} \approx \sigma^{\text{data}}$. It certainly helps to plot both the light curve and the data together! A more physical criterion is the astrophysical plausibility. If astrophysically implausible results are derived, it may be due to a deficiency of the light curve model or due to a fitting caught in a local instead of a global minimum.

Hints for estimating parameters are summarized below; as usual, light curves are assumed to be measured in relative flux units ("light") and not in magnitude.

1. Estimation of the temperature difference: For circular orbits, the ratio of primary to secondary eclipse depths is equal to the corresponding ratio of eclipsed mean star surface brightnesses at the places sampled. It has been argued that surface brightness is sufficient by itself as a modeling parameter and that the use of temperatures as light curve parameters is both unnecessary and, because of stellar atmosphere effects that might not follow the blackbody law [see, e.g., Popper (1993, p. 193)]. Our position on this matter is that the effective temperature is an important *curve-independent* physical parameter because, with a correct understanding of the stellar atmospheres involved, it constrains the relative fluxes in different passbands, and thus it reduces the number of free parameters; moreover, it provides an important handle on the astrophysics of the stellar components. If the atmospheric physics is not correctly understood, then the temperature will vary from curve to curve. Assuming that the atmosphere computation is accurate, the ratio of depths can give the temperature of one component if the temperature of the other one is known, even though stars are not black-bodies. This procedure holds for both partial and complete eclipses, but for only circular orbits. Even so, for small eccentricities they provide an initial guess for one star's temperature as a function of the other's. The temperature of one of the stars needs to be known from spectra, or from color indices if the interstellar reddening is known. We may guess that the total light of the system is dominated by the hotter star. This may be wrong, but we can proceed with the modeling until we confirm it to be so, or prove the opposite – in which case we continue with new T_1 and T_2. Convergence should be reachieved quite quickly.

2. Estimation of the luminosity ratio: For spherical stars, the ratio of the light loss to light remaining at the bottom of a total eclipse is the monochromatic luminosity ratio of the smaller to the larger star. The size ratio can be estimated also, as in the absence of limb darkening we have

$$k^2 = \frac{r_s^2}{r_g^2} = \frac{L_s}{L_g} \frac{1 - l_0^{\text{tr}}}{1 - l_0^{\text{oc}}}, \qquad (4.4.1)$$

where $1 - l_0^{\text{tr}}$ and $1 - l_0^{\text{oc}}$ are the transit and occultation eclipse depths. Wilson (1966) makes a slight modification of (4.4.1) to include limb-darkening. For circular orbits as shown in Fig. 3.10, the two eclipses are of equal duration. The following idea gives only a lower limit of k (because of uncertainties in i).

Although we often find similar formulations in textbooks, this approach is only of limited use because there is no reliable way to know i. The same concerns hold for the following considerations: The duration of the entire eclipse (first contact to last contact) is a measure of the sum of the relative radii, $r_1 + r_2$. If k is known (usually it would not be known a priori), then r_1 and r_2 can also be found. If the eclipse is total, it may be possible to identify contacts 2 and 3 (onset and completion of the totality phase). In this case, the timings provide the radii. If $t_1, ..., t_4$ denote times of first through fourth contact as shown in Fig. 4.8, then k is approximately:

$$k = \frac{(t_4 - t_1) - (t_3 - t_2)}{(t_4 - t_1) + (t_3 - t_2)}. \qquad (4.4.2)$$

Note that (4.4.2) is strictly true only if the stars move along straight lines, not in circular orbits. Which is the larger and which is the smaller star? If the smaller star transits the larger at the time of the primary (deeper) eclipse, the smaller one is star[8] 2; if the smaller star is occulted during primary minimum, it is star 1. Occultation eclipses have a flatter light curve during totality (corresponding to $t_3 - t_2$) than do transits, as a rule. But trial and error may be required to ascertain which of the two possibilities applies. For eccentric orbits, the eclipse which

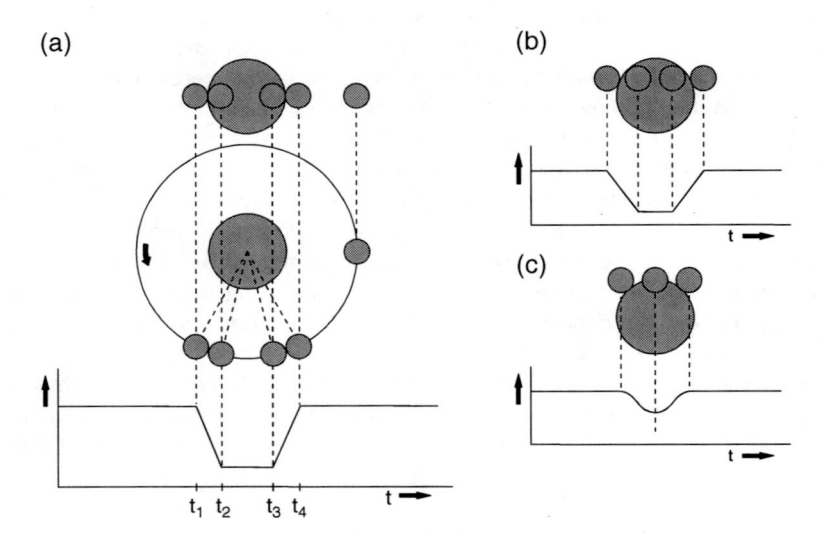

Fig. 4.8 Geometry of contact times. Part (**a**) shows the relative orbit of the secondary around the primary for inclination $i = 90°$. Part (**b**) shows a total eclipse for $i \neq 90°$; note that it is not a central eclipse. Part (**c**) shows a partial eclipse for $i \neq 90°$. Together parts (**a**), (**b**), and (**c**) illustrate that contact times alone cannot determine relative star dimensions, despite the impression given in many elementary textbooks, because the inclination ordinarily is not known

---

[8] See Section 2.8 for details on how to select star 1 and star 2.

4.4 A Priori and A Posteriori Steps in Light Curve Analysis

occurs nearer to apastron is the longer one. However, it is not so easy to make firm estimates of sizes, temperatures, or luminosities as it is for circular orbits because the projected surface area involved in each eclipse is not necessarily the same. Again, for low eccentricities, the estimate is just a bit rougher than for $e = 0$.

3. In eccentric orbit cases, the quantities $e \cos \omega$ and $e \sin \omega$ may be estimated from the separation and durations (widths) of the eclipses, respectively. The relations (3.1.24) and (3.1.26) provide approximations for e and ω separately and give us

$$\tan \omega \approx \frac{\Theta_a - \Theta_p}{\Theta_a + \Theta_p} \frac{2P}{\pi} \left(t^{II} - t^{I} - \frac{P}{2} \right)^{-1} \tag{4.4.3}$$

and

$$e \approx \frac{\pi}{2P \cos \omega} \left(t^{II} - t^{I} - \frac{P}{2} \right). \tag{4.4.4}$$

4. Estimation of the inclination i: The higher the inclination i, the wider and deeper the eclipses. For $i \approx 90°$ there is only a weak dependence of the eclipse structure on the inclination. However, when i becomes low enough so that eclipses are far from central, then larger decreases in width occur. Larger decreases in depth will be seen when the eclipses become partial. We may use an atlas of eclipses like that of Terrell et al. (1992) or make a series of runs with a light curve program keeping all parameters except i and observing the effect. Binary Maker, cf. Bradstreet (1993); Bradstreet & Steelman (2004), provides a convenient and quick way to make such assessments.

5. Estimation of (mean) volume radii for lobe-filling stars with known spectroscopic mass ratio can be found from the relation

$$r = \frac{0.49 q^{2/3}}{0.69 q^{2/3} + \ln(1 + q^{1/3})}, \quad 0 < q < \infty, \tag{4.4.5}$$

derived by Eggleton (1983), which is accurate to about $\pm 1\%$ for all values of the mass ratio q. To compute $r_2 = r$ for the lower mass star, one has to set $q = \mathcal{M}_2/\mathcal{M}_1 \leq 1$, whereas for the higher mass star ($r_2 = r$) one uses $q = \mathcal{M}_1/\mathcal{M}_2 \geq 1$ as input to (4.4.5). Except for Algol type binaries, the problem with (4.4.5) is, however, to know whether a star fills its lobe. More generally, one can often estimate a reasonable r from astrophysical considerations or from published solutions. For well-detached binaries, one can expect relative radii with value of 0.1 or less.

6. Estimating the Roche potential Ω in the circular orbit and synchronous rotation case: If estimated relative radii are available, one could apply (3.1.65) to derive estimations of Ω_1 and Ω_2. If the star is well inside its Roche lobe (detached systems), the relation $\Omega \sim 1/r$ applies. Then there are look-up tables [see,

for instance, Limber (1963, p. 1119)] listing the relative radius r versus Ω. Alternatively, we might use the method described in Appendix E.30 or just use Binary Maker (see Appendix 8.1).

In any case, we urge modelers to produce a synthetic light curve with the initial parameters and to plot calculated against observed light curves. If minima in the computed light curve are less deep than in the observed light curves, i could be increased. The idea is to avoid spending inordinate amounts of time on preparation, yet begin iterations somewhere near the correct minimum.

4.4.2 Criteria for Terminating Iterations

To avoid landing in a relative minimum of parameter hyperspace and for the quantitative determination of light curve parameters in general, we recommend the use of both the Simplex algorithm and (damped) Differential Corrections or another derivative-based method. The former should be used for any initial search. We have found it a helpful tool to explore solution uniqueness and to perform experiments on parameter sets, for example, establishing the directions of convergence of all the parameters by systematically varying the initial mass ratio. Once a parameter set is close to a solution, switching to a derivative-based least-squares solver can increase the rate of convergence and also yield the formal or statistical errors of the parameters.

Independent of the method, the question arises of when to halt iteration. Let us first list some intuitive stopping criteria and discuss whether they make sense:

- By number of iterations — in case no convergence is achieved with the initial parameters (it gives us a chance to monitor the solutions and to interpret the physics);
- by comparison of the standard deviation, σ^{fit}, of the fit with the noise, σ^{data}, in the data;
- by comparison of the parameter corrections with the parameters themselves;
- by comparison of the parameter corrections with the probable or mean standard errors;
- by inspection of the damping constant; and
- by confinement of the adjustments to a limited region (say, an ellipsoid) of parameter space.

All these criteria might appear arbitrary and it is questionable to set up general conventions. The number of iterations needed to converge may be different for different problems. Certainly, σ^{fit} cannot become smaller than σ^{data}. If the model suits the data, we would expect to have $\sigma^{fit} \simeq \sigma^{data}$. Although this probably guarantees a nice looking fit, in a flat valley convergence might not have been achieved. On the other hand, if $\sigma^{fit} > \sigma^{data}$, either the convergence has not been achieved, or the model is deficient.

4.4 A Priori and A Posteriori Steps in Light Curve Analysis

Comparing the parameter corrections Δx_j with the value of the parameter x_j and requiring for all j

$$\Delta x_j \leq \varepsilon_j^P x_j \qquad (4.4.6)$$

with a reasonable relative error ε_j^P does not work for parameters which can take the value 0, and it causes problems for cyclic parameters (angles allowed to any value between 0° and 360°). Hence, this criterion is not recommended.

So, instead, we might consider defining some absolute limits ΔX_j in those cases, and halt iterations if

$$\Delta x_j \leq \Delta X_j \qquad (4.4.7)$$

becomes valid for all j. This idea is again problematic because it introduces some arbitrariness into the problem if it is not possible to scale the parameters.

Very often in the literature on light curve analysis, the criterion

$$\Delta x_j \leq 0.1 \varepsilon_j^P \qquad (4.4.8)$$

is used, that is, convergence is stopped when parameter corrections are substantially, say a factor of 100, smaller than the statistical errors ε_j^P. This criterion has the advantage that it is consistently scaled.

Finally, using a Levenberg–Marquardt scheme, the damping constant λ can be used to terminate the iteration, if the following holds: For a set of parameters $\mathbf{x}_*$ at least one of the conditions (4.4.6), (4.4.7), or (4.4.8) is satisfied. If, for sufficiently small λ the damping factor increases continuously, then $\mathbf{x}_*$ can be accepted as the true solution. In well-defined test cases (Kallrath et al. 1998), λ went down to machine accuracy, i.e., 10^{-15}. Again, this is an empirical rule.

If, for a reasonable number of iterations, the estimated parameter vectors $\mathbf{x}$ stay within a predefined region, then the center of all vectors in that region may define the solution (Wilson 1996). This criterion certainly can detect some secular trends.

Bock (1987) gave a criterion which checks the statistical stability of the solution. This criterion is based on some Lipschitz conditions related to the generalized inverse and considers the quality of the solution, the nonlinearity of the problem, and the number of degrees of freedom. The formalism involved in this criterion is beyond the scope of this book and thus the reader is referred to Bock (1987, pp. 59–73).

It is important to accompany any analysis with a plot of the light curve and the residuals. A careful inspection of the residual plot can reveal any systematic trend. Unfortunately, in most cases the residuals are not normally distributed about zero. Long tails dominate the distribution. The problems associated with such distributions are discussed in the context of robustness and its statistical foundations. For an overview of robustness and robust estimation, we refer the reader to Chap. 15.7 in Press et al. (1992) and references therein.

4.4.3 The Interpretation of Errors Derived from Fitting

A solution $\mathbf{x}_*$ derived from a least-squares problem becomes properly meaningful only if uncertainties are attached. The uncertainty is usually specified by upper and lower bounds x_i^+ and x_i^- which for each parameter lead to the relation

$$x_i^- \leq x_i \leq x_i^+, \quad x_i^\pm := x_i + \delta x_i^\pm. \tag{4.4.9}$$

The errors δx_i include contributions from at least four sources:
- ε^m, the error from approximations and other deficiencies in the light curve model;
- ε^{obs}, the error due to systematic error within the observational data;
- ε^n, the error due to the numerical representation;
- ε^s, the error due to accidental (statistical) error in the observations.

Thus, in combination, if these were random and independent errors, we would get

$$(\delta x_i)^2 = \left(\varepsilon^m\right)^2 + \left(\varepsilon^{obs}\right)^2 + \left(\varepsilon^n\right)^2 + \left(\varepsilon^s\right)^2; \tag{4.4.10}$$

in practice, we have to expect an unknown functional relationship

$$\delta x_i = h\left(\varepsilon^m, \varepsilon^{obs}, \varepsilon^n, \varepsilon^s\right). \tag{4.4.11}$$

Typically, of these sources of error, only the statistical error ε^s is specified in the program output, e.g., derived from the covariance matrix, and the parameters to be determined are given with associated uncertainties in the symmetric form $x_i \pm \Delta x_i$. To be sure, it is very difficult to specify the systematic error ε^m made by approximations in the model. Systematic errors ε^{obs} in the observations are also assumed to be zero. Errors ε^n due to the numerical representation (integration, matrix inversion, round-off, etc.) are usually not discussed.

Within light curve analysis, and in the Wilson–Devinney model in particular, the errors Δx_i of dependent parameters can be determined by applying the Gaußian law of error propagation or by considering total differentials. For instance, in all modes above 0, this relation holds:

$$\Delta L_2 = \frac{L_2}{L_1} \Delta L_1. \tag{4.4.12}$$

If the Differential Correction method is used to determine parameters, then Δx_i follows from the diagonal elements of the covariance matrix (4.3.12), i.e.,

$$S = \mathbf{d}^\mathrm{T} \cdot \mathbf{W} \mathbf{d}, \quad S' = (-\mathbf{A} \cdot \mathbf{W} \cdot \mathbf{d}(\mathbf{x}_0))^\mathrm{T} \Delta \mathbf{x}, \tag{4.4.13}$$

and for the standard deviations,[9] ε^s, of the estimated parameters:

$$\Delta x_i := \varepsilon^s = \frac{S - S'}{n - m}\sqrt{\hat{c}_{ii}}. \qquad (4.4.14)$$

S' can be interpreted as the contribution by which the residuals S corresponding to $\mathbf{x}_0$ are reduced by applying differential corrections to $\mathbf{x}_0$. The errors calculated by (4.4.14) give only the statistical error. Often these statistical errors are much smaller than the realistic uncertainties. The analysis of *V836 Cygni* by Breinhorst et al. (1989), for example, gave an error of the mass ratio, Δq, $\varepsilon^p \approx 5 \cdot 10^{-3}$ but inspection of a $\sigma_{\text{fit}}(q)$ shows that the true error in Δq is at least of order 0.1. Most often this is observed because the correlations among parameters are taken into account to only first order. The off-diagonal elements of the covariance matrix give a measure for the pairwise correlations between parameters but not for higher-order combinations of parameters. Therefore, one should not only specify the errors or uncertainties of the fitted parameters but also add some comments on existence or non-existence of correlations.

Derivative-free least-squares solvers do not produce statistical errors ε^s. Therefore, they should be used only for producing good initial guesses for a derivative-based method.

A completely different approach to estimate $\Delta \mathbf{x}_*$ is the *sensitivity analysis approach* (Appendix B.2). Many simulated observables $\mathcal{O}^{\text{sim}}(\mathbf{x}_*)$ and the corresponding solutions are investigated. This method provides information on how the model reacts on small changes of a model parameter. If a large parameter change leads to only a small or negligible change of model response, the parameter is weakly defined and cannot be determined accurately. For weakly defined parameters, the *grid approach* (Appendix B.3) gives reliable information on the bounds of such parameters.

4.4.4 Calculating Absolute Stellar Parameters from a Light Curve Solution

Curiosa felicitas   *(Painstaking felicity)*

Once all the formal aspects of light curve solution have been reviewed, it is time to discuss the interpretation of the solution. What is the astrophysical significance of the obtained parameter set? Before interpreting the physical state of the binary, we need to compute its absolute parameters, i.e., the temperatures, radii, and luminosities and distance. The radial velocities are necessary to find the semi-major axis

---

[9] The Wilson–Devinney program before 1998 used the probable errors $\varepsilon^p = 0.6745\varepsilon^s$ instead of the standard deviation ε^s used in the 1998 and later versions (Wilson, 1998). Note that the factor 0.6745 is justified only if the errors are normally distributed.

and the absolute masses. In some cases, the semi-major axis can be obtained from astrometry but none of those binaries are known at present to eclipse. From the radii and masses, the surface accelerations (in solar units) are derived. Specific details of the computation depend slightly on the particular light curve parameter set, but there is a general approach to compute stellar parameters from light curve solutions. In order to discuss this issue and interpret the light curve solution, we distinguish between two scenarios: Output obtained from complete data and output obtained from incomplete data.

4.4.4.1 The Complete Data Case

Let us assume that a light curve solution[10] provides the inclination i and the length scale, i.e., the relative orbital semi-major axis, a, in physical units. At first, this enables us to compute the total mass, M, of the binary system

$$M = M_1 + M_2 = \frac{4\pi^2}{G} \frac{a^3}{P^2}. \tag{4.4.15}$$

For some favored cases, the light curve solution also provides the mass ratio

$$q = \frac{M_2}{M_1}. \tag{4.4.16}$$

Combining (4.4.15) and (4.4.16) yields the masses M_1 and M_2 for the stars

$$M_1 = \frac{1}{1+q}M, \quad M_2 = \frac{q}{1+q}M \tag{4.4.17}$$

separately. From (3.1.17) and (3.1.18), we derive

$$a_1 = \frac{q}{1+q}a, \quad a_2 = \frac{1}{1+q}a. \tag{4.4.18}$$

Usually, light curve programs return dimensionless values for each star's surface area, A, and mean surface brightness; we neglect the index indicating the component. The mean radius relative to the semi-major axis can be defined either from the surface area, A, or from volume, V,

$$\bar{r} = \sqrt{\frac{A}{4\pi}}, \quad \bar{r} = \left(\frac{3V}{4\pi}\right)^{1/3}. \tag{4.4.19}$$

---

[10] Here, light curve analysis and light curve solution are used in the extended sense, i.e., each might include photometric data, radial velocities, and other observables.

4.4 A Priori and A Posteriori Steps in Light Curve Analysis

The definitions give slightly different values for $\bar{r}$ but usually the difference is very small. The WD program also provides specific radii for each axis of the star. The arithmetic, geometric, or harmonic mean may be used to average them yielding a mean radius. To the requisite precision, any of these possible estimates of mean radius is probably acceptable. The uncertainty in the surface area or volume is not provided, so an indication of uncertainty in the mean radius is computable only from the individual radii described on page 287. A conservative method is to compute the weighted mean of these radii and to adopt the largest derived uncertainty.

The absolute mean radius and absolute surface may be computed if the semi-major axis, a, is known from astrometry, spectroscopy, or the light curve solution:

$$\mathcal{R} = a\bar{r}, \quad \mathcal{A} = a^2 A. \tag{4.4.20}$$

The acceleration, g', due to gravity at the surface follows as

$$g' = G\frac{M}{\mathcal{R}^2}, \tag{4.4.21}$$

or in dimensionless units as

$$\log g = \log \frac{g'}{g_\odot} = \log \frac{M/M_\odot}{(\mathcal{R}/R_\odot)^2} \quad \text{with} \quad g_\odot = 2.74 \cdot 10^2 \text{ m} \cdot \text{s}^{-2}. \tag{4.4.22}$$

Note that $\log g$ for both components is required if stellar atmospheres are to be incorporated in a light curve model. Note that $\log g_1$ and $\log g_2$ are related by

$$\frac{\log g_2}{\log g_1} = \log \frac{M_2/M_1}{(\mathcal{R}_2/\mathcal{R}_1)^2} = \log \frac{q}{(\bar{r}_2/\bar{r}_1)^2}. \tag{4.4.23}$$

If a light curve program provides the polar surface brightness, it may be used to compute monochromatic luminosities and, by applying a bolometric correction, eventually the bolometric luminosity. However, from the standpoint of calculating the uncertainty, we prefer effective temperatures. Thus, for spherical stars, the bolometric luminosity of each component is

$$\mathcal{L} = \mathcal{A}\sigma T_\text{eff}^4, \tag{4.4.24}$$

and for the more general case we have

$$\mathcal{L} = \sigma \int_{S'} T_\text{eff}^4(\mathbf{r}_s) ds'. \tag{4.4.25}$$

The absolute bolometric magnitude follows:

$$\begin{aligned} M^\text{bol} &= M_\odot^\text{bol} - 2.5\log_{10}(\mathcal{L}/L_\odot) \\ &= M_\odot^\text{bol} - 5\log_{10}(\overline{\mathcal{R}}/R_\odot) - 10\log_{10}(T/T_\odot), \end{aligned} \tag{4.4.26}$$

from which the absolute V magnitude is derived by subtraction of the bolometric correction[11] $B.C.$, i.e.,

$$M_V = M^{bol} - B.C. \tag{4.4.27}$$

Alternatively, with a light curve program available, it might be more consistent to start with the computation of the monochromatic luminosity

$$\mathcal{L}_V = \text{const} \int_{S'} I(\mathbf{r}_s, \cos\gamma; g, T_{\text{eff}}) d\sigma \tag{4.4.28}$$

in absolute units. Then, M_V is computed according to

$$M_V = M_{V\odot} - 2.5 \log_{10}(\mathcal{L}_V/\mathcal{L}_{\odot V}), \tag{4.4.29}$$

a bolometric correction is applied

$$M^{bol} = M_V + B.C., \tag{4.4.30}$$

and, finally, the bolometric luminosity follows from

$$\mathcal{L}/\mathcal{L}_\odot = 10^{-0.4(M^{bol} - M^{bol}_\odot)}. \tag{4.4.31}$$

The color indices of the individual components and of the system may be used to find the interstellar reddening E, especially if the spectral types are sufficiently well known. From this the unreddened distance modulus $(m - M)_0$ is derived. In the Johnson system introduced in Sect. 2.1.1, we have the relations

$$(m_V - M_V)_0 = 5 \log(r/10) \tag{4.4.32}$$

and

$$(m_V - M_V)_0 = m_V - M_V - R E_{B-V}, \tag{4.4.33}$$

where r is the distance in parsecs, and $R = A_V/E_{B-V}$ is the ratio of the attenuation, A_V, of the V light by the interstellar medium to the color excess E_{B-V}. ($N.B.$: $m_V = V$) Although exceptions among determined values of R are well documented, typically, $3.0 \leq R \leq 3.4$.

If masses have been determined, the evolutionary state of the components may be explored. Evolutionary tracks through the two components of known mass are

---

[11] We use the definition $B.C. = M^{bol} - M_V$. Because solar-like stars have their radiation maximum in the visual region of the spectrum, and due to the definition of the zero point of $B.C.$, most stars have negative bolometric corrections. Some care is necessary in the use of $B.C.$ from tables because sometimes the definition $B.C. = M_V - M^{bol}$ is used.

assumed to be coeval, and theoretical models should predict correctly the sizes and luminosities of these components in straight-forward situations. Two examples out of numerous such studies are

- The analysis of the system *DS Andromedae* (Schiller & Milone 1988) in the open cluster NGC 752 showed that the radius of the hotter component is consistent with the age of the cluster and that the Roche radius has not yet been reached.
- The determination of the masses and luminosities of the double-lined spectroscopic and *EB* HD 27130 in the Hyades star cluster by Schiller & Milone (1987) suggests a mass–luminosity relation for these two stars which differs from that of the local field of the Sun.

4.4.4.2 The Incomplete Data Case

If only one component (say, that with mass M_1) has observed radial velocities, the system is called a *single-lined spectroscopic binary*. In this case, besides the orbital period P the radial velocity amplitude,[12] K_1, is known and a useful quantity known as the *mass function* $f(M_1, M_2, i)$ can be obtained:

$$f(M_1, M_2, i) = \frac{M_2^3 \sin^3 i}{(M_1 + M_2)^2} = \frac{M_2 \sin^3 i}{(1/q + 1)^2} = \frac{4\pi^2}{G} \frac{(a_1 \sin i)^3}{P^2}$$

$$= \frac{P}{2\pi G} K_1^3 \left(1 - e^2\right)^{3/2}. \tag{4.4.34}$$

Note that $f(M_1, M_2, i)$ can be calculated based on quantities which can be derived from the spectroscopy only: The period,[13] P, and the eccentricity, e, of the orbit [see Aitken (1964) for details]. If the mass ratio q and the inclination i are known from photometry, then M_2 follows as

$$M_2 = \frac{(1/q + 1)^2}{\sin^3 i} f(M_1, M_2, i) \tag{4.4.35}$$

and M_1 as

$$M_1 = \frac{1}{q} M_2 = \frac{(1+q)^2}{q^3} \frac{1}{\sin^3 i} f(M_1, M_2, i). \tag{4.4.36}$$

---

[12] The radial velocity amplitudes, K_1 and K_2, are useful quantities only if proximity effects are absent. In that case, the mass ratio is just $q = K_1/K_2$. So, if they are used, they should be used carefully.

[13] The period is often more accurately derived from photometry because eclipses make good timing ticks. Also the number of photometric observations tends to be much larger than the number of radial velocity data points. Furthermore, radial velocity observations tend to be taken at the quadratures also, where they have limited use in defining the phases of conjunction.

If we have no photometric light curves (so that the inclination i and mass ratio q are not available), we still can get some information: Combining formulas (4.4.35) and (4.4.36), the sum of masses is

$$M_1 + M_2 = \frac{1}{\sin^3 i} (1 + 1/q)^3 f(M_1, M_2, i). \qquad (4.4.37)$$

For known f, the right-hand side of (4.4.37) takes its minimum value for $q = \infty$ and $i = 90°$. Thus the mass function provides a weak lower bound on the sum of masses, and because $q = \infty$ implies $M_1 = 0$, we get

$$M_2 \geq f(M_1, M_2, i), \qquad (4.4.38)$$

a lower limit of the mass of the star, without radial velocities.

If we are sure that $q \leq 1$, the minimum of $M_1 + M_2$ occurs for $q = 1$ which leads to the tighter bound

$$M_1 + M_2 \geq 8 f(M_1, M_2, i). \qquad (4.4.39)$$

4.5 Suggestions for Improving Performance

Et respice finem   (*Consider the end*)
Sirach 7, 40; *Gesta Romanorum*, c. 103

The determination of light curve parameters from *EB* light curves may require substantial[14] computing time. However, there are several mathematically well-defined alternatives with lower computational cost while still controlling accuracy:

1. Symmetries in the light curve model.
2. Local interpolation of total light between phase grid points, as is done in Linnell's program (Linnell 1989).
3. Choice of lower grid density on the stellar surfaces. When a direct search method is used as the optimizing tool, the required precision (and hence the grid density) does not need to be as high for the calculation of the theoretical light curve by numerical quadrature of the flux over the stellar surface as for a derivative-based method.
4. Use of analytical partial derivatives $\partial \ell / \partial x_i$.
5. Accurate finite difference approximations.

---

[14] We might argue that hardware improvements make these considerations less and less important as time goes on. That is certainly true for existing programs, but models get more and more sophisticated and include more and more details. Also, because the points considered in the next subsections have a sort of general character, it is worthwhile to keep them in mind.

4.5.1 Utilizing Symmetry Properties

Two symmetries can reduce computational time significantly: *Orbital symmetry* and *surface symmetry*. Orbital symmetry leads to symmetry of the light curve w.r.t. either minimum. Expressed w.r.t. the secondary minimum we have

$$l^{\text{cal}}(0.5 - \Delta\Phi) = l^{\text{cal}}(0.5 + \Delta\Phi). \tag{4.5.1}$$

For the computation of light values, l^{cal}, this symmetry can be utilized in the form

$$\Phi \to \begin{cases} l^{\text{cal}}(\Phi), & 0 \le \Phi \le 0.5, \\ l^{\text{cal}}(1 - \Phi), & 0.5 < \Phi \le 1. \end{cases} \tag{4.5.2}$$

Exploiting the orbital symmetry reduces the computational needs by a factor of 2 in simple cases without complicating structures, e.g., circular orbit models without spots. Models including spots or eccentric orbits do not have orbital symmetry in general. Because observations fall wherever they fall, only the solution of the direct problem benefits from this symmetry. Solution of the inverse problem benefits from the symmetry in the use of interpolation techniques described in Sect. 4.5.2.

Symmetry on the stellar surface is present if a given function $f(\theta, \varphi)$, such as the flux emitted from one of the components, is symmetric with respect to the stellar latitude θ and/or longitude φ. Models which are based on spherical stars have the symmetry property

$$\mathcal{S}^0: f(-\theta, \varphi) = f(\theta, \varphi), \quad \theta \in \left[0, \frac{\pi}{2}\right], \quad \varphi \in [0, 2\pi], \tag{4.5.3}$$

and

$$\mathcal{S}^k: f\left(\theta, \frac{\pi}{k} + \Delta\varphi\right) = f\left(\theta, \frac{\pi}{k} - \Delta\varphi\right), \quad k \in \{1, 2\}, \quad \Delta\varphi \in \left[0, \frac{\pi}{k}\right], \tag{4.5.4}$$

whereas the Roche model with circular orbits has only the symmetry properties $\mathcal{S}^0$ and $\mathcal{S}^2$. If the orbital and rotational axes are parallel, the x–y and x–z planes are symmetry planes for each component.

4.5.2 Interpolation Techniques

An efficient method to reduce computation is to use local interpolating polynomials $p(\Phi)$ for determining light curve parameters with derivative-free procedures. Let

$n + 1$ be the number of equidistant grid points

$$x_i = \Phi_i = i\Delta\Phi, \; y_i = f(x_i) = l^{\text{cal}}(\Phi_i), \; i = 0, ..., n; \; y = f(x) = l^{\text{cal}}(\Phi), \quad (4.5.5)$$

with phase grid size

$$\Delta\Phi = \frac{1}{kn}, \quad k = \begin{cases} 1, \text{ no special orbital symmetry is available,} \\ 2, \text{ orbital symmetry w.r.t. phase 0.5.} \end{cases} \quad (4.5.6)$$

The interpolating polynomial may be chosen according to specific needs. Here we choose cubic polynomials based locally on four grid points and evaluated by the Bessel method with central differences [see, e.g., Scarborough (1930)]. This method uses

$$h := x_{i+1} - x_i = \frac{1}{2n}, \quad u := \frac{x - x_0}{h}, \quad v := u - \tfrac{1}{2} \quad (4.5.7)$$

and is based on the scheme

$$\begin{array}{c|c} x_{-1} & y_{-1} \\ x_0 & y_0 \\ x_1 & y_1 \\ x_2 & y_2 \end{array} \begin{array}{c} \Delta y_{-1} \\ \Delta y_0 \\ \Delta y_1 \end{array} \begin{array}{c} \Delta y_{-1}^2 \\ \Delta y_0^2 \end{array} \Delta y_{-1}^3, \quad \Delta y_i^{k+1} := \Delta y_{i+1}^k - \Delta y_i^k. \quad (4.5.8)$$

Local interpolation of the function $f(x)$ by a cubic polynomial $p^3(x)$ is to be understood in the following sense. For an arbitrary $x \in [0, \frac{1}{k}]$, from the set of grid points $\{x_0, x_1, x_2, ..., x_n\}$ a subset of four values $x_{-1}, x_0, x_1,$ and x_2 is chosen to obey $x_0 \leq x < x_1$. In order to achieve the goal, the original set should be extended as needed to $\{x_{-1}, x_0, x_1, x_2, ..., x_{n+1}\}$. Due to the symmetry of the light curve, we have $x_{-1} = x_1$ and $x_{n+1} = x_{n-1}$. This system is indexed as

$$x_i = (i-2)h, \quad 1 \leq i \leq n+3, \quad y_i = f(x_i). \quad (4.5.9)$$

The index i_0 of the local x_0 follows as

$$i_0 = 2 + \left\lfloor \frac{x}{h} \right\rfloor, \quad (4.5.10)$$

and eventually we obtain Bessel's interpolating formulas

$$f(x) = y \approx p^3(x) = \frac{y_0 + y_1}{2} + v\Delta y_0 + \frac{v^2 - \tfrac{1}{4}}{2} \tfrac{1}{2} \left(\Delta^2 y_{-1} + \Delta^2 y_0 \right)$$
$$+ v \left(v^2 - \tfrac{1}{4} \right) \tfrac{1}{6} \Delta^3 y_{-1}. \quad (4.5.11)$$

4.5 Suggestions for Improving Performance

The implementation of this method of interpolation should provide an option to compare the exact light curve with the interpolated one by inspecting the standard deviation and the maximum deviation, thus allowing some control over the error produced by interpolation. Because the standard error of a single observation of observed light curves, σ^{obs}, usually (if light is normalized to unity) is not smaller than 0.005, any error caused by interpolation, say, by a factor of 10 smaller than σ^{obs}, can be safely neglected. This level of accuracy is usually achieved with $n = 50$ to $n = 100$ subintervals. Because well-observed *EB* light curves cover a few hundred data points, say 300–400, the reduction in computational time is significant. The simple interpolation scheme presented above may be replaced by a more sophisticated cubic spline interpolation.

4.5.3 Surface Grid Design

In order to choose appropriate surface grids, we should consider that the time needed to compute $l^{\text{cal}}(\Phi)$ is proportional to the number n_* of integration points on the stellar surface. When constructing a grid of surface points or elements, we should have accuracy, symmetry, and the exploitation of symmetry by mirror imaging in mind. Symmetry properties, as discussed in Sect. 4.5.1, may reduce computation and memory requirements. Let us assume that the x–y and x–z planes are symmetry planes for each component. Notice that there is a potential problem in carrying out the actual mirror imaging, in that equatorial and polar points will be duplicated upon reflection. Accordingly we can eliminate equatorial and polar points. We can do this by placing the first and last latitude curves at half-spaces from the pole, and near-equatorial curves at half-spaces above and below the equator. If we have N of latitude curves, they are located at polar angles

$$\theta_i = \frac{i - 0.5}{N} \frac{\pi}{2}, \quad i = 1, \ldots, N. \tag{4.5.12}$$

A similar argument holds for the distribution of longitude points on the latitude curves. In addition, it improves accuracy if the density of points is reasonably uniform over the entire surface. This goal is obtained, for instance, if the number N_i of longitude points on a latitude curve is proportional to the sine of the polar angle of that curve, i.e.,

$$N_i = \lfloor 1 + 1.3 N \sin \theta_i \rfloor, \tag{4.5.13}$$

as used in the Wilson–Devinney program, and the points are distributed uniformly on this curve. The proportionality factor 1.3 is an artifact of the integration accuracy requirements of the WD program. The total number of grid points on the stars thus depends on the number N of latitude curves on a hemisphere according to the following formula:

$$n_* = n_*(N) = \sum_{i=1}^{N} N_i. \quad (4.5.14)$$

The table below lists this relation for selected values. If Differential Corrections are used to fit the parameters, we should use at least $N \approx 25$ to 30. For derivative-free least-squares solvers, values between 10 and 15 will suffice. The error in flux integration caused by the grid density usually can be reduced to values close to 10^{-4} light units.

N	5	10	15	20	25	30	40	50	100
n_*	23	88	195	342	530	762	1345	2098	8331

4.5.4 Analytic Partial Derivatives

For some of the parameters, partial derivatives, $\partial l^{\text{cal}}/\partial x_i$, involved in the least-squares problem can be calculated analytically, or can be derived from other quantities available when computing a light curve. This is true for the third light (3.4.2), the parallax as in (3.9.3), and in the Wilson–Devinney program for $\partial \ell_j/\partial L_1$.

As has been shown in Chap. 3, the *monochromatic flux* from component j is (3.2.48)

$$\ell_j(\Phi) = \int_{S'} \chi_s(\Phi) I(\mathbf{r}_s, \cos\gamma; g, T, \lambda) \cos\gamma d\sigma, \quad (4.5.15)$$

where χ_s is the characteristic function defined in (3.3.2) ensuring that we integrate over only the visible surface. In spherical polar coordinates, (3.2.48) takes the form (3.2.49)

$$\ell_j(\Phi) = \int_0^\pi \int_0^{2\pi} \chi_s(\Phi) I(\mathbf{r}_s, \cos\gamma; g, T, \lambda) \frac{\cos\gamma}{\cos\beta} r^2 \sin\theta d\varphi d\theta. \quad (4.5.16)$$

Adding all contributions of the binary system, the total emitted flux $\ell(\Phi)$ is (3.2.50)

$$\ell(\Phi) = \ell_1(\Phi) + \ell_2(\Phi) + \ell_3 \quad (4.5.17)$$

with constant third light ℓ_3. Therefore, we have

$$\frac{\partial \ell}{\partial l_3}(\Phi) = 1. \quad (4.5.18)$$

The photospheric parameters such as polar temperature, bolometric albedo, gravity brightening exponent, or limb-darkening coefficient of the model enter only in the function $I(\cos\gamma; g, T, \lambda)$. They do not change the other terms in the integrand of (4.5.16), and they do not change the range of integration or the characteristic

4.5 Suggestions for Improving Performance

function $\chi_s(\Phi)$. Therefore the derivatives of $\ell_j(\Phi)$ with respect to this set $\mathcal{P}_1$ of parameters can be calculated by taking the derivatives of the integrand, i.e., for all $p \in \mathcal{P}_1$ we have

$$\frac{\partial \ell_j}{\partial p}(\Phi) = \int_0^\pi \int_0^{2\pi} \chi_s(\Phi) \frac{\partial I(\mathbf{r}_s, \cos\gamma; g, T, \lambda)}{\partial p} \frac{\cos\gamma}{\cos\beta} r^2 \sin\theta\, d\varphi\, d\theta. \quad (4.5.19)$$

Analytic formulas for the Linnell model are given by Linnell (1989, Appendix). The advantage of this procedure is that the derivative is found by a single integration, rather than by taking the difference of two integrals. This helps to improve accuracy. As an example we consider the albedo A in the Wilson–Devinney model and its monochromatic flux integrand I in formula (6.3.5). Only the reflection factor R depends on the albedo A, and the dependence is described by (3.2.46). The bolometric albedo is by far the easiest parameter because it leads to an analytical expression for $\partial \ell_j / \partial A$. Therefore, we get the simple term

$$\frac{\partial \ell_j}{\partial A}(\Phi) = \int_0^\pi \int_0^{2\pi} \chi_s(\Phi) \frac{F_s}{F_t} \frac{\cos\gamma}{\cos\beta} r^2 \sin\theta\, d\varphi\, d\theta, \quad (4.5.20)$$

where the ratio of bolometric fluxes F_s/F_t is already available from the computation of R itself (see Sect. 3.2.5). In the case of other parameters $p \in \mathcal{P}_1$, the partial derivative $\partial \ell_j / \partial p$ needs to be computed numerically.

The geometric parameters ($p \in \mathcal{P}_2$) are more difficult to treat. Linnell (1989) suggests applying Leibniz's rule. In order to do so, (4.5.16) is rewritten as

$$\ell_j(\Phi) = \int_{\theta_l}^{\theta_u} f_2(\theta, \mathbf{p})\, d\theta, \quad f_2(\theta, \mathbf{p}) := \sin\theta \int_{\varphi_s(\theta)}^{\varphi_n(\theta)} f_3(\varphi, \mathbf{p})\, d\varphi, \quad (4.5.21)$$

with

$$f_3(\varphi, \mathbf{p}) := I(\cos\gamma; g, T, \lambda) \frac{\cos\gamma}{\cos\beta} r^2. \quad (4.5.22)$$

The characteristic function has been replaced by appropriate boundaries or integration limits for longitude φ and colatitude θ. $\varphi_s(\theta)$ is the starting longitude and $\varphi_n(\theta)$ is the ending longitude on a given colatitude θ. θ_l and θ_u represent the lower and upper limits for colatitude. The limits depend on phase Φ and the geometrical parameters. According to Leibniz's rule, the derivatives are now given by

$$\frac{\partial \ell_j}{\partial p}(\Phi) = f_2(\theta_u, \mathbf{p}) \frac{\partial \theta_u}{\partial p} - f_2(\theta_l, \mathbf{p}) \frac{\partial \theta_l}{\partial p} + \int_{\theta_l}^{\theta_u} \frac{\partial f_2(\theta, \mathbf{p})}{\partial p}\, d\theta, \quad \forall\, p \in \mathcal{P}_2, \quad (4.5.23)$$

and

$$\frac{\partial f_2(\theta, \mathbf{p})}{\partial p} = \sin\theta \left[\begin{array}{c} f_3(\varphi_n(\theta), \mathbf{p}) \dfrac{\partial \varphi_n(\theta)}{\partial p} \\ - f_3(\varphi_s(\theta), \mathbf{p}) \dfrac{\partial \varphi_s(\theta)}{\partial p} + \int_{\theta_l}^{\theta_u} \dfrac{\partial f_3(\varphi, \mathbf{p})}{\partial p}\, d\varphi \end{array} \right]. \quad (4.5.24)$$

Since no analytic expression exists for derivatives of $\varphi_s(\theta)$, $\varphi_n(\theta)$, θ_l, and θ_u w.r.t. $\forall\, p \in \mathcal{P}_2$, these derivatives need to be computed numerically. Thus, three numerical derivatives have to be computed instead of one. It is not obvious that some advantage in accuracy is really achieved.

4.5.5 Accurate Finite Difference Approximation

When approximating derivatives $f'(x)$ of a function $f(x)$ by asymmetric finite differences

$$f'(x) \cong \frac{f(x + \Delta x) - f(x)}{\Delta x}, \qquad (4.5.25)$$

or symmetric finite differences

$$f'(x) \cong \frac{f\left(x + \tfrac{1}{2}\Delta x\right) - f\left(x - \tfrac{1}{2}\Delta x\right)}{\Delta x}, \qquad (4.5.26)$$

the question arises as how to choose the size of the increment Δx. As discussed by Press et al. (1992), the optimal choice depends on the curvature, i.e., on the second derivative of $f(x)$. As in most practical cases, generally, as well as in light curve analysis, this information is not available. Instead, for symmetric differences, we may use the heuristic approach

$$\Delta x \approx 2\varepsilon_f^{1/3} x \qquad (4.5.27)$$

described by Press et al. (1992, p. 180) where ε_f is the fractional accuracy with which $f(x)$ is computed. For simple functions, this may be comparable to the machine accuracy, $\varepsilon_f \approx \varepsilon_m$, but for complicated calculations, in which the functions eventually yield the EB observables with additional sources of inaccuracy, ε_f is certainly larger. For this reason, it is wise to use those values suggested by the light curve program developers, who presumably know best the accuracy properties of their program.

4.6 Selected Bibliography

This section is intended to guide the reader to recommended books or articles on least-squares techniques and their statistical foundations.

- *Statistical and Computational Methods in Data Analysis* by Brandt (1976) is a useful resource on basic statistical concepts, e.g., maximum likelihood estimators and least-squares problems.
- Eichhorn's (1993) paper is an excellent review of the methods of least-squares as known and used in astronomy.

- *Solving Least-Squares Problems* by Lawson & Hanson (1974). A classic work, well worth reading.
- *Scientific Data Analysis* by Branham (1990) is an introduction to overdetermined systems on an elementary level.
- *Numerical Recipes – The Art of Scientific Computing* by Press et al. (1992) is already a classic work and a useful source for efficient numerical calculations.

References

Aitken, R. G.: 1964, *The Binary Stars*, Dover Publications, Philadelphia, PA, 3rd edition

Alizamir, S., Rebennack, S., & Pardalos, P. M.: 2008, Improving the Neighborhood Selection Strategy in Simulated Annealing using the Optimal Stopping Problem, in C. M. Tan (ed.), *Global Optimization: Focus on Simulated Annealing*, Chap. 18, pp. 363–382, I-Tech Education and Publication

Baake, E. & Schlöder, J. P.: 1992, Modelling the Fast Fluorescence Rise of Photosynthesis, *Bull. Math. Biol.* **54**, 999–1021

Barone, F., Maceroni, C., Milano, C., & Russo, G.: 1988, The Optimization of the Wilson–Devinney Method: An Application to CW Cas, *A&A* **197**, 347–353

Bock, H. G.: 1981, Numerical Treatment of Inverse Problems in Chemical Reaction Kinetics, in K. H. Ebert, P. Deuflhard, & W. Jäger (eds.), *Modelling of Chemical Reaction Systems*, Series in Chemical Physics, pp. 102–125, Springer, Heidelberg

Bock, H. G.: 1987, *Randwertproblemmethoden zur Parameteridentifizierung in Systemen nichtlinearer Differentialgleichungen*, Preprint 142, Universität Heidelberg, SFB 123, Institut für Angewandte Mathematik, 69120 Heidelberg

Bradstreet, D. H.: 1993, Binary Maker 2.0 – An Interactive Graphical Tool for Preliminary Light Curve Analysis, in E. F. Milone (ed.), *Light Curve Modeling of Eclipsing Binary Stars*, pp. 151–166, Springer, New York

Bradstreet, D. H. and Steelman, D. P., 2004. Binary Maker 3.0. Contact software, Norristown, PA 19087

Brandt, S.: 1976, *Statistical and Computational Methods in Data Analysis*, North-Holland, Amsterdam, 2nd edition

Branham, R. L.: 1990, *Scientific Data Analysis: An Introduction to Overdetermined Systems*, Springer, New York

Code, A. D. & Liller, W. C.: 1962, Direct Recording of Stellar Spectra, in W. A. Hiltner (ed.), *Astronomical Techniques*, Vol. II of *Stars and Stellar Systems*, pp. 281–301, University of Chicago Press, Chicago, IL

Dantzig, G. B.: 1951, Application of the Simplex Method to a Transportation Problem, in T. C. Koopmans (ed.), *Activity Analysis of Production and Allocation*, pp. 359–373, Wiley, New York

Dennis, J. E. & Schnabel, R. B.: 1983, *Numerical Methods for Unconstrained Optimisation and Nonlinear Equations*, Prentice Hall, Englewood Cliffs, NJ

Deuflhard, P. & Apostolescu, V.: 1977, An Underrelaxed Gauss-Newton Method for Equality Constrained Nonlinear Least-Squares Problems, in J. Stoer (ed.), *Proc. 8th IFIP Conf. Würzburg Symposium on the Theory of Computing*, No. 23 in Springer Lecture Notes Control Inform Sci., Springer, Heidelberg

Deuflhard, P. & Apostolescu, V.: 1980, A Study of the Gauss–Newton Method for the Solution of Nonlinear Least-Squares Problems, in J. Frehse, D. Pallaschke, & U. Trottenberg (eds.), *Special Topics of Applied Mathematics*, pp. 129–150, North-Holland, Amsterdam

Djurasevic, G.: 1992, An Analysis of Active Close Binaries (CB) Based on Photometric Measurements. III. The Inverse-Problem Method – An Interpretation of CB Light Curves, *Ap. Sp. Sci.* **197**, 17–34

Eggleton, P. P.: 1983, Approximations to the Radii of Roche Lobes, *ApJ* **268**, 368–369
Eichhorn, H.: 1978, Least-Squares Adjustments with Probabilistic Constraints, *MNRAS* **182**, 355–360
Eichhorn, H.: 1993, Generalized Least-Squares Adjustment, a Timely but Much Neglected Tool, *CMDA* **56**, 337–351
Eichhorn, H. & Standish, M.: 1981, Remarks on Nonstandard Least-Squares Problems, *AJ* **86**, 156–159
Fletcher, R.: 1970, A New Approach to Variable Metric Algorithms, *Comp. J.* **13**, 317–322
Gauß, C. F.: 1809, *Theoria Motus Corporum Coelestium in Sectionibus Conicus Solem Ambientium*, F. Perthes and J. H. Besser, Hamburg
Hill, G.: 1979, Description of an Eclipsing Binary Light Curve Computer Code with Application to Y Sex and the WUMA Code of Rucinski, *Publ. Dom. Astrophys. Obs.* **15**, 297–325
Jefferys, W. H.: 1980, On the Method of Least-Squares, *AJ* **85**, 177–181
Jefferys, W. H.: 1981, On the Method of Least-Squares. II, *AJ* **86**, 149–155
Kallrath, J.: 1993, Gradient Free Determination of Eclipsing Binary Light Curve Parameters – Derivation of Spot Parameters Using the Simplex Algorithm, in E. F. Milone (ed.), *Light Curve Modeling of Eclipsing Binary Stars*, pp. 39–51, Springer, New York
Kallrath, J.: 1999, Least-Squares Methods for Models Including Ordinary and Partial Differential Equations, in R. Dvorak (ed.), *Modern Astrometry and Astrodynamics*, Österreichische Akademie der Wissenschaften, Vienna, Austria
Kallrath, J., Altstädt, V., Schlöder, J. P., & Bock, H. G.: 1999, Analysis of Crack Fatigue Growth Behaviour in Polymers and their Composites Based on Ordinary Differential Equations Parameter Estimation, *Polymer Testing* **18**, 11–40
Kallrath, J. & Kämper, B.-C.: 1992, Another Look at the Early-Type Eclipsing Binary BF Aurigae, *A&A* **265**, 613–625
Kallrath, J. & Linnell, A. P.: 1987, A New Method to Optimize Parameters in Solutions of Eclipsing Binary Light Curves, *Astrophys. J.* **313**, 346–357
Kallrath, J., Milone, E. F., Terrell, D., & Young, A. T.: 1998, Recent Improvements to a Version of the Wilson–Devinney Program, *ApJ Suppl.* **508**, 308–313
Kallrath, J., Schlöder, J., & Bock, H. G.: 1993, Parameter Fitting in Chaotic Dynamical Systems, *CMDA* **56**, 353–371
Khaliullina, A. I. & Khaliullin, K. F.: 1984, Iterative Method of Differential Corrections for the Analysis of Light Curves of Eclipsing Binaries, *Sov. Astron.* **28**, 228–234
Kopal, Z.: 1943, An Application of the Method of Least-Squares to the Adjustment of Photometric Elements of Eclipsing Binaries, *Proc. Am. Philos. Soc.* **86**, 342–350
Kuhn, H. W. & Tucker, A. W.: 1951, Nonlinear Programming, in J. Neumann (ed.), *Proceedings Second Berkeley Symposium on Mathematical Statistics and Probability*, pp. 481–492, University of California, Berkeley, CA
Lawson, C. L. & Hanson, R. J.: 1974, *Solving Least-Squares Problems*, Prentice Hall, Englewood Cliffs, NJ
Legendre, A. M.: 1805, *Nouvelles Methodes pour la Determination des Orbites des Comètes*, Courcier, Paris
Levenberg, K.: 1944, A Method for the Solution of Certain Nonlinear Problems in Least-Squares, *Quart. Appl. Math.* **2**, 164–168
Limber, D. N.: 1963, Surface Forms and Mass Loss for the Components of Close Binaries – General Case of Non-Synchronous Rotation, *ApJ* **138**, 1112–1132
Linnell, A. P.: 1989, A Light Synthesis Program for Binary Stars. III. Differential Corrections, *ApJ* **342**, 449–462
Linnell, A. P. & Proctor, D. D.: 1970, On Weights in Kopal's Iterative Method for Total Eclipses, *ApJ* **162**, 683–686
Lootsma, F. A.: 1972, *Numerical Methods for Nonlinear Optimization*, Academic Press, London
Marquardt, D. W.: 1963, An Algorithm for Least-Squares Estimation of Nonlinear Parameters, *SIAM J. Appl. Math.* **11**, 431–441

Metcalfe, T. S.: 1999, Genetic-Algorithm-Based Light Curve Optimization Applied to Observations of the W UMa star BH Cas, *AJ* **117**, 2503–2510

Metropolis, N., Rosenbluth, A., Rosenbluth, M., Teller, A., & Teller, E.: 1953, Equation of State by Fast Computing Maschines, *Journal of Chemical Physics* **21**, 1087–1092

Milone, E. F., Groisman, G., Fry, D. J. I. F., & Bradstreet, H.: 1991, Analysis and Solution of the Light and Radial Velocity Curves of the Contact Binary TY Bootis, *ApJ* **370**, 677–692

Milone, E. F. & Kallrath, J.: 2008, Tools of the Trade and the Products they Produce: Modeling of Eclipsing Binary Observables, in E. F. Milone, D. A. Leahy, & D. W. Hobill (eds.), *Short-Period Binary Stars: Observations, Analyses, and Results*, Vol. 352 of *Astrophysics and Space Science Library*, pp. 191–214, Springer, Dordrecht, The Netherlands

Moré, J. J.: 1978, The Levenberg–Marquardt Algorithm: Implementation and Theory, in G. A. Watson (ed.), *Numerical Analysis*, No. 630 in Lecture Notes in Mathematics, pp. 105–116, Springer, Berlin, Germany

Murray, W.: 1972, *Numerical Methods for Unconstrained Optimisation*, Academic Press, London, UK

Napier, W. M.: 1981, A Simple Approach to the Analysis of Eclipsing Binary Light Curves, *MNRAS* **194**, 149–159

Nelder, J. A. & Mead, R.: 1965, A Simplex Method for Function Minimization, *Comp. J.* **7**, 308–313

Pardalos, P. M. & Mavridou, T. D.: 2008, Simulated Annealing, in C. A. Floudas & P. M. Pardalos (eds.), *Encyclopedia of Optimization*, pp. 3591–3593, Springer Verlag, New York

Parkinson, J. M. & Hutchinson, D.: 1972, An Investigation into the Efficiency of Variants on the Simplex Method, in F. Lootsma (ed.), *Numerical Methods for Nonlinear Optimisation*, pp. 115–135, Academic Press, London

Plewa, T.: 1988, Parameter Fitting Problem for Contact Binaries, *Acta Astron.* **38**, 415–430

Popper, D.: 1993, Discussion: Comments by D. Popper, in E. F. Milone (ed.), *Light Curve Modeling of Eclipsing Binary Stars*, p. 193, Springer, New York

Powell, M. J. D.: 1964a, A Method for Minimizing a Sum of Squares of Nonlinear Functions without Calculating Derivatives, *Comp. J.* **7**, 303–307

Powell, M. J. D.: 1964b, An Efficient Method for Finding the Minimum of a Function of Several Variables without Calculating Derivatives, *Computer Journal* **7**, 155–162

Press, W. H., Flannery, B. P., Teukolsky, S. A., & Vetterling, W. T.: 1992, *Numerical Recipes – The Art of Scientific Computing*, Cambridge University Press, Cambridge, UK, 2nd edition

Price, W. L.: 1976, A controlled random search procedure for global optimizition, *Comp. J.* **20**, 367–370

Prša, A.: 2005, *Physical Analysis and Numerical Modeling of Eclipsing Binary Stars*, PhD Dissertation, Faculty of Mathematics and Physics, Dept. of Physics, University of Ljubljana, Ljubljana, Slovenia

Prša, A. & Zwitter, T.: 2005b, A Computational Guide to Physics of Eclipsing Binaries. I. Demonstrations and Perspectives, *ApJ* **628**, 426–438

Scarborough, J. B.: 1930, *Numerical Mathematical Analysis*, The Johns Hopkins University Press, Baltimore, MD

Schiller, S. J. & Milone, E. F.: 1987, Photometric Analyses of the Hyades Eclipsing Binary HD 27130, *AJ* **93**, 1471–1483

Schiller, S. J. & Milone, E. F.: 1988, Photometric and Spectroscopic Analysis of the Eclipsing Binary DS Andromedae – A Member of NGC 752, *AJ* **95**, 1466–1477

Schlöder, J. P.: 1988, *Numerische Methoden zur Behandlung hochdimensionaler Aufgaben der Parameteridentifizierung*, Bonner Mathematische Schriften 187, Universität Bonn, Institut für Angewandte Mathematik, 53117 Bonn

Schlöder, J. P. & Bock, H. G.: 1983, Identification of Rate Constants in Bistable Chemical Reactions, in P. Deuflhard & E. Hairer (eds.), *Numerical Treatment of Inverse Problems in Differential and Integral Equations*, Vol. 2 of *Progress in Scientific Computing*, pp. 27–47, Birkhäuser, Boston, MA

Schneider, I.: 1988, *Die Entwicklung der Wahrscheinlichkeitstheorie von den Anfängen bis 1933*, Wissenschaftliche Buchgesellschaft, Darmstadt

Spendley, W., Hext, G. R., & Himsworth, F. R.: 1962, Sequential Application of Simplex Designs in Optimisation and Evolutionary Operation, *Technometrics* **4**, 441–461

Stagg, C. R. & Milone, E. F.: 1993, Improvements to the Wilson–Devinney Code on Computer Platforms at the University of Calgary, in E. F. Milone (ed.), *Light Curve Modeling of Eclipsing Binary Stars*, pp. 75–92, Springer, New York

Stigler, S. M.: 1986, *The History of Statistics*, Belknap, Harvard, Cambridge, MA

Terrell, D., Mukherjee, J. D., & Wilson, R. E.: 1992, *Binary Stars – A Pictorial Atlas*, Krieger Publishing Company, Malabar, FL

Terrell, D., & Wilson, R. E.: 2004, Photometric Mass Ratios of Eclipsing Binary Stars, *Ap. Sp. Sci.* **296**, 221–230

Vassiliadis, V. S. & Conejeros, R.: 2008, Powell Method, in C. A. Floudas & P. M. Pardalos (eds.), *Encyclopedia of Optimization*, pp. 3012–3013, Springer Verlag, New York

Wilson, R. E.: 1966, A Quick Solution Method for Binaries Showing Complete Eclipses, *PASP* **78**, 223–227

Wilson, R. E.: 1978, On the A-Type W Ursae Majoris Systems, *ApJ* **224**, 885–891

Wilson, R. E.: 1979, Eccentric Orbit Generalization and Simulaneous Solution of Binary Star Light and Velocity Curves, *ApJ* **234**, 1054–1066

Wilson, R. E.: 1983, Convergence of Eclipsing Binary Solutions, *Ap. Sp. Sci.* **92**, 229–230

Wilson, R. E.: 1988, Physical Models for Close Binaries and Logical Constraints, in K.-C. Leung (ed.), *Critical Observations Versus Physical Models for Close Binary Systems*, pp. 193–213, Gordon and Breach, Montreux, Switzerland

Wilson, R. E.: 1996, private communication

Wilson, R. E.: 1998, Computing Binary Star Observables (Reference Manual to the Wilson-Devinney Programm, Department of Astronomy, University of Florida, Gainesville, FL, 1998 edition

Wilson, R. E. & Biermann, P.: 1976, TX Cancri - Which Component Is Hotter?, *A&A* **48**, 349–357

Wilson, R. E. & Devinney, E. J.: 1971, Realization of Accurate Close-Binary Light Curves: Application to MR Cygni, *ApJ* **166**, 605–619

Wilson, R. E. & Terrell, D.: 1994, Sub-Synchronous Rotation and Tidal Lag in HD 77581/Vela X-1, in S. S. Holt & C. S. Day (eds.), *The Evolution of X-Ray Binaries*, No. 308 in AIP Conference Proceedings, pp. 483–486, AIP, American Institute of Physics, Woodbury, NY

Wilson, R. E. & Terrell, D.: 1998, X-Ray Binary Unified Analysis: Pulse/RV Application to Vela X1/GP Velorum, *MNRAS* **296**, 33–43

Wyse, A. B.: 1939, An Application of the Method of Least-Squares to the Determination of Photometric Elements of Eclipsing Binaries, *Lick. Obs. Bull.* **496**, 17–27

Yarbro, L. A. & Deming, S. N.: 1974, Selection and Preprocessing of Factors for Simplex Optimization, *Analytica Chimica Acta* **73**, 391–398

Young, A. T., Genet, R. M., Boyd, L. J., Borucki, W. J., Lockwood, G. W., Henry, G. W., Hall, D. S., Smith, D. P., Baliunas, S. L., Donahue, R., & Epand, D. H.: 1991, Precise Automatic Differential Stellar Photometry Analyzing Eclipsing Binaries, *PASP* **103**, 221–242

Chapter 5
Advanced Topics and Techniques

This chapter addresses some of the many improvements and extensions of ideas and techniques in *EB* research over the last decade: The direct distance estimation through the analyses of *EBs* and the derivation of ephemerides and third-body orbital parameters from light and radial velocity curves. The Kepler mission [cf. Koch et al. (2006)] launched on 6th of March 2009, the GAIA mission with a target launch date in 2012, or the ground based survey LSST in discussion for after 2015 will add a new challenge to the field: The analysis of a large number of *EB* light curves from surveys. Detection of extrasolar planets by transit methods is a field where *EB* methods have been used successfully.

5.1 Extended Sets of Observables and Parameters

'Ο θεός έφκιασε τον κόσμο και είπε: 'Οπόχει μυαλό ας πορεύεται.'
God made the world and said: "*He who has a brain will go on.*"
(Cephalonian proverb)

In *EB* analysis, the observables are usually the light curves and radial velocity curves, whereas the adjustable parameters are typically dimensionless quantities such as the mass ratio, Roche potential, and inclination, and also a few parameters in physical units such as the semi-major axis, and mean temperatures. In this section, we discuss absolute masses, temperatures, and distance as adjustable parameters and treat, for instance, the mass ratio as an additional observable. An important consequence and advantage is that the two-step procedure is replaced by *one* consistent least-squares analysis. We also learn that adjustable parameters can have observables as direct counterparts.

After outlining a general concept of extended sets of observables and parameters, especially useful in the context of direct absolute parameters estimation (Sect. 5.1.1), in Sect. 5.1.2.1 we focus on the following problem: Most light curve models require the temperature of at least one star as input. One of the main difficulties of modeling *EBs* is the accurate determination of the individual temperatures. In Sect. 5.1.2.3 we describe how to evaluate light curves in at least two passbands to determine both temperatures *and* the distance (Sect. 5.1.4). This overcomes some limits of the traditional two-step approach to estimate distances of *EB*s.

The section is ended by a discussion of using main sequence constraints to reduce correlations among adjustable parameters and the incorporation of intrinsic variables into light curve models.

5.1.1 Inclusion of Absolute Parameters in Light Curve Analysis

In Sect. 4.4.4.1 the absolute parameters and, especially, the distance of a star were computed a posteriori. Here we outline some ideas of how absolute parameters can be made part of the least-squares analysis providing their consistent standard errors, and thus improving on the standard two-step scheme in which the absolute parameters are computed only after the light curve solution has been performed. We introduce the absolute masses, M_1 and M_2, as examples. Later we shall present recent ideas that derive, for instance, both temperatures and the distance as part of the overall least-squares algorithm.

To illustrate the addition of adjustable parameters (not only absolute parameters, but any new ones) and relations to an existing model that already includes adjustable parameters p_k, we show how to add M_1 and M_2 as free parameters to an existing model, in which the semi-major axis a, and the mass ratio q are already adjustable (there might be others as well). Based on the relationships (4.4.14), (4.4.15), and (4.4.16), we obtain two equations

$$M_1 = \frac{1}{1+q}\frac{4\pi^2}{G}\frac{a^3}{P^2}, \quad M_2 = \frac{q}{1+q}\frac{4\pi^2}{G}\frac{a^3}{P^2} \qquad (5.1.1)$$

for the two new parameters M_1 and M_2. In the least-squares analysis, we need to consider the extended equation of condition

$$o_\nu^o - o_\nu^c = \frac{\partial o_\nu^c}{\partial M_1}\delta M_1 + \frac{\partial o_\nu^c}{\partial M_2}\delta M_2 + \sum_{k=1}^{K}\frac{\partial o_\nu^c}{\partial p_k}\delta p_k \qquad (5.1.2)$$

and the new partial derivatives $\partial o_\nu^c/\partial M_1$ and $\partial o_\nu^c/\partial M_2$, where o_ν^o and o_ν^c denote observed and calculated values of some observable at time t_ν. As (5.1.1) and (4.4.15) are explicit equations, this is rather simple and is given by

$$\frac{\partial o_\nu^c}{\partial M_j} = \frac{\partial o_\nu^c}{\partial a}\frac{\partial a}{\partial M_j} + \frac{\partial o_\nu^c}{\partial q}\frac{\partial q}{\partial M_j}, \quad j = 1, 2. \qquad (5.1.3)$$

The q-derivatives are given by

$$\frac{\partial q}{\partial M_1} = -\frac{1}{M_1^2}M_2 = -\frac{q}{M_1} \quad , \quad \frac{\partial q}{\partial M_2} = \frac{1}{M_1}. \qquad (5.1.4)$$

5.1 Extended Sets of Observables and Parameters

Computing the a-derivatives leads to

$$\frac{4\pi^2}{G}\frac{1}{P^2}\frac{\partial\left(a^3\right)}{\partial M_j} = 1, \quad j = 1, 2 \tag{5.1.5}$$

and thus

$$\frac{\partial a}{\partial M_j} = \frac{G}{4\pi^2}\frac{P^2}{3a^2}, \quad j = 1, 2. \tag{5.1.6}$$

This procedure has the advantage that we do not need to compute any additional numerical derivatives with respect to M_1 or M_2, as the derivatives $\partial o_v^c/\partial a$ and $\partial o_v^c/\partial q$ are known already.

The situation becomes more complicated if the period P is also an adjustable parameter, as in that case we have no explicit relationship $P(M_j)$. Therefore, we need to resort to numerical derivatives for $\partial o_v^c/\partial M_j$, although the ones for q and a are available. Thus, to conclude this illustration, we summarize that for adding additional adjustable parameters one has to check whether full explicit relations are available (that is, no further numerical derivatives are needed), or whether the additional relationships do not allow derivation of analytic derivatives.

There are three major advantages of a consistent one-step approach:

1. It is impersonal, driven only by the input data *and* their standard deviations.
2. The least-squares analysis automatically provides standard errors of the estimated new adjustable parameters.
3. It allows one to include estimations *and* standard deviations of these parameters from other sources, e.g., from published data.

We conclude this section by stressing the advantage of including parameter estimations from other data sources when they are specified with reliable standard errors. Distance or mass ratio serves as examples. In such cases, an adjustable parameter p can be compared to a direct counterpart observable p_m^o with standard error ε_m. Index m, $m = 1, \ldots, M$, refers to several published values, where M is usually a small number, say, $M \leq 3$. To consider the observable p_m^o in the least-squares analysis, we rename p into p^c and obtain M equations of condition

$$p_m^o - p^c = \delta p_m, \quad m = 1, \ldots, M, \tag{5.1.7}$$

or the corresponding least-squares term

$$\sum_{m=1}^{M} w_m \left(p_m^o - p^c\right)^2, \tag{5.1.8}$$

where the weights w_m are functions of ε_m.

If the light curve analysis provides p with standard error $\varepsilon < \varepsilon_m$ *without* considering p_m^o, the values p_m^o will have little effect as their weights are small and they are typically not numerous. In the opposite case, $\varepsilon \geq \varepsilon_m$, they may have a strong effect. Part of the reason, in this case with large ε, is that the standard errors of other light curve parameters could also be large due to correlations. In this case, use of the external estimations p_m^o and their standard errors, ε_m, has a correlation-reducing influence on the overall light curve solution. As an example of a parameter-observable relation, we consider the mass ratio, q, in a detached binary. For detached binaries, q is strongly correlated with the Roche potentials. In this case, all is fine if the radial velocity curves are made part of a simultaneous least-squares analysis. However, in many publications, q is just fixed to a value q_* found in other publications. By contrast, exploitation of the parameter-observable relation and (5.1.8), the published q_* value *and* its standard error ε_{m*} are consistently embedded into the analysis.

Another interesting case has not yet been considered in light curve analysis. In Sect. 5.2.2 we discuss whole-curve fitting as an *alternative* to traditional times-of-minima analysis for deriving an *EB* ephemeris. With introduction of the observable T_m^o and its parameter counterpart T_m^c, the observed times of minima, T_m^o, could be made an integral part of the analysis with T_m^c being a function of the ephemeris parameters T_0 (time of conjunction), P_0 (period at reference epoch), dP/dt (period time derivative), and in principle the other orbital parameters.

5.1.2 Determining Individual Temperatures

One of the main difficulties of modeling *EB*s is the accurate determination of the individual temperatures. Frequent practice in the literature is to *assume* the temperature of *one star*, or better to *obtain it from spectra or color indices* in an a priori step as described on page 199, after which the other star's temperature follows by fitting the light curve model to the data. In this two-step approach, the accuracy of the estimated temperature depends either on the how well the individual spectra can be extracted from the composite binary spectrum or the reliability of deriving that temperature from the composite time- or phase-dependent color index. As it is difficult to estimate accurately the contribution of only one star in advance, in Sect. 5.1.2.1 we exploit extra information to derive estimates of the temperatures, Sect. 5.1.2.2 uses color indices to determine the temperature, while in Sect. 5.1.2.3 we describe how to evaluate light curves in at least two passbands to determine both temperatures *and* the distance (Sect. 5.1.4) which overcomes some limits of the traditional two-step approach to estimate distances of *EB*s (Sect. 5.1.4).

5.1.2.1 Temperature Estimations

One way to estimate individual temperatures is to exploit extra information about the system. If the stars under consideration are known members of a stellar ensemble, for example, and the ensemble has been well studied, the condition is met. By well

5.1 Extended Sets of Observables and Parameters

studied, we mean that the color-magnitude (CMD) is well established, the distance has been found, and there is a well-fitted isochrone available. With these criteria, a method has been devised by Milone et al. (2004b) which utilizes the properties of the isochrone to determine the properties of both components of an *EB*, in the case that no spectroscopy and only two-passband photometry is available. In the case of the 47 Tuc survey for extrasolar planets, cf. Gilliland et al. (2000); Milone et al. (2004b), this proved to be the case. The idea is basically this: From the Russell–Merrill treatment in Sect. 6.2.1, we obtain a relation for eclipse depths. The depth of mid-eclipse for primary (p) and secondary (s) minimum can be written as

$$d_p = 1 - \ell_p = (L_1 + L_2) - [L_1 + (1 - \alpha_0)L_1] = \alpha_0 L_1 d_s \quad (5.1.9)$$
$$= 1 - \ell_s = (L_1 + L_2) - [L_1 + (1 - \alpha_0)L_2 k^2] = \alpha_0 k^2 L_2, \quad (5.1.10)$$

where ℓ_j is the luminosity of component j, α_0 is the fraction of light lost at mid-eclipse, and $k = R_2/R_1$. The passband luminosity is the product of the surface brightness and the area of the disk, so the ratio of depths in any passband is

$$\frac{d_p}{d_s} = \frac{L_1}{L_2 k^2} = \frac{\sigma_1}{\sigma_2}, \quad (5.1.11)$$

where σ_j is the surface brightness of component j. Comparing now the ratio of depth ratios for two different passbands

$$Q = \frac{(d_p/d_s)\lambda_1}{(d_p/d_s)\lambda_2} = \frac{(\sigma_1/\sigma_2)\lambda_1}{(\sigma_1/\sigma_2)\lambda_2}. \quad (5.1.12)$$

Rearranging this

$$Q = \frac{[\sigma_1(\lambda_1)]/[\sigma_1(\lambda_2)]}{[\sigma_2(\lambda_1)]/[\sigma_2(\lambda_2)]}. \quad (5.1.13)$$

But this ratio is directly related to the color index difference between components:

$$\Delta C = C_1 - C_2 = -2.5 \log Q. \quad (5.1.14)$$

What is observed, however, is the net color index of the system, C_{12}, and, if the cluster has been well-enough studied, the well-known cluster ambiguities of interstellar and cluster reddening and metallicity have been resolved, so that the intrinsic color indices and the absolute visual magnitude can be assumed known for the system. This is where the isochrone comes in: In the case of the detached systems, from which the independent properties of the system may, indeed, represent those of independently evolved stars, we assume that one of the two components lies on the isochrone, if the system is a cluster member. If the eclipse is total, the occultation eclipse provides an individual CI; if partial, one may assume initially that either component is on the isochrone. Proximity of the system to either the isochrone or to the "binary main sequence" indicates that the components are either dominated by one component or are equally luminous. Thus the difference between the system's M_V and that of the isochrone provides vital clues for the modeling of the light curves. The full procedure for the determination of the full set of parameters and the preliminary results for the 47 Tuc cluster are further discussed in Milone et al. (2004b). The bottom line of this method is that the accuracy and precision of the derived parameters, including the temperatures of the components, are strongly dependent on the accuracy and precision of the model for the cluster.

5.1.2.2 Color Indices as Individual Temperature Indicators

Prša & Zwitter (2005b) explicitly discuss an approach that is often taken by light curve modeling astronomers, to determine the individual effective temperatures, T_1 and T_2, from standard photometry observations without perhaps fewer a priori assumptions than other approaches. This approach involves the system's *binary effective temperature*, T_B. Observationally, an unresolved binary may be regarded as a point-source with a time- or phase-dependent effective temperature $T_B = T_B(t)$ that could be compared to an observed color index curve. Both components contribute to $T_B(t)$ according to their sizes and individual temperatures, T_1 and T_2, and inclination. If a model is to accurately reproduce observations, the composed contributions of both components must match this behavior.

An initial value of $T_B(t)$ may be obtained from a color-temperature calibration. For the Johnson $B - V$ color, Prša & Zwitter (2005b) in their program PHOEBE compute the empirical color-based effective temperature, $T_{\text{eff}}(B - V)$, by a polynomial of degree 7,

$$T_{\text{eff}}(B - V) = \sum_{i=0}^{7} C_i (B - V)^i \qquad (5.1.15)$$

with the coefficients from updated Flower (1996) tables,

Coefficient:	V, IV, III, II	I
C_0	3.979145	4.012560
C_1	−0.654992	−1.055043
C_2	1.740690	2.133395
C_3	−4.608815	−2.459770
C_4	6.792600	1.349424
C_5	−5.396910	−0.283943
C_6	2.192970	–
C_7	−0.359496	–

where the second column applies to main sequence stars (V), sub-giants (IV), giants (III), and bright giants (II), the third one (I) to supergiants. To make use of this scheme, CIs from other passbands, e.g., Johnson V and Cousins I, need to be transformed into $B - V$ in order to use Flower (1996), e.g., by exploiting Caldwell et al. (1993) who provide different color index dependencies on $B - V$. As these calibrations serve only to obtain an initial value of $T_B(t)$, there is no need to worry about accuracy. As we see below the individual temperatures reproducing the color index are derived in an iterative scheme exploiting the synthetic energy distribution of the binary.

The Wilson–Devinney (WD) model used in the following discussion computes the observable flux scaled to an arbitrary level. The model adapts to this level by determining for each passband i the corresponding WD passband luminosity, L_{1i}, not considering any color information that might have been present in the data. Because T_B is observationally revealed by its $B-V$ (or any other suitable) color index,[1] some of the relevant temperature information is lost. One solution to this problem is to couple the passband luminosity by exploiting the observed color index (the method proposed by Prša & Zwitter 2005b), another solution is to resort to Wilson's (2007a, b, 2008) temperature–distance theorem matching discussed in Sect. 5.1.2.3. Both approaches depend on reliable photometric calibrations [cf. Landolt (1992) covering celestial equator regions, Henden & Honeycutt (1997) and Bryja & Sandtorf (1999) covering fields around cataclysmic variables, Henden & Munari (2000) covering fields around symbiotic binaries].

In Prša & Zwitter's (2005b) light curve program PHOEBE, described in Sect. 8.2, the parameters L_{1i} are initially regarded as simple level-setting quantities – physical context comes in only after the color index relationship is exploited. For the sake of simplicity, consider that input observational data are supplied in magnitudes rather than fluxes, without any arbitrary scaling of the data: Colors must be preserved and the photometry must be absolute, i.e., not relative, and fully transformed to a standard system, for which the zero magnitude flux is known accurately. PHOEBE, built on WD, inherently works with flux using a single, passband-independent parameter m_0 to transform all light curves from magnitudes to fluxes. The value of m_0 is chosen so that the fluxes of the dimmest light curve are of the order of unity. It is a single quantity for all passbands, which immediately implies that the magnitude difference, now the flux ratio, is preserved; hence, the color index is preserved. If the distance to the binary is also known (e.g., from astrometry), m_0 immediately yields observed luminosities in physical units; intrinsic luminosities in physical units are obtained if the color excess $E(B - V)$ is also known. This is where physics comes in: From such a set of observations, the calculated L_{1i} are indeed passband luminosities, the ratios of which are the constraints we need. Passband luminosities of light curves are now connected by the corresponding color indices. Note that the temperatures can be derived without the distances.

Once the color index relationship is set, only a single passband luminosity L_{1i} is adjusted, while the remaining L_{1i} are computed from the color index constraints. This way color indices are preserved and effective temperatures of the binary may

---

[1] Useful relations among color indices are given in Caldwell et al. (1993). Although color indices depend also on $\log g/g_0$, metallicity, and rotational velocity, their effect is much smaller than the temperature.

be obtained. For a given model solution, a synthetic spectrum of the binary may be computed from the current values of $T_{\rm eff}$, $\log(g)$, $v_{\rm rot}$, and metallicity, [M/H]. As the stars are not spherical, this spectrum is constructed by integrating over local emergent intensities computed for each surface element. That spectrum is then "corrected" for interstellar extinction (by multiplying it with a wavelength-dependent interstellar extinction curve), Rayleigh scattering due to Earth's atmosphere, and any other intrinsic wavelength-dependent corrections (circumbinary attenuation, clouds, etc.) that are included in the model (Prša 2009). The yielded spectral energy distribution is then multiplied by the CCD, optics and filter response functions (this combination is usually referred to as the passband transmission function). By this one obtains the theoretical spectrum as it would be detected. To obtain the theoretical flux for the given passband, we need to integrate this spectrum over wavelength. By repeating the same process for different passbands, we can obtain theoretical fluxes for all passbands at all phases, the ratios of which – as one might have guessed from the discussion in the preceding section – are the color indices and these need to be constrained by observations.

A historical definition of color indices, dating back to the time of visual photometry and differential photometry, requires that their values are set to 0.0 for Vega (spectral type A0V). Stellar energy distribution functions do not provide us with that a priori, so one needs to offset the theoretical color indices. This is achieved easily by computing passband fluxes for Vega (by the same procedure sketched above), deriving color indices and setting them to 0. This provides color index *offsets* that are then applied to binary star color indices.

Applying (5.1.15) yields $T_{\rm B}(t)$ at some observational time t or phase. For binaries with well-determined Roche surface potentials, T_1 and T_2 follow from the least-squares analysis. For wide detached binaries, this is somewhat hindered by the fact that the Roche potentials and temperatures are fully correlated via the surface brightness ratio (the eclipse depth ratio is directly proportional to the surface brightness ratio, which is in turn a function of $[(T_1, R_1), (T_2, R_2)]$). This renders individual T_1 and T_2 uncertain to the extent of the degeneracy between surface brightness parameters. Regardless, color constraining projects out only those combinations of parameters that *preserve* $T_{\rm B}(t)$ and hence the color index. Because the relation between effective temperatures of individual components is, in cases where parameter correlations are weaker, fully determined by the light curve (predominantly by the primary-to-secondary eclipse depth ratio) and because the sum of both components' contributions must match the effective temperature of the binary, the color-constrained least-squares method yields effective temperatures of individual components.

5.1.2.3 Both Temperatures from Absolute Light Curves

The essential requirement for direct distance estimation described in Sect. 5.1.4 and temperature determination follows from Wilson's (2007a, b, 2008) *EB* temperature–distance theorem: *Eclipsing binary light curves can yield temperatures of both stars and distance if and only if the data are standardized, the absolute geometry*

5.1 Extended Sets of Observables and Parameters

is determined, and two or more substantially different photometric passbands are fitted. Here temperature means *effective mean surface temperature* excluding the reflection effect.

In this context, the "standardization" requirement can be fulfilled by calibration from standard[2] magnitudes to physical units as described by Wilson (2007a, b) and discussion in Wilson (2008, Sect. 5). Absolute flux calibrations for the Strömgren *uvby* photometric system have been derived by Fabregat & Reig (1996) and Gray (1998). Two calibrations suitable for converting standard magnitudes in the Johnson system to standard physical units are Johnson (1965, 1966) and Bessell (1979), designated by subscripts *J* and *B*, respectively, below. They have been converted by Wilson (2007a, b) and lead to the following table:

Band $\lambda_{\text{eff}}(u)$	f_J	f_B	f_J/f_B
U	360 $4.35 \cdot 10^{-1}$	$4.19 \cdot 10^{-1}$	1.038
B	440 $6.87 \cdot 10^{-1}$	$6.60 \cdot 10^{-1}$	1.041
V	550 $3.78 \cdot 10^{-1}$	$3.61 \cdot 10^{-1}$	1.047
K	2200 $3.92 \cdot 10^{-3}$	$4.02 \cdot 10^{-3}$	0.975

This table, for various passbands u with effective wavelength $\lambda_{\text{eff}}(u)$ measured in nm gives the absolute flux values for zero apparent magnitude, e.g., $f_J(V) = 0.378$ erg/cm^3/s is the Johnson V flux for $m_V = 0$. The conversion from arbitrary apparent magnitude m_V to absolute flux $f(m_V)$ is then given by

$$f(m_V)/f_J = 10^{-0.4 m_V}. \tag{5.1.16}$$

Scaling using both calibrations differs by only 4% which seems to be sufficiently accurate given that the error in the distance varies with the square root, e.g., for Johnson V with $\sqrt{1.04} \approx 1.02$, or $\sqrt{0.975} \approx 0.98$, respectively. In addition to the Johnson or Bessell calibration one needs to know the interstellar extinction A in magnitudes.

A nonstandard or non-calibrated light curve essentially contains information about the components' surface brightness via the ratio of eclipse depths yielding a relation

$$T_2 = f(T_1). \tag{5.1.17}$$

As this provides no further temperature information, we have one relation for two unknown temperatures. Therefore, it has been common practice to adopt one temperature, usually T_1, from spectra or color indices and solve for the other one. Following Wilson (2007a, b) one can also understand why, in principle, additional light curves in other passbands do not help to determine both temperatures. In an

---

[2] Standardized means U, B, V, etc. – not only on a standard system but with magnitudes as opposed to magnitude differences. In astronomical photometry, this is referred to as "absolute photometry."

ideal situation where all system parameters are known and the radiative emission is described fully and correctly, the photometry and spectroscopy are reproduced correctly – only the pair (T_1, T_2) is not yet resolved. Theory then will correctly predict T_2 for given T_1, for *every passband* from the primary to secondary depth ratios, and because all the computed T_2s are correct, they will be band independent and the resulting relations $T_2 = f(T_1)$ will just repeat that found from any one curve.

With standard light curves convertible to absolute units, the situation changes. Only one of the infinite pairs (T_1, T_2) satisfying (5.1.17) will also *reproduce the absolute system flux* for a given distance D. As we do not know D, we add the two constraints

$$F_1^{cal}(T_1, T_2, D) = F_1^{obs} \quad (5.1.18)$$

$$F_2^{cal}(T_1, T_2, D) = F_2^{obs} \quad (5.1.19)$$

and two passbands. The three equations (5.1.17), (5.1.18), and (5.1.19) are needed and sufficient to determine all three unknown quantities T_1, T_2, and D as stated above in the temperature–distance theorem. As a consequence, one standard light curve suffices to find (T_1, T_2) if D is independently known.

One has to keep in mind that this is the ideal situation. If the absolute flux calibration is not accurate enough or the observed data are too noisy, the temperatures and distance can appear larger or smaller than they really are, leading to an error in the temperature perhaps by several hundred degrees. Thus, it may well continue to be the case that light curves in more passbands improve precision and that spectroscopic temperature determinations prove to be more accurate.

5.1.3 Traditional Distance Estimation

Traditional distance estimation is an a posteriori step that follows the light curves analyses in Sect. 4.4.4.1. Here we summarize the underlying idea. As pointed out in Chap. 1 (page 22) the distance D or parallax π of a binary can (in favorable cases) be derived if both light and radial velocity curves are available. Although the orbit size, relative dimensions, and temperature difference are the results of a simultaneous least-squares analysis, it is useful to remember that these quantities strongly relate to the data sources as listed by Wilson (2008):

1. The orbit size, i.e., the semi-major axis a, and thus the linear scale of the system follow from the radial velocity curves.
2. The star dimensions relative to a (i.e., their mean radii) follow from the form of the light curve that establishes a one-dimensional family of mean surface temperatures $[T_1, T_2]$ through the two eclipse depths.
3. If one mean temperature is obtained from spectra or color indices (as is usually the case), the other temperature follows implicitly from the $[T_1, T_2]$ family through the least-squares analysis. Together with the surface gravity distribution, irradiation from the other star, and possibly spots, a temperature distribution

5.1 Extended Sets of Observables and Parameters

can be computed. If this distribution is entered into an atmosphere model, or a simplified model involving a blackbody law with limb darkening, we obtain the emission per unit surface area on both stars.

4. The above information determines the absolute passband luminosities, as well time-dependent observable flux for any assumed distance.

The traditional way to compute the distance from the four data types above in most "distance deriving" publications prior to 2007 is described in Sect. 4.4.4.1 and finds D from the unreddened distance modulus, for instance in the Johnson system introduced in Sect. 2.1.1

$$(m_V - M_V)_0 = 5\log(D/10) \qquad (5.1.20)$$

with

$$(m_V - M_V)_0 = m_V - M_V - RE_{B-V},$$

where D is the distance in parsecs, and $R = A_V/E_{B-V}$ is the ratio of the passband extinction, A_V, of the V light by the interstellar medium to the color excess E_{B-V}. Note that the apparent magnitude of the binary needs to be known in a standard system, e.g., Johnston UBV or Strömgren *uvby*. Differential photometry alone is not sufficient. If distance were derived at all, we further note the following.

1. The distance derivation using the separate follow-up step is strictly valid only for spherical stars as otherwise the local physics is not treated properly.
2. Formal standard errors of the derived distance are not provided, although other kinds of error estimates may be given.

Therefore, below we argue in favor of Wilson's (2008) self-consistent approach that exploits photometric light curves in standard physical units.

5.1.4 Direct Distance Estimation

Direct distance estimation means that the distance, D, is derived in a least-squares analysis without the a posteriori step discussed in Section 5.1.3. If D is determined in this way, it is consistent with all local physical surface quantities, including proximity effects and any other phenomena supported by the model, avoiding the need to resort to mean radii. The only model requirement is the ability to compute distance-dependent flux, F_d^{ph}, in standard physical units, e.g., *cgs* units. In the analysis, these fluxes are then compared to the observed fluxes, F^{obs}, by exploiting the relation

$$D_{pc} = a_{pc}\sqrt{\frac{F_d^{ph}}{F^{obs}}}. \qquad (5.1.21)$$

Thus, observed light needs to be available also in these standard units, or should be properly converted from standard magnitudes or relative flux to these units. As discussed already in Sect. 5.1.2.3 for the Johnson UBV system, F^{obs} has been calibrated to standard physical units.

In the case of the WD program, there has always been an explicit absolute coupling between 4π sr passband luminosity, L, and calculated flux, F. Use of the normal emergent intensity, I_{pole}, at a reference point (usually, a pole) as a scaling factor, made it easy and required only minor changes to compute light curves in any standard system of physical units by introducing a second scaling step from user-defined flux units to physical units by the relation

$$L^{ph} = \left(\frac{I^{ph}_{pole}}{I^{mod}_{pole}}\right) L^{mod}, \tag{5.1.22}$$

where superscript "mod" refers to WD intrinsic model quantities. Thus, absolute I^{ph}_{pole} yields absolute L^{ph} and F^{ph}_d, given that the geometry is correctly followed. The distance-dependent flux, F^{ph}_d, in physical standard units, in the WD program, is computed as

$$F^{ph}_d = 10^{-0.4A} \left(\frac{a}{D}\right)^2 \left[F^{mod}_{a1}\left(\frac{I^{ph}_1}{I^{mod}_1}\right) + F^{mod}_{a2}\left(\frac{I^{ph}_2}{I^{mod}_2}\right)\right], \tag{5.1.23}$$

where superscript "mod" refers to WD intrinsic model quantities, a is the orbital semi-major axis, and A is interstellar passband extinction in magnitudes, with subscripts 1 and 2 for the binary components. The polar normal intensities I^{ph}_1 and I^{ph}_2 are computed by an interpolation routine approximating the radiation by a stellar atmosphere.

Utilization of a stellar atmosphere, or an approximation to it, requires a good temperature estimate for one star from spectra or color indices, or derivation of both temperatures from two light curves as described in Sect. 5.1.2.3. As stated in Wilson's *EB* temperature–distance theorem, calibrated light curves in two passbands are needed and are, in principle, sufficient to determine both temperatures and the distance. Note that the light curve model involves and reproduces mean surface temperature, temperature at a definite reference point (usually, the pole), at any local surface point, and observed (aspect-dependent) temperature.

Distances derived from *EB*s can be very accurate and can be determined for nearby galaxies [cf. Wilson (2004)]. However, we should keep in mind that they depend on extinction, A, and they could also be contaminated by third light. The light, ℓ_3, of a third star or planet makes a system appear brighter and thus apparently nearer, and thus influences *EB* solutions in other ways. It dilutes all variations (eclipses, tides, reflection, etc.) and thus reduces light curve amplitudes.

5.1 Extended Sets of Observables and Parameters 233

Interstellar extinction dims light curves so as to increase distance estimates, whereas its associated reddening decreases temperature estimates. Reduced theoretical temperatures reduce predicted absolute fluxes and thus decrease distance estimates. Thus, in regard to distance determined from this light curve analysis approach, extinction and reddening partly offset one another and, accordingly, the overall effect of extinction on distance determination is less than one might suppose (Wilson 2008, Sect. 7). Wilson (2008) also investigated the possibility of determining A through the least-squares analysis. Although this is indeed possible given absolutely scaled light curves in three passbands, it is not very practical as the sensitivity with respect to the calibration of the light curves is typically strong. Small deviations in the calibration lead to significantly different A values.

In addition to these interstellar extinction or third light problems specific to the analysis and observation of a particular binary, Wilson (2008, Section 5) carefully discusses the radiative model, photometric response functions, absolute photometric calibration, and photometric transformations as sources of systematic errors of the overall procedure.

5.1.5 Main-Sequence Constraints

The *EB* parameter estimation problem often suffers from a lack of uniqueness. Reducing parameter space may help but needs to be justified and carried out carefully. Some degrees of freedom can be eliminated by making an assumption about the morphological type of a binary star, e.g., that it is semi-detached or a contact binary. With more *EB* software becoming available, the set of special assumptions is increasing. Prša & Zwitter (2005b) introduced main sequence constraints (*MSC*) assuming that either or both components are main sequence stars. Because a significant percentage of all stars are on the main sequence, there is a fair chance that this assumption is correct. In principle, applying *MSCs* to one component or both components of the modeled binary means imposing relations among mass, luminosity, temperature, and radii of main sequence stars (see, e.g., Malkov 2003 for such relations specific to *EB*s) which in turn relates to the computation of absolute dimensions. Consequentially, given a single parameter (e.g., a component's effective temperature), all other parameters (its mass, luminosity, and radius) are calculable. This in turn implies that in the case of circular and nearly circular orbits, the effective potential of the constrained component is fully determined.

However, the situation is not as ideal as described above. The main sequence is not a line in the Hertzsprung–Russell diagram (luminosity or absolute magnitude versus spectral type, stellar temperature or color index: Basically brightness plotted against color) but rather a strip with a non-zero width. Depending on the position selected, rather different values for the derived absolute quantities could follow. Prša & Zwitter (2005b) account for the width of the main sequence by treating the main sequence relations not as strict constraints but rather as a user-defined penalty cost terms.

Although *MSC*s break the degeneracy, at best they could be used for testing whether either or both stars can plausibly be main sequence stars: Depending on the behavior of the standard deviation of the fit, such a hypothesis can be accepted or rejected. *MSC*s are of a different nature than morphological constraints and may lead to a circular argument: *EB*s provide absolute parameters for stars, which can then be used to establish various calibrations. *MSC*s, on the other hand, use calibrations. Thus, main sequence constrained solutions must never be used to establish calibrations of any kind.

5.1.6 Intrinsic Variability of Eclipsing Binaries' Components

In Chap. 1 we introduced extrinsic and intrinsic variables stars; in the latter group, we find pulsating stars. As more than 50% of all stars are members of binary or multiple systems, we should expect to find pulsating stars, or more generally, intrinsic variables in binary and multiple systems, and thus in *EB*s. As pulsation indicates instability, we might be concerned with stellar evolution effects if such pulsating stars happen to be a binary member. However, at least for well-detached *EB*s we should expect that each star evolves independently. Pigulski (2006) presented a brief overview of the present knowledge of intrinsic variability in binary and multiple systems and discusses various examples of *EB*s in the literature. Adding intrinsic variability to a component of an *EB*, the stellar parameters of a binary component become further constrained.

Pigulski & Michalska (2007) reported detection of pulsating components in 11 *EB*s in the ∼11,000 stars from the public ASAS-3 database. Among them are three classical Algols, MX Pav, IZ Tel, and VY Mic, with δ Scuti-type primary components. In six other *EB*s, the short-period variability can also be interpreted in terms of δ Scuti-type pulsations, where both components are probably main sequence stars. The pulsation mode in *HD 99612* shows significant amplitude decrease during the observed interval. In addition, one component of the eclipsing and double-lined spectroscopic O-type binary ALS 1135 shows β Cep-type pulsation. Finally, *Y Cir* is a good candidate for a slowly pulsating B-star in an *EB*.

Given that the physical processes that cause pulsations lead to time-dependent temperatures and radii, it is not surprising that no light curve program yet exists that can determine both the normal *EB* parameters and the parameters describing the intrinsic variability. However, Wilson (2009) has already started the first numerical experiments on the subject. Overall, as discussed by Lampens (2006), the field has many open questions – among them whether binarity can influence pulsations or whether binarity can explain the exotic behavior of some pulsating stars.

5.2 Multiple Star Systems and their Dynamics

Ὅταν κάνεις ὅτι μπορείς, κάνεις ὅτι πρέπει.

(*When you do what you can, you do what you must.*)

5.2.1 Third-Body Effects on Light and Radial Velocity Curves

*EB*s are sometimes members of multiple star systems [cf. Mayer (2005) and references therein]. Among 728 multiple systems with 3–7 components contained in an extended 6/1999 edition of the Tokovinin (1997) catalogue of physical multiple stars, there are 83 eclipsing binaries. In the simplest case a third body orbits with the binary around the system barycenter. Observational evidence of third bodies comes from spectroscopy (disturbances of the radial velocity curve), or from analysis of times of minima. A linear relation between *Observed–Computed* times of minima and time indicates constant period[3]; a parabola implies a constant rate of period change; and a sinusoid implies *apsidal motion*[4] or variation in arrival time (*light-time effect*) due to orbital motion of the close binary around the system barycenter (binary plus third body). *EB* programs ideally should consider three short-term effects imposed by the third body: Third light as described in Sect. 3.4.1, the light-time effect to all observable phenomena, and radial velocities.

Although they are not short-term effects, here we want to summarize a few results on the dynamics and the effects of the third body on the binary, referring the reader to Eggleton (2006) for a more detailed discussion.

Apsidal motion can be due to gravitational perturbations by other bodies, finite nonspherical mass distributions of the stars, and general relativistic effects [cf. Quataert et al. (1996)]. If due to a third body, apsidal motion may be accompanied by precession. Both, apsidal motion and precession, result from the rotation of the binary's orbital frame around the barycenter. If the third body orbit is the outer orbit, both effects are on a timescale (Eggleton 2006, p. 203)

$$\tau_{3b} = \frac{h}{a^2 C_3} \approx \frac{M_1 + M_2}{M_3} \frac{P_3^2}{2\pi P} \sqrt{(1-e_3)^3}, \qquad (5.2.1)$$

where P and P_3 are the period of the inner and outer orbit, e_3 is the eccentricity of the outer orbit, h is the angular momentum of the third body, and

---

[3] The period is constant but incorrect if the slope of the line is different from zero.

[4] The term *apsidal motion* refers to the rotation of lines of apsides of an eccentric binary orbit, or the rotation of the periastron. Apsidal motion is caused by perturbations to the $1/r$ gravitational potential and indicates deviations from Keplerian elliptic motion.

$$C_3 = \frac{GM_3}{2a_3^3\sqrt{(1-e_3)^3}}. \tag{5.2.2}$$

Neither precession nor apsidal motion is expected to have a significant effect on the long-term orbital evolution of a binary. Given that P and a are purely functions of the orbital energy but not of angular momentum, they remain constant in the lowest approximation of the Keplerian orbit perturbation. The dynamic cause is that the third body's force is derivable from a potential and thus does no work around a closed curve. It thus has no net effect on the binary's orbital energy during a complete cycle. As the binary eccentricity and angular momentum can fluctuate, a third body can, nevertheless, strongly influence a binary's orbital evolution. One would not intuitively expect this behavior, particularly if the third body moves in a wide orbit with a period of the order of 10^4 years or has only very low mass. If the inclination between the outer and inner orbit is larger than about $39°$, large cyclic variations in e become possible at high inclinations (Kozai 1962). These cycles are known as *Kozai cycles* and are related to Kozai's (1962, Sect. 4.8) analysis of the interaction between Jupiter and solar system asteroids with high inclination. The cyclic eccentricity variations, coupled with the approximate constancy of a, means that tidal friction can become important at periastron during part of the cycle, even if it is unimportant during the small eccentricity part of the cycle. Due to this friction, over many Kozai cycles the inner orbit will shrink as well as become circularized, the final period being roughly the period when the stars are close enough for apsidal motion due to their distortion to dominate over apsidal motion due to the third body. As other perturbations of the third body can reduce the effect of Kozai cycles, there is a maximum outer orbit size that can generate Kozai cycles, but this may still be several thousand times larger than the inner orbit with periods up to 10^3–10^4 years. Eggleton & Kiseleva-Eggleton (2001) applied this theory to *SS Lac*, a binary, now lacking eclipses due to a rotating orbital plane.

Having discussed the dynamic consequences of a third body qualitatively, we present a few quantitative relations from Van Hamme & Wilson (2005, 2007) involving six third-body parameters that can be also derived from light and radial velocity curve fitting. These parameters are semi-major axis, a', of the outer relative orbit; eccentricity, e_3; the argument of periastron, ω', of the close orbit's center of mass; period, P'; inclination, i' (angle between the plane of sky and the third-body orbit), and superior conjunction, T_c', of the *EB* center-of-mass with the barycenter. Note that T_c' is well defined, whereas the *EB* periastron passage, T_{peri}', is undefined for circular orbits and weakly defined for small eccentricities.

As the binary star components orbit the barycenter, the periodically varying light-travel time leads to the light-time effect, i.e., the difference, $\Delta t = t_{obs} - t_{sys}$, between the barycentric time, t_{obs}, and the *EB* center-of-mass time, t_{sys}, which sets the orbital phase in the third-body orbit. The light-time effect time difference affects all observables and is given by

$$\Delta t = \mathcal{A}_{\Delta t} \frac{1-e_3^2}{1+e_3\cos v}\sin(v+\omega'), \tag{5.2.3}$$

5.2 Multiple Star Systems and their Dynamics

with the true anomaly, v, and the light-time semi-amplitude for circular third-body orbits

$$\mathcal{A}_{\Delta t} = \frac{a' \sin i'}{c} A, \qquad (5.2.4)$$

and

$$A = 1 - \left(\frac{a}{a'}\right)^3 \left(\frac{P}{P'}\right)^2, \qquad (5.2.5)$$

where $a' = a_{12} + a_3$ is the semi-major axis of the outer relative orbit, and c is the vacuum speed of light. The true anomaly v and the eccentric anomaly E are related by (3.1.27), which here gives

$$\tan \frac{v}{2} = B \tan \frac{E}{2}, \quad B := \sqrt{\frac{1+e_3}{1-e_3}}. \qquad (5.2.6)$$

The mean anomaly

$$M = \frac{2}{P'} \left(t_{sys} - T'_{peri} \right) \qquad (5.2.7)$$

is related to the eccentric anomaly E through Kepler's equation (3.1.28)

$$E - e_3 \sin E = M. \qquad (5.2.8)$$

Alternatively, the mean anomaly can be expressed as

$$M = M_c + \frac{2\pi}{P'} \left(t - T'_c \right), \qquad (5.2.9)$$

with the mean anomaly M_c at the time of superior conjunction, T'_c, obtained from Kepler's equation with $E = E_c$. The latter can be obtained from the true anomaly at conjunction

$$v_c = \frac{\pi}{2} - \omega', \qquad (5.2.10)$$

with

$$\tan \frac{E_c}{2} = \frac{1}{B} \tan \frac{v_c}{2}. \qquad (5.2.11)$$

Third bodies lead also to a change of the radial velocities due to the eclipsing pair's systemic velocity. As the EB orbits the barycenter, its velocity varies and is given by

$$V_\gamma = V_{\gamma 0} + \frac{A_\gamma}{\sqrt{1-e_3^2}}\left[e_3 \cos\omega' + \cos(\upsilon_c + \omega')\right] \quad (5.2.12)$$

with the semi-amplitude

$$A_\gamma = \frac{2\pi}{P'} a' \sin i' A \quad (5.2.13)$$

of the V_γ reflex motion for circular third bodies.

The least-squares approach to estimate the third-body parameters a', e_3, ω', P', i', and T'_c, requires the calculation of the derivatives $d(\Delta t)/dp$, where p denotes one of the third-body parameters. Van Hamme & Wilson (2007) give analytic relations for these derivatives.

5.2.2 Ephemerides Derived from Whole Light Curves and Radial Velocity Curves

In most publications on *EBs*, we find that ephemerides (epoch, period, and period changes) are derived from the times of minima. This usually requires a long-time coverage of the binary, which is a useful basis to establish the period and period changes. However, if the *EB* has just been discovered, the times of minima are only a fraction of the data available when complete light curves and radial velocity curves have been observed. An alternative method has been proposed by Wilson (2005) and by Van Hamme & Wilson (2007). It does not depend on traditional timing diagrams but derives an ephemeris from whole light and radial velocity curves. The time coverage is improved as there may be epochs with only light curves and other epochs with only radial velocity curves. Wilson, in Elias et al. (1997), had used this technique already in an analysis of *AX Monocerotis*, a (K giant, Be giant) binary that lacks eclipses, with radial velocities from two epochs being the only reliable means to an ephemeris. Because "time markers" such as light curve eclipses are absent in radial velocity curves, an algorithm was needed for treating whole curves, and the idea for *AX Mon* was to work within a full binary star observables program (in this case, WD), rigorously compute phases from time, and let WD take care of all sophistications of the synthesized velocity curves. The analysis involves the orbit ephemeris parameters T_0 (reference epoch), P_0 (period at reference epoch), $\dot{P} = dP/dt$ (period time derivative), and $d\omega/dt$ (orbit rotation, i.e., apsidal motion). For $\dot{P} = 0$ and thus constant $P_0 = P$, phase and time are connected by

$$\Delta\phi = \frac{\Delta t}{P}, \quad (5.2.14)$$

5.2 Multiple Star Systems and their Dynamics

where $\Delta\phi$ denotes the length $\Delta\phi = \phi - \phi_0$ of a phase interval $[\phi_0, \phi]$, and Δt the corresponding time interval. Thus we can compute phase as a function of time by

$$\phi(t) = \phi_0 + \frac{t - T_0}{P}. \qquad (5.2.15)$$

For constant, but nonzero $\dot{P} \neq 0$ we need to derive[5] a relation between between phase, $\phi(t)$, and time for nonconstant period. We start with

$$d\phi = \frac{dt}{P} \qquad (5.2.16)$$

and integrate this

$$\Delta\phi = \int_{T_0}^{t} \frac{d\tau}{P} = \int_{T_0}^{t} \frac{d\tau}{P_0 + (\tau - T_0)\dot{P}}. \qquad (5.2.17)$$

Using the integral relation

$$\int \frac{dt}{a + bt} = \frac{1}{b} \ln(a + bt) + \text{const} \qquad (5.2.18)$$

with $a = P_0 + T_0 \dot{P}$ and $b = \dot{P}$ we obtain

$$\Delta\phi = \frac{1}{\dot{P}} \left\{ \ln\left[P_0 + (t - T_0)\dot{P}\right] - \ln[P_0] \right\} \qquad (5.2.19)$$

$$= \frac{1}{\dot{P}} \ln \frac{P_0 + (t - T_0)\dot{P}}{P_0} = \frac{1}{\dot{P}} \ln \left(1 + \frac{(t - T_0)dP/dt}{P_0}\right) \qquad (5.2.20)$$

or, eventually,

$$\phi(t) = \phi_0 + \frac{\ln\left[1 + \frac{(t-T_0)dP/dt}{P_0}\right]}{dP/dt}. \qquad (5.2.21)$$

To evaluate (5.2.21) numerically for small values of dP/dt or in the limit $dP/dt \to 0$, we provide the Taylor series expansion (5.2.22)

$$\Delta\phi = \frac{(t - T_0)}{P_0} \sum_{k=0}^{\infty} \frac{(-1)^k}{k+1} \left(\frac{(t - T_0)}{P_0} \frac{dP}{dt}\right)^k, \qquad (5.2.22)$$

for small dP/dt, which in the limit $dP/dt \to 0$ yields the phase–time relation

---

[5] This derivation is reproduced from R. E. Wilson's personal notes of Dec. 1995 and Feb 1997.

$$\phi(t) = \phi_0 + \frac{(t - T_0)}{P_0} \qquad (5.2.23)$$

for constant P. Except for the new formulae (5.2.21) and (5.2.22) to compute phase, radial velocities and light curve flux may be computed in the usual way from an EB model. Terms in T_0, P_0, dP/dt, and $d\omega/dt$ were added to the DC equation of condition in the WD model so, with f the general symbol for an observable quantity (here radial velocities or flux), the equation becomes

$$f_o - f_c = \frac{\partial f_c}{\partial T_0}\delta T_0 + \frac{\partial f_c}{\partial P_0}\delta P_0 + \frac{\partial f_c}{\partial(dP/dt)}\delta(dP/dt) + \frac{\partial f_c}{\partial(d\omega/dt)}\delta(d\omega/dt)$$
$$+ \sum_{k=1}^{K} \frac{\partial f_c}{\partial p_k}\delta p_k. \qquad (5.2.24)$$

Subscripts o and c are for "observed" and "computed," and the δp_k values are corrections to input parameters p_k in the iterative scheme. The partial derivatives in T_0, P_0, dP/dt, and $d\omega/dt$ can each be expressed as the product of a numerical derivative and an analytic derivative. For instance, $\partial f_c/\partial T_0$ is given by how the basic observables depend on phase, $\partial f_c/\partial\phi$, and an analytic derivative that quantifies how phase depends on a given parameter (e.g., $\partial\phi/\partial T_0$). Accordingly, each of the four derivatives

$$\begin{array}{ll} \frac{\partial f_c}{\partial T_0} = \frac{\partial f_c}{\partial \phi}\frac{\partial \phi}{\partial T_0} & \frac{\partial f_c}{\partial P_0} = \frac{\partial f_c}{\partial \phi}\frac{\partial \phi}{\partial P_0} \\ \frac{\partial f_c}{\partial(dP/dt)} = \frac{\partial f_c}{\partial \phi}\frac{\partial \phi}{\partial(dP/dt)} & \frac{\partial f_c}{\partial(d\omega/dt)} = \frac{\partial f_c}{\partial \phi}\frac{\partial \phi}{\partial(d\omega/dt)} \end{array} \qquad (5.2.25)$$

includes a numerical factor, $\partial f_c/\partial\phi$, that needs to be computed only once (not four times) per data point, and also an analytic factor. The analytic factors require negligible computing time and the numerical factor is the same for all four parameters, so four derivatives can be generated for the computational price of one. To compute, for instance, $\partial\phi/\partial T_0$, we provide the inverse relation to (5.2.21)

$$t - T_0 = e^{(\Delta\phi dP/dt - 1)}\frac{P_0}{dP/dt} \qquad (5.2.26)$$

and its Taylor series expansion

$$t - T_0 = P_0 \Delta\phi \sum_{k=0}^{\infty} \frac{1}{(k+1)!}\left(\Delta\phi\frac{dP}{dt}\right)^k. \qquad (5.2.27)$$

Exploiting the analytic derivatives eliminates the need to specify appropriate increments for T_0, P_0, dP/dt, or $d\omega/dt$ in computing the four derivatives $\partial f_c/\partial T_0$, etc.; we only need to think about one increment in ϕ.

Traditional timing plots and the ephemeris solutions described here are complementary in that timing diagrams are naturally visual and intuitive, while "multiple whole-curve" solutions potentially access a larger body of information. Ideally, one should include the times of minima as another set of observables into the light curve analysis exploiting the ideas outlined in Sect. 5.1.1.

5.3 Analyzing Large Numbers of Light Curves

Ultra posse nemo obligatur.   (*Nobody is obliged to do the impossible.*)

Thousands of *EB*s have already been identified in the light curve database of the MACHO projects [cf. Faccioli et al. (2007)]. The discovery rates, for example, of variable stars (including eclipsing binaries), are increasing rapidly as more powerful ground-based telescopes and new satellites come on line, and we could expect discoveries in the millions. To extract astrophysical information from even a small fraction of these, humans cannot be in the data-reduction loop and new techniques are needed to eliminate this requirement. Here we discuss some ideas on analyzing large number of *EB* light curves from surveys.[6]

It should be understood that modeling eclipsing binaries and solving inverse problems in such a context is a major research effort and requires expertise to use the software effectively. Wherever possible, great effort has been invested to make the software as stable as possible, but in some places careful user interaction is needed.

For the analysis of large numbers of *EB* light curves obtained from surveys, detailed investigations need to be replaced by a highly automated procedure. There is a price to be paid for doing this in terms of accuracy. Nevertheless, such an approach should produce good approximate results and may indicate interesting *EB* stars for detailed analysis.

At the time of this writing, a few attempts have already been made, but there is a significant amount of work needed to support efficiently the analysis of, for instance, the data expected from survey missions such as OGLE, MACHO, TrES, HAT, or Kepler, as well as from the *Large Synoptic Survey Telescope* (LSST).

5.3.1 Techniques for Analyzing Large Numbers of Light Curves

The critical issues are speed and stability. Speed is obviously necessary to analyze large numbers of data. Stability is required to automate the procedure. Automation is required if the user is to analyze large sets of data produced by surveys. Approaches to automating light curve solutions have taken various forms to date.

---

[6] Methods on how eclipsing binary stars are detected in surveys and distinguished from other variable stars are found, for instance, in Eyer & Blake (2005) and references therein.

1. *Matching approach* [Wyithe & Wilson 2002a,b and Wilson & Wyithe (2003)]: Match one or several light curves to a large test set of pre-computed light curves. This set is used in the first step to analyze observed light curves.
2. *Rule-based approaches*: These follow an a priori rule-based procedure to extract relevant light curve information to produce a good initial parameter set for further light curve fitting. They are helpful to non-experts and can be combined with the matching approach.
3. *Simplified physical models*: Devor (2005) developed an automated pipeline for a simple spherical star model without tidal or reflection physics, whose starting values are guessed and then refined with a downhill simplex method followed by simulated annealing. This approach identifies detached binaries which can be described sufficiently accurate by spherical models. Tamuz et al. (2006) employed the EBOP ellipsoidal model (Popper & Etzel 1981). Using this engine, they arrived at initial solutions after a combination of grid search, gradient descent, and geometrical light curve analysis.
4. *Artificial neural network* (Prša et al. (2008)): The neural network approach can be understood as a formal mathematical approximation technique in which amplifiers between input and output signals are adjusted.

The first two approaches have in common that they can exploit the correct binary star physics that gives rise to the light curves. The neural network approach, as a formal approximation technique, may lack this facet without extra information. If spherical or ellipsoidal stars are used to approximate semi-detached or over-contact binaries, it is not clear if the initial information obtained from such simplifications will be useful. All of these treatments have the advantage that they allow non-experts to produce reasonable initial parameter sets for further detailed study.

A prerequisite for the first two approaches is to know the ephemerides. In principle, the period may be extracted by a power series analysis. Surveys may not have a long-enough duration to derive period changes. However, period changes might be an issue for non-expert users when analyzing *EB* light curves obtained at different epochs.

5.3.2 The Matching Approach

In their work to establish the best distance indicators among detached and semi-detached binaries in the Small Magellanic Cloud, Wyithe & Wilson (2002a,b) and Wilson & Wyithe (2003) obtained starting parameters for the rigorous WD model by comparing each light curve with a set of archived model light curves, and then sending the best match to an automated version of the WD differential corrector program DC.

In ongoing work, Kallrath & Wilson (2057) are extending this approach by an inner linear regression loop, incorporating a priori information, adding interpolation techniques, and increasing storage and numerical efficiency. This approach now supports all WD parameters.

5.3 Analyzing Large Numbers of Light Curves

For a given binary system, let ℓ_{ic}^o be any observed value for observable c, $c = 1\ldots C$, at phase θ_i. Correspondingly, ℓ_{ick}^o denotes the computed value at the same phase θ_i for the *archive*[7] curve k, $k = 1\ldots K$. Note that K might easily be a large number such as 10^{10}. The matching approach returns the number of the best fitting archive light curve, a scaling parameter, a, and a shift parameter, b, by solving the following nested minimization problem:

$$\min_k \left\{ \min_{a,b} \sum_{i=1}^{I} w_i \left[\ell_{ic}^o - \left(a\ell_{ick}^c + b \right) \right]^2 \right\}. \quad (5.3.1)$$

Note that the inner minimization problem only requires solution of a linear regression problem. Thus, for each k, there exists an analytic solution for the unknown parameters a and b. Note that the ℓ_{icn}^c values are obtained by interpolation. The archive light curves are generated in such a way that they are well covered in the eclipses, while a few points will do in those phases that show only small variation. Thus, there is a non-equidistant distribution of grid points that is well interpolated by cubic polynomials.

5.3.2.1 Solving Linear Regression Problems

Although solving linear regression problems is not difficult as such, one should exploit a priori knowledge of light curve parameters when looping over k. If a priori knowledge is available, for instance, on the mass ratio, q, or the temperatures T_1 and T_2, then certain k values can be excluded. The analysis of C observables (radial velocity curves and light curves) requires to solve C linear regression problems. If the observable is a radial velocity curve, the additive constant b gives the systemic velocity γ. For light curves, a returns the WD scaling quantity L_1 and b is third light, ℓ_3

5.3.2.2 Generation and Storage of the Archive Curves

Archive generation requires appropriate looping and proper interfacing to subroutine LC of the WD program. Special attention should be paid to the way the stored sets can be accessed. If a priori knowledge is available in connection with the Roche potentials Ω_1 and Ω_2, for instance, on the mass ratio, q, we should exclude unphysical configurations and ensure that certain values of k can be excluded.

An additional aspect is the storage of the computed archive light curves. For each light curve and observable (wavelength), we need $r_k = 4 \times C \times I_k$ bytes, where we consider 4 bytes, I_k phases and C bands. Note that we may have different numbers of phases depending on the shape and amplitude of the light curve (used in our interpolation scheme). The total memory requirement is then $R := \sum_{k=1}^{K} r_k$. Note that R may easily reach the order of $4C \cdot 10^8$ light curves if all reasonable

---

[7] Synonomously, we use the terms *stored*, *library*, or *template* light curves.

combinations of the photometric parameters e, ω, i, q, Ω_1, Ω_1, T_1, T_2, and $\log g$ are considered.

The choice of unadjusted parameters A_1, A_2, g_1, and g_2 depends on T_1 and T_2. L_1 can be set arbitrarily to $L_1 = 1$ because the matching problem involves the scaling parameter anyway. L_2 follows as a function of L_1. ℓ_3 is covered by the linear regression in the matching problem. Limb-darkening parameters also can be chosen, from, for instance, Van Hamme's (1993) limb-darkening coefficients. As the computation of limb-darkening coefficients depends on $\log g$, we have added this as a parameter. Great care is necessary when involving the eccentric orbit parameters. Both eccentricity, e, and length of the perihel, ω, need a very fine grid.

In addition to the sets generated automatically, we add all the light curve parameters sets for those *EB*s for which a light curve solution is available. This way, when we find a match to an observed light curve, we are able to provide not only some reasonable light curve parameters but, in addition, also a candidate similar to the current *EB*.

One might think to store the library light curves in a type of database. However, database techniques become very poor when talking about 10^{10} light curves. Therefore, a flat storage scheme is used. In the simplest case, for each k we store the physical and geometric parameters, then those parameters describing observable c, and then the values of the observable.

5.3.3 The Expert Rule Approach

Among other topics this approach depends on pattern recognition in graphs. The expert rule approach requires that we are able to derive morphological features of the light curves from data. Although humans can easily identify straight-line segments in graphs, pattern recognition in graphs is complicated. Light curve minima can easily be identified qualitatively on a plot. In that way, Algol type light curves are detected readily. If the minima are not arranged symmetrically, this immediately indicates an eccentric binary orbit.

5.3.4 Simplified Physical Models

Devor (2005) developed an automated pipeline for a simple spherical star model without tidal or reflection physics, and the starting values of which are first guessed and then refined with a downhill simplex method followed by simulated annealing. This approach identifies detached binaries which can be described with sufficient accuracy by spherical models. Devor & Charbonneau (2006a, b) extend this approach by utilizing theoretical models of stellar properties to estimate the orbital parameters as well as the masses, radii, and absolute magnitudes of the stars. This approach requires only a light curve and an estimate of the binary's combined color.

5.3.5 Artificial Neural Networks

The concept of neural networks (cf. Hertz et al. 1990) has it roots in brain research but has a counterpart in artificial intelligence, where it also involves elements of approximation theory and mathematical optimization. During the training phase, the artificial neural network is adjusted to a set of observations by a learning algorithm. In approximation theory, this corresponds to the computations of a set of weighting coefficients associated with some basis functions. After this phase, the tuned network computes output based on further input.

Neural networks have been used by Sarro et al. (2006) to separate pulsating stars from *EB*s and to classify *EB*s into four categories according to their characteristics such as eclipse depth and widths. The classification is performed by a Bayesian ensemble of neural networks trained with HIPPARCOS data of seven different categories including eccentric binary systems and two types of pulsating light curve morphologies. In a follow-up step, these four categories are related to the configurations detached, semi-detached and over-contact binaries.

Whereas Sarro et al. (2006) use artificial neural networks for automatic classification, Devinney et al. (2006) used it in their project *Eclipsing Binary Artificial Intelligence* (EBAI) for deriving starting parameters. They employed a neural network to "map" observational data to approximate model elements. Observational data for an *EB* are presented to the network's input nodes, and its output nodes yield starting model elements. The network is first "trained" on many observational data-model element pairs.

Prša et al. (2008) constructed a three-layer back-propagation neural network that solves nonlinear regression problems. They describe the basic concepts and procedures for applying artificial neural networks to detached *EB*s. Their neural network was trained with 33,235 WD-generated light curves and applied to a set of 10,000 synthetic detached *EB*s, to 50 detached binaries from the *Catalog and AtLas of Eclipsing Binaries* (CALEB1) and to the set of 2,580 OGLE LMC binaries (Wyrzykowski et al. 2003) classified as detached.

5.4 Extrasolar Planets

Adde parvum parvo magnus acervus erit.
(Add little to little and there will be a big pile. Ovid)

Extrasolar planet research has similarities with *EB* studies in the sense that similar data and analyzing methods are used because a star–planet (or other low-luminosity object) system, with transits and radial velocities for the star only, is in many respects analogous to a single-lined spectroscopic and detached *EB*. As the number of detected transiting planets is continually increasing (on July 1, 2009, the *Extrasolar Planet Encyclopedia*[8] contained 59 transiting planets), we have included

---

[8] URL http://exoplanet.eu/catalog.php (or, http://exoplanet.eu/catalog-transit.php).

this section on extrasolar planets and analyzing methods. For the *EB* community, we provide some introduction to the field of extrasolar planets.

There has been a long history of claims of detection of extrasolar planets, but only in recent decades have such claims been verifiable and, indeed, verified. Campbell et al. (1988) tentatively reported the detection of a planet in orbit around the star γ Cephei, but this was not confirmed conclusively until 15 years later (Hatzes et al. 2003). The first confirmed detection of extrasolar planets was by Wolszczan & Frail (1992) around the pulsar PSR 1257+12. In 1995, Mayor & Queloz (1995) made the first unambiguous discovery of a planet around a main sequence star: 51 Pegasi. The planet turned out to be more massive than Jupiter, and in close proximity to the star, a characteristic that earned it and subsequently discovered similarly placed objects the nickname "hot Jupiters." Thus far, this and most of the extrasolar planets discovered to date have been identified through radial velocity variations of the orbited stars. The discovery of planets through this technique required critical increases of precision to 10 m/s and better. The technical improvements have been in spectrograph stability, spectral comparison techniques, and in analysis methods. As a result, at present writing, uncertainties have been reduced to $\sim$3 m/s. In addition, long-term averaging of data is yielding longer-period, lower-amplitude effects in the radial velocity variations of the stars – the effects of planets several AU or more from the star. Still further improvement to $\sim$1 m/s or less appeared to be unproductive initially, because at this level the noise arising from chromospheric network motions in solar analogue stars is likely to mask any periodic effects due to planets. The noise arises from localized velocity variations, modulated by stellar rotation and activity cycles. Such modulation would produce periodicities in the noise level that could be misinterpreted as due to planets. However, over long periods of time (say years or tens of years), the strict orbital periodicity may be recoverable in the noise, because the latter is essentially stochastic and will be superimposed on the rotation (which is measurable for the star through high-resolution spectroscopy) and on a merely *cyclic* phenomenon – the star's activity or star spot cycle.

Another method for finding extrasolar planets is based on photometry, and an increasing number of planets are being detected through transit eclipses and through gravitational lensing spikes in the lensing signatures of their parent stars. Because the analysis of planetary transits has a lot in common with *EB* modeling, we discuss this method and successful applications of it in this chapter. Extrasolar planets that transit their stars provide tight constraints on the orbital inclination and, when coupled with radial velocity data, allow the determination of planetary masses and sizes. Parameters of the parent star still need to be determined independently. The temperature can be estimated from the star's spectral characteristics, whereas the mass and size can be derived from stellar evolutionary models (Cody & Sasselov 2002; Sasselov 2003).

Astrometry is also beginning to yield planetary detections, even though such methods are strictly distance limited in two ways: Extrasolar "hot Jupiter" systems must be close to the Sun to be observed in this way, but planets may be seen in systems that are more distant from the Sun if those planets are at greater distances from their parent stars, and yet are luminous enough to be visible. The latter is a nontrivial requirement; to be seen, the substellar object must be glowing in the infrared from its own internal heat sources, if it is too far from its parent to receive and reflect sufficient visible light.

5.4.1 General Comments About Substellar Objects

One of the best-studied extrasolar planets is that about the star *HD 209458*. *HD 209458b*, as the planet is designated, was discovered through radial velocity measurements of its parent, *HD 209458a*, but light curves were obtained and the eclipses were observed (Charbonneau et al. 2000) permitting the radius, the inclination, and therefore, the mass and absolute dimensions of the planet and orbit to be determined. Table 16.2 in Milone & Wilson (2008) provides a list of known extrasolar planets.

Some fraction of the objects designated planets may be brown dwarfs, which are typically larger, more massive, and have larger internally generated radiative flux than planets. This may be expected of substellar objects detected through the radial velocity variation of their more massive parent stars, because the single-lined curve yields only projected masses. Nevertheless, not all of them can be brown dwarfs, because the distribution of projected masses with semi-major axis shows few of the yet higher mass objects that would be expected if most of those detected were brown dwarfs (Mayor et al. 1998).

Brown dwarfs have been shown to be absent in binary combinations with normal stars, the so-called brown dwarf desert . However, near-infrared surveys such as 2MASS, DENIS, and Sloan show them to be present both individually and in binaries with other low-mass, low-luminosity objects. If the "desert" is not due to a selection effect, the survey results imply that the dearth in binary systems is, at least, as Basri (2000) called it, a brown-dwarf "desert island." As the number of survey results continue to accumulate, however, the dearth looks more and more like a *bonafide* desert.

5.4.2 Methods to Find "Small"-Mass Companions

Several of the search methods for extrasolar planets are similar to those employed to search for eclipsing and other periodic variable stars, but need to be more exacting because of the low amplitudes involved. At present there are five direct and a few indirect methods available in the search for extrasolar planets:

- astrometric variations;
- (direct) imaging/spectroscopy of planets;
- gravitational lensing;
- radial velocity variations;
- transits; and
- indirect effects of planets on (O–C) diagrams of *EB*s, and on stellar disks such as warps, gaps, and clumps.

5.4.2.1 Astrometry Variations

Periodic and nonlinear proper motions indicate binarity. Astrometric binaries involve low-mass companions detected through proper motion variation. Their usefulness in determining the properties of unseen companions has already been discussed on page 11. The process for extracting the mass, M_{inv}, and barycentric distance of the non-visible, low-mass object is described in Milone & Wilson (2008). If the astrometric precision is high enough, the method can work for brown dwarfs (defined roughly as having masses between 13 and 75 Jupiter masses (M_J) if they have solar composition and up to 90 if extremely metal-poor, or even for planets (objects with $M_{\text{inv}} \leq 13\ M_J$).

On the basis of astrometry, Gatewood (1996; 2000) suggested the presence of planets in the systems Lalande 21185 and ε Eri, but the former, at least, still requires confirmation. Benedict et al. (2002) have made astrometric measurements of a planet previously detected from radial velocities of stellar reflective motions, *Gliese 876b*, planet of an M4 dwarf, also known as *Ross 780*. The measurements were made using the Fine Guidance Sensor on the HST. With radial velocity data, the mass is not in doubt. Together, the data yield an unprojected mass of $1.89 \pm 0.34 M_J$ for the planet.

Advanced astrometry space missions may be capable of finding variations due to precise and frequent astrometric measurements. These missions include NASA's SIM (Space Interferometry Mission) and ESA's GAIA, both of which are expected to achieve several micro-arc-seconds of positional precision. SIM is a pointed mission, while GAIA will be a survey instrument. GAIA will be equipped with a radial velocity spectrometer (resolution $\sim$11,200) and from which photometric fluxes may be integrated across any number of photometric passbands. The spectrometer resolution may be insufficient to detect variations due to planets, but the astrometric resolution will be and transits may be detected in the integrated flux.

5.4.2.2 Direct Imaging and Spectroscopy

In both optical and infrared spectral regions, one can look for faint companions to nearby stars, but true planets are rarely likely to be luminous enough to be seen directly, if they are as close to their parent stars as are the "hot Jupiter," typified by *51 Peg b* or *HD 209458b*. It is even more difficult to obtain high spectral resolution to discern identifying features in the spectrum of any such candidate objects. The difficulty is that the overwhelming light of the parent star makes it difficult to separate the flux of the planet from its star. Coronagraphic (Lyot & Marshall 1933) and diffraction techniques are beginning to yield results as new generation instruments come into play, as have high-resolution techniques on existing telescopes. Such techniques include median averaging of rotating fields to produce clean flats for background subtraction. IR surveys are turning up very red and faint objects, and a number of these have been confirmed to be brown dwarfs through subsequent spectroscopy. Some very red objects in clusters are also turning out to be substellar objects (see Basri 2000 for a still useful summary).

5.4 Extrasolar Planets

In 2004, an apparent companion to the brown-dwarf 2MASS J12073346-3932539 (or "2M1207", for short) was observed in the infrared with the Very Large Telescope (VLT) in Chile. If it is at the same distance as the primary, the companion is 55 AU from the star (Chauvin et al. 2004). Observations made 4 months later with the Near Infrared Spectrometer Camera and Multi-Object Spectrometer (*NICMOS*) on the HST showed no relative proper motion, and from color indices confirmed its temperature at 1250K. The brown-dwarf candidate is in the TW Hydrae cluster, thought to be only 8 million years old. If it is indeed gravitationally associated with the brown-dwarf, the fainter and cooler object is modeled to have a mass equal to $5 M_J$. In 2005, a faint companion to the variable star *GQ Lupi* was observed in visible light; the objects display common proper motion, but is very far from the parent star, and its mass is uncertain within a factor of 10 (Neuhäuser et al. 2005), so it could, in fact, be a brown-dwarf star. Most recently, Kalas et al. (2008) has demonstrated orbital motion in HST images of Fomalhaut (α Piscis Austrini). Finally, a system of substellar objects has been imaged by Marois et al. (2008) with the VLT around *HR 8799*; the estimated masses of these objects are between 7 and 10 M_J, and so may prove to be brown dwarfs. In any case, this is an interesting system.

5.4.2.3 Radial Velocity Variations of the Visible Component

Periodic variations in the Doppler shift of the star as seen in its spectrum are a dead giveaway for something pulling the star around. Because masses of planets are much smaller than those of stars, the orbital motion of the star around the common centre of mass is small also. Therefore high accuracy and precision are required: tens of meters per second or better.

Detection of planets through stellar radial velocity variations has been the major method of detection thus far. The method is illustrated in Fig. 16.3 in Milone & Wilson (2008). Technical improvements in spectrograph stability, in spectral comparison techniques, and in analysis methods have now reduced the uncertainties to $\sim$3 m/s. In addition to detection improvement, long-term averaging of data is beginning to yield long-period, low-amplitude effects in the radial velocity signatures of the parent stars — the effects of planets several AU or more away from the star, in other words, the searches have begun to probe the region occupied by giant planets in our own solar system.

Further improvements to $\sim$1 m/s means investigators must enter a realm dominated by noise effects in the atmospheres of solar analogue stars. Solar-like activity may generate localized velocity variations that will be modulated by both stellar rotation and magnetic activity cycle intervals. The separation of these effects from the effects of multiple low-amplitude planetary periodicities will become a major problem.

5.4.2.4 Gravitational Lensing

The gravitational field of a star causes light from more distant objects lying in nearly the same direction to be bent. Thus the star acts as a lens. The passage of a single star (lens) in front of a more distant one causes varying brightness resulting in two peaks. If the stars are in syzygy with the observer, so that there is an exact match in direction on the sky, an "Einstein ring" is seen instead. From the first detection in 1993, hundreds of events have been seen. Usually there is no consensus of the distance of the star that is acting as the lens, but when the lens turns out to be a binary star, the additional lensing action of the second star and an assumption about the motion of the lensing system in the plane of the sky permit a distance estimate to be made. See Fig. 16.5 in Milone & Wilson (2008), where the crossing by the "caustic" (a surface of maximum brightness created by spherical aberration in a spherical lens/mirror) lasted 8 1/2 h; for comparison, a corresponding event in the galactic halo, 15 Kpc away, would have taken only $\sim$1/2 h. If there is a planet in the system, a sharp spike will be seen, in addition to the star's effects.

The ground-based OGLE (for Optical Gravitational Lensing Experiment) survey of the galactic bulge region of the galaxies has now detected some of these. The OGLE program also has detected apparent planetary transits, as we note below.

5.4.2.5 Transit Eclipses

The presence of a planet can be established through an eclipse of the star's light by a planetary transit of the star's disk. This is now a proven technique, but it requires highly precise photometry, relatively small stars and/or large planets, or very long monitoring intervals; selection effects favor planets close to their parent stars with occultations on timescales of hours.

In 2000, radial velocity variation detected with the 1.5-m telescope at the Harvard College Observatory revealed a planetary candidate around the field star *HD 209458*. D. Charbonneau monitored the star for photometric evidence and, with the help of other observers at Texas and Hawaii, succeeded in observing it (Charbonneau et al. 2000, Henry et al. 2000). It has subsequently been observed with the HST and limitations on perturbations due to moons, and rings have been established (none have been seen). Subsequent investigation shows the detection of such transits in Hipparcos satellite data. Finally, through a careful analysis of the HST data set, the spectral signature of sodium has been detected from the absorption of the star's light as it passed through the planet's atmosphere (Charbonneau et al. 2002). This is the first such identification!

The OGLE lensing survey has revealed tens of potential transit-like events of very low depth, many of these are repeating, suggesting planetary transits. Several of these have been followed-up with radial velocities studies on large telescopes. Three cases that have proven to be planetary transits are *TR-56b*, *TR-113b*, and *TR-132b* (Konacki et al. 2004). The planets in these systems are even closer to their parent stars than the previously found "hot Jupiters."

More recently, large-field surveys of brighter stars have been revealing transits. The first such detection, of *TrES-1*, was announced by Alonso et al. (2004). It has an orbit similar to that of *HD 209458b*, and similar mass, but smaller radius ($\sim$1.08 R_J).

5.4 Extrasolar Planets

5.4.2.6 Indirect Effects: O–C Variation

A team headed by E. F. Guinan (Villanova Univ.) claimed detection of one or more planets in the *CM Draconis* system, an eclipsing M-dwarf binary. The Villanova group claims only one photometric event (which another group disputes) but it has studied the timing of the mutual eclipses and has compiled an O–C (for Observed–Computed instants of mid-eclipse) curve of the eclipsing system, which, Guinan feels, furnishes evidence of the gravitational effect on the orbits of the two stars. At present, a planet in this system remains unconfirmed.

5.4.2.7 Effects on Disks

Protoplanetary disks have been seen around several stars, including β *Pictoris* and Vega, and remnants of disks have been seen around older stars. Gaps and warping have been attributed to the presence of planets or protoplanets in some of these systems. Gorkavyi et al. (2004) summarize the case for a $10M_E$ planet orbiting β *Pictoris*, the first star discovered to have a disk around it. Subsequent direct imaging in the infrared has revealed the existence of a possible planet well within the disk at a distance comparable to Saturn's from the Sun (Lagrange et al. 2009).

5.4.3 Star–Planet Systems and Eclipsing Binary Models

In *EB* models or programs we need to characterize planets by those parameters usually used to describe stars. The fundamental parameters are mass, radius, and temperature. A star–planet (or other low-luminosity object) system, with transits and radial velocities for the star only, is analog to a single-lined spectroscopic and detached *EB*. The orbital period, P, can be obtained from either radial velocities or light curves of the system and is usually the most precisely determined quantity. The radial velocity curve provides the eccentricity, e, and the radial velocity amplitude, K_*, of the parent star. From the transit light curves one can derive the inclination, i, and relative radii r_* and r_p of star and planet with respect to the semi-major axis, a.

5.4.3.1 Comparing Stars, Brown Dwarfs, and Planets

To estimate reasonable initial values of the planet's parameters we start by comparing stars, brown dwarfs, and planets. Deuterium burning begins at a mass $\sim 13 M_J$, making this a convenient dividing line for planets and low-mass stellar objects (Saumon et al. 1996). Objects of mass greater than this are assumed to be brown dwarfs or stars; objects with masses greater than $\sim 75 M_J$ (or 0.072 $M_\odot$), with solar composition, stars. For stars with no metals, this limit increases to $90 M_J$. One may distinguish among these three types of objects, namely a planet from a brown dwarf, or a brown dwarf from a star, at least partly through spectral characteristics. One criterion to distinguish between a brown dwarf and a star is the presence of lithium (Li), which is easily destroyed in stars through large-scale convection. Basri (2000,

p. 494) argues that any object with spectral class later than M7, in which Li is detected, must be substellar. Chabrier & Baraffe (2000) summarize the characteristic spectral features with temperature for the cool end of the sequence as follows (the temperature limits are approximate only).

- ≤4000 K: M dwarfs. Most of the hydrogen is in the form of H_2; and most of the carbon in CO. O is bound mainly in TiO, VO, and H_2O, some in OH and in monoatomic O, and metal oxides. Metal hydrides (e.g., CaH, FeH, MgH) are also present. In optical spectra, TiO and VO dominate; in the IR, H_2O and CO features are seen.
- ≤2800 K: O-rich compounds condense in the atmosphere; possibly perovskite ($CaTiO_3$) may be present.
- ≤2000 K: L dwarfs. [Example: GD 165B]. Some TiO remains, but metal oxides and hydrides disappear from the spectra. Alkali metals are present in atomic form. Some methane may be seen.
- ≤1800 K: Refractory elements (e.g., Al, Ca, Ti, Fe, V) condense into grains. Corundum (Al_2O_3), perovskite condense. Depending on the pressure, rock-forming elements such as Mg, Si, Fe may condense as metallic iron, forsterite (Mg_2SiO_4), or enstatite ($MgSiO_4$).
- ~1700 K (to ~1000 K): Cross-over to methane or T dwarfs. [Example: *Gliese 229B*]. Methane absorption strong in H (1.7:m), K (2.4 μm), and L (3.3 μm), giving rise to steep spectrum at shorter wavelengths, with $J - K \leq 0$, but with $I - J \geq 5$.

It appears that objects with temperatures below about 1300 K or so may qualify as planets, but such limits are not without controversy.

Consequently it is more prudent to use the mass range rather than spectral classification to distinguish planets from brown dwarfs.

We can conclude that we could model a star–planet pair in an *EB* program by a cool secondary putting its temperature T_2 to, say, 500–1,000 K, a small mass ratio, say, $0.01 \leq q \leq 0.1$, and a small ratio, r_2/r_1, of radii. For optical light curves of a transit, the planetary color is effectively black relative to the star, so the planetary temperature is not critical. This is *not* true, however, if thermal infrared light curves are available. Indeed, occultations (eclipse of the planet by the star) have now been observed in the thermal infrared, thanks to the Spitzer Space Telescope.

5.4.3.2 Transit Geometry and Modeling Approaches

The transit geometry is identical to that of a small star transiting a large one. The plan and elevation views of the geometry can be seen in Figs. 5.1 and 5.2 taken from Williams (2001). The orbital radius is a, the planetary and stellar radii are, respectively, r and R, and the angles subtended by the planet and by the star are α_1 and α_2, respectively. The star's distance from the Sun is d. From Fig. 5.2, it is seen that the quantity $\alpha = \alpha_1 + \alpha_2$ is related to the planet's longitude at first contact, θ, by the expression

5.4 Extrasolar Planets

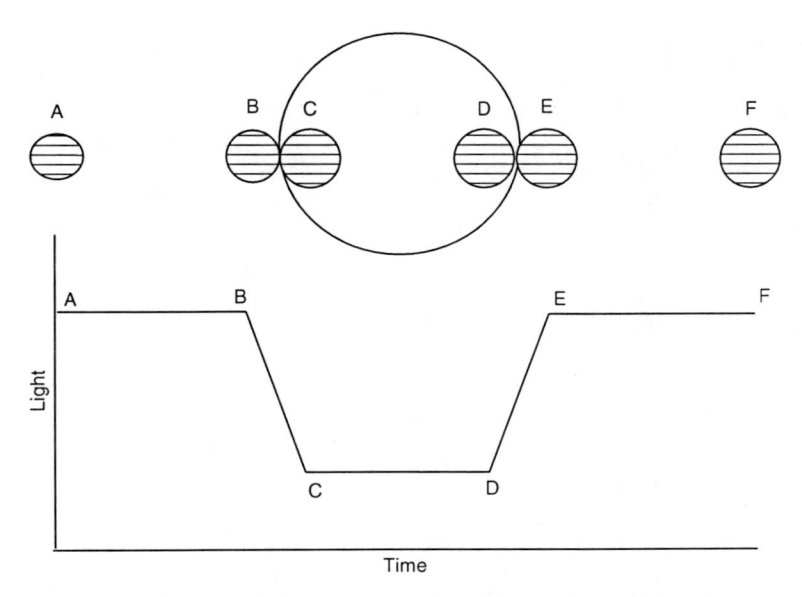

Fig. 5.1 The *top part* of the figure, Fig. 1.14 in Williams (2001), shows the path of the planet as it transits the disk of the star. The *bottom part* of the figure is the corresponding light curve for the transit. Courtesy Michael D. Williams, University of Calgary, Calgary, AB, Canada

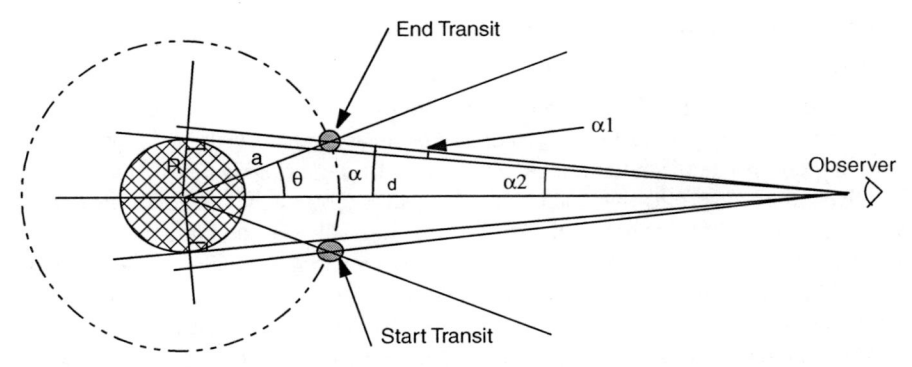

Fig. 5.2 The star has radius R. The planet's orbit has radius a and period P. The observer is located at distance d away, and the star has an angular radius α_2. Courtesy Michael D. Williams, University of Calgary, Calgary, AB, Canada

$$\frac{a}{\sin \alpha} \cong \frac{d-a}{\sin \theta}. \qquad (5.4.1)$$

As $d \gg a$ we get for sufficiently small angles $\theta \leq 10°$

$$\theta \approx \frac{R+r}{a}. \qquad (5.4.2)$$

Based on the transit geometry with its accompanying requirement of sphericity, one could get approximate solutions by exploiting the analytic formulae, as outlined, for instance, by Seager & Mallén-Ornelas (2003) or Kipping (2008). Although, analytic transit models cannot cover more complicated physics such as distortion of the star due to its rotation, or distortion of the planet if it is close to the star, nonlinear limb darkening, or stellar atmospheres to name a few, their solutions can provide initial parameter estimations to light curve programs.

The next step could be to use a simple light curve program such as EBOP, which simulates the components of an *EB* using biaxial ellipsoids. As EBOP can also be restricted to spherical objects, so systematic effects arising from the assumption of a physical shape can be easily quantified. The EBOP model has been shown to work well for transiting extrasolar planetary systems by a number of authors, including Giménez (2006), Wilson et al. (2006), Shporer et al. (2007), and Southworth et al. (2007b). Southworth (2008) provided homogeneous studies of 14 well-observed transiting extrasolar planets, among them also *HD 209458*, based on a modified version of EBOP. His study focused on the effect of limb darkening on the accuracy and error limits of the solution. To include more detailed physics of the star, the WD program is an appropriate candidate. As it uses Roche geometry, it very accurately reproduces distorted surfaces if the Roche surface grid are carefully generated. In addition it can model the star's radiation properly.

5.4.3.3 Representing Planets in the WD Model

In this section we review a few literature cases in which a star–planet system has been modeled with *EB* models and the WD program, see also Milone et al. (2004a). The first problem to be resolved in modeling planetary transits using *EB* models is to find an appropriate discretization or resolution of time, or equivalently, phase. The relative sizes of the transiting and transited object determine the scale. For the transit of a Jupiter-sized object across the disk of the Sun, 0.1 is a good approximation to r_p. Adequate phase sampling in such a case requires longitudinal grid elements of order 30. This is quite similar to that usually used in stellar eclipse modeling, and holds for both occultation and transit eclipses. One may model the transit of a white dwarf, with characteristic radius of the Earth, across the disk of a red dwarf (i.e., a red main sequence star), with characteristic radius of, say, $0.5 R_\odot$, appropriate for an M2V star. But in this case, one would need a resolution of 0.006 or about 165 grid elements to be able to sample ingress and egress adequately. The occultation eclipse is not observable in the visible (we will discuss the infrared situation in the next section), so the transit eclipse alone must determine the period and the geometry sufficiently well to yield the planet's parameters.

The second problem arises from the relative surface brightnesses. In visible light the contrast between the stellar surface brightness of the two objects determines the relative eclipse depths. In the case of a white dwarf transiting a red dwarf, visible light curves may not even register a dip, because the eclipsed surface brightness of the red star may be too low. In such a case, infrared photometry is necessary to observe a sufficient depth for analysis. In the case of a cataclysmic variable (CV), the red star has filled its inner lobe so the infrared light curve may reveal sufficient

5.4 Extrasolar Planets

curvature in the light curve to demonstrate the shape of the cool secondary. But in this case, the occultation eclipse will be very deep in the visible. For the transit of a much cooler star across a solar-type star, however, the cooler star's contribution to system light may yield a negligible dip at primary minimum. This is the case for planetary transits, and for most purposes, at least for optical light curves, we may assume zero surface brightness of the planet during the transit.

5.4.3.4 HD 209458b: Transit Analysis of an ExtraSolar Planet

This planet was the first extrasolar planet to have its radius determined from transit analysis and the first to have had a constituent of its atmosphere detected. The HST observations were so precise that both extensive rings and satellites of this object can be ruled out (Brown et al. 2004). The HST and previous light curves were analyzed with new light curve analysis programs in use at the University of Calgary. Williams (2001) assumed the values $M/M_\odot = 1.09 \pm 0.01$ and $R/R_\odot = 1.145 \pm 0.003$ for the star *HD 209458a*, and $a = 10.06 R_\odot$. From a simultaneous analysis of transit light curve (Fig. 5.3) and radial velocity curve (Fig. 5.4) he derived the following best-fit parameters: $M_p/M_J = 0.69 \pm 0.01$, $R_p/R_J = 1.37 \pm 0.01$, $P/\text{days} = 3.52478 \pm 0.00005$, $E_0 = 2451254.587 \pm 0.002$, $e = 0.0000 \pm 0.0001$, and $i = 86.54 \pm 0.02$. These results were more precise than previous published results.

The radius of this planet is larger than expected for a planet less massive than Jupiter because of its proximity to the star, which increases its equilibrium temperature and therefore the pressure scale height of the atmosphere:

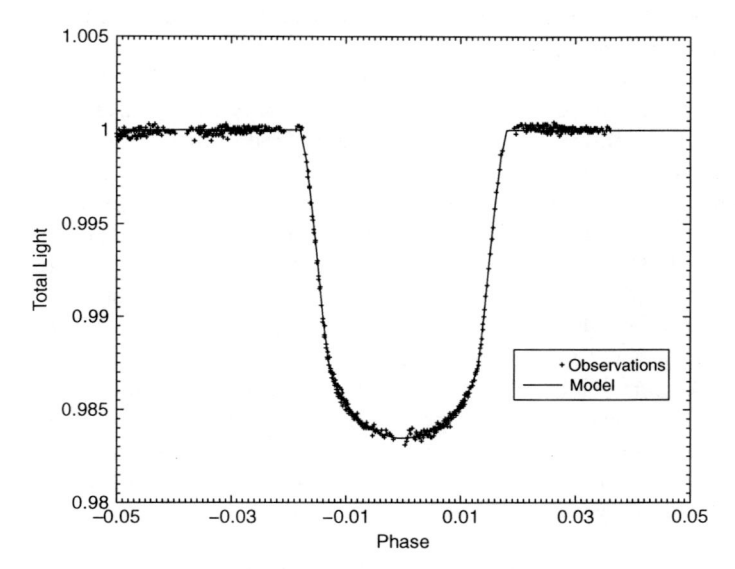

Fig. 5.3 This figure, Fig. 4.3 in Williams (2001), shows the HST transit light curve and best-fit model obtained by Williams. Courtesy Michael D. Williams, University of Calgary, Calgary, AB, Canada

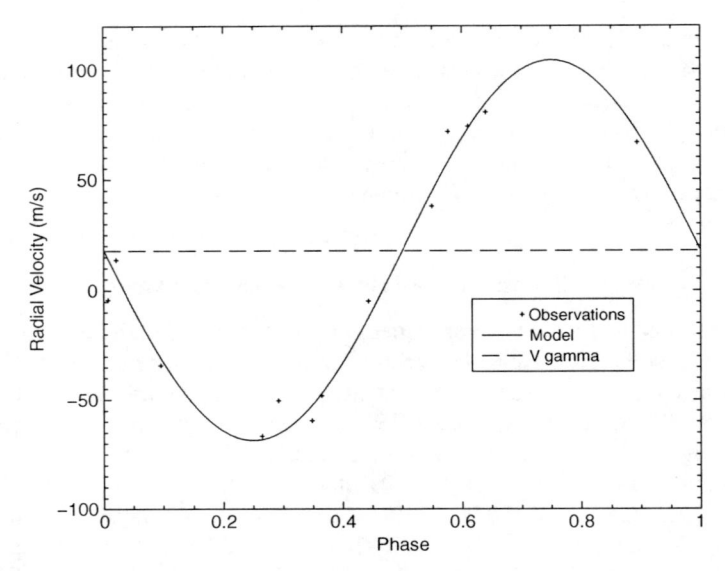

Fig. 5.4 This figure, Fig. 4.1 in Williams (2001), shows the radial velocity curve of *HD 209458* and best-fit model obtained by Williams. Courtesy Michael D. Williams, University of Calgary, Calgary, AB, Canada

$$H = \frac{kt}{\mu m_u g}, \qquad (5.4.3)$$

where H is the pressure scale height, k is the Boltzmann constant, μ is the mean molecular weight, m_u is the atomic mass unit, basically the mass of the proton, and g is the acceleration of gravity.

The mean density of *HD 209458b* from Williams's work is found to be 330 ± 10 kg/m^3, less than half that of Saturn. The atmosphere is clearly distended and a reasonable explanation has been advanced to explain it. Observations (Vidal-Madjar et al. 2004) suggest the existence of a trailing cloud of hydrogen, carbon, and oxygen, indicating hydrodynamic loss of atmosphere from this planet. If the observations are fully confirmed, nothing could better illustrate the dynamic character of such a planet. These authors suggest that planets older and closer to their parent stars than *HD 209458b* may have been deprived of their atmospheric envelopes and become a new class of planets (chthonian). The confirmed OGLE planets do constitute a closer and therefore hotter class of hot Jupiters, but their sizes appear to be smaller than that of *HD 209458b*. The system *TrES-1* similarly appears to be smaller. Recent planetary modeling also suggests that *HD 209458b* is anomalously large, but it is not alone. It is being joined by a growing number of low-density extrasolar planets. *HAT P-1b* and *WASP-1b*, from two of the many surveys now being undertaken, are two examples.

5.4.3.5 The OGLE-TR-56 Star Planet System

Vaccaro & Van Hamme (2005) simultaneously fitted light and velocity data for the star–planet system *OGLE-TR-56* with the WD program. They solved for orbital and planet parameters, along with the ephemeris using all currently available observational data: one photometric light curve and the star's radial velocity curve. The mass, $\mathcal{M}_s$, and temperature, T_1, for the star (*OGLE-TR-56a*) were kept fixed at values derived from spectral characteristics and stellar evolutionary tracks (Cody & Sasselov, 2002; Sasselov, 2003), namely $\mathcal{M}_s = 1.04 \mathcal{M}_\odot$ and $T_1 = 5900$ K. A logarithmic limb-darkening law was adopted with coefficients x, y from Van Hamme (1993). The star's rotation rate F_1 was set to 0.06 corresponding to a rotational period of 20 days (Sasselov, 2003).

In the WD program, adjustable radial velocity-related parameters are the semi-major axis, a, systemic velocity, V_γ, and mass ratio, q. As in a single-lined binary, the mass ratio q cannot be determined from the velocity curve, they proceeded as follows: From an initially assumed planetary mass, $\mathcal{M}_p$, they adopted $q = \mathcal{M}_p/\mathcal{M}_s$. An initial value of the semi-major axis, a, follows from known period, P, and Kepler's third law

$$P^2 = \frac{4\pi^2 a^3}{G\mathcal{M}_s(1+q)} \qquad (5.4.4)$$

follow q and the planet's mass $\mathcal{M}_p = q\mathcal{M}_s$.

Time instead of phase was the independent variable, and ephemeris parameters (reference epoch T_0, period P, and possibly rate of period change dP/dt) were fitted together with the other parameters (inclination i, a and systemic velocity, V_γ, Roche potentials, Ω_1 and Ω_2). Note that q was kept fixed. The result of this fit were values for a, i, and other adjustable parameters consistent with the fixed mass ratio. However, a, P, and q in Kepler's law (5.4.4) could lead to a different stellar mass, $\mathcal{M}_s$. Therefore, for given $\mathcal{M}_s$ from (5.4.4) a new mass ratio is computed from

$$q = \frac{4\pi^2 a^3}{G\mathcal{M}_s} \frac{1}{P^2} - 1. \qquad (5.4.5)$$

The data fitting procedure was repeated until the triple $(a, P, q, \mathcal{M}_s)$ converged to values yielding a stellar mass from (5.4.4) consistent with the pregiven value of $\mathcal{M}_s$.

Their results are in good agreement with parameters obtained by other authors but have significantly smaller errors for i and V_γ, slightly smaller errors for period, and larger errors for $\mathcal{M}_s$ and $\mathcal{M}_p$. Especially, the value and error of the stellar radius, found by fitting the light curve data, agree very well with the value found by other authors who fit evolutionary tracks. The authors found no significant change in orbital period that may be due to orbital decay.

5.5 Selected Bibliography

Utilia et delectabilia (*useful and delightful*)
This section is intended to guide the reader to recommended books or articles related to the advanced modeling topics covered in this chapter.

- *Semi-detached Binaries as Probes of the Local Group* by Wilson (2004) – a review article on the role of *EB*s in distance estimation.
- The proceedings of IAU Symposium No. 240 (2006) under the title *Binary Stars as Critical Tools & Tests in Contemporary Astrophysics* edited by Hartkopf et al. (2007) provide an excellent overview on state-of-the and ongoing activities in close binary research. They review major advances in instrumentations and techniques, and new observing techniques and reduction methods including surveys.
- Seager & Mallén-Ornelas (2003) and Kipping (2008) solved a set of nonlinear equations to obtain the parameter of star–planet systems with spherical stars and planets. These papers could serve as a starting point to become familiar with the geometry of the problem.

References

Alonso, R., Brown, T. M., Torres, G., Latham, D. W., Sozzetti, A., Mandushev, G., Belmonte, J. A., Charbonneau, D., Deeg, H. J., Dunham, E. W., O'Donovan, F. T., & Stefanik, R. P.: 2004, TrES-1: The Transiting Planet of a Bright K0 V Star, *ApJ Letters* **613**, L153–L156

Basri, G.: 2000, Observations of Brown Dwarfs, *Annual Review of Astronomy and Astrophysics* **38**, 485–519

Benedict, G. F., McArthur, B. E., Forveille, T., Delfosse, X., Nelan, E., Butler, R. P., Spiesman, W., Marcy, G., Goldman, B., Perrier, C., Jeffreys, W. H., & Mayor, M.: 2002, A Mass for the Extrasolar Planet Gliese 876b Determined from the Hubble Space Telescope Fine Guidance Sensor 3 Astrometry and High-Precision Radial Velocities, *ApJ* **581**, L115–L118

Bessell, M. S.: 1979, UBVRI photometry. II – The Cousins VRI System, its Temperature and Absolute Flux Calibration, and Relevance for Two-dimensional Photometry, *PASP* **91**, 589–607

Brown, T. M., Charbonneau, D., Gilliland, R. L., Albrow, M. D., Burrows, A., Cochran, W. D., Baliber, N., Edmonds, P. D., Frandsen, S., Bruntt, S., Guhathakurta, P., Choi, P., Lin, D. N. C., Vogt, S. S., Marcy, G. W., Mayor, M., Naef, D., Milone, E. F., Stagg, C. R., Williams, M. D., Sarajedini, A., Sigurdsson, S., & Vandenberg, D. A.: 2004, HST Photometry of 47 Tucanae: Time Series Analysis and Search for Giant Planets, in A. J. Penny, P. Artymowicz, A.-M. Lagrange, & S. Russell (eds.), *Planetary Systems in the Universe: Observation, Formation and Evo lution*, Vol. 214 of *ASP Conference Series*, pp. 66–68, Astronomical Society of the Pacific, San Francisco

Bryja, C. & Sandtorf, J. R.: 1999, Secondary Standard Field Star Calibration and BVRI Photometric Monitoring of Selected Cataclysmic Variables, in *American Astronomical Society Meeting Abstracts*, Vol. 194 of *American Astronomical Society Meeting Abstracts*, p. 115.03

Caldwell, J. A. R., Cousins, A. W. J., Ahlers, C. C., van Wamelen, P., & Maritz, E. J.: 1993, Statistical Relations between the Photometric Colours of Common Types of Stars in the UBV(RI)c, JHK, and uvby Systems, *South African Astronomical Observatory Circular* **15**, 1–+

Campbell, B., Walker, G. A. H., & Yang, S.: 1988, A search for substellar companions to solar-type stars, *ApJ* **331**, 902–921

Chabrier, G. & Baraffe, I.: 2000, Theory of Low-Mass Stars and Substellar Objects, *Ann. Rev. Astron. Astrophys.* **38**, 337–377

Charbonneau, D., Brown, T., Latham, D., Mayor, M., & Mazeh, T.: 2000, Detection of Planetary Transits Across a Sun-Like Star, *ApJ* **529**, L45–L48

Charbonneau, D., Brown, T. M., Noyes, R. W., & Gilliland, R. L.: 2002, Detection of an Extrasolar Planet Atmosphere, *ApJ* **568**, 377–384

Chauvin, G., Lagrange, A.-M., Dumas, C., Zuckerman, B., Mouillet, D., Song, I., Beuzit, J.-L., & Lowrance, P.: 2004, A giant planet candidate near a young brown dwarf. Direct VLT/NACO observations using IR wavefront sensing, *A&A* **425**, L29–L32

Cody, A.M., & Sasselov, D.D.: 2002, HD 209458: Physical Parameters of the Parent Star and Transiting Planet, *ApJ* **569**, 451–458.

Devinney, E. J., Guinnan, E. F., Bradstreet, D., DeGeorge, M., Giammarco, J., Alcock, C., & Engle, C.: 2006, Observe the Model: On Inverse Problems in Astronomy, *Bulletin of the American Astronomical Society* **37**, 1212

Devor, J.: 2005, Solutions for 10,000 Eclipsing Binaries in the Bulge Fields of OGLE II Using DEBiL, *ApJ* **628**, 411–425

Devor, J. & Charbonneau, D.: 2006a, MECI: A Method for Eclipsing Component Identification, *ApJ* **653**, 647–656

Devor, J. & Charbonneau, D.: 2006b, A Method For Eclipsing Component Identification in Large Photometric Datasets, *Ap. Sp. Sci.* **304**, 351–354

Eggleton, P.: 2006, *Evolutionary Processes in Binary and Multiple Stars*, Vol. 40 of *Cambridge Astrophysics Series*, Cambridge University Press, Cambridge, UK, 2nd edition

Eggleton, P. P. & Kiseleva-Eggleton, L.: 2001, Orbital Evolution in Binary and Triple Stars, with an Application to SS Lacertae, *ApJ* **562**, 1012–1030

Elias, N. M., Wilson, R. E., Olson, E. C., Aufdenberg, J. P., Guinan, E. F., Gudel, M., Van Hamme, W., & Stevens, H. L.: 1997, New Perspectives on AX Monocerotis, *ApJ* **484**, 394–411

Eyer, L. & Blake, C.: 2005, Automated Classification of Variable Stars for All-Sky Automated Survey 1–2 Data, *MNRAS* **358**, 30–38

Fabregat, J. & Reig, P.: 1996, The Absolute Flux Calibration of the UVBY Photometric System, *PASP* **108**, 90–+

Faccioli, L., Alcock, C., Cook, K., Prochter, G. E., Protopapas, P., & Syphers, D.: 2007, Eclipsing Binary Stars in the Large and Small Magellanic Clouds from the MACHO Project: The Sample, *AJ* **134**, 1963–1993

Flower, P. J.: 1996, Transformations from Theoretical Hertzsprung-Russell Diagrams to Color-Magnitude Diagrams: Effective Temperatures, B-V Colors, and Bolometric Corrections, *ApJ* **469**, 355–365

Gatewood, G.: 1996, Lalande 21185, *Bulletin of the American Astronomical Society* **28**, 885

Gatewood, G.: 2000, The Actual Mass of the Object Orbiting Epsilon Eridani, *Bulletin of the American Astronomical Society* **32**, 1051

Gilliland, R. L., Brown, T. M., Guhathakurta, P., Sarajedini, A., Milone, E. F., Albrow, M. D., Baliber, N. R., Bruntt, H., Burrows, A., Charbonneau, D., Choi, P., Cochran, W. D., Edmonds, P. D., Frandsen, S., Howell, J. H., Lin, D. N. C., Marcy, G. W., Mayor, M., Naef, D., Sigurdsson, S., Stagg, C. R., Vandenberg, D. A., Vogt, S. S., & Williams, M. D.: 2000, A Lack of Planets in 47 Tucanae from a Hubble Space Telescope Search, *ApJ* **545**, L47–L51

Giménez, A.: 2006, Equations for the Analysis of the Light Curves of Extrasolar Planetary Transits, *A&A* **450**, 1231–1237

Gorkavyi, N., Heap, S., Ozernoy, L., Taidakova, T., & Mather, J.: 2004, Indicator of Exo-Solar Planet(s) in the Circumstellar Disk around Beta Pictoris, in A. J. Penny, P. Artymowicz, A.-M. Lagrange, & S. Russell (eds.), *Planetary Systems in the Universe: Observation, Formation and Evolution*, Vol. 214 of *ASP Conference Series*, pp. 331–332, Astronomical Society of the Pacific, San Francisco

Gray, R. O.: 1998, The Absolute Flux Calibration of Strömgren UVBY Photometry, *AJ* **116**, 482–485

Hartkopf, W. I., Guinan, E. F., & Harmanec, P. (eds.): 2007, *Binary Stars as Critical Tools and Tests in Contemporary Astrophysics*, No. 240 in Proceedings IAU Symposium, Dordrecht, Holland, Kluwer Academic Publishers

Hatzes, A. P., Cochran, W. D., Endl, M., McArthur, B., Paulson, D. B., Walker, G. A. H., Campbell, B., & Yang, S.: 2003, A Planetary Companion to γ Cephei A, *ApJ* **599**, 1383–1394

Henden, A. & Munari, U.: 2000, UBV(RI)_C Photometric Comparison Sequences for Symbiotic Stars, *A&A Suppl.* **143**, 343–355

Henden, A. A. & Honeycutt, R. K.: 1997, Secondary Photometric Standards for Northern Cataclysmic Variables and Related Objects, *PASP* **109**, 441–460

Henry, G. W., Marcy, G. W., Butler, R. P., & Vogt, S. S.: 2000, A Transiting "51 Peg-like" Planet, *ApJ Letters* **529**, L41–L44

Hertz, J., Palmer, R. G., & Krogh, A. S.: 1990, *Introduction to the Theory of Neural Computation*, Perseus, New York

Johnson, H. L.: 1965, The Absolute Calibration of the Arizona Photometry, *Comm. Lunar Planet. Lab.* **3**, 73–77

Johnson, H. L.: 1966, Fundamental Stellar Photometry for Standards of Spectral Type on the Revised System of the Yerkes Spectral Atlas, *Ann. Rev. Astron. Astrophys.* **4**, 193–206

Kalas, P., Graham, J. R., Chiang, E., Fitzgerald, M. P., Clampin, M., Kite, E. S., Stapelfeldt, K., Marois, C., & Krist, J.: 2008, Optical Images of an Exosolar Planet 25 Light-Years from Earth, *Science* **322**, 1345–1348

Kipping, D. M.: 2008, Transiting Planets – Light-curve Analysis for Eccentric Orbits, *MNRAS* **389**, 1383–1390

Koch, D., Borucki, W., Basri, G., Brown, T., Caldwell, D., Christensen-Dalsgaard, J., Cochran, W., Dunham, E., Gautier, T. N., Geary, J., Gilliland, R., Jenkins, J., Kondo, Y., Latham, D., Lissauer, J., & Monet, D.: 2006, The Kepler Mission: Astrophysics and Eclipsing Binaries, *Ap. Sp. Sci.* **304**, 391–395

Konacki, M., Torres, G., Sasselov, D. D., Pietrzynski, G., Udalski, A., Jha, S., Ruiz, M. T., Gieren, W., & Minniti, D.: 2004, The Transiting Extrasolar Giant Planet Around the Star OGLE-TR-113, *ApJ* **609**, L37–L40

Kozai, Y.: 1962, Secular Perturbations of Asteroids with High Inclination and Eccentricity, *AJ* **67**, 591–598

Lagrange, A.-M., Gratadour, D., Chauvin, G., Fusco, T., Ehrenreich, D., Mouillet, D., Rousset, G., Rouan, D., Allard, F., Gendron, É., Charton, J., Mugnier, L., Rabou, P., Montri. J., Lacombe, F.: 2009, A Probable Giant Planet Imaged in the β Pictoris Disk. VLT/NaCo Deep L'-Band Imaging. *A&A* **493**, L21–L25

Lampens, P.: 2006, Intrinsic Variability in Multiple Systems and Clusters: Open Questions, in C. Aerts & C. Sterken (eds.), *Astrophysics of Variable Stars*, Vol. 349 of *Astronomical Society of the Pacific Conference Series*, pp. 153–+

Landolt, A. U.: 1992, UBVRI Photometric Standard Stars in the Magnitude Range 11.5–16.0 around the Celestial Equator, *AJ* **104**, 340–371

Lyot, B. & Marshall, R. K.: 1933, The Study of the Solar Corona without an Eclipse, *J. R. Astrono. Soc. Can.* **27**, 225–+

Malkov, O. Y.: 2003, Eclipsing Binaries and the Mass-luminosity Relation, *A&A* **402**, 1055–1060

Marois, C., Macintosh, B., Barman, T., Zuckerman, B., Song, I., Patience, J., Lafrenière, D., & Doyon, R.: 2008, Direct Imaging of Multiple Planets Orbiting the Star HR 8799, *Science* **322**, 1348–1352

Mayer, P.: 2005, Triple and Multiple Systems, *Ap. Sp. Sci.* **296**, 113–119

Mayor, M. & Queloz, D.: 1995, A Jupiter-Mass Companion to a Solar-Type Star, *Nature* **378**, 355–*

Mayor, M., Queloz, D., & Udry, S.: 1998, Mass Function and Orbital Distributions of Substellar Companions (invited review), in R. Rebolo, E. L. Martin, & M. R. Zapatero Osorio (eds.), *Brown Dwarfs and Extrasolar Planets*, Vol. 134 of *Astronomical Society of the Pacific Conference Series*, pp. 140–+

Milone, E. F., Kallrath, J., Stagg, C. R., & Williams, M. D.: 2004a, Modeling Eclipsing Binaries in Globular Clusters, *RevMexAA (Serie de Conferencias)* **21**, 109–115

Milone, E. F., Williams, M. D., Stagg, C. R., McClure, M. L., Desnoyers Winmil, B., Brown, T., Charbonneau, D., Gilliland, R. L., Henry, G. W., Kallrath, J., Marcy, G. W., Terrell, D., & Van Hamme, W.: 2004b, Simulation and Modeling of Transit Eclipses by Planets, in A. J. Penny, P. Artymowicz, A.-M. Lagrange, & S. Russell (eds.), *Planetary Systems in the Universe: Observation, Formation and Evo lution*, Vol. 214 of *ASP Conference Series*, pp. 90–92, Astronomical Society of the Pacific, San Francisco

Milone, E. F. & Wilson, W. J. F.: 2008, *Solar System Astrophyics: Planetary Atmospheres and the Outer Solar System*, Astronomy and Astrophysics Library, Springer, New York

Neuhäuser, R., Guenther, E. W., Wuchterl, G., Mugrauer, M., Bedalov, A., & Hauschildt, P. H.: 2005, Evidence for a Co-moving Sub-stellar Companion of GQ Lup, *A&A* **435**, L13–L16

Pigulski, A.: 2006, Intrinsic Variability in Multiple Systems and Clusters: an Overview, in C. Aerts & C. Sterken (eds.), *Astrophysics of Variable Stars*, Vol. 349 of *Astronomical Society of the Pacific Conference Series*, pp. 137–152

Pigulski, A. & Michalska, G.: 2007, Pulsating Components of Eclipsing Binaries in the ASAS-3 Catalog, *Acta Astronomica* **57**, 61–72

Popper, D. M. & Etzel, P. B.: 1981, Photometric Orbits of Seven Detached Eclipsing Binaries, *ApJ* **86**, 102–120

Prša, A.: 2009, *Spectral Energy Distribution and Calibrated Color Indices*, private communication

Prša, A., Guinan, E. F., Devinney, E. J., DeGeorge, M., Bradstreet, D. H., Giammarco, J. M., Alcock, C. R., & Engle, S. G.: 2008, Artificial Intelligence Approach to the Determination of Physical Properties of Eclipsing Binaries. I. The EBAI Project, *ApJ* **687**, 542–565

Prša, A. & Zwitter, T.: 2005b, A Computational Guide to Physics of Eclipsing Binaries. I. Demonstrations and Perspectives, *ApJ* **628**, 426–438

Quataert, E. J., Kumar, P., & Ao, C. O.: 1996, On the Validity of the Classical Apsidal Motion Formula for Tidal Distortion, *ApJ* **463**, 284–296

Sarro, L.~ M., Sánchez-Fernández, C., & Giménez, .: 2006, Automatic Classification of Eclipsing Binaries Light Curves using Neural Networks, *A&A* **446**, 395–402

Sasselov, D. D.: 2003, The New Transiting Planet OGLE-TR-56b: Orbit and Atmosphere, *ApJ* **596**, 1327–1331

Saumon, D., Hubbard, W. B., Burrows, A., Guillot, T., Lunine, J. I., & Chabrier, G.: 1996, A Theory of Extrasolar Giant Planets, *ApJ* **460**, 993–+

Seager, S. & Mallén-Ornelas, G.: 2003, A Unique Solution of Planet and Star Parameters from an Extrasolar Planet Transit Light Curve, *ApJ* **585**, 1038–1055

Shporer, A., Tamuz, O., Zucker, S., & Mazeh, T.: 2007, Photometric Follow-up of the Transiting Planet WASP-1b, *MNRAS* **376**, 1296–1300

Southworth, J.: 2008, Homogeneous Studies of Transiting Extrasolar Planets – I. Light-curve Analyses, *MNRAS* **386**, 1644–1666

Southworth, J., Wheatley, P. J., & Sams, G.: 2007, A Method for the Direct Determination of the Surface Gravities of Transiting Extrasolar Planets, *MNRAS* **379**, L11–L15

Tamuz, O., Mazeh, T., & North, P.: 2006, Automated Analysis of Eclipsing Binary Light Curves – I. EBAS – a new Eclipsing Binary Automated Solver with EBOP, *MNRAS* **367**, 1521–1530

Tokovinin, A. A.: 1997, MSC – A Catalogue of Physical Multiple Stars, *A&A Suppl.* **124**, 75–84

Vaccaro, T. & Van Hamme, W.: 2005, The OGLE-Tr-56 Star Planet System, *Ap. Sp. Sci.* **296**, 231–234

Van Hamme, W.: 1993, New Limb-Darkening Coefficients for Modeling Binary Star Light Curves, *AJ* **106**, 2096–2117

Van Hamme, W. & Wilson, R. E.: 2005, Estimation of Light Time Effects for Close Binaries in Triple Systems, *Ap. Sp. Sci.* **296**, 121–126

Van Hamme, W. & Wilson, R. E.: 2007, Third-Body Parameters from Whole Light and Velocity Curves, *ApJ* **661**, 1129–1151

Vidal-Madjar, A., Désert, J.-M., Lecavelier des Etangs, A., Hébrard, G., Ballester, G. E., Ehrenreich, D., Ferlet, R., McConnell, J. C., Mayor, M., & Parkinson, C. D.: 2004, Detection of

Oxygen and Carbon in the Hydrodynamically Escaping Atmosphere of the Extrasolar Planet HD 209458b, *ApJ Letters* **604**, L69–L72

Williams, M. D.: 2001, In the Shadows of Unseen Companions: Modeling the Transits of Extra-Solar Planets, Msc thesis, The University of Calgary, Calgary, Canada

Wilson, D. M., Enoch, B., Christian, D. J., Clarkson, W. I., Collier Cameron, A., Deeg, H. J., Evans, A., Haswell, C. A., Hellier, C., Hodgkin, S. T., Horne, K., Irwin, J., Kane, S. R., Lister, T. A., Maxted, P. F. L., Norton, A. J., Pollacco, D., Skillen, I., Street, R. A., West, R. G., & Wheatley, P. J.: 2006, SuperWASP Observations of the Transiting Extrasolar Planet XO-1b, *PASP* **118**, 1245–1251

Wilson, R. E.: 2004, Semi-detached Binaries as Probes of the Local Group, *New Astronomy Reviews* **48**, 695–671

Wilson, R. E.: 2005, EB Light Curve Models – What's Next?, *Ap. Sp. Sci.* **296**, 197–207

Wilson, R. E.: 2007a, Close Binary Star Observables: Modeling Innovations 2003–2006, in W. I. Hartkopf, E. F. Guinan, & P. Harmanec (eds.), *Binary Stars as Critical Tools and Tests in Contemporary Astrophysics*, No. 240 in Proceedings IAU Symposium, pp. 188–197, Kluwer Academic Publishers, Dordrecht, Holland

Wilson, R. E.: 2007b, Eclipsing Binary Flux Units and the Distance Problem, in Y. W. Kang, H.-W. Lee, K.-S. Cheng, & K.-C. Leung (eds.), *The Seventh Pacific Rim Conference on Stellar Astrophysics*, No. 362 in ASP Conf. Ser., pp. 3–14, Astronomical Society of the Pacific, San Francisco, CA

Wilson, R. E.: 2008, Eclipsing Binary Solutions in Physical Units and Direct Distance Estimation, *ApJ* **672**, 575–589

Wilson, R. E.: 2009, *Modeling Intrinsic Variable Stars into Eclipsing Binary Programs*, private communication

Wilson, R. E. & Wyithe, S. B.: 2003, Toward Optimal Processing of Large Eclipsing Binary Data, in U. Munari (ed.), GAIA Spectroscopy: Science and Technology, Vol. 298 of Astronomical Society of the Pacific Conference Series, pp 313–322, San Francisco

Wolszczan, A. & Frail, D. A.: 1992, A planetary system around the millisecond pulsar PSR1257 + 12, *Nature* **355**, 145–147

Wyithe, J. S. B. & Wilson, R. E.: 2002a, Photometric Solutions for Detached Eclipsing Binaries: Selection of Ideal Distance Indicators in the Small Magellanic Cloud, *ApJ* **559**, 260–274

Wyithe, J. S. B. & Wilson, R. E.: 2002b, Photometric Solutions for Semi-detached Eclipsing Binaries: Selection of Distance Indicators in the Small Magellanic Cloud, *ApJ* **571**, 293–319

Wyrzykowski, L., Udalski, A., Kubiak, M., Szymanski, M., Zebrun, K., Soszynski, I., Wozniak, P. R., Pietrzynski, G., & Szewczyk, O.: 2003, The Optical Gravitational Lensing Experiment. Eclipsing Binary Stars in the Large Magellanic Cloud, *Acta Astronomica* **53**, 1–25

Part III
Light Curve Programs and Software Packages

Chapter 6
Light Curve Models and Software

This chapter describes the characteristics and details of implementation for the major light curve models and programs. The purpose is to provide an overview of existing capabilities.

6.1 Distinction Between Models and Programs

Utile dulci   (*The useful with the agreeable*)
Horaz (65–8 B.C.), *ars poetica*, 343

In the context of light curve analysis, a *model* (see footnote on page 77 also) is a set of mathematical and physical relations which enables the mapping of a set of EB parameters $\mathbf{x}$ to a light curve L^{cal} for a given set of phase values. A light curve *code* or *program* is the software implementation of such a model. The output is typically in digital form suitable for graphic visualization of both light curves and a representation of the binary model itself. Whereas a model is abstract and generic and relates a stellar system's physical attributes (gravitational potential, eclipse conditions, etc.), the program requires a choice of coordinates, integration or summation procedures, and matrix inversion routines. A light curve program's most important ingredient is the physical model. The degree of realism of the model fundamentally determines the reliability of the predicted light curve. However, the program itself constrains the accuracy of the result as well as the efficiency with which the result is reached. Therefore, it seems reasonable that those who develop light curve models maintain close contact with those who write and upgrade the program. In the past, model and program developer often have been one and the same person. This may change in the future, as we note in Sect. 7.3. The "models versus programs" topic is discussed more extensively in Wilson (1994).

A desirable feature of a light curve program is an ability to incorporate additional astrophysics. There is a continuing need to improve the model physics, as we become more aware of the observational properties of stars. This is also very important when we use the light curve program as a diagnostic tool and thereby derive astrophysical properties of both system and component stars from the modeling process itself. This requires the light curve software to be structurally well defined but especially expandable, as we describe in Chap. 7.

Some (`EBOP`, `LIGHT2`, `PGA-E,WD`) of the most frequently used light curve modeling programs are reviewed by their authors in Milone (1993). Others, such as the Russell–Merrill procedure and Kopal's frequency domain method, are dealt with extensively in the literature. Thus, we give only a brief overview here. In McNally (1991, p. 485) and in Milone (1993), tables are given which list the percentage use of different light curve models. We make the distinction between basic geometric models in which orbit determination is the primary goal, and more elaborate models in which the radiative properties of the stars are explored also.

We begin with the mature treatment by Russell & Merrill (1952), an evolved version of the model that Russell began developing at the beginning of the twentieth century. We may call this work the Genesis of light curve modeling – an appropriate phrase because Russell was the first to travel the "royal road" of eclipses. It is also the model with the simplest geometry and is certainly not "state of the art." But we cover it in this chapter for historical reasons because many light curve solutions were determined with this method and because, even now, it is useful in providing basic views of the geometric properties of many binary star systems. Additional details of the method and the procedures used to find solutions and compute light curves are given in Appendix D.1.

6.2 Synthetic Light Curve Models

Reculer pur mieux sauter   (*To go back in order to take a better leap*)

6.2.1 The Russell–Merrill Model and Technique

In limine   (*On the threshold; at the beginning*)

The basic assumption in this model is that stars are spherical or, in successor versions, ellipsoidal in shape. The original exposition and the notation for the Russell–Merrill model are given in a series of papers by Russell (1912a, b) and by Russell & Shapley (1912a, b). The technique is best described in Russell & Merrill (1952). Light curves of stars which show evidence of tidal distortion and reflection are transformed by the rectification process into those of spherical stars. The geometrical model for distorted stars is a tri-axial ellipsoid. The physics is limited to Planckian radiation, linear limb-darkening law (3.2.23), gravity brightening, and a simple treatment of the reflection effect. In principle, only systems that conform to the "rectifiable model" are to be treated. In practice, it is not always obvious if a given binary conforms, except, of course, that some systems may not yield solutions.

The basic Russell model was implemented not by a computer program but by manual techniques, augmented in the 1950s and 1960s by computation of "intermediate" or final orbit least-squares techniques [see, e.g., Irwin (1962)]. Manual methods continued to be used until recently (although at a declining rate since the 1970s) because of simplicity, and where computing power was unavailable,

6.2 Synthetic Light Curve Models

for convenience. The value of the basic technique is that we can obtain and test a solution, with a relatively straightforward procedure, and with the help of either a set of tables or nomograms. In the precomputer era, this method was the most elaborated and powerful technique available to *EB* astronomers. Nowadays, the use of tables and nomograms is decreasing. Another disadvantage of the Russell–Merrill technique is that it allows for only limited augmentation of its astrophysical content. Why do we describe it so extensively, then, here, and in Appendix D.1? Even though the basic method is less and less favored by light curve modelers, it still may be in use, and it was the method of choice for most of the twentieth century. In the early 1970s, there were computerized versions [Jurkevich (1970), Proctor & Linnell (1972)] of the Russell–Merrill approach available but they are no longer maintained. Although the programs EBOP and WINK cannot be considered as successor codes of the Russell–Merrill approach, the early versions of these codes benefited from ideas in the Russell–Merrill model. In order to be able to evaluate the results of light curve analyses carried out with this and similar methods, it may help modelers to be familiar with both its concepts and practices, even if most of those solutions have been redone, or at least retried.

The Russell–Merrill method uses the geometric phase introduced in Sect. 3.1.2. One of its basic ingredients is the function $\alpha = \alpha(\delta, k, x_1, x_2)$, defining the light lost at any phase compared to that lost at internal tangency (i.e., at second or third contact). As shown in Fig. 6.1, the quantity α depends on the quantities introduced in Sect. 3.1.3: The projected separation of centers δ, the ratio of radii $k = r_s/r_g$, and the assumed limb-darkening coefficients x_g and x_s.[1] The light curve is assumed to be in units of normalized flux, such that[2] $\ell = L_g + L_s = 1$ at maximum light, which is assumed to be flat. For total eclipses, it is possible to derive the ratio of the components' luminosities from the light levels during total phases. For given L_g the light loss of an eclipsed star is then found in a table containing a set of empirical α values for observations of a given minimum. Equation (3.1.10) would then provide, for different values during eclipse, a set of equations of conditions for the radius of the eclipsed star and the inclination. The following discussion is an abbreviated version of Russell & Merrill's (1952) description of the procedure.

In general, the light observed at any phase may be written as

$$\ell^{oc} = 1 - L_s \tau^{oc} \alpha^{oc} = 1 - L_s \alpha^{oc} \tag{6.2.1}$$

during an occultation eclipse, and

$$\ell^{tr} = 1 - L_g \tau \alpha^{tr} \tag{6.2.2}$$

---

[1] Russell and Merrill refer to the components as "greater" and "smaller" stars, abbreviated g and s, respectively.

[2] Note that this Russell notation equates light and luminosity and is thus an inconsistency in theory. This problem did not cause trouble in the old models because ℓ is proportional to L for spherical stars.

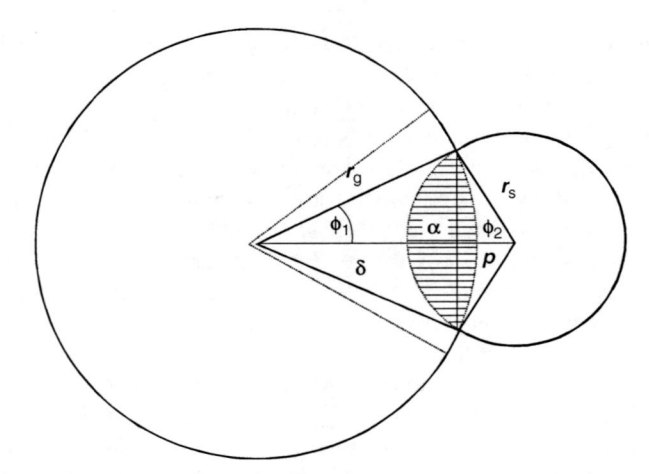

Fig. 6.1 The relation among δ/r_g, $k = r_s/r_g$, and α. The geometrical depth $p = (\delta - r_g)/r_s$ represents the degree of overlap of the disks. The hatched area α is the same independent of the particular star covered; adapted from figure 134 in Binnendijk (1960, p. 266). Note that the meaning of α here is slightly different from the α used in the Russell–Merrill model, where it is defined as a ratio of light lost at any phase to that lost at internal tangency. Thus the hatched area is $\tau\alpha$, if expressed as a unitless fraction of the eclipsed star's disk area

during a transit eclipse. Here, τ is the ratio of the light lost at internal tangency to that of the entire eclipsed star:

$$\alpha = \frac{1-\ell}{1-\ell_{\text{int}}}, \quad \tau = \frac{1-\ell_{\text{int}}}{L_g}. \tag{6.2.3}$$

For an occultation, $\tau^{\text{oc}} = \dfrac{1-\ell_{\text{int}}}{L_s} = 1$, whence the right-hand side of (6.2.1):

$$\ell^{\text{oc}} = 1 - L_s \alpha^{\text{oc}}. \tag{6.2.4}$$

To study complete eclipses, Russell and Merrill defined a conveniently normalized photometric measure of projected separation, ψ, a function of k, α, and the limb-darkening coefficient, x:

$$\psi(x, k, \alpha) := \frac{\sin^2\theta - \sin^2\theta_1}{\sin^2\theta_1 - \sin^2\theta_2} = \frac{\delta_1^2 - \delta^2}{\delta_2^2 - \delta_1^2}, \tag{6.2.5}$$

where θ_1 corresponds to one fixed α on a branch of the minimum and θ_2 to another; Russell & Merrill's (1952) prescription was to look up θ_1 at $\alpha_1 = 0.6$, and θ_2 at $\alpha_2 = 0.9$, with hand-drawn curves through the data of the minimum for interpolation. Defining and computing quantities $A = \sin^2\theta_1$, $B = \sin^2\theta_1 - \sin^2\theta_2$, (6.2.5) may be written as

6.2 Synthetic Light Curve Models

$$\sin^2\theta = A + B\psi, \quad \psi = \frac{\sin^2\theta - A}{B}. \tag{6.2.6}$$

For a range of values of α, θ is read from the plot, and ψ is computed. For each value of α, a value of ψ is found, so that a table can be constructed. The ratio of radii, k, can be found by interpolation in tables for ψ^{oc} and for ψ^{tr}, appropriate for occultation and transit eclipses, respectively. It may not be known in advance whether a given eclipse is an occultation or a transit, and both sets of tables can be consulted. If total, the occultation eclipse is flat-bottomed and in the case of EA light curves tend to be the deeper eclipse, whereas the bottom of the transit eclipse is usually gently curved because of the limb darkening of the larger star.

If the external contact (first or fourth contact) phase can be identified, then from (3.1.10),

$$\sin^2\theta_{\text{ext}} = A + B\psi(x, k, \alpha = 0) = \frac{r_g^2}{\sin^2 i}(1 + k)^2 - \cot^2 i \tag{6.2.7}$$

and, from the phase of internal contact,

$$\sin^2\theta_{\text{int}} = A + B\psi(x, k, \alpha = 1) = \frac{r_g^2}{\sin^2 i}(1 - k)^2 - \cot^2 i, \tag{6.2.8}$$

and from these two equations, both r_g and i can be found. The equations are independent of whether the eclipse is a transit or an occultation, but we must consult different tables for the two cases.

In the partial eclipse case, both the shape and depth must be evaluated; for complete eclipses information from either the shapes or depths is sufficient in principle (of course, in this case also it is safer to use both kinds of information). A new variable n is defined in terms of the relative light lost in the minimum, α, and that at mid-eclipse, α_0, so that

$$n = \frac{\alpha}{\alpha_0} = \frac{1 - \ell}{1 - \ell_0}. \tag{6.2.9}$$

This is the measured relative light loss fraction. The predicted fraction will be different for occultations and transits; both hypotheses must be tried in any solution unless good spectral information is available. Often one or the other can easily be excluded. A function χ is defined such that

$$\chi(x, k, \alpha_0, n) = \frac{\sin^2\theta(n)}{\sin^2\theta(n = \frac{1}{2})}. \tag{6.2.10}$$

The procedure is that n is calculated for a range of values of ℓ on both descending and ascending branches (often folded by reflection about an axis through the mid-minimum point). Initially, neither α_0 nor α is known. The corresponding phase,

θ, is then read from the light curve, and χ is computed. For a given χ, a tabular relationship between k and α_0 can be produced from χ tables. Several such curves, generated by selecting a new n, computing the corresponding χ, and returning to the tables to find the new set of values of (k, α_0), will intersect in a region defining the "best" values of k and α_0. The tables for both χ^{tr} and χ^{oc} must be tested because the relationships will be different for the occultation and transit cases. These provide the "shape relations" for partial eclipses.

The "depth relations" provide indispensable information about the solution in partial eclipse cases. To discuss these, Russell introduced the q function

$$q := \tau \frac{\alpha^{tr}}{\alpha^{oc}} = \frac{\dfrac{1 - \ell^{tr}}{L_g}}{\dfrac{1 - \ell^{oc}}{L_s}} = \frac{\tau \alpha^{tr} + \ell^{tr} - 1}{1 - \ell^{oc}}, \qquad (6.2.11)$$

where the last equivalence results from the condition that $L_g + L_s = 1$, the equivalence that

$$\alpha^{oc} = 1 - \ell^{oc} + \frac{1 - \ell^{tr}}{q}, \qquad (6.2.12)$$

and substitution of equations (6.2.1) and (6.2.2). The evaluation of q thus depends on the availability of data from both eclipses. Again, both hypotheses about the eclipses must be tested: First, one of the eclipses is assumed to be an occultation; the computation is done, the same eclipse is then assumed to be a transit, and the computation redone. Russell and Merrill recommended as a first assumption that the occultation is the deeper eclipse. After q is computed, α_0 is computed from equation (6.2.12), and the table of $k(x_g, x_s, \alpha^{oc}, q)$ is used to establish a table of relations between k and α_0. The elements are then computed, and from these a theoretical light curve is constructed and plotted against the observations. At this point, if the observations were good enough, improvements could be made.

The analysis depends on the assumption that the stars are well represented by the spherical model. Therefore, in order to use the Russell–Merrill method, the light curve first must be transformed into what it would be if the stars were spherical. This process is called *rectification*, and in principle can be applied only to "rectifiable cases." These rectifiable stars are similar prolate ellipsoids with limb- and gravity-darkening, and "gray-body" radiators (i.e., not perfectly efficient blackbodies). The contributions due to "ellipticity" (called elsewhere the "oblateness effect"), gravity and limb-darkening, and "reflection" are computed theoretically and applied semi-empirically, because they depend on the surface brightnesses, sizes, and separations of the components. The rotation of a prolate ellipsoid could cause a sinusoidal variation of the surface area, and thus of the brightness. If stars were true ellipsoids, such a correction would be exact (with additional corrections for expected limb darkening and gravity darkening). However, stars are not ellipsoids. Russell and Merrill recognized that the ellipsoidal assumption was only an

6.2 Synthetic Light Curve Models

approximation to the true shape of a binary star component. It was a question of the adequacy of the approximation weighed against the tedium of exact calculation – a formidable and daunting task in the early 1950s, and a difficult one, in any case, until the 1970s.

Rectification begins with a Fourier analysis of the light curve outside of eclipse. The representation is of the following kind:

$$\ell(\theta) = \sum_{m=0}^{n} [A_m \cos m\theta + B_m \sin m\theta], \qquad (6.2.13)$$

where n was usually taken as 2 although some practitioners used as many significant terms as necessary to fit the data satisfactorily. Any term other than A_0, A_1, and A_2 was considered "unexplained" (if the minima are centered on phases 0 ;and 0.5 – otherwise such terms are generated even if there are no asymmetries in the light curve) and had to be dealt with in an arbitrary way. The A_1-term, which peaks at $\theta = 0°$ and reaches minimum at $180°$, and the A_2-term, which peaks at $0°$ and $180°$ and reaches minimum at $90°$ and $270°$, are the contributors to the reflection effect. The $\cos 2\theta$ term may be at least partially due to what has been called the "oblateness effect," the effect of tidal distortion on the light curves. The presence of a B_1-term indicates a difference between the maxima of the light curve – the *O'Connell effect* [see page 6, footnote 37 on page 135, and Davidge & Milone (1984)], – and higher-order B-terms indicate other light curve perturbations. Russell and Merrill recognized that it was "against human nature" to defer light curve analysis data solely because of unexplained asymmetries, and so, to permit analysis, they provided a general prescription for full rectification (Russell & Merrill 1952, p. 53).

Further details of the Russell–Merrill method can be found in Appendix D.1. Here we merely cite one example of the utility of the method applied to an apparently intractable – but rectifiable – problem.

As an example of the Russell–Merrill approach, we illustrate the method with the plot of the depth and shape relations for the secondary (occultation) eclipse of the system *RT Lacertae* (see Fig. 6.2). Although this is not an ideal system for application of the Russell–Merrill technique (in fact its many complications make it a truly difficult challenge for any light curve analysis program), nonetheless this technique at least permits a solution, and so *RT Lac* provides a suitable example. Precisely because of the complications in this binary, the Russell–Merrill method succeeds at least as well as other methods, and thereby demonstrates the continued usefulness of the method. The binary is a $5^{\rm d}07$-period system, with a superimposed sinusoidal wave moving through the light curve ~ 10 years. The primary minimum is the more negative (suggesting higher temperature for the eclipsing star) in both (B-V) and (U-B) color indices. Hα, Ca II H and K, and far-ultraviolet spectroscopy reveal signatures of gas streams associated with Algol systems and thus mass exchange [cf. Huenemoerder (1985); Huenemoerder & Barden (1986a, b)]. Infrared photometry indicates the presence of a phase-dependent infrared excess.

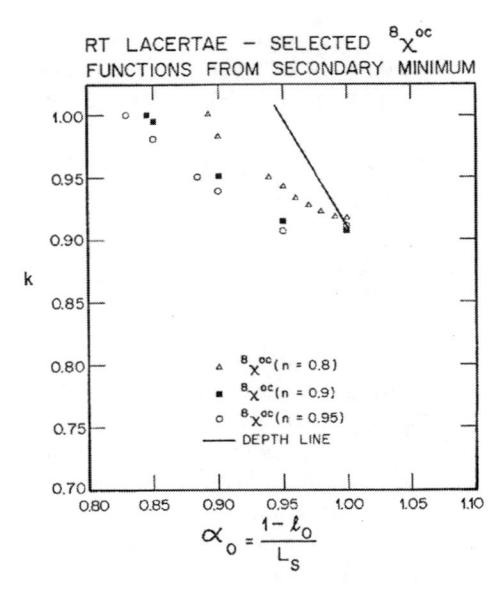

Fig. 6.2 Geometry of the secondary eclipse of *RT Lacertae*. This figure [taken from Fig. 3 in Milone (1977, p. 1003)], shows the shape and depth relations of the secondary eclipse, i.e., the interception of the χ functions and the depth line for the fully rectified minimum

RT Lac seems to be both an RS CVn-type variable and also what Plavec (1980) called a "W Serpentis" type[3] variable, where a hot spot on the accretion disk of one of the stars excites far-ultraviolet emission. Huenemoerder (1988) discovered just such variable ultraviolet emission in *RT Lac*. If sufficiently thick, an accretion disk can resemble a shroud enveloping the hotter component, thereby dimming its light but not necessarily altering its color significantly. In the Russell–Merrill analysis by Milone (1976, 1977), *both* eclipses are shown to be occultation eclipses, because the envelope around the hotter star occults light but does not radiate it. This kind of model cannot be explored with most of the modern light curve programs – at present. Yet, because the Russell model permits the primary and secondary eclipses to be treated independently, the apparent paradoxes of the system can be resolved, at least to a degree.

The infrared light curves are shown in Fig. 6.3. The Milone (1976, 1977) solution is not unique, however. Eaton & Hall (1979) analyzed the system under the assumption that the light curve and color index anomalies can be explained by fortuitous combinations of large numbers of small spot regions. Such a model agrees with the results of a study by Crawford (1992), who found that a particular spectroscopic

---

[3] According to Wilson (1989), these stars can be described as a group of long-period Algol-like mass-transferring binaries characterized by very substantial disks around the more massive components, strange and poorly repeating light curves, prominent optical emission lines, and large secular period changes.

6.2 Synthetic Light Curve Models

feature is better matched with a spot model, at a particular epoch of observations. However, the analysis still fails to account for all the light curve features, despite the increase in the number of parameter-fitting elements.

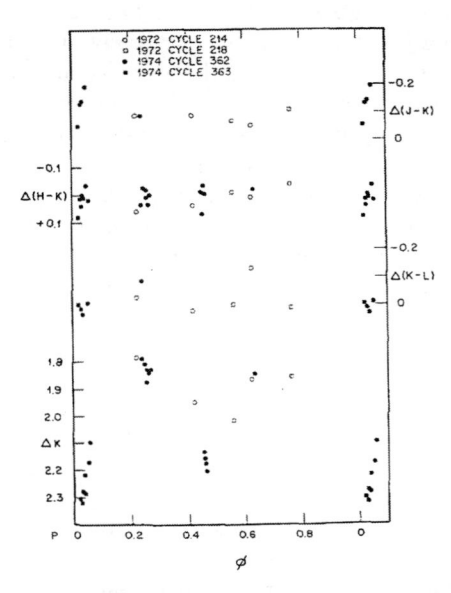

Fig. 6.3 The infrared light and color curves of *RT Lacertae*. This figure [taken from Milone (1976, p. 101, Fig. 3)] shows the differential *J-K*, *H-K*, *K-L*, and *K* light curves of *RT Lacertae* relative to the star BD +43° 4108

In summary, it is quite likely that the components of *RT Lac* are spotted, but the presence of a transient stream and a thick disk in the system is also likely. It is probable that these various features may alternate in importance with epoch.

For those readers who are interested in the nature of such a curious system, a description of the investigation history and the properties of *RT Lac* can be found in Milone (2002). Lanza et al. (2002) carried out spot modeling on a long series of synoptic observations by Turkish observers. Popper (1991, 1992) identified it as a *cool Algol* (an Algol system in which both components are relatively cool and thus are evolving on similar timescales; physically, the $M - T_{\text{eff}}$ relation for these objects is opposite that of normal Algols). Finally, its evolutionary state has been discussed and compared to those of other such systems by Eggleton & Kiseleva-Eggleton (2002).

6.2.2 The "Eclipsing Binary Orbit Program" EBOP

Etzel's (1981) Fortran program EBOP is based on the Nelson & Davis (1972) spheroidal model called the *NDE model*. It is an efficient software for the analysis of detached binary systems with minimal shape distortion due to proximity effects.[4] It is not appropriate for modeling significantly deformed components. The NDE model and its assumptions are close to those in the rectification model by Russell & Merrill (1952). However, as EBOP computes light curves directly, it is much more flexible and it provides options to implement more physics. It is not necessary to rectify for proximity effects. Nevertheless, the model is purposely tied to light curve defined geometrical rather than "astrophysical" parameters (Etzel 1993).

EBOP makes use of spheroidal stars moving in circular or eccentric orbits; here we explain only the circular case. It uses the linear limb-darkening law (3.2.23). The eclipsed area, a, and the light loss during an eclipse is integrated semi-analytically using some basic formulas for circular disks, rings, and sectors. The stellar disk of the covered star is partitioned into concentric rings of radius $r \sin \vartheta$ and width $\Delta \gamma = r \cos \vartheta d\vartheta$, where ϑ is the angular distance to the disk center. Integration over the entire disk

$$\pi \frac{\int I(\vartheta) ds'}{\int ds'} = \frac{1}{r^2} \int_0^{\pi/2} I(\vartheta) 2\pi r \sin \vartheta r \cos \vartheta d\vartheta = \int_0^{\pi/2} I(\vartheta) \cos \vartheta d\vartheta = \mathcal{F}$$
(6.2.14)

yields the averaged flux $\mathcal{F}$. The accuracy of this "eclipse function" depends on the width, $\Delta \gamma$, of the rings and the accuracy of the computations involved. Choosing $\Delta \gamma = 5°$ yields a relative error of 10^{-4} which is usually sufficient. The advantage of this semi-analytic integration procedure is its efficiency. It is faster and more accurate than the standard procedure based on elliptic integrals or purely numerical procedures (two-dimensional Gauß quadrature or direct summation over a stellar grid). The major parameters to compute light curves of spherical stars with EBOP are

- relative surface brightness at the disk center of the secondary component, J_s;
- relative radius r_p of the primary component;
- ratio $k = r_s/r_p$ of the radii r_s and r_p of the secondary and primary components, respectively (note that these radius definitions differ from those of Russell–Merrill);
- inclination i; and
- limb-darkening coefficients x_p, x_s;

and the associated parameters:

- Eccentric orbit characterization ($e \cos \Omega$, $e \sin \Omega$);
- third light $L_3 = l - L_p - L_s$;
- ephemeris phase correction $\Delta \phi$;
- normalization of the light curve m_q; and
- size of the integration rings $\Delta \gamma$.

---

[4] Here we discuss only the program's application to spherical stars. For the extensions including slightly deformed components modeled as ellipsoids, the reader is referred to Etzel (1981); the implementation is based on the evaluation of oblateness as described in Binnendijk (1974) and requires the mass ratio.

6.2 Synthetic Light Curve Models

It should be stressed that in EBOP the central surface brightness, J_s, is used and not the stellar temperature. This is inherited from the Russell–Merrill method. Note that J_s is the *relative* surface brightness at the disk center of star 2 because the value of J_p is defined as unity. In the NDE spheroidal star model, J_s is directly connected to the ratio of depths of minima, whereas temperature has only an indirect influence on the light curve – one that is related to the physical model. The advantage of J_s over effective temperatures or the ratio of bolometric luminosities is that, to a large degree, it can be determined empirically from the light curve whereas stellar temperatures are related less directly to the light curve, through the many assumptions of the radiative model. Therefore, for checking as well as comparative purposes, computation of this quantity in other light curve models is desirable.

Based on the linear limb-darkening law (3.2.23) in EBOP, the unnormalized luminosities[5] of each spherical component follow as

$$l_s = \pi J_s r_s^2 \left[1 - \frac{x_s}{3}\right], \quad l_p = \pi J_p r_p^2 \left[1 - \frac{x_p}{3}\right], \quad (6.2.15)$$

according to (E.29.16). The luminosity ratio

$$\frac{l_s}{l_p} = k^2 J_s \frac{1 - x_s/3}{1 - x_p/3} \quad (6.2.16)$$

depends only on the ratio of radii, ratio of surface brightness, and a correction term for the limb darkening. The relative luminosities

$$L_s = \frac{l_s}{l_s + l_p}, \quad L_p = \frac{l_p}{l_s + l_p} \quad (6.2.17)$$

in normalized units are used in the absence of third light. If third light has to be considered, EBOP uses a modified definition for the luminosities and imposes the normalization

$$L_s + L_p + L_3 = 1. \quad (6.2.18)$$

EBOP does not support direct modeling of the proximity effects caused by distortion of the components and the reflection effect. However, terms for some perturbations, such as oblateness, are considered (Etzel, 1981, pp. 114–115); they are basically derived from Binnendijk (1960, 1974, 1977).

The perturbation terms describing the reflection effect are based on the assumption of a point source (the illuminating star) that illuminates the facing stellar hemisphere of the other star. Quantitatively, this is described by the simple bolometric phase law found in Russell & Merrill (1952, p. 44) and Binnendijk (1960, p. 119):

---

[5] NB: in the EBOP notation, ℓ and L mean something different than in the general *EB* literature.

$$f(\phi) = 0.2 + 0.4\cos\phi + 0.2\cos^2\phi, \quad s_x = \cos\phi = \sin i \cos\theta, \quad (6.2.19)$$

where $\cos\phi$ is the direction cosine of the line-of-sight w.r.t. the line joining both stars, and θ is the true phase angle calculated according to (3.1.19) or (3.1.37). The light of the primary component varies according to

$$L'_p = L_p + \Delta L_p, \quad \Delta L_p = S_p f(\phi), \quad S_p = 0.4 A_p L_s r_p^2, \quad (6.2.20)$$

where S_p is the contribution of the illuminated hemisphere of the primary and A_p is the bolometric albedo introduced in Sect. 3.2.5. A similar expression holds for the secondary component. Due to the phase shift of π, a negative cosine term occurs.

The total brightness variation caused by reflection is obtained by adding the contributions of the stars. Because the contribution of the star in front is very small, and only increases when the eclipsed area, a, becomes small, the observer sees the brightness variation

$$R = \Delta L_s + \Delta L_p - a\Delta L_e \quad (6.2.21)$$
$$\cong (1-a)(\Delta L_s + \Delta L_p)$$
$$= (1-a)\left[\tfrac{1}{2}(S_s + S_p) - (S_s - S_p)\cos\phi + \tfrac{1}{2}(S_s + S_p)\cos^2\phi\right],$$

where ΔL_e is the light reflected from the eclipsed star. Obviously, the reflection effect produces a brightness variation outside eclipse. The Fourier series representation of light variation outside eclipse generally contains the terms $\cos\phi$ and $\cos^2\phi$. According to (6.2.21), the term $\cos\phi$ vanishes if the surface brightnesses S_s and S_p are equal. In the case $S_s > S_p$, the effect of the $\cos\phi$ term is that the brightness of the binary system increases when approaching the secondary minimum.

In order to reproduce the brightness at quadrature ($\theta = 90°$), a free normalization parameter m_q is added to the least-squares problem. In a system with no significant reflection effect, m_q is identical with the luminosity corresponding to the brightness at quadrature.

In practice, EBOP seems to be sufficiently accurate for relatively uncomplicated detached systems with an average oblateness, $\bar{\varepsilon} \leq 0.04$. This program was attractive in the 1970s and 1980s because of its high integration accuracy, and a computational time which saves a factor of 15–40 when compared with the more sophisticated Wood (1972) WINK program. Therefore it is possible to use it for sensitivity analysis (see Appendix B.2) of small parameter changes.

EBOP is still popular and frequently used [cf. Devor (2005), Devor & Charbonneau (2006a), Tamuz et al. (2006) for analyzing large number of light curves, or Southworth (2008) studying transiting extrasolar planets]. It has been enhanced by individual users [cf. Southworth et al. (2004a, b, c), and Southworth et al. (2007a, b) who have incorporated the Levenberg–Marquardt algorithm (MRQMIN; Press et al. 1992, p. 678), an improved treatment of limb darkening, and extensive error analysis techniques].

6.2.3 The Wood Model and the WINK program

The Wood (1972) WINK program is based on the Wood (1971) model; the program is no longer maintained, so its use is decreasing. The model assumes the components of the binary system to be tri-axial ellipsoids. It is well suited for systems with moderate oblateness and reflection effect. WINK computes the total flux of both components received by the observer on Earth by integration over the visible parts of the stars without considering eclipses at first for each phase Φ. The integrals

$$\ell_j(\Phi) = \int I_j \cos\gamma d\sigma, \quad j = 1, 2 \qquad (6.2.22)$$

represent the emitted flux from component j; I_j is the local intensity according to the limb and gravity darkening law and modified by reflection. WINK uses the Russell–Merrill expressions (D.1.3) and (D.1.4) leading to

$$\ell_j(\Phi) = \int I_{0j} \left[1 - y + y\left(\frac{g}{g_0}\right)\right](1 - x + x\cos\gamma)\frac{\cos\gamma}{\cos\beta}r\sin\theta d\theta d\phi. \qquad (6.2.23)$$

For the primary, the mid-disk intensity I_{01} is normalized to $I_{01} = 1$ which leads to $I_{02} = J_2/J_1$ for the secondary component, where J_1 and J_2 are the (central) surface brightnesses of the stars. The numerical integration uses the Gauß quadrature. In most cases, a 16 by 16 point integration is sufficient and leads to an accuracy of 0.0003 light units for spherical stars.

The total light ℓ of the system is computed by adding the light of the components and subtracting the light lost during eclipse

$$\ell(\Phi) = \ell_1(\Phi) + \ell_2(\Phi) - \ell^E(\Phi). \qquad (6.2.24)$$

Light is normalized to quarter phase, i.e., $\ell(\Phi = 0.25) = 1$.

6.3 Physical Models: Roche Geometry Based Programs

Per angusta ad augusta (*Through trials and tribulations to honor and glory*)

In the early 1970s, the Roche model became the foundation for most of the light curve models and programs used thereafter. In this section we will describe, in alphabetical order, only a selection of the most frequently used models. There are others, e.g., that of Yamasaki (1981, 1982), which are not treated in this book but are mentioned in a review by Wilson (1994).

6.3.1 Binnendijk's Model

Binnendijk's model was coded by Nagy (1975) and lead to a synthetic light curve program for contact binaries, based on the Roche model with cylindrical coordinates. A full discussion is found in Binnendijk (1977), which contains also a good exposition of the physical principles behind the models and a fine summary of the synthetic light curve programs to the end of 1975. So, for example, the differences between the forms of the potential, which were (and still are) widely used, are well described.

Nagy and Binnendijk used step sizes of 0.04 along the axis joining the components, in units of the separation of centers, and a 5° angle about this axis. Surface brightness was computed by a method based on Mochnacki & Doughty (1972a). Binnendijk stressed the importance of *surface brightness*, rather than *effective temperature* as a radiation parameter, an appropriate emphasis at a time when stellar atmospheres were just beginning to be well characterized. Binnendijk emphasized the importance of radial velocity, line profile, and spectrophotometric data, and therefore foresaw many of the modern advances that have been made since.

6.3.2 Hadrava's Program FOTEL

FOTEL (Hadrava 1997, 2004) is a code developed at the Ondřejov observatory by Hadrava for the simultaneous solution of light curves and radial-velocity curves of *EB*s. It uses (in its version from 1995) a very simplified model of the physics[6] and geometry for the flux calculations, but it has some options rarely available in more physically sophisticated codes. It is designed to handle a large number of original observational data (typically up to several thousand points) from up to 30 data sets of radial velocities and different passband magnitudes and is designed to run efficiently on a PC. The zero point for each data set is determined by fitting, i.e., the program corrects for systematic instrumental shifts or for different comparison stars. The mean quadratic error for each data set is calculated and the weight of the set can be chosen according to its estimated reliability. The code can take into account a third body or component including observations of its radial velocities and the corresponding light-time effect. Secular changes of several parameters (e.g. , the periastron advance, change of period, or amplitude of radial velocities) can be fitted. An option is also included to look for a change of radius (and volume) due to the variations of tidal force in an eccentric orbit. The code uses a Simplex method (for up to 10 parameters in one run) combined with a direct least-squares solution of multiplicative parameters (i.e., K-velocity and zero points of data sets). Errors and cross-correlations of all these fitted parameters are also calculated.

---

[6] The version FOTEL 3, allows the user to choose between the approximation of stellar shapes by triaxial ellipsoids (which is faster and usually satisfactory) and by Roche equipotentials (which is more accurate).

6.3.3 Hill's Model

The program LIGHT2 [Hill (1979), Hill & Rucinski (1993)] is the result of mating Hill's previous modeling program called LIGHT, which combined the Roche model with Wood's (1971, 1972) Gauß–Legendre quadrature scheme, and Rucinski's WUMA3 (see Sect. 6.3.5) model which was derived from Lucy (1968). It achieves an accurate representation of the system brightness while dealing with horizons and eclipses. The LIGHT2 program has the following characteristics:

- Roche model;
- blackbody/semi-empirical hybrid of calculated blackbody, color-index based, and theoretical atmosphere fluxes;
- irradiation computed in sectors and rings [as per Hutchings (1968)];
- differential corrections via CURFIT (Bevington 1969) based on the Marquardt method;
- multiple (≤ 10), elliptical spot structures; location and ΔT may be unknowns;
- line profiles are calculated.

The program itself uses a command system of keywords with defaults and is available to potential users. The integration scheme for LIGHT is based on the Gauß–Legendre method and that of WUMA3 on Gauß–Legendre–Chebyshev quadrature. Improvements by Wade & Rucinski (1985) have involved limb darkening based on Kurucz (1979) stellar atmosphere models, reference to a line profile database for early-type stars, and use of the Simplex algorithm for uniqueness tests.

The line-profile analysis tool is very powerful and has provided much more precise determinations of masses than has hitherto been possible. See Holmgren (1988) for several examples, and Hill et al. (1990) for an excellent demonstration of how the program can uncover the source of system variations (e.g., movement of spots caused by differential rotation or asynchronous rotation) through line profile analysis.

Among unique features is the option to model elliptical spot regions. Although this may well be an improvement over conventional circular spots of all other light curve modeling codes, elliptical spots require more free parameters and thus, according to Sect. 3.4.2, may lead to overparameterization. Semi-empirical stellar atmospheres are used to simulate stellar fluxes. For a more complete review of LIGHT2 and its capabilities see Hill & Rucinski (1993).

6.3.4 Linnell's Model

Linnell's model as described in Linnell (1984, 1993) is also based on Roche geometry. The initial 1984 version assumed circular orbits, but in the current version, this has been generalized to include eccentric orbits. The 1984 version required the input data specifying the photospheric potentials to be in physical units. The purpose was to permit some model other than the Roche model (such as a polytropic model) to be used; this was effected by an input switch to program PGA that computes the shapes and surfaces of the stars. A built-in alternative assumes a spherical model and permits testing of integration accuracy.

To begin the computations, the program CALPT accepts dimensionless Roche potentials and calculates the corresponding physical potentials. A component's rotation may be nonsynchronous. The software package consists of a series of programs. PGA defines the geometric properties of the photospheres, including direction cosines and values of photospheric grid radii, gravity values, and related quantities. PGB accepts an orbital inclination and either a single orbital longitude or an array of them. It projects the components onto the plane of the sky and fits an array of overlapping parabolas to the horizon points of each component. These quantities are used in turn to calculate shadow boundaries on a given component during an eclipse. PGB determines direction cosines for the line-of-sight to the observer and calculates the zenith distance and its cosine, for that line-of-sight, at all photospheric grid points. PGC determines the radiation characteristics of each component. The program interfaces with an external data file for evaluation of limb-darkening coefficients. PGC calculates linear or quadratic limb-darkening coefficients appropriate to each grid point by fast interpolation in an external file tabulated in (λ, g, T). The data file reference is passed to PGC by name, permitting easy change to an updated table of limb-darkening coefficients. The external file of limb-darkening coefficients is based on Kurucz model atmospheres.

The effective temperature T_{eff} is distinguished from the boundary temperature T_0. According to Chandrasekhar (1950, Chap. XI, Eq. (31)), T_{eff} and T_0 are related by

$$T_0^4 = \frac{\sqrt{3}}{4} T_{\text{eff}}^4 \quad \Leftrightarrow \quad T_0 = 0.8112 T_{\text{eff}}. \tag{6.3.1}$$

The temperatures stored for the grid points follow the specification of (3.2.16), but the equation for grid point intensities uses T_0. The reflection effect is modeled with high geometric accuracy which increases not only the reliability of the results but also the computing time.

PGC provides several alternatives for calculating continuum intensities. The least accurate, but fastest, alternative uses the Planck law. A model atmospheres option interfaces to external files of continuum fluxes tabulated by $(\lambda, g, T_{\text{eff}})$, and interpolates, as with limb darkening. In the mid-1990s, Linnell & Hubeny (1994) developed a much more accurate self-consistent procedure that interfaces directly to a spectrum synthesis program. In this procedure, an accurate spectrum synthesis is carried out. It encompasses distorted, irradiated binary star components and permits the computation of intensities at all local grid points at the effective wavelengths of the observational data. This information, in turn, permits determination of limb-darkening coefficients for the same data set that represents the component spectrum.

In common with other light synthesis procedures, this approach calculates monochromatic light curves. As noted earlier, photometric observations involve an integration over the passband of the product of the composite stellar spectrum, the transmission function of the Earth's atmosphere, and the response function of the optics and detector. Monochromatic light curves represent only an approximation

6.3 Physical Models: Roche Geometry Based Programs

to the photometric quantities actually obtained observationally. Buser (1978) has studied this problem from the standpoint of single stars and has introduced the term *synthetic photometry* to describe his representation of single star colors. Linnell et al. (1998) have extended the use of synthetic spectra for binary stars to include synthetic photometry.

PGD integrates over the components to calculate surface areas, total emergent flux, and flux toward the observer. The program also integrates between shadow boundaries to determine light lost by eclipse. The integration uses Simpson's rule rather than summations of light contributions from discrete surface elements.

SPT, originally called PGE, calculates theoretical light values at the times of observation. It does this in the following way. Starting with PGB the program uses an array of fiducial orbital longitudes. The output of PGD is an array of theoretical light values at the same orbital longitudes. SPT uses an accurate nonlinear interpolation algorithm, with the output of PGD, to calculate theoretical light values at the times of observation. The array of fiducial longitudes is chosen so that no interpolation is required across phases where the light derivative is discontinuous. The reason for this arrangement is that it is possible to calculate theoretical light values for comparison with an indefinitely large data set while preserving the basic accuracy of the light synthesis calculation. This in turn means that it is unnecessary to combine observations into normal points (NB: This is true for all light curve programs now that computers have gotten so fast). The reason for the change in name, from PGE to SPT, is that SPT permits placement of dark (or bright) spots on the component photospheres (Linnell, 1991). The accurate light curve computations consider a variety of horizon and eclipse problems, including over-contact "self-eclipses."

Linnell's differential correction program DIFCORR (Linnell 1989) produces corrections for the parameters i, Ω_1, Ω_2, q, A_1, A_2, g_1, g_2, T_1, T_2 (polar temperatures), S_1, S_2 (limb-darkening scaling coefficients), and U, a light level normalizing factor to fit the observed light curve to the calculated light curve. The program can handle multicolor light curves and so, in principle, can determine both T_1 and T_2. For the sake of self-consistency, the values of L_1 and L_2 are defined as derived quantities, calculated in PGD. The program makes no provision to decouple L_1 and L_2 from T_1 and T_2, as the Wilson–Devinney model does. A separate differential correction program determines star spot parameters to fit residual light curve effects. It is also possible to adjust the entire primary minimum by a shift-parameter t_p. This parameter should not be confused with a shift due to possible orbital eccentricity.

A major difference exists between the calculation of light derivatives in Linnell's program and other light synthesis programs. In Linnell's program, there is a central reference set of parameters which determine a central reference light curve at each observed wavelength. For each parameter, a displacement for that parameter from the central reference value is chosen and two outlying parameter sets are established, symmetric with the central reference set, and displaced by the chosen value. Care must be exercised to avoid physically impossible parameters. The three light curves for each parameter determine two first differences and one second difference, and these remain fixed for successive iterations. Only the coefficient of the second difference changes in the calculation of new first derivatives. See Linnell (1989, Eq. (30)).

At the cost of substantial initial calculations, the computing time for successive iterations is greatly reduced. In addition, the accuracy of the first derivatives is quite high.

A justification for the separate, multiple program approach (rather than subroutines in a single program) is flexibility. A diagnostic run with a modified T_2, say, is possible without recomputing the geometry. The running of PGB is the most time consuming part of the calculation. A call to a batch Fortran program, that oversees the whole process, runs all the separate programs in sequence. Implementation of the synthetic photometry program (Linnell et al. 1998) would have been impractical without the division of the entire project into separate programs.

6.3.5 Rucinski's Model

Rucinski's WUMA3 uses sky-grid integration and hence has low integration efficiency, requiring at least 10,000 integration points. The intention was to have a program which would be totally free of systematic errors related to uneven distribution of integration points. Other properties are

- Roche model (only for equipotentials between the inner and outer critical common envelopes);
- interpolation between semi-empirical, color-index-based fluxes;
- differential corrections via Rucinski's least-squares program;
- no spots are modeled but provision to change temperature or surface brightness along (or on one side of) the common equipotential;
- line profiles calculated in WUMA3's descendants (WUMA5 and WUMA6);
- input through a separate parameter file;
- Fortran code; and
- extreme simplicity and modular structure permit easy modification.

This code is not supported generally and has restricted distribution. According to Rucinski, there is no intention at present to continue development of the WUMAn codes.

6.3.6 Wilson–Devinney Models

6.3.6.1 The 1998 Wilson–Devinney Model

The original Wilson–Devinney model and program has been extended in many publications and software releases [Wilson & Devinney (1971), Wilson (1979, 1990, 1993)]. All versions released after 1982 treat elliptic orbits with eccentricity, e, and semi-major axis, a. The constant distance between the stars is replaced by the phase-dependent distance $d = d(\Phi)$. Nonsynchronous rotation of components is described by a parameter F which is the ratio of rotational angular velocity to mean

6.3 Physical Models: Roche Geometry Based Programs

orbital angular velocity. The surfaces of the components are derived from the Roche equipotentials, defined as

$$\Omega(\mathbf{r}; q, F, d) = \frac{1}{r} + q\left[\frac{1}{\sqrt{d^2 - 2d\lambda r + r^2}} - \frac{\lambda r}{d^2}\right] + \frac{q+1}{2}F^2 r^2 (1 - \nu^2). \quad (6.3.2)$$

The structure of the software is as follows. At first, based on (6.3.2), a discrete representation of the stellar surface is computed. For both stars, for a given distribution of grid points (θ_i, φ_j), the corresponding radial vectors $r_{ij} = r_{ij}(\theta_i, \varphi_j; q, F, \Omega, d)$ are determined, and a system of surface points $\mathbf{r}_s$ is established. Note that both Ω_1 and Ω_2 are measured in the coordinate system of component 1. To generate the surface of component 2, it is necessary to transform to the coordinate system of component 2.

$$\Omega'_2 = q'\Omega_2 + \tfrac{1}{2}(1 - q'), \quad q' = \frac{1}{q}. \quad (6.3.3)$$

The inverse transformation to (6.3.3) is

$$\Omega_2 = q\Omega'_2 + \tfrac{1}{2}(1 - q). \quad (6.3.4)$$

The Wilson–Devinney program uses the following geometrical conventions: The normal vectors point into the interior of the stars, and the line-of-sight vector originates at the binary and points to the observer. This implies that only points satisfying the condition $\cos\gamma < 0$ contribute to the flux seen by the observer.

In the WD program, the local flux function contribution $d\ell_l(\cos\gamma; g_l, T_l, \lambda)$ in the integrand of (3.2.49) is computed for each component j according to

$$d\ell_l(\mathbf{r}_s, \cos\gamma; g_l, T_l, \lambda) = G_j D_j R_j I_j \cos\gamma\, d\sigma, \quad (6.3.5)$$

where the dimensionless ratios $G_j = G_j(\mathbf{r}_s)$, $D_j = D_j(\mathbf{r}_s)$, and $R_j = R_j(\mathbf{r}_s)$ account for gravity brightening, limb darkening, and reflection effect, and I_j is a reference intensity. The computation of these factors is performed according to the formulas (6.3.8), (3.2.23) or one of the other limb-darkening laws in Sect. 3.2.4, (3.2.46), and (6.3.7). Some further details are discussed below.

Because an analysis of photometric data allows the derivation of only relative dimensions of the components (such as the ratio of radii, masses, or luminosities), many quantities and parameters in the model are dimensionless. This has the following consequences for intensities and luminosities. Because the flux at the poles is unknown, a scaling factor I_j, $j = 1, 2$, is introduced. I_j is the normal surface intensity at the pole of component j, which is computed to reproduce the luminosity L_j:

$$\mathcal{D}_j \int_{S'_j} I_j G_j(\mathbf{r}_s) d\sigma = L_j, \quad \mathcal{D}_j = \pi\left(1 - \frac{x_j}{3}\right), \quad j = 1, 2 \quad (6.3.6)$$

and[7]

$$I_j := \frac{L_j}{\mathcal{D}_j} \int_{S'_j} G_j(\mathbf{r}_s) d\sigma, \quad j = 1, 2, \tag{6.3.7}$$

where $G_j(\mathbf{r}_s)$ gives the ratio of normal local to polar intensity due to gravity brightening:

$$G_j(\mathbf{r}_s) := \frac{\alpha_\lambda(T_l)}{\alpha_\lambda(T_j)} \frac{\wp(\lambda, T_l)}{\wp(\lambda, T_j)}, \tag{6.3.8}$$

where T_j and T_l denote the polar and local effective temperatures. The function $\alpha_\lambda(T)$ allowed the introduction of a model atmosphere into the Wilson–Devinney model. To elaborate, $F_\nu^{CG}(T)$ represents the monochromatic flux according to the *Carbon–Gingerich* model atmosphere (see Appendix E.1) tabulated in units of erg/cm^2/s/ster/Hz in Carbon & Gingerich (1969). The implementation followed according to the definition

$$\alpha_\lambda(T) := F_\nu^{CG}(T)/B_{\nu(\lambda)}(T). \tag{6.3.9}$$

WD95, as a spin-off of the more recent version WD93K93 (see page 305), also includes the Kurucz stellar atmospheres models.

The WD model considers a simple *reflection effect* in a form close to that described in Sect. 3.2.5 and Appendix E.25, and, alternatively, a more detailed one which even considers multiple reflection developed by Wilson (1990), as described in Appendix E.25.

The present version of the WD program approximates the integral (3.2.49) by a simple sum but with fractional area corrections for eclipses, i.e., using (6.3.5) we get

$$\begin{aligned}\ell(\Phi) &= \int_0^\pi \int_0^{2\pi} \chi_s(\Phi) I(\cos\gamma; g, T, \lambda) \frac{\cos\gamma}{\cos\beta} r^2 \sin\theta d\varphi d\theta \\ &= I \sum_\varphi \sum_\theta \left\{ GDR \frac{\cos\gamma}{\cos\beta} r^2 \sin\theta \Delta\varphi \Delta\theta \right\}, \end{aligned} \tag{6.3.10}$$

where the sums are evaluated for angles φ and θ associated with visible points, denoting the grid size in φ and θ. In the Wilson–Devinney program, the discretization is chosen as

---

[7] In the Wilson–Devinney program (1993 and later versions), there is a flag LD which gives a choice among linear, logarithmic, and square-root limb-darkening laws. The effective limb-darkening factor $\mathcal{D}_j$ then changes and needs to be recomputed as outlined in Sect. 3.2.4. In all versions of the program, only x is an adjustable parameter, not y (Wilson 1998).

6.3 Physical Models: Roche Geometry Based Programs

$$\theta_i := \frac{\pi}{2} \frac{i - 0.5}{N} \Delta\theta, \quad \Delta\theta := \frac{\pi/2}{N}, \quad i = 1, \ldots, N, \quad (6.3.11)$$

where N denotes the number of latitude circles on each hemisphere, and

$$\varphi_k := \pi \frac{k - 0.5}{M(\theta_i)} \Delta\varphi, \quad \Delta\varphi := \frac{\pi}{M(\theta_i)}, \quad k = 1, \ldots, M(\theta_i), \quad (6.3.12)$$

with [compare term on page 213]

$$M(\theta_i) := \left\lfloor 1 + 1.3N \sin\left(\frac{\pi}{2} \frac{i - 0.5}{N}\right) \right\rfloor. \quad (6.3.13)$$

Note that I, the normal emergent intensity at the pole, does not depend on φ and θ, and thus is extracted from the sum in (6.3.10). Due to (6.3.7) we note, in addition, the relation

$$\frac{\partial \ell}{\partial L_j}(\Phi) = \frac{\ell_j(\Phi)}{L_j}, \quad j = 1, 2. \quad (6.3.14)$$

As already mentioned in Sect. 4.1.1.4, the WD model provides several modes to specify the geometry of the binary system or to add constraints or functional relations among parameters. The WD operations are summarized in the following table (the vector $\mathbf{p}^G$ contains the geometrical parameters):

−1	satisfy X-ray eclipse duration	$\Theta_e = \Theta_e(i, \Omega, q, e, \Omega, F)$
0	approximate the R–M model	$L_1, L_2, T_1,$ and T_2 uncoupled
1	over-contact binary with $T_2 = f(T_1, \mathbf{p}^G)$	$\Omega_2 = \Omega_1 > \Omega^c, \quad T_2 = f(T_1)$
2	L_2 is a function of (L_1, T_1, T_2)	$L_2 = L_2(L_1, T_1, T_2)$
3	over-contact binary (circular case only)	$\Omega_2 = \Omega_1, x_2 = x_1, g_2 = g_1$
4	primary star fills its limiting lobe	$\Omega_1 = \Omega^c$
5	secondary star fills its limiting lobe	$\Omega_2 = \Omega^c$
6	double-contact binaries	$\Omega_1 = \Omega^{c_1}, \quad \Omega_2 = \Omega^{c_2}$

The function f (applied in mode 1 but not in mode 3)

$$T_2 = f(T_1) = T_1 \left(\frac{g_{2p}}{g_{1p}}\right)^\beta \quad (6.3.15)$$

relates the mean polar effective temperatures T_1 and T_2 such that the local effective temperatures of the stars are equal at their interface on the neck region (Wilson & Devinney, 1973, p. 542). g_{1p} and g_{2p} are the surface gravity accelerations at the poles, and β is the exponent defined in (3.2.16); for over-contact binaries we have $\beta = \beta_1 = \beta_2$.

The model described so far is called *mode 0*, in which the luminosities L_1 and L_2, as well as the temperatures T_1 and T_2, are mutually independent. This is similar to the assumptions by Russell & Merrill (1952) in the Russell–Merrill model. In all other modes, except mode -1, L_2 is a function of L_1 and the temperatures T_1 and T_2:

$$L_2 = \frac{\alpha_\lambda(T_2)}{\alpha_\lambda(T_1)} \frac{\wp(\lambda, T_2)}{\wp(\lambda, T_1)} \frac{3 - x_2}{3 - x_1} \frac{\int_{S_2} G_2 d\sigma}{\int_{S_1} G_1 d\sigma} L_1. \qquad (6.3.16)$$

From (6.3.16), it follows that the luminosity ratio L_2/L_1 is a constant depending explicitly on T_1, T_2, L_1, x_1, and x_2, and implicitly on geometric parameters. Due to (6.3.16) the total flux scales with L_1 and thus we can compute the partial derivative as

$$\frac{\partial \ell}{\partial L_1}(\Phi) = \frac{\ell_1(\Phi) + \ell_2(\Phi)}{L_1}. \qquad (6.3.17)$$

According to (6.3.16), the luminosity L_2 is no longer a free parameter. The condition expressed by (6.3.17) allows the interpretation that the free parameter L_1 acts as a scaling factor to couple the calculated light curve with the observed data. Although many modelers publish only the monochromatic luminosity ratio,[8] L_2/L_1, the individual luminosities are coupled self-consistently to the output fluxes. This point is made obvious by the units of $L_{1,2}$ and $\ell(\Phi)$. For example, the user may understand an input L_1 to be in the unit 10^{33} erg/s/μm. Thus an entered L_1 of 6.0000 would mean $6 \cdot 10^{33}$ erg/s/μm. Corresponding units for $\ell(\Phi)$ would then be 10^{33} erg/s/μm/d^2/cm^2, where d is the distance of the binary in centimeters. In the program, d is assumed to be equal to the semi-major axis a which in turn is measured in solar radii. Naturally, when dealing with arbitrarily scaled data, we need not worry about these absolute meanings.

The extension of mode 0 by the relation (6.3.16) is realized in all modes above 0. Additional constraints are added in *mode 1* and *mode 3*,[9] appropriate for overcontact systems, where $\Omega_{1,2} \geq \Omega(\mathbf{r}_c; q, F)$,

$$\Omega_2 = \Omega_1, \quad g_2 = g_1, \quad A_2 = A_1, \quad x_2 = x_1, \qquad (6.3.18)$$

and $\mathbf{r}_c$ is the coordinate vector of the equilibrium point L_1^p. In Cartesian coordinates, $\mathbf{r}_c = (x_{L_1^p}, 0, 0)^T$, i.e., $x_{L_1^p}$ follows from the condition

---

[8] Note that L_2/L_1 is the monochromatic luminosity ratio and not the bolometric luminosity ratio L_2^{bol}/L_1^{bol} needed in the computation of the reflection effect. Appendix E.5 shows how to compute L_2^{bol}/L_1^{bol} as a function of L_2/L_1.

[9] Note that from the 2003 version on the WD program, the equalities $g_2 = g_1$, $A_2 = A_1$, and $x_2 = x_1$ are not enforced any longer in *mode 3*.

6.3 Physical Models: Roche Geometry Based Programs

$$0 = \frac{\partial \Omega}{\partial x}(x = x_{L_1^p}, y = z = 0; q, F). \quad (6.3.19)$$

Appendix E.12 shows in detail how to compute $x_{L_1^p}$. In the coordinate system of the secondary star, the equilibrium point L_1^p has the coordinate

$$x'_{L_1^p} = 1 - x_{L_1^p}. \quad (6.3.20)$$

In contrast to mode 3, mode 1 requires, in addition to (6.3.18), the condition that $T_2 = f(T_1, \mathbf{p}^G)$ which allows the modeling of an over-contact system in thermal contact. $\mathbf{p}^G$ denotes the geometrical parameters. Note that there are different temperatures on the two stars, but the local temperatures are equal where their surfaces meet on the neck of the binary (for idealized thermal contact).

Explicit modeling of semi-detached systems is forced by *mode 4* and *mode 5* which apply the constraints $\Omega_1 = \Omega_c$ or $\Omega_2 = \Omega_c$, respectively. In *mode 4* the primary and in *mode 5* the secondary fill its limiting lobe – that is, the Roche lobe – assuming circular orbits and synchronous rotation.

Let $r_j^{\text{pole}}, r_j^{\text{point}}, r_j^{\text{side}}$, and r_j^{back} denote, respectively, the local radii of component j, at its pole, in the direction of the other component: In the direction perpendicular to that direction and in the plane of the orbit; and in the direction opposite to the other component. A useful mean radius $\bar{r}_j$ can be defined by means of the volume:

$$\bar{r}_j^V = \left[\frac{3}{4\pi} V_j\right]^{1/3}, \quad V_j = 4 \int_0^\pi \int_0^{\pi/2} \left(\int_0^{r(\theta,\varphi)} r^2 dr\right) \sin\theta d\theta d\varphi. \quad (6.3.21)$$

In (6.3.21) the symmetry of the Roche geometry is reflected by the factor 4. For detached and semi-detached systems, the function $r(\theta, \varphi)$ is well defined for all θ and φ. The volumes computed according to (6.3.21) are dimensionless. Multiplication by $\mathcal{R}^3$ yields the physical volume. For particular cases other means, such as harmonic, may be used also. As described in Appendix E.30, relation (6.3.21) also can be used to compute the associated Roche potential Ω_j for an estimated radius r_j^* by solving the equation

$$\bar{r}_j^V(\Omega_j) = r_j^*. \quad (6.3.22)$$

The WD program consists of two main programs named LC and DC, and approximately two dozen subroutines further discussed in Appendix D.1.

6.3.6.2 New Features in the 1999–2007 Models

The 2003 version of the WD program followed the first edition of this book. It provided stellar atmosphere approximation functions by Van Hamme & Wilson (2003). The functions are based on model stellar atmospheres by Kurucz (1993). The radiative atmosphere implementation is in terms of photometric bands, with 25 bands now accommodated.

Input for the radiative treatment includes log g (to allow for the handling of giants, subgiants, etc., in addition to main sequence stars) and 19 chemical compositions in addition to effective temperature, T_{eff}. Influences of T_{eff} and log g on radiative output are applied *locally* on all surface elements (not merely with one correction for the entire star as in some programs). See Van Hamme & Wilson (2003) for information about smooth transitions from atmosphere to blackbody treatment at the limits of the atmosphere tables and for other specifics of the radiative atmosphere application.

In versions of 1998 and earlier using the old stellar atmosphere routine, one could enter any period, P, or orbit size, a, without affecting light curves, as the scaling of observable light (output) from luminosities (input) involved only temperature but not log g or chemical composition. In the newer versions, this is different as log g is derived from GM/R^2 (strictly speaking, from local conditions, including effects of rotation and the other star's gravity) with M and R dependent on period and absolute size. Thus, realistic guesses for P, a, and [M/H]s should be used with the new radiative treatment (pure blackbody computations remain unaffected by absolute masses and dimensions). In a non-simultaneous light-velocity solution, the final semi-major axis a from the velocities should be the same as used for the light curves. That condition will be satisfied automatically in a simultaneous solution. The programs LC and DC are made to be mutually consistent but will have different L_2s and light if absolute dimensions and masses differ between the two programs. Naturally the foregoing warnings do not apply for blackbody computations, where the programs' light curves are unaffected by absolute masses and dimensions.

Another significant change from earlier versions concerns MODE=3 operations. The parameters A_2, g_2, x_2, and y_2 are free parameters, not set equal to A_1, g_1, x_1, and y_1. The reason for this relaxation is that T_2 may differ considerably from T_1; *TU Muscae* is a good example of a hot over-contact binary with different temperatures [cf. Andersen & Grønbech (1975)]. As A_2, g_2, x_2, and y_2 depend on temperature, it would be hard to argue why these parameters should be set equal to A_1, g_1, x_1, and y_1.

The WD model in its most recent version (2007, at this writing) includes three major new features (Van Hamme & Wilson 2007):

- an alternative method to derive the ephemeris by considering whole light and radial velocity curves yields the time of conjunction, period, rate of period change, and orbital rotation (apsidal motion)
- light-time and velocity shifts due to a third body (or several of them) lead to six additional adjustable *third-body parameters* (heliocentric reference time or epoch $T_{0,3b}$ (time of conjunction) and period $P_{0,3b}$, eccentricity e_{3b}, argument of periastron Ω_{3b}, orbital semi-major axis a_{3b}, and inclination i_3 of the third-body orbit relative to plane of sky)
- LC and DC now can interpolate (locally) in [T_{eff}, log g] for x and y limb-darkening coefficients from the Van Hamme (1993) tables for any of 19 compositions ([M/H]). Negative values of the control integers LD1 and LD2 activate interpolation while positive values enforce the use of fixed limb-darkening coefficients.

Each of the features is a major step in completing the picture of *EB* analysis. The essential basis of the third-body procedure [cf. Wilson (2007, Sect. 3.1) and Van Hamme & Wilson (2007)] is to fit multiple curves in *time* rather than phase. The most problematic part of the analysis is the determination of the orbit period of the third body.

Direct distance estimation as described in Sect. 5.1.4 and Wilson (2008) will be available in the 2009 version of the WD program. This is based on absolute, calibrated light curves which allow us to determine the temperatures of both components and to compare absolute, computed distance-dependent flux with the observed flux.

6.4 Cherepashchuk's Model

Et ex oriente lux   *(And from the East, Light)*

Cherepashchuk's approach is widely used in Russia, and we might even apply the term "Russian school" to those who use the light curve modeling procedures described by Cherepashchuk. Since the work by Cherepashchuk (1966), there is a strong focus on *EBs* with extended atmospheres[10] such as the Wolf–Rayet binary *V444 Cygni* [Cherepashchuk (1975), Cherepashchuk & Khaliullin (1976)] and X-ray binaries. The light curve models are mostly spherical models with enhanced features for atmospheric eclipses, which are interpreted to include disks. The least-squares problem is solved with great attention to the ill-posed character of the general light curve analysis problem (Cherepashchuk et al. 1967). The methods are based on regularization algorithms by Tikhonov (1963a, b). They are more recently described in Tikhonov & Arsenin (1979). An English translation outlining the method is given in Tsesevich (1973, pp. 237–244). There is a strong mathematical background in the Russian school on ill-posed problems in the sense of Hadamard, and the reader should not be surprised to find Fredholm integral equations of the first kind involved in the light curve analysis as, for example, in Cherepashchuk et al. (1975).

The models describe semi-transparent emitting disks surrounded by nonemissive atmospheres. In the simplest case, the stars are uniform, circular disks without reflection effect enhancements. In the case of extended atmospheres, a distinction is made between the radii of the luminous disks and those of the larger absorbing disks. In Cherepashchuk's notation, in which components 1 and 2 are identified with the notation for the distances ξ and ρ, respectively, from the disk centers, the luminosity of the system is

$$L = L_\xi + L_\rho = 2\pi \int_0^{r_{\xi c}} I_c(\xi)\xi \, d\xi + 2\pi \int_0^{r_{\rho c}} I_c(\rho)\rho \, d\rho = 1, \qquad (6.4.1)$$

---

[10] Readers interested in extended atmospheres may also consult Wehrse (1987) and Wolf (1987).

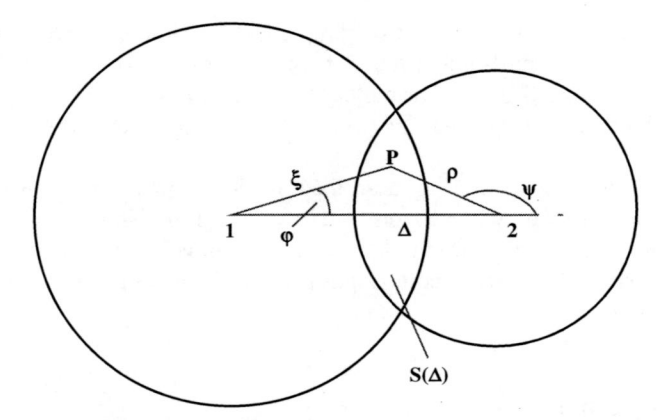

Fig. 6.4 Basic eclipse geometry. The figure shows the distances ξ and ρ of a point P to components 1 and 2, the distance Δ between the centers of the stars, and $S(\Delta)$

where r_c denotes the radius of the luminous disk. The luminosity of the system is normalized to unity. See Fig. 6.4 for the geometry. During the eclipse of star ρ, the flux $I_c(\rho)$ originating from surface area element $d\sigma$, and beamed into solid angle element $d\Omega$, is absorbed by the extended atmosphere of star ξ to the extent

$$I_c(\rho)\left[1 - e^{-\tau(\xi)}\right] d\sigma \, d\Omega, \tag{6.4.2}$$

where $\tau(\xi) = \int_{-\infty}^{+\infty} a(r) dx$, where $a(r)$ is the absorption coefficient per unit volume in star ξ, and r is the distance from the center of this star. The absorbed flux must be reradiated in the extended atmosphere of star ξ, and this is asserted to be the equivalent of the reflection effect [Goncharsky et al. (1978)]. The light loss due to eclipse is determined by the integration over the overlapping surface S_Δ

$$L_\xi + L_\rho - \ell_1(\Delta) = 1 - \ell_1(\Delta) = \int_{S_\Delta} I_c(\rho)\left[1 - e^{-\tau(\xi)}\right] d\sigma, \tag{6.4.3}$$

where $\ell_1(\Delta)$ is the light seen when the star centers are separated by plane-of-sky distance $\Delta^2 = \cos^2 i + \sin^2 i \sin^2 \theta$, and S denotes the overlap region. During the secondary minimum, the light absorbed by the atmosphere of star ρ is written, analogously, as

$$L_\xi + L_\rho - \ell_2(\Delta) = 1 - \ell_2(\Delta) = \int_{S_\Delta} I_c(\xi)[1 - e^{-\tau(\rho)}] d\sigma. \tag{6.4.4}$$

Cherepashchuk refers to the quantities $[1-e^{-\tau(\xi)}]$ and $[1-e^{-\tau(\rho)}]$ as $I_a(\xi)$ and $I_a(\rho)$, respectively. The relations among I_c and I_a require two a priori relations in addition to (6.4.3) and (6.4.4). These relations must involve the choice of a model of particular structure for each component. Models of stars of unknown structure will not

6.4 Cherepashchuk's Model

suffice. Cherepashchuk suggests two types of models: Classical and semi-classical. The following brief exposition of the treatment of these two cases is intended to convey the flavor of the Russian school's concern about the determinability of light curve solutions. The *classical model* assumes spherical stars with opaque disks and thin atmospheres. The functions in this model are

$$I_a(\xi) := \begin{cases} 1, & \text{if } 0 \leq \xi \leq r_{\xi a}, \\ 0, & \text{if } \xi > r_{\xi a}, \end{cases} \quad I_a(\rho) := \begin{cases} 1, & \text{if } 0 \leq \rho \leq r_{\rho a}, \\ 0, & \text{if } \rho > r_{\rho a}. \end{cases} \quad (6.4.5)$$

It is the "standard model" with arbitrary limb-darkening laws described by the functions $I_c(\xi)$, and $I_c(\rho)$, found from solving the integral equations (6.4.3) and (6.4.4).

The *semi-classical model* comprises a classical model star and a "peculiar star" with extended atmosphere. In Cherepashchuk's formulation, the functions $I_c(\rho)$ and $I_a(\rho)$ describing the "normal" star are known. The radiative and absorption properties of the peculiar component, viz., $I_c(\xi)$ and $I_a(\xi)$, are not assumed a priori but are determined by solving the equations (6.4.3) and (6.4.4). The latter equations determine only two functions which depend on the parameters of the light curve: The stellar radii and the orbital inclination. To permit the determination of other elements, the luminosity normalization equation (6.4.1) must be solved as well. The number of additional equations required to provide system parameters depends on the maximum value of the overlap region, i.e., on whether the eclipses are partial or total. Cherepashchuk considers two cases:

(a) $\cos i > r_\rho$: where r_ρ is the radius of the normal component. In this case, at the moment of conjunction ($\vartheta = 0$), the limb of star ρ does not reach the center of the disk of component ξ and the light curve does not contain information about the functions $I_c(\xi)$ and $I_a(\xi)$ for the central regions of the ξ-component, described by the expression: $\xi < \cos i - r_\rho$. In such a case, even though it is possible in principle to derive the parameters of the normal star, a unique solution of (6.4.1), (6.4.3), and (6.4.4) is not possible for the peculiar binary. In the second case, on the other hand, where

(b) $\cos i < r_\rho$: all parts of component ξ are eclipsed, and functions $I_c(\xi)$ and $I_a(\xi)$ are determined by the troika of equations for all values of ξ : $0 \leq \xi \leq r_{\xi_{a,c}}$. The accuracy of the light curve and the specific values for r_ρ and i determine the efficiency with which (6.4.3) and (6.4.4) can determine the parameters.

On the basis of a "determinability" analysis for the classical and semi-classical models, Cherepashchuk (1971) concluded that unique solutions are determinable only in the following cases:

1. Classical models:
   - total eclipses;
   - partial eclipses when for each minimum, $\cos i < r_{\text{occ}}$, where r_{occ} is the radius of the eclipsing star.

2. Semi-classical models:

   - total eclipse of the peculiar star by the normal component;
   - partial eclipses with the condition: $\cos i < r_\rho$, where r_ρ is the radius of the normal component.

3. Semi-classical models including opaque cores:

   - total eclipses of the peculiar star by the normal component;
   - total eclipses of the normal star by the core of the peculiar star;
   - partial eclipses when $\cos i < r_\rho$.

Cherepashchuk recommends that external sources of information, such as spectrophotometric light ratios, be used to determine if the $\cos i < r_\rho$ condition holds. In order to keep the problem well-posed and therefore to be able to solve the light curve of extended atmosphere systems, it is assumed that any unknown functions, such as the center-to-limb variation, are monotonic and nonnegative.

Perhaps the line of attack is best illustrated by an analysis of the light curves of the WN5+O6 system *V444 Cygni,* studied in detail by Cherepashchuk and his colleagues. It is described as a typical semi-classical system with an extended atmosphere (disk) around the Wolf–Rayet WN5 component, which is in front at primary minimum. The absorption of the O6 star's light by the disk of the WN5 star is expressed as

$$I_a^\rho(\xi) = I_\rho^0 \left[1 - e^{-\tau(\xi)}\right], \tag{6.4.6}$$

where I_ρ^0 is the brightness of the O6 component at the disk center, and $\tau(\xi)$ is the optical depth of the disk of the Wolf–Rayet component. The data are assumed to be in the form of normal points in a rectified light curve. The adopted formalism requires solutions of a series of equations:

$$1 - \ell_1(\vartheta) = \int_0^{R_{\xi_a}} K_1(\xi, \Delta, r_\rho) I_a^\rho(\xi) d\xi, \quad \text{if} \quad \cos i \leq \Delta \leq R_{\xi_a} + r_\rho, \tag{6.4.7}$$

$$0 \leq I_a^\rho(\xi) \leq I_a^\rho(0), \tag{6.4.8}$$

$$1 - \ell_2(\vartheta) = \int_0^{R_{\xi_c}} K_2(\xi, \Delta, r_\rho) I_c(\xi) d\xi, \quad \text{if} \quad \cos i \leq \Delta \leq R_{\xi_c} + r_\rho, \tag{6.4.9}$$

$$0 \leq I_c(\xi) \leq C_2, \tag{6.4.10}$$

$$I_a^\rho(0) = \frac{1 - 2\pi \int_0^{R_{\xi_c}} I_c(\xi) \xi d\xi}{\mathcal{D} r_\rho^2}, \quad \mathcal{D} := \pi \left(1 - \frac{x}{3}\right). \tag{6.4.11}$$

6.4 Cherepashchuk's Model

The upper limit of $I_c(\xi)$ in (6.4.10), C_2, is not known but is found by trial and error. The computation of the quantities on the left-hand side of (6.4.7) and (6.4.9) is to be done so as to achieve the least-squares minima of the two sets of data of $|O - C|$, here defined as:

$$|O - C| = \sum_{j=1}^{N} w_j \left\{ [1 - \ell_{1,2}^{obs}(\theta_j)] - [1 - \ell_{1,2}^{cal}(\theta_j)] \right\}^2, \qquad (6.4.12)$$

where the weight of a normal point, $w_j = k/\epsilon_j^2$; k is arbitrary and ϵ_j is the standard deviation of the jth normal point. Computer codes to accomplish the fitting and extraction of the unknowns are described in Cherepashchuk (1973), Cherepashchuk et al. (1973), and Goncharsky et al. (1985). The steps are done as follows:

- the condition $\cos i < r_\rho$ is established;
- the quantities r_ρ and i are assumed, and (6.4.9) is solved for $I_c(\xi, r_\rho, i)$ and the corresponding values of $|O - C|_{r_\rho, i}$; and
- substitution of $I_c(\xi, r_\rho, i)$ in (6.4.8) and the solution of (6.4.7), which then provides the function $I_a^\rho(\xi, r_\rho, i)$ and corresponding values of $|O - C|_{r_\rho, i}$.

The absolute minima of the two sets of (6.4.12) yield the parameters r_ρ, i and the functions $I_c(\xi)$ and $I_a^\rho(\xi)$. These two functions are used to determine physical characteristics of the WN5 component of *V444 Cygni* such as the temperature and absorption coefficient over the disk and core radius, r_0. The data which were analyzed were taken in a rectified light curve in a 7.5 nm passband centered at $\lambda = 424.4$ nm in the continuum. The results by Cherepashchuk (1975) are

$$r_\rho = 0.25 \pm 0.02, \; i = 78° \pm 1°, \; \mathcal{L}_{WR} = 0.197 \pm 0.03, \; 2.2 \leq \frac{r_0}{R_\odot} \leq 2.6.$$

Cherepashchuk notes that the relatively small value of r_0 suggests that the WN5 component is the helium remnant core of a star formed as a result of extensive mass exchange in an interacting, massive binary.

Treatment of more complicated cases is illustrated by the solution of the inverse problem for *SS433*, an interesting, interacting X-ray binary with relativistic jets. Extensive fitting of a model with a thick, precessing accretion disk with an oblate spheroid shape to an extensive body of optical (*V*-passband) data consisting of 10 light curves representing a range of precessional phases failed to satisfy the level of significance (1% by the accuracy and precision of the data). The results, described in detail by Antokhina & Cherepashchuk (1987) and in Goncharsky et al. (1991), are as follows: The "normal" star fills its Roche lobe; and the equatorial radius of the accretion disk is equal to the maximum radius of the compact object's Roche lobe. The choice of the optimum mass ratio was uncoupled to the temperature of the "normal" star. The model for $q \geq 0.25$ is shown in Fig. 6.5 along with a still more complicated model with $q = 0.20$ (lower three figures).

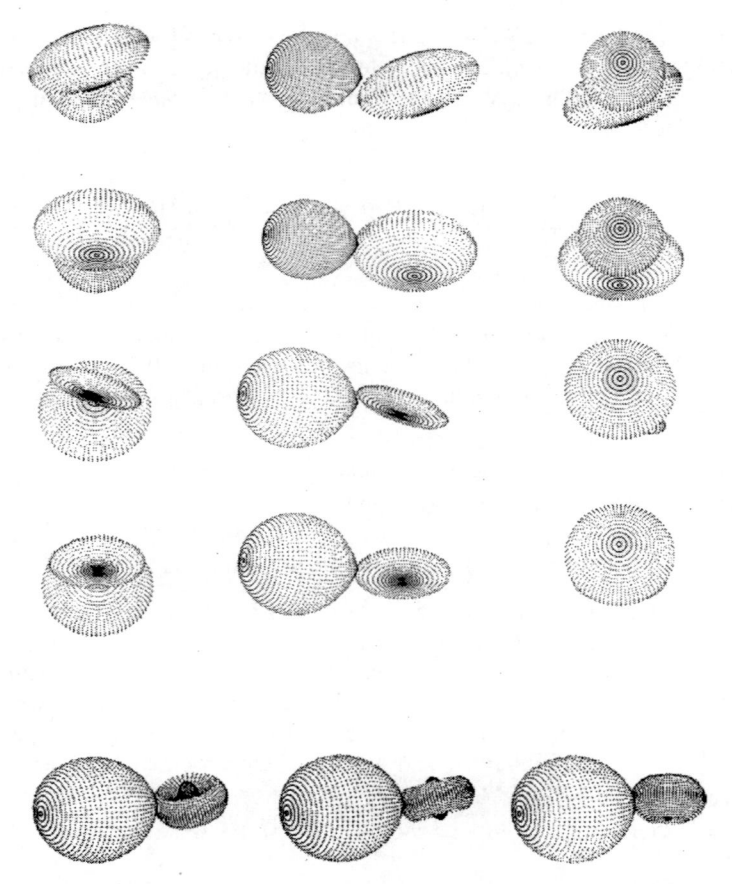

Fig. 6.5 More complex models: Three-dimensional representatives of the interacting X-Ray system *SS433*. Courtesy, A. M. Cherepashchuk

The mass ratio of the *SS433* system is not directly observed. Instead, the mass function

$$f_x(m) = \frac{m_1^3 \sin^3}{(m_x + m_1)^2} \qquad (6.4.13)$$

has been determined to be somewhere between 2 $M_\odot$ (cf. D'Odorico et al. 1991) and 10 $M_\odot$ (Crampton & Hutchings, 1981), with a third determination of 7.7 $M_\odot$ (Fabrica & Bychkova, 1990). X-ray systems are among the principal binaries studied by the "Russian School." Goncharsky et al. (1991) and Antokhina et al. (1992, 1993) discuss the more recent studies. A more recent review on this field is by Cherepashchuk (2005).

6.5 Other Approaches

Et sic de similibus   *(And so of like kind)*

6.5.1 Budding's Eclipsing Binary Model

Budding (1993, Chap. 8) discusses in didactic detail his "Standard Eclipsing Binary Star Model" (SEBM) to analyze circular orbit systems. It is basically a spherical model including circular spots, and thus, often applied to analyze spotted stars; cf. Budding & Zeilik (1987). Note that the original light curves are "cleaned" of the effects of spots.

6.5.2 Kopal's Frequency Domain Method

Kopal's contributions to the field of modern light curve analysis were crucial; his early works prepared the way for the new era. Here, however, we discuss a technique that he developed in his later years, the "Frequency Domain Method."

The basic aspects of this technique are described in Kopal (1979) and also in Kopal (1990, pp. 41–69). Kopal discusses, in great mathematical depth, the representation of the fractional light loss in symmetric *EB* light curves in terms of integral transforms, especially Hankel and Fourier transforms, and the asymptotic properties of finite sums of the latter.

The method concentrates on the determination of quantities such as the area, A_{2m}, under the function of measured light, ℓ, plotted against $\sin^{2m}\theta$. A_{2m} is the *moment of the eclipse* of index *m*. This empirically determined quantity, A_{2m}, may be defined as

$$A_{2m} = \int_0^{\theta_1} (1 - \ell)\mathrm{d}(\sin^{2m}\theta), \qquad (6.5.14)$$

with phase of first contact, θ_1. A theoretical expression for A_{2m} is

$$A_{2m} = mL_1 \csc^{2m} i \int_{\delta_2^2}^{\delta_1^2} \left(\delta^2 - \delta_0^2\right)^{m-1} \alpha \mathrm{d}\delta^2, \qquad (6.5.15)$$

where α is the fractional loss of light, δ indicates the separation of the components so that $\delta_0 = \cos i$, namely, the value of δ at closest approach (at $\theta = 0$), and $\delta_{1,2}$ mark separations at first and second contact, respectively. The analysis also requires the determination of the coefficients c_j of the light curve representation

$$L_1 + L_2 = 1 + c_0 - \sum_{j=1}^{4} c_j \cos^j \psi. \qquad (6.5.16)$$

Although his approach and the notation used above is based mostly on uniform, circular disks, some corrections are applied for limb- and gravity-darkened, tidally distorted nonspherical stars with light curve perturbations (Kopal, 1989, 1990). The analysis of the moments yields the elements of the system. For perturbed light curves, the prescriptive procedure is as follows:

1. determine the necessary number of moments of the light curves (at least 4);
2. evaluate the necessary number of constants $c_{j=1,n}$ (at least 4) from the light curve maxima and take their weighted sum;
3. evaluate the (normalized) moments;
4. evaluate the $r_{1,2}$, i, and $L_{1,2}$ from the moments;
5. compute the light curves from the elements and evaluate the perturbations, $\mathcal{P}_{2m}$;
6. use the perturbations to obtain improved moments; and
7. iterate Steps 3–6 to the final set of elements.

Kopal (1979, p. 193) asserts that no "time-domain" analyses, by which he characterizes all the other methods described here, can make use of empirical data to convert observed light curves into a form from which the elements can be derived directly; keeping in mind that the whole idea of least-squares procedures is to extract parameters indirectly, it is not clear why the direct determination of parameters should be an advantage. Interpreting the assertion to mean that a closed analytic solution cannot be found for such cases, the question remains whether *this* method can do so either. Its claimed superiority has yet to be demonstrated.

The Fourier analysis frequency domain method is a rather formal mathematical approach suffering from the following disadvantages:

1. the components are assumed to be spheres, or at best, corrected spheres (Kopal 1989, 1990);
2. it is doubtful that a few Fourier coefficients can reveal fully the information in a light curve, generally. The Fourier analysis method uses an integral moment[11] of the observations, thus masking the contribution of each point to the value of A_{2m}, particularly for small $\sin^{2m} \theta$ as m increases. That means that perturbation terms for proximity effects are evaluated essentially from the uneclipsed portion of the light curve;
3. it is necessary to specify the luminosity scaling;
4. the value of the angle θ_1 of external tangency must be specified; and
5. it is not easy to introduce additional physical effects into the model in this analysis.

---

[11] This process smoothes the observations. Strictly speaking, information is thrown away in the process.

Budding (1993, p. 211) notes similarly that although the method is straightforward, in practical situations complications arise, especially because the proximity effect representation must be very precise, but the c_j quantities "are not well suited to accurate numerical derivation." He also notes that the procedure loses the advantage of simplicity for partial eclipses.

Light curve analysis based on the solution of an inverse problem with an underlying physical model as its core is not limited by these disadvantages. Physical models are open to the implementation of improved physics.

6.5.3 Mochnacki's General Synthesis Code, GENSYN

The GENSYN code was developed by Mochnacki and Doughty in the early 1970s, of Mochnacki & Doughty (1972a, b), and further improved around 1983. Mochnacki's work began by using Lucy's (1968) code, which he found to be reliable but slow due to its use of a "ray-tracing" algorithm. The GENSYN code was intended to do both light curve and line profile synthesis. It used a cylindrical coordinate scheme making the radius vector a single-valued function for all configurations: contact and noncontact. This might be an advantage, although not a crucial one, because there is no problem finding the correct surface in spherical coordinates either. The program was designed to be compact, fast, and numerically stable but, unlike the Wilson–Devinney LC program, did not incorporate an accurate correction scheme for horizon visibility. GENSYN was applied originally to totally eclipsing A-type contact systems (such as *AW Ursae Majoris* and *V566 Ophiuchi*). According to Mochnacki, it was the first Roche geometry-based light curve code to incorporate full mutual irradiation by mapping each surface element to all those illuminating it.

6.5.4 Collier–Mochnacki–Hendry GDDSYN Spotted General Synthesis Code

Andrew Collier and Stefan Mochnacki combined their programs SPOTTY and GENSYN in 1987 to analyze spotted eclipsing systems. An improved geodesic grid system with triangular elements was introduced by Paul Hendry as part of his MSc thesis at the University of Toronto under Mochnacki's direction. Mochnacki states that the resulting code, GDDSYN, is both faster and more accurate than WD93. All grid elements have comparable surface areas and there is little aliasing due to projected symmetries (Hendry & Mochnacki, 1992). In addition, Hendry wrote a program to determine the most likely spot distribution which makes use of a maximum entropy algorithm, and combined it with GDDSYN and SPOTTY to determine spot distributions in contact systems (Hendry et al. 1992). Finally, Hendry more recently wrote a program to fit both orbital elements and spot distributions to photometric and spectroscopic data. Mochnacki notes that this code was used to analyze several years of DDO observations as part of Hendry's PhD thesis.

6.5.5 Other Spot Analysis Methods

A long-standing tradition at the Osservatorio Astrofisico di Catania, on the island of Sicily, has been the study of the effects of star spots on light curves. Their work has focused on the RS CVn (e.g., Rodonò et al. 1995) and to the similar systems *AR Lac*, *II Peg*, and *SZ Psc*. This work was then extended to the somewhat more complicated system *RT Lac*, discussed earlier. The spot modeling technique is described in several papers; see Lanza et al. (1998, 2002) and references contained therein. Their goal has been to observe the spot distribution as a function of longitude on the star's surface and to study spot migrations over longer intervals of time and modulation over shorter intervals. Information of this sort can lead to an understanding of magnetic structures and their relationship to the rotational and orbital dynamics of the system. Eclipsing systems provide the necessary spatial resolution to explore the brightness distribution on the surface of the eclipsed component, which is divided into several hundred pixels, each of which is allowed to vary within constrained limits. Fittings were obtained with the use of a priori assumptions supplied by a maximum entropy criterion, and, in an independent check on the results, a Tikhonov regularization criterion. According to the authors, the use of such criteria leads to stable and unique solutions of the distribution over the surfaces of the stars. A lengthy series of observations extending over decades provided the database, and the maximum brightness of the system observed during that time served as a kind of unspotted calibration for the pixels. In the case of *AR Lac*, data were available for 20 annual light curves in the interval 1968–1992. The system is subject to variations exceeding 0.05 magnitudes in a single day, possibly due to flares on timescales of tens of minutes. The reference level was established for the system as observed in 1987 when V = $6^{m}_{.}030 \pm 0^{m}_{.}005$ at orbital phase $0^{d}_{.}7395$. The *EB* model for the system, at least for the earlier studies, involved tri-axial ellipsoids for the stellar shapes but also Kopal's (1959) geometric treatment of ellipsoids, reflection, and gravity darkening (Lanza et al. 1998). The resulting longitude distributions are obtained for each component and show general agreement for particular years for each of the two restrictive criteria.

6.6 Selected Bibliography

This section is intended to guide the reader to recommended books or articles on light curve models and programs.

- The review article by Wilson (1994) gives an excellent overview on light curve models. It provides a historical view and discusses the underlying astrophysics.
- An early overview on modern *Light Curve Modeling of Eclipsing Binaries* is provided by Milone (1993). In this collection, several authors briefly describe or comment on the latest versions of some of the better-known light curve models and programs.

- The *Determination of the Elements of Eclipsing Binaries* by Russell & Merrill (1952) is probably the best and most complete description of material related to the Russell–Merrill model.
- In his review *Close Binary Star Observables: Modeling Innovations 2003–06*, Wilson (2007) summarizes the developments of the WD program during 2003 and 2006.
- The journal publication *Eclipsing Binary Solutions in Physical Units and Direct Distance Estimation* by Wilson (2008) is a pleasant-to-read article on the assumption and implications of direct distance estimation.

References

Andersen, J. & Grønbech, B.: 1975, The Close O-type Eclipsing Binary TU Muscae, *A&A* **45**, 107–115

Antokhina, Éh. A. & Cherepashchuk, A. M.: 1987, SS 433: Parameters of the Eclipsing System with a Thick Precessing Accretion Disk, *Sov. Astron.* **31**, 295–307

Antokhina, Éh. A., Pavlenko, E. P., Cherepashchuk, A. M., & Shugarov, S. Y.: 1993, The Best Candidate for a Black Hole – X-Ray Nova V404 Cygni: the Light Curve and Parameters, *Astron. Rep.* **37**, 407–411

Antokhina, Éh. A., Seyfina, E. V., & Cherepashchuk, A. M.: 1992, Analysis of X-Ray Eclipses in SS 433, *Sov. Astron.* **36**, 143–146

Bevington, P. R.: 1969, *Data Reduction and Error Analysis for the Physical Sciences*, McGraw-Hill, New York

Binnendijk, L.: 1960, *Properties of Double Stars*, University of Pennsylvannia Press, Philadelphia, PA

Binnendijk, L.: 1974, The Intrinsic Light Variations in Very Close Eclipsing Binary Systems, *Vistas* **16**, 61–83

Binnendijk, L.: 1977, Synthetic Light Curves for Binaries, *Vistas* **21**, 359–391

Budding, E.: 1993, *An Introduction to Astronomical Photometry*, Cambridge University Press, Cambridge, UK

Budding, E. & Zeilik, M.: 1987, An Analysis of the Light Curves of Short-Period RS CVn stars: Starspots and Fundamental Properties, *ApJ* **319**, 827–835

Buser, R.: 1978, A Systematic Investigation of Multicolor Photometric Systems, *A&A* **62**, 411–424

Carbon, D. F. & Gingerich, O.: 1969, A Grid of Model Stellar Atmospheres from 4000 to 50,000 K, in O. Gingerich (ed.), *Theory and Observation of Normal Stellar Atmospheres*, Proc. of the Third Harvard-Smithonian Conference on Stellar Atmospheres, pp. 377–400, MIT University Press, Cambridge, MA

Chandrasekhar, S.: 1950, *Radiative Transfer*, Oxford University Press, Oxford, UK

Cherepashchuk, A. M.: 1966, Determination of the Elements of Eclipsing Systems Containing a Component with an Extended Atmosphere, *Sov. Astron.* **10** (3), 227–252

Cherepashchuk, A. M.: 1971, Eclipses of Spherical Stars (Arbitrary Limb Darkening Law), in V. P. Tsesevich (ed.), *Eclipsing Variable Stars*, pp. 225–269, Nauka, Moscow, (English translation, 1973, New York)

Cherepashchuk, A. M.: 1973, The Direct Method of Light Curve Solution of the Eclipsing Binary System with an Extended Atmosphere. Calculation and Application of Weights. Computer Programmes, *Peremennye Zvezdy (Variable Stars)* **19**, 227–252

Cherepashchuk, A. M.: 1975, Photometric Elements of the Eclipsing Binary V444 Cygni, and the Nature of the Wolf–Rayet Component, *Sov. Astron.* **19**(1), 47–57

Cherepashchuk, A. M.: 2005, Atmospheric Eclipses in WR-O Binaries: From Kopal and Shapley to Present Days, *Ap. Sp. Sci.* **296**, 55–65

Cherepashchuk, A. M., Goncharsky, A. V., & Jagola, A. G.: 1973, The Algorithm and Computer Programme for Light Curve Solution of an Eclipsing Binary System Containing the Component with an Extended Atmosphere, *Peremennye Zvezdy (Variable Stars)* **18**, 535–569

Cherepashchuk, A. M., Goncharsky, A. V., & Jagola, A. G.: 1975, A Class of Monotonic Functions as a Solution of Eclipsing Binary Light Curves, *Sov. Astron.* **18(4)**, 460–463

Cherepashchuk, A. M., Goncharsky, A. V., & Yagola, A. G.: 1967, An Interpretation of Eclipsing Systems as an Inverse Problem of Photometry, *Sov. Astron.* **11(6)**, 990–999

Cherepashchuk, A. M. & Khaliullin, K. F.: 1976, Nature of the Wolf–Rayet Component of the Binary System V444 Cygni, *Sov. Astron.* **19(6)**, 727–733

Crampton, D. & Hutchings, J. B.: 1981, The SS 433 Binary System, *ApJ* **251**, 604–610

Crawford, J.: 1992, A Photometric and Spectroscopic Analysis of the Chromospherically Active Binary RT Lacertae, Dissertation, San Diego State University, Department of Astronomy

Davidge, T. J. & Milone, E. F.: 1984, A Study of the O'Connell Effect in the Light Curves of Eclipsing Binaries, *ApJ Suppl.* **55**, 571–584

Devor, J.: 2005, Solutions for 10,000 Eclipsing Binaries in the Bulge Fields of OGLE II Using DEBiL, *ApJ* **628**, 411–425

Devor, J. & Charbonneau, D.: 2006a, MECI: A Method for Eclipsing Component Identification, *ApJ* **653**, 647–656

D'Odorico, S., Oosterloo, T., Zwitter, T., & Calvani, M.: 1991, Evidence that the Compact Object in SS 433 is a Neutron Star and not a Black Hole, *Nature* **353**, 329–331

Eaton, J. A. & Hall, D. S.: 1979, Starspots as the Cause of the Intrinsic Light Variations in RS Canum Venaticorum Type Stars, *ApJ* **227**, 907–922

Eggleton, P. P. & Kiseleva-Eggleton, L.: 2002, The Evolution of Cool Algols, *ApJ* **575**, 461–473

Etzel, P. B.: 1981, A Simple Synthesis Method for Solving the Elements of Well-Detached Eclipsing Systems, in E. B. Carling & Z. Kopal (eds.), *Photometric and Spectroscopic Binary Systems*, pp. 111–120, D. Reidel, Dordrecht, Holland

Etzel, P. B.: 1993, Current Status of the EBOP Code, in E. F. Milone (ed.), *Light Curve Modeling of Eclipsing Binary Stars*, pp. 113–124, Springer, New York

Fabrica, S. N. & Bychkova, L. V.: 1990, The Mass Function of SS 433, *A&A Letters* **240**, L5–L7

Goncharsky, A. V., Cherepashchuk, A. M., & Yagola, A. G.: 1978, *Numerical Methods for Solving Inverse Problems of Astrophysics*, Nauka, Moscow (in Russian)

Goncharsky, A. V., Cherepashchuk, A. M., & Yagola, A. G.: 1985, *Ill-Posed Problems of Astrophysics*, Nauka, Moscow (in Russian)

Goncharsky, A. V., Romanov, S. Y., & Cherepashchuk, A. M.: 1991, *Finite-Number Parametric Inverse Problems of Astrophysics*, Moscow University Press, Moscow, Russia (in Russian)

Hadrava, P.: 1997, FOTEL 3 – User's Guide, Technical report, Astronomical Institute of the Academy of Sciences of the Czech Republic, 25165 Ondrejov, Czech Republic

Hadrava, P.: 2004, FOTEL 4 - User's guide, *Publications of the Astronomical Institute of the Czechoslovak Academy of Sciences* **92**, 1–14

Hendry, P. D. & Mochnacki, S. W.: 1992, The GDDSYN Light Curve Synthesis Method, *ApJ* **388**, 603–613

Hendry, P. D., Mochnacki, S. W., & Cameron, A. C.: 1992, Photometric Imaging of VW Cephei, *ApJ* **399**, 246–264

Hill, G.: 1979, Description of an Eclipsing Binary Light Curve Computer Code with Application to Y Sex and the WUMA Code of Rucinski, *Publ. Dom. Astrophys. Obs.* **15**, 297–325

Hill, G., Fisher, W. A., & Holmgren, D.: 1990, Studies of Late-Type Binaries. IV. The Physical Parameters of ER Vulpeculae, *A&A* **238**, 145–159

Hill, G. & Rucinski, S. M.: 1993, LIGHT2: A light-curve modeling program, in E. F. Milone (ed.), *Light Curve Modeling of Eclipsing Binary Stars*, pp. 135–150, Springer, New York

Holmgren, D. E.: 1988, *The Absolute Dimensions of Ten Eclipsing Binary Stars with Components of Early Spectral Type*, PhD Dissertation, University of Victoria, Department of Physics and Astronomy, University of Victoria

Huenemoerder, D. P.: 1985, Hydrogen Alpha Observations of RS Canum Venaticorum Stars. IV. Gas Streams in RT Lacertae, *AJ* **90**, 499–503

References

Huenemoerder, D. P.: 1988, Optical and Ultraviolet Activity in RT Lacertae in 1985 and 1986, *PASP* **100**, 600–603

Huenemoerder, D. P. & Barden, S. C.: 1986a, Optical and UV Spectroscopy of the Peculiar RS CVn System, RT Lacertae, in M. Zeilik & D. M. Gibson (eds.), *Cool Stars, Stellar Systems, and the Sun. Proc. 4th Cambridge Workshop, Santa Fe*, pp. 199–201, Springer, New York

Huenemoerder, D. P. & Barden, S. C.: 1986b, Optical and UV Spectroscopy of the Peculiar RS CVn System RT Lacertae, *AJ* **91**, 583–589

Hutchings, J. B.: 1968, Expanding Atmospheres in OB Supergiants. I., *MNRAS* **141**, 219–249

Irwin, J.: 1962, Tables Facilitating the Least-Squares Solution of an Eclipsing Binary Light-Curve, *ApJ* **106**, 380–426

Jurkevich, I.: 1970, Machine Solutions of Light Curves of Eclipsing Binary Systems, in A. Beer (ed.), *The Henry Norris Russell Memorial Volume*, Vol. 12 of *Vistas*, pp. 63–116, Pergamon Press, Oxford, UK

Kopal, Z.: 1959, *Close Binary Systems*, Chapman & Hall, London

Kopal, Z.: 1979, *Language of the Stars*, D. Reidel, Dordrecht, Holland

Kopal, Z.: 1989, *The Roche Problem*, Kluwer Academic Publishers, Dordrecht, Holland

Kopal, Z.: 1990, *Mathematical Theory of Stellar Eclipses*, Kluwer Academic Publishers, Dordrecht, Holland

Kurucz, R. L.: 1979, Model Atmospheres for G, F, A, B, and O Stars, *ApJ Suppl.* **40**, 1–340

Kurucz, R. L.: 1993, New Atmospheres for Modelling Binaries and Disks, in E. F. Milone (ed.), *Light Curve Modeling of Eclipsing Binary Stars*, pp. 93–102, Springer, New York

Lanza, A. F., Catalano, S., Cutispoto, G., Pagano, I., & Rodonò, M.: 1998, Long-term Starspot Evolution, Activity Cycle and Orbital Period Variation of AR Lacertae, *A&A* **332**, 541–560

Lanza, A. F., Catalano, S., Rodonò, M., İbanoğlu, C., Evren, S., Taş, G., Çakırlı, Ö., & Devlen, A.: 2002, Long-term Starspot Evolution, Activity Cycle and Orbital Period Variation of RT Lacertae, *A&A* **386**, 583–605

Linnell, A. P.: 1984, A Light Synthesis Program for Binary Stars, *ApJ Suppl.* **54**, 17–31

Linnell, A. P.: 1989, A Light Synthesis Program for Binary Stars. III. Differential Corrections, *ApJ* **342**, 449–462

Linnell, A. P.: 1991, A Light Synthesis Study of W Ursae Majoris, *ApJ* **374**, 307–318

Linnell, A. P.: 1993, Light Synthesis Modeling of Close Binary Stars, in E. F. Milone (ed.), *Light Curve Modeling of Eclipsing Binary Stars*, pp. 103–111, Springer, New York

Linnell, A. P., Etzel, P. B., Hubeny, I., & Olson, E. C.: 1998, A Photometric and Spectrophotometric of MR Cygni, *ApJ* **494**, 773–782

Linnell, A. P. & Hubeny, I.: 1994, A Spectrum Synthesis Program for Binary Stars, *ApJ* **434**, 738–746

Lucy, L. B.: 1968, The Light Curves of W Ursae Majoris, *ApJ* **153**, 877–884

Marquardt, D. W.: 1963, An Algorithm for Least-Squares Estimation of Nonlinear Parameters, *SIAM J. Appl. Math.* **11**, 431–441

McNally, D. (ed.): 1991, *Reports on Astronomy Symposium on the Theory of Computing*, No. XXIA in Close Binary Stars, The Netherlands, IAU

Milone, E. F.: 1976, Infrared Photometry of RT Lacertae, *ApJ Suppl.* **31**, 93–109

Milone, E. F.: 1977, Preliminary Solution for RT Lacertae, *AJ* **82**, 998–1007

Milone, E. F. (ed.): 1993, *Light Curve Modeling of Eclipsing Binary Stars*, Springer, New York

Milone, E. F.: 2002, A Reprise of the Properties of the Exotic Eclipsing Binary RT Lacertae, in C. A. Tout & W. van Hamme (eds.), *Exotic Stars as Challenges to Evolution*, Vol. 279 of *Astronomical Society of the Pacific Conference Series*, pp. 65–71

Mochnacki, S. W. & Doughty, N. A.: 1972a, A Model for the Totally Eclipsing W Ursae Majoris System AW UMa, *MNRAS* **156**, 51–65

Mochnacki, S. W. & Doughty, N. A.: 1972b, Models for Five W Ursae Majoris Systems, *MNRAS* **156**, 243–252

Nagy, T. E.: 1975, The Binary System V566 Oph Revisited, *Bull. Amer. Astr. Soc.* **7**, 533

Nelson, B. & Davis, W. D.: 1972, Eclipsing-Binary Solutions by Sequential Optimization of the Parameters, *ApJ* **174**, 617–628

Plavec, M. J.: 1980, IUE Observations of Long Period Eclipsing Binaries: A Study of Accretion onto Non-Degenerate Stars, in M. J. Plavec, D. M. Popper, & R. K. Ulrich (eds.), *Close Binary Stars: Observations and Interpretation*, pp. 251–261, D. Reidel, Dordrecht, Holland

Popper, D. M.: 1991, Orbits of Close Binaries with CA II H and K in Emission. IV – Three Systems with Mass Ratios far from Unity, *AJ* **101**, 220–229

Popper, D. M.: 1992, The Cool ALGOLS, in Y. Kondo, R. Sistero, & R. S. Polidan (eds.), *Evolutionary Processes in Interacting Binary Stars*, Vol. 151 of *IAU Symposium*, pp. 395–398

Press, W. H., Flannery, B. P., Teukolsky, S. A., & Vetterling, W. T.: 1992, Numerical Recipes – The Art of Scientific Computing, Cambridge University Press, Cambridge, UK, 2nd edition

Proctor, D. D. & Linnell, A. P.: 1972, Computer Solution of Eclipsing-Binary Light Curves by the Method of Differential Corrections, *ApJ Suppl.* **24**, 449–477

Rodonò, M., Lanza, A. F., & Catalano, S.: 1995, Starspot Evolution, Activity Cycle and Orbital Period Variation of the Prototype Active Binary RS Canum Venaticorum., *A&A* **301**, 75–88

Russell, H. N.: 1912a, On the Determination of the Orbital Elements of Eclipsing Variable Stars. I, *ApJ* **35**, 315–340

Russell, H. N.: 1912b, On the Determination of the Orbital Elements of Eclipsing Variable Stars. II, *ApJ* **36**, 54–74

Russell, H. N. & Merrill, J. E.: 1952, The Determination of the Elements of Eclipsing Binary Stars, *Princeton. Obs. Contr.* **26**, 1–96

Russell, H. N. & Shapley, H.: 1912a, On Darkening at the Limb in Eclipsing Variables. I, *ApJ* **36**, 239–254

Russell, H. N. & Shapley, H.: 1912b, On Darkening at the Limb in Eclipsing Variables. II, *ApJ* **36**, 385–408

Southworth, J.: 2008, Homogeneous Studies of Transiting Extrasolar Planets - I. Light-curve Analyses, *MNRAS* **386**, 1644–1666

Southworth, J., Bruntt, H., & Buzasi, D. L.: 2007a, Eclipsing Binaries Observed with the WIRE Satellite. II. β Aurigae and Non-linear Limb Darkening in Light Curves, *A&A* **467**, 1215–1226

Southworth, J., Maxted, P. F. L., & Smalley, B.: 2004a, Eclipsing binaries in Open Clusters - I. V615 Per and V618 Per in h Persei, *MNRAS* **349**, 547–559

Southworth, J., Maxted, P. F. L., & Smalley, B.: 2004b, Eclipsing Binaries in Open Clusters - II. V453 Cyg in NGC 6871, *MNRAS* **351**, 1277–1289

Southworth, J., Smalley, B., Maxted, P. F. L., & Etzel, P. B.: 2004c, Accurate Fundamental Parameters of Eclipsing Binary Stars, in J. Zverko, J. Ziznovsky, S. J. Adelman, & W. W. Weiss (eds.), *IAU Symposium*, pp. 548–561

Southworth, J., Wheatley, P. J., & Sams, G.: 2007b, A Method for the Direct Determination of the Surface Gravities of Transiting Extrasolar Planets, *MNRAS* **379**, L11–L15

Tamuz, O., Mazeh, T., & North, P.: 2006, Automated analysis of eclipsing binary light curves – I. EBAS – a new Eclipsing Binary Automated Solver with EBOP, *MNRAS* **367**, 1521–1530

Tikhonov, A. N.: 1963a, Regularization of Incorrectly Posed Problems, *Dokl. Akad. Nauk USSR* **153**, 49–52

Tikhonov, A. N.: 1963b, Solution of Incorrectly Formulated Problems and the Regularization Method, *Dokl. Akad. Nauk USSR* **151**, 501–504

Tikhonov, A. N. & Arsenin, V. Y. (eds.): 1979, *Methods for Solving Ill-Posed Problems*, Nauka, Moscow

Tsesevich, V. P. (ed.): 1973, *Eclipsing Variable Stars*, A Halsted Press Book, Wiley, New York

Van Hamme, W.: 1993, New Limb-Darkening Coefficients for Modeling Binary Star Light Curves, *AJ* **106**, 2096–2117

Van Hamme, W. & Wilson, R. E.: 2003, Stellar Atmospheres in Eclipsing Binary Models, in U. Munari (ed.), *GAIA Spectroscopy: Science and Technology*, Vol. 298 of *Astronomical Society of the Pacific Conference Series*, pp. 323–328, San Francisco

Van Hamme, W. & Wilson, R. E.: 2007, Third-Body Parameters from Whole Light and Velocity Curves, *ApJ* **661**, 1129–1151

Wade, R. A. & Rucinski, S. M.: 1985, Linear and Quadratic Limb-darkening Coefficients for a Large Grid of LTE Model Atmospheres, *A&A Suppl.* **60**, 471–484

Wehrse, R.: 1987, Theory of Circumstellar Envelopes, in I. Appenzeller & C. Jordan (eds.), *Circumstellar Matter*, IAU Symposium 122, pp. 255–266, Kluwer Academic Publishers, Dordrecht, Holland

Wilson, R. E.: 1979, Eccentric Orbit Generalization and Simultaneous Solution of Binary Star Light and Velocity Curves, *ApJ* **234**, 1054–1066

Wilson, R. E.: 1989, The Relation of Algols and W Serpentis Stars, *Space Sci. Rev.* **50**, 191–203

Wilson, R. E.: 1990, Accuracy and Efficiency in the Binary Star Reflection Effect, *ApJ* **356**, 613–622

Wilson, R. E.: 1993, Computation Methods and Organization for Close Binary Observables, in J. C. Leung & I.-S. Nha (eds.), *New Frontiers in Binary Star Research*, Vol. 38 of ASP Conference Series, pp. 91–126, Astronomical Society of the Pacific, San Francisco, CA

Wilson, R. E.: 1994, Binary-Star Light-Curve Models, *PASP* **106**, 921–941

Wilson, R. E.: 1998, *Computing Binary Star Observables (Reference Manual to the Wilson–Devinney Programm*, Department of Astronomy, University of Florida, Gainesville, FL, 1998 edition

Wilson, R. E.: 2007, Close Binary Star Observables: Modeling Innovations 2003-06, in W. I. Hartkopf, E. F. Guinan, & P. Harmanec (eds.), *Binary Stars as Critical Tools and Tests in Contemporary Astrophysics*, No. 240 in Proceedings IAU Symposium, pp. 188–197, Kluwer Academic Publishers, Dordrecht, Holland

Wilson, R. E.: 2008, Eclipsing Binary Solutions in Physical Units and Direct Distance Estimation, *ApJ* **672**, 575–589

Wilson, R. E. & Devinney, E. J.: 1971, Realization of Accurate Close-Binary Light Curves: Application to MR Cygni, *ApJ* **166**, 605–619

Wilson, R. E. & Devinney, E. J.: 1973, Fundamental Data for Contact Binaries: RZ Comae Berenices, RZ Tauri, and AW Ursae Majoris, *ApJ* **182**, 539–547

Wolf, B.: 1987, Some Observations Relevant to the Theory of Expending Envelopes, in I. Appenzeller & C. Jordan (eds.), *Circumstellar Matter*, IAU Symposium 122, pp. 409–425, Kluwer Academic Publishers, Dordrecht, Holland

Wood, D. B.: 1971, An Analytic Model of Eclipsing Binary Star Systems, *AJ* **76**, 701–710

Wood, D. B.: 1972, *A Computer Program for Modeling Non-Spherical Eclipsing Binary Star Systems*, Technical Report X-110-72-473, GSFC, Greenbelt, MD

Yamasaki, A.: 1981, Light Curve Analysis of Contact Binaries: Characteristic Quantities of the Light Curve, *Ap. Sp. Sci.* **77**, 75–109

Yamasaki, A.: 1982, A Spot Model for VW Cephei, *Ap. Sp. Sci.* **85**, 43–48

Chapter 7
The Wilson–Devinney Program: Extensions and Applications

Because the Wilson–Devinney program is the most widely used of all the light curve modeling tools, it is appropriate to describe its features, capabilities, and continuing development in some detail. The WD program itself has seen continual improvements, and the current version (briefly summarized in Chap. 6) with its powerful features provides the opportunity to extract a maximum of information from a variety of observational data. As a side-effect, publications on the WD model and on the WD program[1] have stimulated the development of new programs, which in their kernel use the WD program. Several such programs with new innovative features or added functionality now coexist with the WD program. Some of these features were developed independently in several programs, including WD.

Starting in the 1980s, several extensions and enhancements to the original WD program have been implemented in the programs LCCTRL and LC83KS by Kallrath (1987), WD83K83 by Stagg & Milone (1993), WD83K93 by Milone et al. (1992b), WD93K93 by Stagg & Milone (1993)), and LC93KS by Kallrath (1993). The successor to these programs is WD95 (Kallrath et al. 1998) and versions up to WD2007 (Milone & Kallrath, 2008). The naming is now unified to WDx2007 with the WDx indicating extensions to the WD program. WD has now also become the physical model engine of PHOEBE, a simulation and analysis tool developed by Prša & Zwitter (2005b) and described in Sect. 8.2 that offers an attractive graphical user interface.

The first section of this chapter describes the functionality of WDx2007, the second focuses on the Kurucz atmospheres option in WDx2007 and WD, the third gives a brief review of research performed with the WD program and its offspring versions (LC83KS, WD93K93, LC93KS, and WDx2007), which include enhanced features by other authors. The third section also discusses analyses of astrophysically interesting systems such as X-ray binaries and *EB*s in globular and open clusters. Finally, in the fourth section, some future prospects for the WD program are considered.

---

[1] The WD program itself continues to be developed by R. E. Wilson.

7.1 Current Capabilities of WDx2007

Ad vitam aut culpam   (*For life – or, until you mess up*)

The stand-alone Fortran77 program WDx2007[2] enables the user to make use of the Wilson–Devinney (WD) program[3] to compute *EB* light and radial velocity curves and to analyze data, i.e., to fit light curves or merely to compute a synthetic light curve. WDx2007 contains the WD program but provides several features around it, which enables the user to

- carry out computations using the original WD programs LC and DC;
- convenient preprocessing of the input data to LC and DC;
- use the Kurucz atmospheres and to model better the wavelength-dependent stellar flux (Milone et al. 1992b);
- on-the-fly computation of limb-darkening coefficients using an interpolation scheme based on the Van Hamme (1993) tables (during the iterations of the least-squares iteration the limb-darkening coefficients are automatically computed as a function of temperature, log g and wavelength);
- fit *EB* observables using the simplex method (Kallrath & Linnell 1987), differential corrections (Wilson & Devinney 1971), a Levenberg–Marquardt scheme (Kallrath et al. 1998), or simulated annealing as described in Milone & Kallrath (2008);
- do automatic iterations (Kallrath 1987);
- produce gnuPlot graphics files;
- obtain best-fit solutions for grids or tables over fixed parameters (Kallrath & Kämper 1992) which otherwise may be poorly determined; and
- develop and test new features, e.g., the analysis of large number of light curves using the matching approach described in Sect. 5.3.2.

WDx2007 couples directly to the WD code which may cause some delays in keeping up with newly issued versions of the WD program. It runs under the operating system LINUX (especially, Ubuntu Linux), CygWin, as well as in Windows Command Shell under Win95, Win98, WinNT, Win2000, and WinXP. WDx2007 has been developed upon the framework of the preceding programs LCCTRL (Kallrath 1987), WD93, WD95, WD98 (Kallrath et al. 1998), and WD2002, the first that combined all previous developments and included all stellar atmosphere improvements by Milone et al. (1992b) and local limb-darkening coefficients. These additional atmosphere and limb-darkening features are now an integral part of the WD program.

---

[2] This program, maintained and further developed by Josef Kallrath, and its documentation is available at http://www.astro.ufl.edu/˜kallrath/

[3] The Wilson–Devinney program is distributed by Robert E. Wilson (University of Florida) and available at ftp://ftp.astro.ufl.edu/pub/wilson/lcdc2007

7.2 Atmospheric Options

In nubibus   *(In the clouds)*

The WD program has seen two enhancements for stellar atmospheres. The earlier one was implemented in WD95 (and carried through WDx2007); in 2003 another was integrated directly into the WD code by Van Hamme & Wilson (2003). Both are based on Kurucz's stellar atmospheres models contained on a CD-ROM[4] made available by Robert L. Kurucz.

Kurucz's stellar atmospheres incorporate computed opacities from a comprehensive list of 58 million lines. The opacities were computed for 56 temperatures in the range 2,000 to 200,000 K, 21 values of the log of the pressure (measured in *cgs* units) over the range −2 to +8, and 5 values of microturbulent velocity (0, 1, 2, 4, and 8 km/s). They include ranges of abundances: 20 models in the range [Fe/H][5] = +1.0 to −5.0, and for 0.0 with no helium. The new line opacities and added continuous opacities were all included in the new atmosphere calculations. Temperature ranges depend on log g and abundance [M/H], with the largest range from 3,500 to 50,000 K. Temperature limits and the 19 abundances are given in Kurucz (1993).

7.2.1 Kurucz Atmospheres in WDx2007

WD95 (as well as WD93K93, which has been used for light curve modeling of *EBs* since 1993) and the current version, WDx2007, make use of Kurucz's stellar atmosphere models. The atmospheres were integrated over the standard system passbands $UBVR_JI_J$, R_CI_C, and *uvby*, the nominal extended Johnson infrared passbands *JHKLMN*, the improved infrared passbands, iz, iJ, iH, iK, iL', iM, iN, in, and iQ, and for a range of narrow, square-edged passbands centered on wavelengths in the far-ultraviolet, appropriate for IUE, HST, or other space platforms with far-ultraviolet detectors. The raisons d'être and other details of infrared passband optimization may be found in Young et al. (1994).

The flux ratio files for a specified filter can be computed as described below. This procedure was developed and applied by C.R. Stagg at the University of Calgary. At first we tabulate the filter transmission function. Then create a file of wavelength versus transmission for the filter.

```
line 1        FACTOR, 1, WLBEGIN (nm), WLEND (nm), NW
line 2        WLBEGIN, TRANSMISSION (WLBEGIN)
...
last line     WLEND, TRANSMISSION (WLEND)
```

---

[4] Kurucz CD-ROM No. 13 ATLA S9—Stellar Programs and 2 km/s grid. Robert L. Kurucz, Aug 22, 1993, Smithsonian Astrophysical Observatory. Cambridge, MA 021138, USA. Copyright Smithsonian Astrophysical Observatory, 1993.

[5] The symbol [Fe/H] denotes the logarithm of the ratio of iron and hydrogen abundances relative to the same ratio for the Sun. Thus, stars of solar composition have $[Fe/H] = 0$.

FACTOR is the factor to multiply the wavelengths to convert them to nm, i.e., WLBEGIN*FACTOR = WLBEGIN (nm). The wavelengths (in nm) must correspond to those in the file FPHEAD. NW is the number of wavelengths listed. All lines are in open format.

```
1.0 1 485.0 743.0 130
  485.00   0.0000
  487.00   0.0032
  ...
  743.00   0.0000
```

The program FLUX2000 prompts the user for the name of the Kurucz flux file (e.g., e:\cdromfm\fm05k2.pck), the name of the filter transmission file, (e.g., Rtrans), and the name of the output file (e.g., Rm05)?

Here are the extra lines the header that must be added to the top of the LC or DC input to run it with the revised Kurucz atmospheres program:

```
line 1: 1 for LC, 2 for DC
line 2: g(star1), g(star2)
line 3: flux file for first filter listed
line 4: flux file for second filter listed
...
last line: END
```

Example of lines to be added to DC input:

```
2
5.0 5.0
Uflux
Bflux
Vflux
END
```

Below you will find the program listing.

```
program read2000
character*2 a2
character*18 a18
character*22 a22
character*65 a65
character*2 char(35)
DATA CHAR/
&8*' ',' A',' E',' W','F1','F2','PH','VG',' I',
&'G1','G2','T1','T2','A1','A2','P1','P2',' Q',
&'T0',' P','DP','DW',' ','L1','L2','X1','X2','L3'/
```

```
          imax=20
          open(4,file='parm')
          open(7,file='pout')
        9 write(7,35)
          do 6 n=1,2
        1 read(4,2,end=10) a2,a18,a22
        2 format(a2,25x,a18,18x,a22)
          if(a2.ne.'No') go to 1
          if(n.eq.1) write(7,25) a2,a18,a22
       25 format(a2,1x,a18,1x,a22)
        5 read(4,3) i,a18,a22
        3 format(i2,25x,a18,18x,a22)
          if(i.eq.0) go to 6
          if(n.eq.1) write(7,25) char(i),a18,a22
          go to 5
        6 write(7,35)
       35 format(1x)
          do 8 n=1,imax
          read(4,7) a65
        7 format(a65)
        8 write(7,7) a65
          imax=6
          go to 9
       10 close(4)
          close(7)
          stop
          end
```

7.2.2 Kurucz Atmospheres in WD

Van Hamme & Wilson (2003) implemented the following Kurucz atmosphere approximation in the WD program. It requires only a few additional a priori computed files and increases the computational time only slightly. The scheme models temperature dependence of passband intensities[6] through Legendre polynomial approximation functions; the only remaining interpolation is in $\log g$. The Legendre approximations reproduce normal emergent stellar atmosphere passband intensities in absolute units with errors typically smaller than astrophysical uncertainties in the original atmospheres and are used to compute the integrated response for the 25 passbands listed in the WD program documentation.[7] Among these

---

[6] Note that the Legendre approximation is not used for a specific wavelength but directly over the whole integrated passband response. This is more accurate than a posteriori integration over all wavelengths of the passband.

[7] Monograph *ebdoc2007* available at *ftp://ftp.astro.ufl.edu/pub/wilson/lcdc2007*

are Johnson standard $UBVR_JI_J$, R_CI_C, and Strömgren *uvby*, the nominal extended Johnson infrared passbands *JHKLMN*, and the HIPPARCOS *hip* and TYCHO B_T and V_T passbands.

Their computations are based on the normal emergent intensities provided on CD-ROMs 16 and 17 described in Kurucz (1993) for microturbulent velocity 2 km/s. Intensities are given at 1,221 wavelengths from 9 to 160,000 nm and 11 $\log g$'s from 0.0 to 5.0 (cgs).

For each atmosphere model Van Hamme & Wilson integrated intensities over each passband, weighted by response function, and similarly integrated blackbody intensities. Irrespective of abundance or surface gravity, Legendre polynomials represent passband intensities accurately as functions of T_{eff} over four subintervals, with passband-dependent beginning and end points. Accordingly, four T_{eff} subintervals were bounded by lower (T_l) and upper (T_h) effective temperatures with T_{eff} scaled according to $\phi_T = (T_{\text{eff}} - T_l)/(T_h - T_l)$. The coefficients ϕ_T result from least-squares Legendre fits of degree m with $m \leq 9$ based on the number of points in a subinterval, which leads to a maximum of 10 Legendre coefficients. Use of a subinterval end point as the starting point of the next subinterval greatly reduces discontinuities at subinterval boundaries.

In many close binaries, especially those with tidally distorted components, at least one of the stars has part of its surface outside the range of available atmosphere models. Van Hamme & Wilson (2003) give one example; late-type W UMa overcontact binaries with low gravity connecting necks are provide other examples. It would be perverse to abandon atmosphere models for the entire star because of such range limitations, yet a simple blackbody patch would impose an artificial discontinuity that could introduce very bad numerical effects in light curve computation. To avoid radiative discontinuities in very high and very low temperature regions, Van Hamme & Wilson developed polynomial ramping functions in T_{eff} and $\log g$ that smoothly transition from atmosphere to blackbody regions. If a $(T_{\text{eff}}, \log g)$ pair is outside the range of atmosphere applicability, the program smoothly connects atmosphere model intensities to blackbody band intensities over built-in ranges in $\log g$ and T_{eff} whose limits can easily be changed. This strategy allows atmosphere computations of spotted stars with surface parts hotter or cooler than existing atmosphere models.

7.3 Applications and Extensions

Abeunt studia in mores   (*You are what you learn*)

This section describes specific astrophysically interesting applications and resulting experiences with the most recent versions of the Wilson–Devinney program, or special extensions (pulse arrival times, line profile analysis, LC93KS, and WD95) of the Wilson–Devinney program. The intention is to provide hints about what can be achieved and how various extensions can be used to derive astrophysical results. The binaries discussed also demonstrate how interesting binary star astrophysics can be. *HD 77581/GP Velorum* is an X-ray pulsar which makes it a very rich data source. Some binaries in the ancient globular cluster *NGC 5466* and some others in *M71* provide useful information both on binaries and clusters, although they are very faint stars. The binary *H235* in the open cluster *NGC 752* is an interesting example of the evolution of binary stars in clusters. The 24.6-day period *EB AI Phoenicis* is one of the best-studied field *EB* systems. Finally, the chapter provides some results on the analysis of fast and slowly rotating Algols based on line profile fitting.

7.3.1 The Eclipsing X-Ray Binary HD 77581/Vela X-1

This binary contains an X-ray pulsar. The pulse arrival times are an additional observable used in a simultaneous analysis described in Sect. 4.1.1.6 and are modeled as outlined in Sect. 3.8. For more background on pulse arrival time modeling, we refer the reader to Wilson & Terrell (1998) on the methodology of incorporating pulse arrival times in light curve modeling.

The ellipsoidal variable*HD 77581/GP Velorum* is the optical counterpart of the pulsed, eclipsing X-ray source *Vela X-1*. A B0.5 supergiant and a neutron star move in an eccentric orbit of about $e = 0.1$ and a period of 8.96 days. Spectral line broadening indicates that the optical star rotates subsynchronously with $0.5 \leq F_2 \leq 0.75$ (Zuiderwijk 1995). The analyses by Wilson & Wilson & Terrell (1994, 1998), performed with the original WD program, demonstrate the fruitfulness of a simultaneous least-squares analysis of *all* available data. The analysis included the data shown in Fig. 7.1:

1. B and V light curves observed by Van Genderen (1981);
2. optical (He I) radial velocity curves by Van Paradijs et al. (1977), Petro & Hiltner (1974), Van Kerkwijk et al. (1995), and Stickland et al. (1997);
3. pulse arrival times measured by Rappaport et al. (1980) and other sources given in Wilson & Terrell (1998); and
4. estimations of the X-ray eclipse duration[8] by Watson & Griffiths (1977).

The light curves were used only in preliminary experiments, not in the final solution. The known X-ray eclipse duration was included in the analysis as an additional constraint with the formalism implemented in subroutine DURA (see Appendix E.11). In the simultaneous fitting of many types of data, appropriate weighting is very important (see Sect. 4.1.1.6). Wilson & Terrell (1994, 1998) commented on which observations contribute most to specific output parameters:

X-ray eclipse $\Rightarrow$ relative size of the B star,
pulse arrival times $\Rightarrow$ orbital eccentricity,
radial velocities and pulse arrival times $\Rightarrow$ absolute dimensions.

Neither the X-ray nor optical variations can yield the inclination. They indicate only that the orbit is close enough to edge-on to produce broad eclipses. The important point is that all relationships are used simultaneously to ensure a self-consistent solution.

---

[8] The 1994 analysis used the eclipse duration data by Avni (1976), Ögelman et al. (1977), and also by Van der Klis and Bonnet-Bidaud (1984).

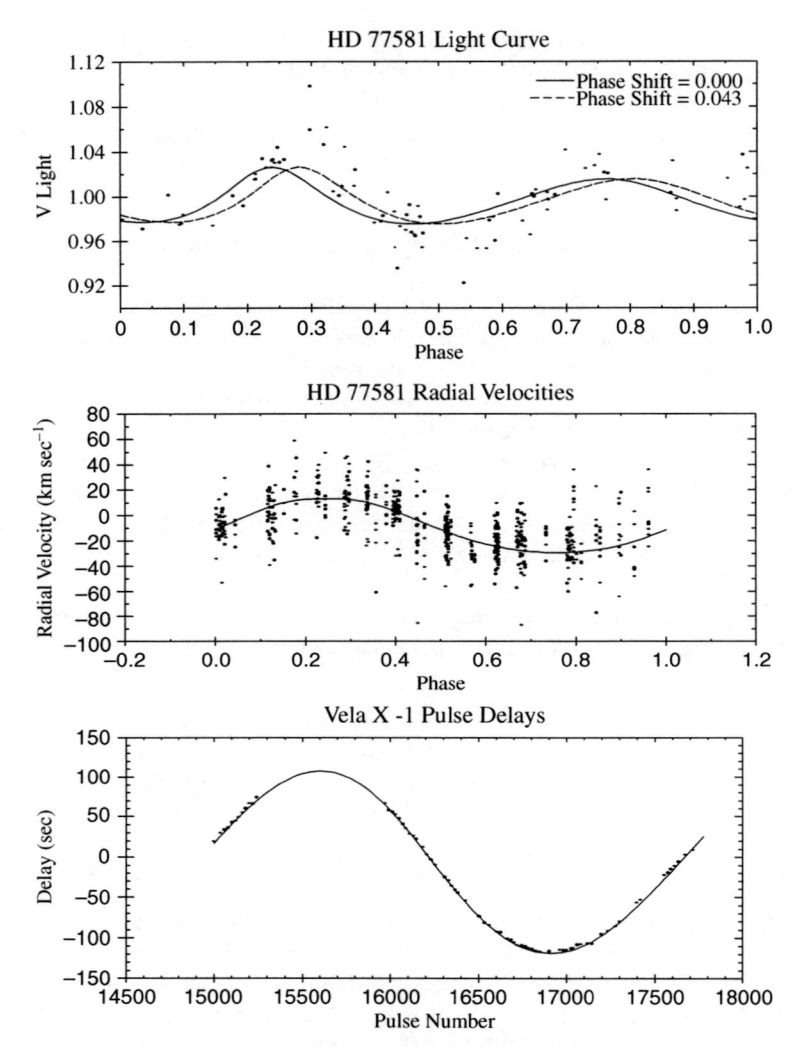

Fig. 7.1 *HD 77581* light curve and radial velocities, Vela X-1 pulse delays. This figure, taken from Wilson & Terrell (1994), shows the light curve, radial velocity, and pulse delay data involved in the simultaneous analysis and the curves fitted to the data. Courtesy D. Terrell

7.3.2 The Eclipsing Binaries in NGC 5466

The *EB* systems in the ancient globular cluster *NGC 5466* (see Fig. 2.3) were included in the University of Calgary binaries-in-clusters program. The objective was to analyze *EB*s in well-studied clusters in a boot-strap program to increase knowledge of both binaries and clusters. The systems were discovered to eclipse by Mateo et al. (1990), who provided the only published light curves. These were

7.3 Applications and Extensions 313

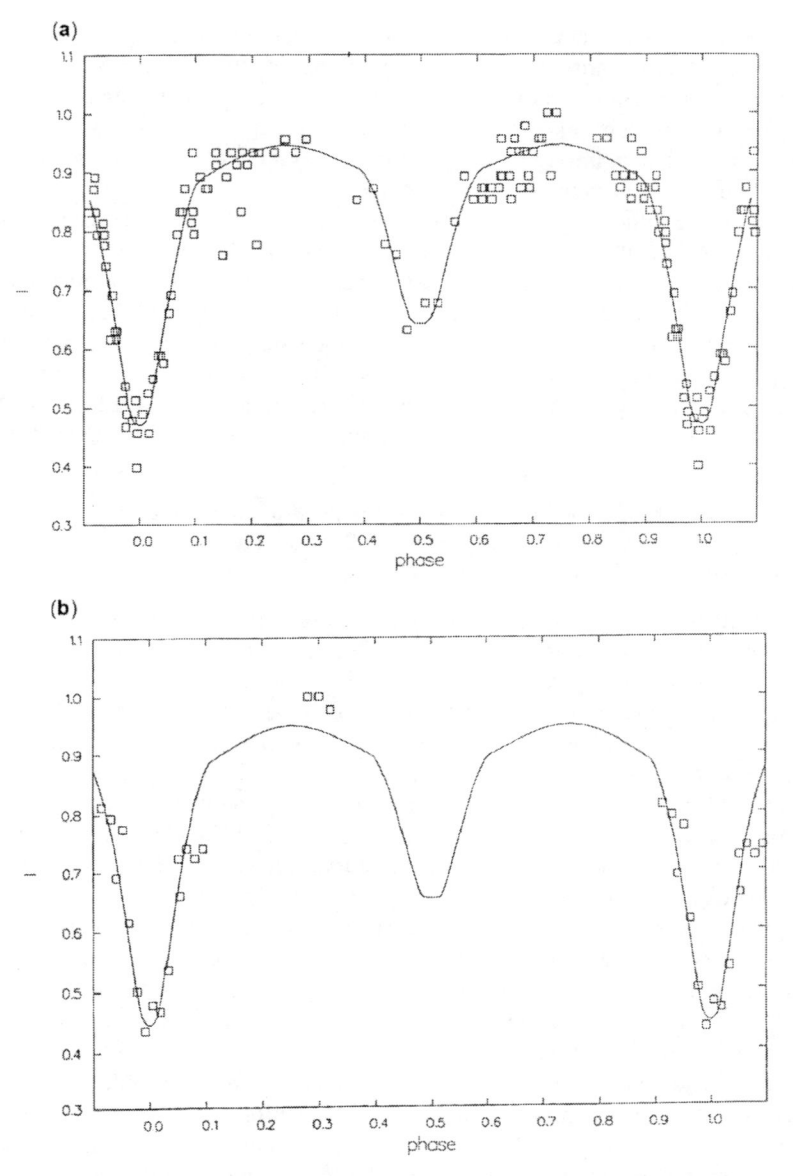

Fig. 7.2 Light curves of *NH31* in the globular cluster *NGC 5466*. Observed data and computed *V* and *B* light curves of the detached (**a**) transit, (**b**) semi-detached, and (**c**) occultation models. This is Fig. 1 in Kallrath et al. (1992)

the first binaries modeled with an updated version of LC83KS in which spot parameters could be adjusted automatically. The modeling work to date is summarized in Kallrath et al. (1992) and in Milone et al. (1992a). The models for the components of the over-contact systems [*NH19* and *NH30* in the notation of Nemec & Harris (1987)] and the two most likely models for the non-contact system *NH31* (its light curves and fits are shown in Fig. 7.2) indicate all the stars to be among the large blue straggler population of this cluster. Since blue stragglers are brighter and bluer than the isochrones drawn through the bulk of the stars on the color-magnitude diagram, they appear to be far younger than the other cluster stars. Since most over-contact systems are not blue stragglers, the two cases in *NGC 5466* are very important. A current theory for the origin of blue stragglers is that they are merely unresolved binaries which appear brighter and bluer as a result of a dredge up of material from the interior because of some sort of binary interaction. Another, not necessarily independent, theory is that they represent merged stars. The over-contact systems appear to have contact parameters exceeding 0.90, whereas the non-contact system has a period of only $\sim 0\overset{d}{.}5$, making it apparently one of the shortest-period Algol systems. Therefore, all three seem to be heading toward mergers. To summarize, the three main facts about binaries and blue stragglers in *NGC 5466* are

1. each star of these blue straggler systems is itself a blue straggler;
2. these three systems are the only known *EB*s in the cluster; and
3. there are many blue stragglers in this cluster.

These facts argue for an ongoing merger process during which the three systems have not yet merged because they are themselves the remnants of four-body systems, all previous two-body systems having merged long ago.

The conclusions are dependent on the light curve solutions for the *EB*s, but it is legitimate to ask how reliable the solutions are. The systems are faint: The V magnitudes range from 18.4 for *NH19* at maximum to 19.6 for *NH30* at minimum, and only V and fragmentary B light curves have been published. Two of the three show a clear *O'Connell effect* (defined on pages 6 and 135). One way to overcome this problem is to include spots in the light curve modeling, although in such cases spots serve only as an artifice to "save the phenomena." As we have noted elsewhere, the general acceptance of a spot model depends on a satisfactory representation of the observations, and on evidence beyond the light curve. Important tracers of the physical existence of star spots are the following: variability of the O'Connell effect, phase migration of the light curve asymmetry with time, the presence of molecular bands in spectra taken at phases for which large, cool spots are predicted, magnetometry evidence of strong Zeeman splitting, and line profile perturbations which are interpreted as rotational velocity fields. Unfortunately, most of these tracers require very high spectral resolution and light gathering power. In the case of the nineteenth magnitude systems in *NGC 5466*, this evidence is hardly forthcoming. Therefore we cannot show in any conclusive way that the O'Connell effect is correctly modeled by spots; neither can we show that it was not. The other major difficulty with the solutions is the lack of radial velocity measurement. Milone et al. (1987, 1991) and Milone (1993, pp. 195–202) demonstrated that radial velocity curves provide an important discriminant among models, especially if light curve phase coverage is incomplete or if eclipses are only partial. Unfortunately, it is difficult to convince telescope allocation committees (for telescopes of sufficient size to provide the data) that this kind of project is feasible and worthy of telescope time. The best that can be done at present is to obtain data in passbands across a large range of wavelength. This will make it easier to find the relative surface fluxes; it will also permit exploration of the flux distribution of any light curve perturbations. But for the final discrimination among occultation and transit models in the detached case, and critical confirmation in others, we must continue to seek radial velocity data. Hopefully, some "telescope allocation committees" will sufficiently appreciate the importance to astronomy of fundamental properties of stars to risk three days of telescope time.

7.3.3 The Binary H235 in the Open Cluster NGC 752

For the over-contact system *H235* in the open cluster *NGC 752* both radial-velocity (Fig. 7.4) and light curves are available. The system was modeled with the WD93K93 code (the enhanced 1993 Wilson–Devinney program with 1993 Kurucz atmospheres), and the upgraded Simplex version LC93KS, which permitted the simultaneous fitting of radial velocity and light curves, with spots for the first time.

The solution (Milone et al. 1995) assumed a radiative envelope because of the relatively early spectral type. The analysis was successful but evidence of a variable O'Connell effect precluded the modeling of an additional, incomplete CCD light curve, and necessitated the use of three spots to secure a satisfactory, but not perfect fit (Fig. 7.3) to the photoelectric light curve. Subsequently it was found that the assumption of nonlinear limb darkening, a two-iteration reflection effect treatment, and the use of albedos and gravity brightening coefficients appropriate for convective envelopes improved the fit by nearly 8%, most of which occurred with the use of the two-pass reflection option. The most important changes in the parameters of the fittings (Milone & Terrell 1996) are

1. the temperature difference is almost zero with $\Delta T = -1 \pm 42$, so that $T_2 = 6501 \pm 156$, compared to $T_1 = 6500 \pm 150$ (*est.*);
2. the components are slightly larger, increasing both luminosities and fill-out factor;
3. the "fill-out factor" changed from 0.214 to 0.583; and
4. the absolute magnitude changed from $M_V = 3.27$ [*not 3.41* as given in Milone et al. (1995)] to 3.18 ± 0.11.

Although the changes in modeling elements may appear to be modest, they lead to important differences in the interpretation of *H235*. It appears to be more evolved and thus closer to a merger system than was originally thought. The system is important in our understanding of the dynamical processes in open clusters because the star density of open clusters is sufficiently low that collisional interactions were considered not frequent enough to be important. The case of *H235* may require a reassessment of the effects of many weak interactions on binary star evolution in such clusters.

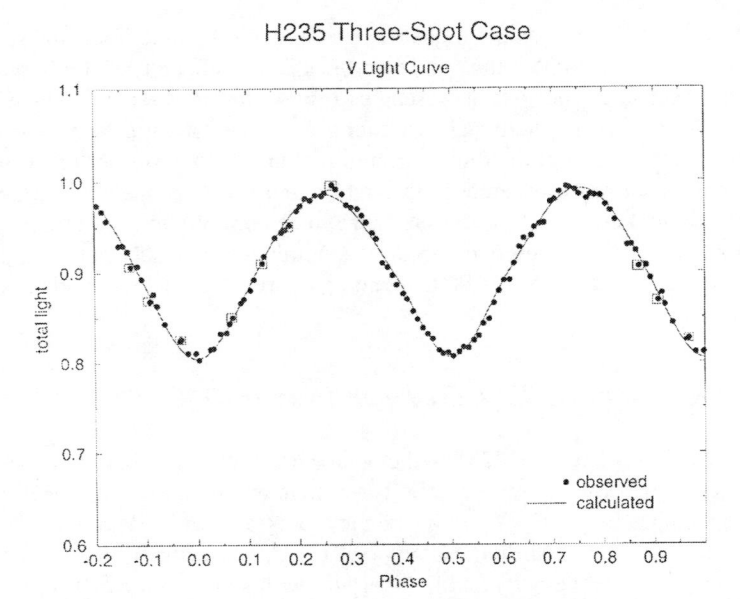

Fig. 7.3 Light curve of the over-contact system *H235*. This plot, part of Fig. 2 in Milone et al. (1995), shows the *V* light curve (observed data and fit in the three-spot case) of *H235* in the open cluster *NGC 752*

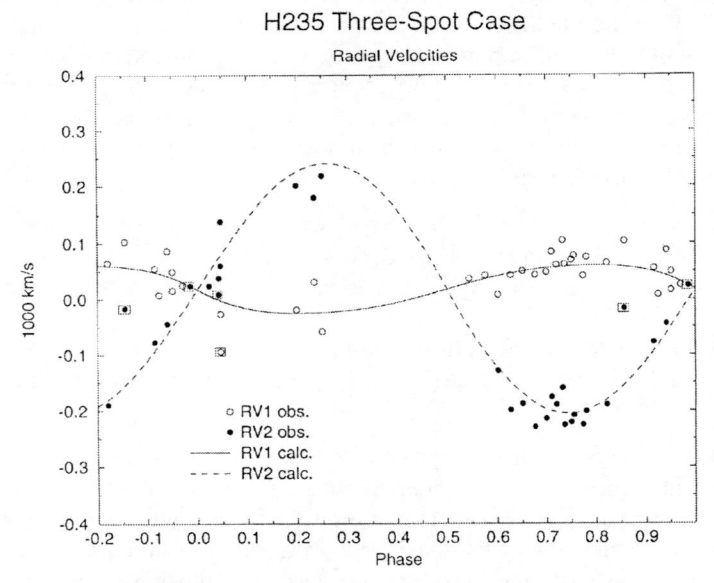

Fig. 7.4 Radial velocities of the over-contact system *H235*. This plot, part of Fig. 5 in Milone et al. (1995), shows the observed and calculated radial velocity curves (three-spot case) of *H235* in the open cluster *NGC 752*

7.3 Applications and Extensions

7.3.4 The Field Binary V728 Herculis

Not every application is automatically successful. The system *V728 Herculis* was modeled with both WD93K93 and LC93KS, but for a long time the latter consistently yielded physically unrealistic temperatures for the secondary which was found to be the hotter component. These earlier runs were carried out under the assumption that both envelopes were radiative – a reasonable assumption because the spectral type is F3-5, and the minima are of similar depth. However, subsequent modeling indicated that deeper valleys existed in parameter space if *convective* atmospheres were assumed. When gravity brightening and albedo coefficients appropriate for this case were adopted, the Simplex and the WD programs converged to quite similar results. These later trials were concluded with WD93K93 in a two-iteration reflection effect treatment, with nonlinear limb darkening. Both radiative and convective solutions were obtained with binned and unbinned data sets. The results for the convective modeling solution indicate that binning sometimes is significant to the final solution: For the radiative modeling solution, it did not matter (both data sets converged to the same solution). For the convective runs, the results were 1σ or more different for some parameters. The observations and analytical results are discussed in Nelson et al. (1995). The final mass ratio was determined to be $q = 0.1786 \pm 0.0023$ and the contact parameter, $f = 0.71 \pm 0.11$. Work by (1995) and Rasio & Shapiro (1994, 1995) suggests that over-contact systems with deep convective envelopes and small mass ratios ($q < 0.45$) may be unstable and enter into the final merger stage. One of the interesting results of the Nelson et al. (1995) study, however, is that there was no evidence of significant period change that might be expected to accompany such a situation, although more recent times of minimum obtained by Nelson et al. suggest at least a different period if not a variable one is needed to satisfy them. However, for *V728 Her*, one index of instability, the radius of gyration as defined by Rasio (1995), is found to be in the stable region of Rasio's Fig. 1: $k_1^2 = 0.16$. The instability of this over-contact system is, therefore, not demonstrated observationally – at least not yet.

7.3.5 The Eclipsing Binaries in M71

There are five known *EB*s in the globular cluster*M71* (Yan & Mateo 1994, Mateo & Yan 1996). The systems are around eighteenth magnitude in *V*, with $V - I \approx 1$ but the Yan & Mateo CCD light curves are relatively smooth. These have been modeled with WD93K93 by J. McVean for an MSc thesis at the University of Calgary. The results are summarized in McVean et al. (1997). He assumed[9] initial masses of 1.7 $M_\odot$, roughly twice the turn-off mass for the cluster, from the models of Bergbusch & Vandenberg (1992). Three of the binaries (V1, V2, and V5) are over-contact

---

[9] The radial velocity curves by Mateo & Yan (1996) were not yet included in this analysis.

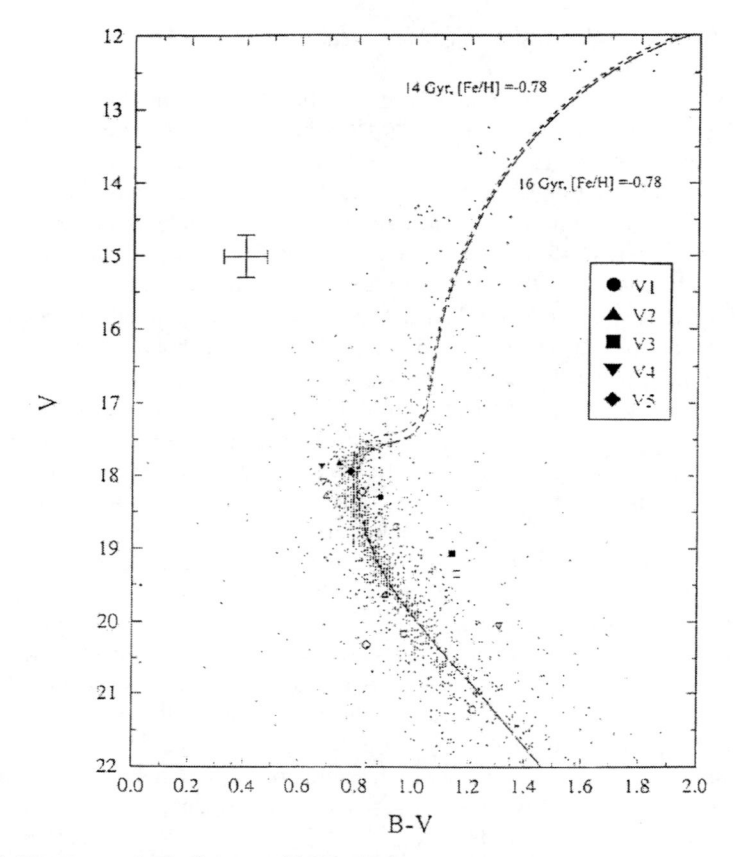

Fig. 7.5 Cluster magnitude diagram of *M71* with isochrones. It also contains the positions of the binaries and their individual components (*open symbols*); from Fig. 2 in McVean et al. (1997)

systems; the other two appear to be detached systems according to the best fit models, but the photometric secondary star of V4 has a contact parameter of −0.11, indicating that it nearly fills its Roche lobe. Due to O'Connell effects in the light curves of V1 and V3, a spot group was required on one of the components of both systems. All five systems have derived distances, within errors, in agreement with those (3.6 ± 0.5 kpc) determined by Cudworth (1985) but V3, has an unusual location on the color-magnitude diagram of the cluster (see Fig. 7.5), and for this reason Yan & Mateo (1994) doubted its membership. The interesting results of the analysis were first that a slight preference was found for the [Fe/H]= −0.3 Kurucz atmosphere fluxes and second that the uncertainties, especially for the mass ratios, were unusually low for systems of this kind, where radial velocities are not available. As in most investigations involving the WD93K93 program, the probable errors for the full set of the final run were used and cited, but in some cases, the formal probable error was of order ∼2%. This is a purely internal error, of course, and refers only to the uncertainty in the deepest minimum in each of the modeling runs, because uncertainties in the DC output file refer only to the values near the minimum. The true errors in such parameters are certainly higher, but the accuracy of the result nevertheless is very high, mainly on the basis of the extensive Simplex as well as perturbed WD modeling.

7.3.6 The Eclipsing Binaries in 47 Tuc

A HST search for "hot Jupiters" around the stars of the globular cluster NGC 104, otherwise known as 47 Tuc, has been described by Gilliland et al. (2000). A by-product of the 8-day continual imaging of a portion of the core region of the cluster produced a large number of variables, only some of which were previously known. Subsequent ground-based imaging of wider regions of the cluster produced even more variable star discoveries and more light curves on those that were obtained by Gilliand's experiment, but here we describe only a few results from the HST experiment alone.

Although the data for each light curve are numerous (∼1290) in each of pass-bands transformed to the V and I_C band, respectively), the information is not always as complete as one would desire. Two passbands are the minimal information needed to argue about temperatures of the components from the standpoint of light curve data alone. The faintness of these stars and the crowdedness of the field effectively eliminates the possibility of obtaining radial velocity dispersion spectra for these objects. The only way one can proceed to obtain fundamental data from *EB* light curves in this cluster is to rely on the very careful studies that have been carried out of the cluster itself. Fortunately, Bergbusch & VandenBerg (2001) have been able to obtain consistent models for the cluster isochrones (curves of constant age on the magnitude-color diagram). With these, Milone et al. (2004) were able to devise and use a method to bootstrap this information and the results of modeling the two passband curves to obtain a consistent set of results for both component stars in a handful of the systems of the cluster. Initial parameter guesses were made from a set of light curve properties (eclipse depths, widths, contact phases) provided by simulated light curves for systems with 47 Tuc metallicity and with the brightness and color of the VandenBerg (2000) isochrone [see also Bergbusch & VandenBerg (2001) and VandenBerg et al. (2002)]. This work is described briefly by Milone et al. (2004). The method to use only the two light curves and the most reliable isochrone for the cluster has been described in Chapter 5. Since the initial modeling work, R. Gilliland (private correspondence to EFM) suggested that the 47 Tuc photometry should be redone. This has been carried out by R. Guhathakurta (2009, private communication) for constant stars; R. Stagg and Milone are examining the impact of revised photometry on the properties of the eclipsing binary star components.

7.3.7 The Well-Studied System AI Phoenicis

This 24.6-day period binary is one of the best-studied field *EB*. Its discovery on sky patrol plates taken at a Remeis-Sternwarte Bamberg Southern Station was announced by Strohmeier (1972). Subsequently, Reipurth (1978) obtained *uvby* data and Imbert (1979) obtained the first radial velocities and analyzed the system. Hrivnak & Milone (1984) obtained the first *UBVRI* light curves and performed the first light curve analyses with the WD program using the blackbody option. This was one of the earlier uses of the program to study an eccentric orbit binary. They found the components to be a mid-F dwarf of mass (1.12 ± 0.03) $M_\odot$ and (1.77 ± 0.03) $R_\odot$ and a late G subgiant of (1.16 ± 0.03) $M_\odot$ and (2.85 ± 0.03) $R_\odot$, respectively. Vandenberg & Hrivnak (1985) investigated the evolutionary state of the system. From the color indices, they deduced the bounds of the metallicity ($Z = 0.0169$ to 0.04) and used these to establish bounds for the helium content and the age: For $Y = 0.33$, $\tau = (4.3 \pm 0.3)$ Gyr, whereas for $Y = 0.43$, $\tau = (2.9 \pm 0.2)$ Gyr. Subsequently, Andersen et al. (1988) made use of additional *uvby* photometry, new radial velocities, and high resolution ($R \approx 50{,}000$) spectra obtained at good S/N ($\sim$200) to reanalyze all existing data with the NDE model, although they do not specify the program they used (presumably therefore a version of EBOP). Andersen et al. (1988) do not show the computed light curves but the radial velocity predictions are seen to be in excellent agreement with the data, and they reported good fittings of the separate passband solutions to the light curve data. They found somewhat larger uncertainties for the elements at least partially because they rejected two points on the rising portion of the light curve that had been accepted by Hrivnak & Milone (1984) and improved precision for the *uvby* light curves. Andersen et al. (1988) concluded from an analysis of their high- resolution spectra that [Fe/H] = -0.14 ± 0.10, and $Z = 0.012 \pm 0.003$; from this they derived $Y = 0.27 \pm 0.02$ and $\tau = (4.1 \pm 0.4)$ Gyr.

Thus, even before the modeling with improved atmospheres options, the elements of *AI Phe* were among the best determined of all evolved systems. Milone et al. (1992b) reanalyzed all previously published data and IUE ultraviolet observations obtained in the primary minimum. The latter had been obtained in order to investigate the limb darkening in the hotter component, which, on the basis of earlier work by Imbert (1979) was thought to be a solar analogue with a spectral type of G2V, and suitable for direct comparison with the solar center-to-limb variation studied by Kjeldseth-Moe & Milone (1978). Milone et al. (1992b) made use of the Kurucz (1979) atmospheres option in the University of Calgary version of the WD program, WD83K83, and obtained a multiwavelength solution consistent with individual passband elements. This modeling used the empirical corrections of Wade & Rucinski (1985) to fit the ultraviolet light curves better than any previous modeling. They found the mass and radius for the hotter and cooler star, respectively, to be (1.190 ± 0.006) $M_\odot$ and (1.762 ± 0.007) $R_\odot$; and (1.231 ± 0.005) $M_\odot$ and (2.931 ± 0.007) $R_\odot$. With the temperature of the hotter component taken as 6310 K ($\pm$150 K, assumed), that of the secondary was determined to be (5151 ± 150) K. Subsequently, modeling with WD93K93 was carried out to test the effects of multiple reflections and slight changes in metallicity with flux files that made use of the atmosphere models by Kurucz (1993). This modeling has produced little improvement at the present writing, but this was expected given the small range of metallicity of the models attempted so far ([Fe/H] = −0.02, 0, +0.02). In the later 1990s *Al Phe* was analyzed with WD95, as a test for this new package. Finally, in Sect. 8.4 we briefly indicate how the distribution of residuals in the *Al Phe* modeling process have been analyzed.

It is useful to summarize why this system has been of such interest. The main reason is that the components have evolved from the main sequence, and at different rates. Their intrinsic properties thus provide the means to test evolution models. A precise determination of the elements is possible in this case because the system is double-lined, and the eclipses are total ($i = 88°.45 \pm 0°.01$) despite the long period of the system, so that the shape of the minima can be observed in detail. The improved modeling has provided more precise values of the radiative and at least as precise values of the nonradiative properties as previous studies, but with improved confidence. Additional reasons for modeling this system are the significant differences between the component's colors and magnitude, due to the relatively cool temperature of the secondary component. The modeling of the flux of the cool secondary is a challenge for stellar atmosphere theory. The far-ultraviolet limb darkening of the hotter star can be studied more easily because of the lack of contribution from the cooler component. Finally, effects seen in other systems involving a late-type subgiant are absent, namely the RS CVn-type behavior, which can complicate the determination of fundamental stellar properties.

Current modeling involves the exploration of the effects of multiple reflection, of nonlinear limb darkening, and of chemical composition. The models tested have thus far not included stars as deficient as −0.1 and −0.2 but this is planned for the near future.

The full sweep of the importance of systems such as *Al Phe* and binaries in clusters is beyond the scope of this book, but we refer to Andersen (1991) for an extensive discussion of the former, and the contributions detailed in Milone & Mermilliod (1996) for the latter.

7.3.8 HP Draconis

The $10^d.76$-day period, eccentric orbit double-lined eclipsing binary *HP Draconis* was among the systems used to test the capability of the GAIA mission to yield fundamental stellar data; cf. Milone et al. (2005). The system was discovered as variable in the HIPPARCOS mission albeit with an incorrect period ($6^d.67$). The test involved using HIPPARCOS and TYCHO photometric data, which were of somewhat lower precision than GAIA was expected to provide, and radial velocities measured from echelle spectra obtained at Asiago Observatory. The mean standard errors for the HIPPARCOS and TYCHO light curves were 0.012 and 0.11 magn., respectively; the spectral resolution is 20,000 and the mse of the RV curve is 3 km/s. The latter is higher than expected for the GAIA spectroscopy but the latter were to include more observations in compensation. The main problem with the photometry was the paucity of data in the minima: only three data points in the primary minimum and seven in the secondary minimum of the *hip* light curve. The TYCHO data were even worse: the minima could not be identified. They were, however, useful for determining colors and thus temperatures; T_1 was taken as 6,000 K. The modeling code was the package WD2002, in which the simplex program and self-iterating damped least-squares routines were used. The converged solution was able to yield very good results for some parameters and excellent if preliminary results, for others. Among the preliminary results were a significant dω/dt term.

Independently, Kurpinska–Winiarska et al. (2000) with photometry from Cracow Observatory determined the system to be eccentric. The Cracow photometry consisted of *B* and *V* light curves with complete coverage of the minima and this group obtained additional radial velocities from the ELODIE spectrograph on a telescope at the Haute Provence Observatory.

In 2008–2009 all available data except the TYCHO set were analyzed with the Wilson–Devinney program, 2007 version. Nineteen parameters were simultaneously adjusted, 13 of which are curve independent: a, ϵ, ω, γ, T_2, i, $\Omega_{1,2}$, q, t_0, P, dP/dt, dω/dt, L_1(B,V,hip), ℓ_3(B,V,hip). Initial values were taken from the solutions obtained by Milone et al. (2005) and Kurpinska–Winiarska et al. (2000). This program is not self-iterating and the method of non-correlating subsets was used to improve fittings from run to run. Solar composition was assumed for the stellar atmospheres corrections, which are completely internal in this version. Several thousand runs were carried out in several series. The lowest SSRs were obtained with a set of elements that included a significant third light component ineach of the *B*,

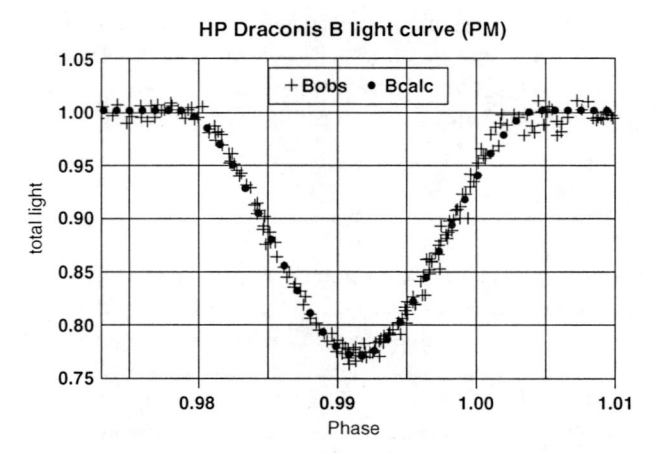

Fig. 7.6 The fitting of the *B* primary minimum of the *HP Draconis* light curve. The data are from Cracow Observatory. Courtesy, M. Kurpinska-Winiarska and E. Oblak.

7.3 Applications and Extensions

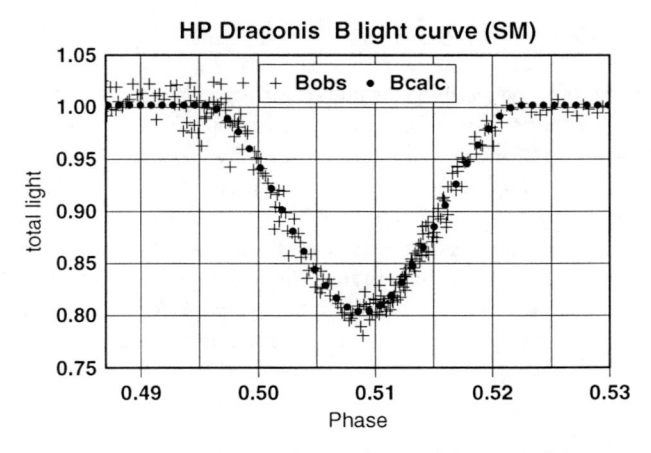

Fig. 7.7 The fitting of the B secondary minimum of the *HP Draconis* light curve. The data are from Cracow Observatory. Courtesy, M. Kurpinska-Winiarska and E. Oblak

V, and hip passbands and there is a marginally significant $d\omega/dt$ term. Figures 7.6 and 7.7 show the final B light curve fitting at primary and secondary minimum, respectively.

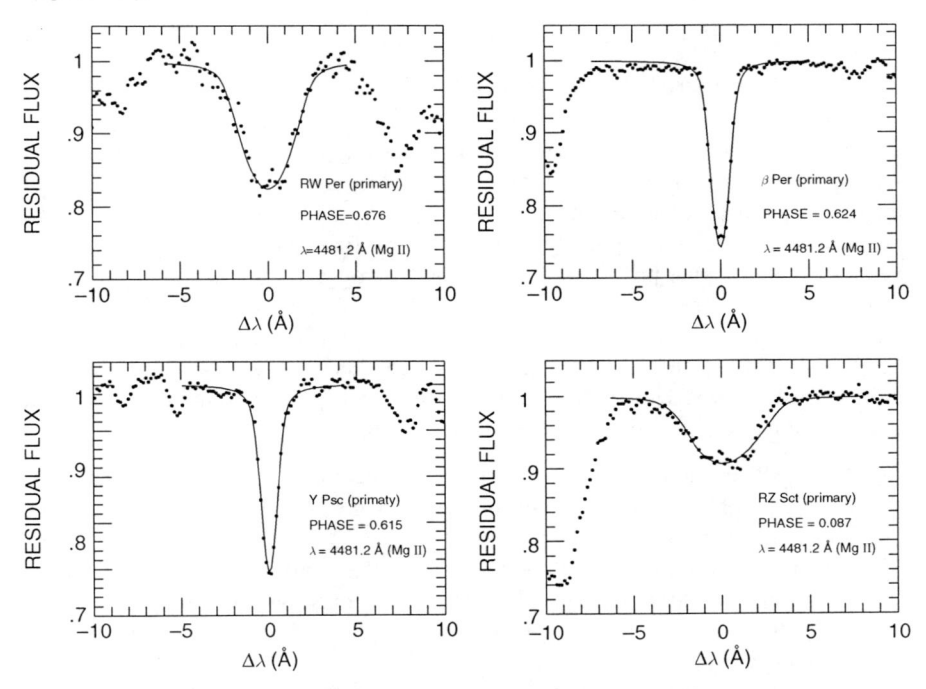

Fig. 7.8 Fitting line profiles. This figure, Fig. 6 in Mukherjee et al. (1996), shows fitted versus observed profiles. Courtesy J. D. Mukherjee

7.3.9 Fitting of Line Profiles

Although it never became available as public software, Mukherjee et al. (1996) combined the theory of stellar line broadening for local profiles with the WD program and used it to estimate rotation rates of Algol binaries by fitting line profiles to observed data. They used the Simplex algorithm and the method of Differential Corrections to adjust the damping constant Γ, number N_f of absorbers along the line-of-sight, turbulent velocity v^{tur}, absorption versus scattering parameter ε, and the rotation parameter F_1 for the primary star. Figure 7.8 shows the observed and fitted line profiles for some rapidly (*U Cep, S Cnc*) and slowly (*RZ Cas, TV Cas*) rotating Algols. For the full analysis we refer the reader to Mukherjee et al. (1996).

7.4 The Future

In futuro ...

This section looks into the future of the WD program. We keep the *Future as Seen in 1999* (the year when the first edition was printed), add comments [in square brackets] where significant progress has been made, and then lend a current perspective on future developments.

7.4.1 "The Future" as Envisioned in 1999

The history of the WD program has been one of various special purpose versions that were developed for particular problems, followed by absorption of their capabilities into the general program. For example, it is anticipated that the 1998 version computes light curves, radial velocity curves, spectral line profiles, and images, while versions that compute polarization curves and X-ray pulse arrival times now exist separately and eventually will be absorbed. Generalizations that a user need not worry about are embedded invisibly wherever practical. For example, computational shortcuts for many special case situations speed execution without compromising more intricate cases. The 1998 version (Wilson 1997; private communication) will have the following changes/additions vis-à-vis 1992:

1. The model can have circumstellar scattering regions (see Sect. 3.4.4) that attenuate star light by several scattering mechanisms. A first application is to *AX Monocerotis* (Elias et al. 1997).
2. Spectral line profiles can be computed, with the various proximity and eclipse effects and blending. Line profiles can be associated with specific regions on a star. This feature permits analysis of chromospheric fluorescence and also lines from star spots.
3. Either time or phase can be the independent variable. We can thereby solve for ephemeris parameters, including time derivatives of the period and argument of periastron, and can combine data from several epochs.

7.4 The Future

4. Simulated observational error can now be added to light curves so as to facilitate solution tests on synthetic data.
5. Output needed to make pictures of a binary, including spots, is now provided. The data can be used as input to any commercial or private plotting program.
6. The Levenberg–Marquardt procedure can be applied by entering a nonzero value for the damping constant λ.
7. The program is now entirely in double precision.
8. Input and output formats have been revised so as to assure entirely adequate numbers of digits, even for rare and extreme cases. Some quantities that previously were in floating format are now in exponent format.
9. Because of the options to compute additional kinds of quantities, a control integer now determines the LC input/output format and triggers certain decisions about what will be computed. The output is thereby easier to read because LC does not try to squeeze everything into the page width.
10. Because of a preference in the literature for standard deviations, as opposed to probable errors, the parameter error estimates are now standard deviations.

Several improvements await incorporation:

- Polarimetry is a ripe field for exploitation in light curve analysis. The data are scarce but the means to incorporate them into analytical methods could encourage further observational progress in this demanding field. It is expected that in a few years the WD light curve program will support the analysis of polarimetry data. [NB.: The development and deployment of ESPaDOnS (for *Echelle SpectroPolometric Device for the Observation of Stars*) on the Canada–France–Hawaii Telescope (CFHT) will be a boon for polarimetric studies of *EB*s, both hot and cool systems. The instrument is capable of producing spectra from 0.37 to 1.00 μm at a resolution of 50,000. The study of stellar magnetic properties has experienced vigorous growth over the past two decades, and with an instrument such as this on a major facility at a site with excellent seeing, continued growth of the field seems assured (assuming that practitioners are granted time by telescope allocation committees). A number of investigations of rapidly rotating early- and late-type stars have been studied with this instrument, yielding information about magnetic field structures. It will be interesting to see such investigations carried out on binary stars.]
- Atmospheric eclipses for components with extended atmospheres.
- Improved accuracy, e.g., light curve quadrature and spot geometry.

An accuracy improvement for WD is on the way, but is intricate and not yet debugged. At present the work is a "back burner" project. The accuracy improvement will be major and might be used either to reduce error or to reduce execution time for the same level of precision. On a shorter timescale, we can expect line profile improvements such that several broadening mechanisms (e.g., thermal and turbulent Doppler broadening, damping) will be included or the capability to fit spectral line profile parameters. Also, the circumstellar scattering regions mentioned above will be made to scatter starlight into the line-of-sight, in addition to their present function as attenuating clouds. Clouds might be treated on the basis of rigid hydrodynamics. The effects of streams in Algol systems on light curves have been modeled outside WD by several workers, cf. Terrell (1994), who developed a hydrodynamic code. This work could be combined with cloud modeling in a more extended module (see Sect. 9.1.1 for a definition of modular structure and modules). Having clouds in the models also requires us to model Thomson scattering and other types of scattering and absorption. Although the program has a simple stellar atmosphere capability (main sequence stars only), the atmosphere provisions of some other programs [e.g., Linnell (1991), Milone et al. (1992b)] are major improvements and very important. Therefore it is anticipated that a relatively general atmosphere routine will be added to the WD program at some point. [NB.: The improvement of the least-squares engine to a damped least-squares one has been a major advance. Eclipses caused by discrete clouds in atellar atmospheres can now be carried out in WD programs. Kurucz atmospheres are now standard in the most recent WD programs and have been directly incorporated into the code instead of relying on auxiliary files of model atmosphere to blackbody fluxes as had to be the case in the WD98k93 and W98 package programs developed by Milone, Kallrath, and collaborators. There still need to be auxiliary files for the individual metallicities, however. Wilson's own 2007 version of WD incorporates ramp functions to move between regions for which Kurucz models are applicable and those where black bodies must be used. The atmospheric models are available for a wide variety of metallicities and can be applied for a large number of passbands.]

7.4.2 The Future (as Seen in 2009)

In futuro ...

In addition to the improvements described above that are only partially implemented at present, a number of additional improvements are in the process of being implemented and still others can be foreseen. These include the following:

Direct distance estimation as described in Sect. 5.1.4 and Wilson (2008) will be available in the 2009 version of the WD program. This is based on absolute calibrated light curves which permits the determination of the temperatures of both components and the comparison of computed absolute, distance-dependent flux with the observed flux. This feature also allows the fitting of the interstellar extinction attenuation, A_λ, i.e., for specific passbands at effective wavelength λ.

Inverse distance estimation exploits distances obtained, for example, by the HIPPARCOS and GAIA missions to constrain the solution and to reduce correlations among parameters.

Analysis of an eclipsing binary component with pulsations: Future editions of the WD program may include the physics of *pulsating variable stars*. Progress toward such a goal is in preparation by Wilson (2009).

Discontinuous period changes (Wilson 2005) also discusses possible implementation of discontinuous period changes.

Continuum and spectral line polarization modeling: This feature is available in a new version of WD under development by Wilson.

Addition of other adjustable parameters: There is provision in the 2007 version of WD to include more adjusted parameters. One of the most useful of these could be di/dt. It has been found in several eclipsing systems that the inclination varies with time, due to the presence of a third body in the dynamical system of the binary. This situation proved to be the case with the system *SS Lacertae*, for example. A group of parameters that could be useful are those that stipulate the shape of spotted regions on the stars. In his program LIGHT2 described in Sect. 6.3.3, Graham Hill incorporated elliptical spots. Spot groups may be represented better by such a shape than a circular set of spots. The major axis size could replace the current spot radius, but the ellipticity and major axis orientation would have to be adjustable also, or at least they would need to be specifiable.

Incorporation of stellar tomography into WD codes. The discussion of the SHELLSPEC software in Sect. 2.5 indicated how powerful a tool this can be in the case where specific spectral features can be associated with specific regions in a binary star system. It would be very useful to be able to incorporate such a tool into general light curve analysis software.

Facilitation of Passband additions. In the wd98k93 program and the WD98 package, instructions were provided to create new "flux files," so that additional passband data could be modeled. In the current WDx2007 program, however, the passbands are "hard-wired" so that alterations to the program are required to add new passbands. This cannot be done frequently, and pressure to do so would create a nightmare for Robert Wilson and Walter Van Hamme who have already spent a great deal of time and effort to make the program as self-contained as possible with regard to stellar atmosphere application. Therefore, it would be useful to allow additional passbands to be added by the user.

References

Andersen, J.: 1991, *Accurate Properties of Normal Stars: Determination and Applications*, PhD Dissertation, University of Copenhagen, Copenhagen University Observatory

Andersen, J., Clausen, J. V., Gustafsson, B., Nordström, B., & Vandenberg, D. A.: 1988, Absolute Dimensions of Eclipsing Binaries. XIII. AI Phoenicis: A Case Study in Stellar Evolution, *A&A* **196**, 128–140

Avni, Y.: 1976, The Eclipse Duration of the X-Ray Pulsar 3U 0900-40, *ApJ* **209**, 574–577

Bergbusch, P. A. & Vandenberg, D. A.: 1992, Oxygen-Enhanced Models for Globular Cluster Stars. II. Isochrones and Luminosity Functions, *ApJ Suppl.* **81**, 163–220

Bergbusch, P. A. & VandenBerg, D. A.: 2001, Models for Old, Metal-poor Stars with Enhanced α-Element Abundances. III. Isochrones and Isochrone Population Functions, *ApJ* **556**, 322–339

Cudworth, K. M.: 1985, Photometry, Proper Motions, and Membership in the Globular Cluster M71, *AJ* **90**, 65–73

Elias, N. M., Wilson, R. E., Olson, E. C., Aufdenberg, J. P., Guinan, E. F., Gudel, M., Van Hamme, W., & Stevens, H. L.: 1997, New Perspectives on AX Monocerotis, *ApJ* **484**, 394–411

Gilliland, R. L., Brown, T. M., Guhathakurta, P., Sarajedini, A., Milone, E. F., Albrow, M. D., Baliber, N. R., Bruntt, H., Burrows, A., Charbonneau, D., Choi, P., Cochran, W. D., Edmonds,

P. D., Frandsen, S., Howell, J. H., Lin, D. N. C., Marcy, G. W., Mayor, M., Naef, D., Sigurdsson, S., Stagg, C. R., Vandenberg, D. A., Vogt, S. S., & Williams, M. D.: 2000, A Lack of Planets in 47 Tucanae from a Hubble Space Telescope Search, *ApJ* **545**, L47–L51

Hrivnak, B. J. & Milone, E. F.: 1984, Observations, Analyses, and Absolute Parameters of the Evolved Binary AI Phoenicis, *ApJ* **282**, 748–757

Imbert, M.: 1979, Orbite Spectroscopique et Dimensions de la Binaire à Éclipse AI Phe, *A&A* **36**, 453–456

Kallrath, J.: 1987, Analyse von Lichtkurven enger Doppelsterne, Diploma thesis, Rheinische Friedrich-Wilhelms Universität Bonn, Sternwarte der Universität Bonn

Kallrath, J.: 1993, Gradient Free Determination of Eclipsing Binary Light Curve Parameters – Derivation of Spot Parameters Using the Simplex Algorithm, in E. F. Milone (ed.), *Light Curve Modeling of Eclipsing Binary Stars*, pp. 39–51, Springer, New York

Kallrath, J. & Kämper, B.-C.: 1992, Another Look at the Early-Type Eclipsing Binary BF Aurigae, *A&A* **265**, 613–625

Kallrath, J. & Linnell, A. P.: 1987, A New Method to Optimize Parameters in Solutions of Eclipsing Binary Light Curves, *Astrophys. J.* **313**, 346–357

Kallrath, J., Milone, E. F., & Stagg, C. R.: 1992, Modeling of the Eclipsing Binaries in the Globular Cluster NGC 5466, *ApJ* **389**, 590–601

Kallrath, J., Milone, E. F., Terrell, D., & Young, A. T.: 1998, Recent Improvements to a Version of the Wilson-Devinney Program, *Astrophys. J.* **508**, 308–313

Kjeldseth-Moe, O. & Milone, E. F.: 1978, Limb Darkening 1945-3245 Å for the Quiet Sun from *Skylab* Data, *ApJ* **226**, 301–314

Kurpinska–Winiarska, M., Oblak, E., Winiarski, M., & Kundera, T.: 2000, Observations of Two HIPPARCOS Eclipsing Variables, *Information Bulletin on Variable Stars* **4823**, 1–3

Kurucz, R. L.: 1979, Model Atmospheres for G, F, A, B, and O Stars, *ApJ Suppl.* **40**, 1–340

Kurucz, R. L.: 1993, New Atmospheres for Modelling Binaries and Disks, in E. F. Milone (ed.), *Light Curve Modeling of Eclipsing Binary Stars*, pp. 93–102, Springer, New York

Linnell, A. P.: 1991, A Light Synthesis Study of W Ursae Majoris, *ApJ* **374**, 307–318

Mateo, M., Harris, H. C., Nemec, J., & Olszewski, E. W.: 1990, Blue Stragglers as Remnants of Stellar Mergers: The Discovery of Short-Period Eclipsing Binaries in the Globular Cluster NGC 5466, *AJ* **100**, 469–484

Mateo, M. & Yan, L.: 1996, Errata: Primordial Main Sequence Binary Stars in the Globular Cluster M71, *AJ* **111**, 567

McVean, J. R., Milone, E. F., Mateo, M., & Yan, L.: 1997, Analyses of the Light Curves of the Eclipsing Binaries in the Globular Cluster M71, *ApJ* **481**, 782–794

Milone, E. F. (ed.): 1993, *Light Curve Modeling of Eclipsing Binary Stars*, Springer, New York

Milone, E. F. & Kallrath, J.: 2008, Tools of the Trade and the Products they Produce: Modeling of Eclipsing Binary Observables, in E. F. Milone, D. A. Leahy, & D. W. Hobill (eds.), *Short-Period Binary Stars: Observations, Analyses, and Results*, Vol. 352 of *Astrophysics and Space Science Library*, pp. 191–214, Springer, Dordrecht, The Netherlands

Milone, E. F. & Mermilliod, J.-C. (eds.): 1996, *The Origins, Evolution, and Destinies of Binaries in Clusters*, No. 90 in PASP Conference Series, San Francisco, Astronomical Society of the Pacific

Milone, E. F. & Terrell, D.: 1996, Analysis of the Contact System H235 in the Intermediate-Age Open Cluster NGC 752, in E. F. Milone & J.-C. Mermilliod (eds.), *The Origins, Evolution, and Destinies of Binary Stars in Clusters*, pp. 283–285, A.S.P. Conference Series, Provo, UT

Milone, E. F., Stagg, C. R., & Kallrath, J.: 1992a, The Eclipsing Binaries in NGC 5466 and Implications for Close Binary Evolution, in Y. Kondo (ed.), *Evolutionary Processes in Interacting Binary Stars*, pp. 483–486, IAU, The Netherlands

Milone, E. F., Stagg, C. R., & Kurucz, R. L.: 1992b, The Eclipsing Binary AI Phoenicis: New Results Based on an Improved Light Curve Analysis Program, *ApJ Suppl.* **79**, 123–137

Milone, E. F., Wilson, R. E., & Hrivnak, B. J.: 1987, RW Comae Berenicis. III. Light Curve Solution and Absolute Parameters, *ApJ* **319**, 325–333

Milone, E. F., Groisman, G., Fry, D. J. I. F., & Bradstreet, H.: 1991, Analysis and Solution of the Light and Radial Velocity Curves of the Contact Binary TY Bootis, *ApJ* **370**, 677–692

Milone, E. F., Kallrath, J., Stagg, C. R., & Williams, M. D.: 2004, The Modeling of Binaries in Globular Clusters, *Revista Mexicana AA (SC)* **21**, 109–115

Milone, E. F., Munari, U., Marrese, P. M., Williams, M. D., Zwitter, T., Kallrath, J., & Tomov, T.: 2005, Evaluating GAIA performance on eclipsing binaries: IV. Orbits and stellar parameters for SV Cam, BS Dra and HP Dra, *Astronomy and Astropyhsics* **441**, 605–613

Milone, E. F., Stagg, C. R., Sugars, B. A., McVean, J. R., Schiller, S. J., Kallrath, J., & Bradstreet, D. H.: 1995, Observations and Analysis of the Contact Binary H235 in the Open Cluster NGC 752, *AJ* **109**, 359–377

Mukherjee, J. D., Peters, G. J., & Wilson, R. E.: 1996, Rotation of Algol Binaries – A Line Profile Model Applied to Observations, *MNRAS* **283**, 613–625

Nelson, R. H., Milone, E. F., VanLeeuwen, J., Terrell, D., Penfold, J. E., & Kallrath, J.: 1995, Observations and Analysis of the Field Contact Binary V728 Herculis, *AJ* **110**, 2400–2407

Nemec, J. M. & Harris, H. C.: 1987, Blue Straggler Stars in the Globular Cluster NGC 5466, *ApJ* **316**, 172–188

Ögelman, H., Beuermann, K. P., Kanbach, G., Mayer-Hasselwander, H. A., Capozzi, D., Fiordilino, E., & Molteni, D.: 1977, Increase in the Pulsational Period of 3U0900-40, *A&A* **58**, 385–388

Petro, L. D. & Hiltner, W. A.: 1974, Optical Observations of HD 77581 and a Model for the System HD 77581-2U 0900-40, *ApJ* **190**, 661–666

Prša, A. & Zwitter, T.: 2005b, A Computational Guide to Physics of Eclipsing Binaries. I. Demonstrations and Perspectives, *ApJ* **628**, 426–438

Rappaport, S., Joss, P. C., & Stothers, R.: 1980, The Apsidal Motion Test in 4U 0900-40, *ApJ* **235**, 570–575

Rasio, F. A.: 1995, The Minimum Ratio of W Ursae Majoris Binaries, *ApJ* **444**, L41–L43

Rasio, F. A. & Shapiro, S.: 1994, Hydrodynamics of Binary Coalescence. I. Polytropes with Stiff Equations of State, *ApJ* **432**, 242–261

Rasio, F. A. & Shapiro, S.: 1995, Hydrodynamics of Binary Coalescence. I. Polytropes with $\Gamma = 5/3$, *ApJ* **438**, 887–903

Reipurth, B.: 1978, Photometry of AI Phe, *Inform. Bull. Variable Stars* **1419**, 1–2

Stagg, C. R. & Milone, E. F.: 1993, Improvements to the Wilson–Devinney Code on Computer Platforms at the University of Calgary, in E. F. Milone (ed.), *Light Curve Modeling of Eclipsing Binary Stars*, pp. 75–92, Springer, New York

Stickland, D., Lloyd, C., & Radziun-Woodham, A.: 1997, The Orbit of the Supergiant of Vela-X1 derived from *IUE* Radial Velocities, *MNRAS* **286**, L21–L24

Strohmeier, W.: 1972, Three New Bright Eclipsing Binaries, *Inform. Bull. Variable Stars* **665**, 1–3

Terrell, D.: 1994, Circumstellar Hydrodynamics and Spectral Radiation in Algols, PhD thesis, Department of Astronomy, University of Florida, Gainesville, FL

Van der Klis, J. & Bonnet-Bidaud, J. M.: 1984, The Orbital Parameters and the X-Ray Pulsation of Vela X-1 (4U 0900-40), *A&A* **135**, 155–170

Van Genderen, A. M.: 1981, A Discussion on VBLUW Photometry of the X-Ray Binary HD 77581 (=Vela X-1 =3 U 0900-40) and on the Overluminosity of the Primaries in X-Ray Binaries. The Optical Micro Variability of the Hot Supergiant Primaries HD 77581 and HD 153919, *A&A* **96**, 82–90

Van Hamme, W.: 1993, New Limb-Darkening Coefficients for Modeling Binary Star Light Curves, *AJ* **106**, 2096–2117

Van Hamme, W. & Wilson, R. E.: 2003, Stellar Atmospheres in Eclipsing Binary Models, in U. Munari (ed.), *GAIA Spectroscopy: Science and Technology*, Vol. 298 of *Astronomical Society of the Pacific Conference Series*, pp. 323–328, San Francisco

Van Kerkwijk, M. H., Van Paradijs, J., Zuiderwijk, E. J., Hammerschlag-Hensberge, G., Kaper, L., & Sterken, C.: 1995, Spectroscopy of HD77581 and the Mass of Vela X-1, *A&A* **303**, 483–496

Van Paradijs, J., Zuiderwijk, E. J., Takens, R. J., Hammerschlag-Hensberge, G., Van den Heuvel, E. P. J., & Lorre, C. D.: 1977, The Spectroscopic Orbit and the Masses of the Components of the Binary X-Ray Source 3U0900-40/HD 77581, *A&A Suppl.* **30**, 195–211

VandenBerg, D. A.: 2000, Models for Old, Metal-Poor Stars with Enhanced α-Element Abundances. II. Their Implications for the Ages of the Galaxy's Globular Clusters and Field Halo Stars, *ApJ Suppl.* **129**, 315–352

Vandenberg, D. A. & Hrivnak, B. J.: 1985, The Age and Helium Content of the Eclipsing Binary AI Phoenicis, *ApJ* **291**, 270

VandenBerg, D. A., Richard, O., Michaud, G., & Richer, J.: 2002, Models of Metal-poor Stars with Gravitational Settling and Radiative Accelerations. II. The Age of the Oldest Stars, *ApJ* **571**, 487–500

Wade, R. A. & Rucinski, S. M.: 1985, Linear and Quadratic Limb-darkening Coefficients for a Large Grid of LTE Model Atmospheres, *A&A Suppl.* **60**, 471–484

Watson, M. G. & Griffiths, R. E.: 1977, Ariel V Sky Survey Instrument: Extended Observations of 3U 0900-40, *MNRAS* **178**, 513–524

Wilson, R. E.: 2005, EB Light Curve Models – What's Next? *Ap. Sp. Sci.* **296**, 197–207

Wilson, R. E.: 2008, Eclipsing Binary Solutions in Physical Units and Direct Distance Estimation, *ApJ* **672**, 575–589

Wilson, R. E.: 2009, *Modeling Intrinsic Variable Stars into Eclipsing Binary Programs*, private communication

Wilson, R. E. & Devinney, E. J.: 1971, Realization of Accurate Close-Binary Light Curves: Application to MR Cygni, *ApJ* **166**, 605–619

Wilson, R. E. & Terrell, D.: 1994, Sub-Synchronous Rotation and Tidal Lag in HD 77581/Vela X-1, in S. S. Holt & C. S. Day (eds.), *The Evolution of X-Ray Binaries*, No. 308 in AIP Conference Proceedings, pp. 483–486, AIP, American Institute of Physics, Woodbury, NY

Wilson, R. E. & Terrell, D.: 1998, X-Ray Binary Unified Analysis: Pulse/RV Application to Vela X1/GP Velorum, *MNRAS* **296**, 33–43

Yan, L. & Mateo, M.: 1994, Primordial Main Sequence Binary Stars in the Globular Cluster M71, *AJ* **108**, 1810–1827

Young, A. T., Milone, E. F., & Stagg, C. R.: 1994, On Improving IR Photometric Passbands, *A&A Suppl.* **105**, 259–279

Zuiderwijk, E. J.: 1995, The Rotation Period of HD 77581 (Vela X-1), *A&A* **299**, 79–83

Chapter 8
Light Curve Software with Graphical User Interface and Visualization

ad unguem (*to a fingernail; exactly; nicely done*)

In this chapter we summarize the approach and current contributions to this area by a number of authors. Here, however, direction rather than specific packages must be emphasized because this subfield is rapidly changing.

Graphics and visual support[1] include the plotting of light curves, graphing of the fit and residuals, providing projected views of the components, and sometimes the distribution of such physical quantities as surface brightness, spots, and prominences. Three-dimensional and virtual reality visualizations will be discussed.

8.1 Binary Maker

Binary Maker is a commercially available software package developed by David Bradstreet (1993); Bradstreet and Steelmans (2004) (Eastern College, Pennsylvania) to visualize light and radial velocity curves and the appearance of the system itself with varying phase. It permits to plot the Roche potentials and the outer and inner Lagrangian surfaces in either Kopal or modified Kopal potentials and the calculation of several quantities among them: the radii of the components in back, side, pole and point facings, surface areas and volumes, mean densities, locations of the Lagrangian points L_1^p and L_2^p, and the fill-out factor. Most importantly, it computes and plots synthetic light and radial velocity curves. It can also be used to derive the mass ratio from observed radial velocity curves, and other parameters, through trial approximations of light curve fittings to data.

Binary Maker can plot observed light and radial velocity data for comparison with computed light and radial velocity curves. It also computes and plots spectral line profiles at selected phases. Generated light curve and radial velocity curve data points can be simultaneously displayed with the projected three-dimensional view of the system model replete with star spots and in the correct orientation

---

[1] In the late 1990s Dirk Terrell (University of Florida, now at the Southwest Research Institute, Boulder, CO) distributed a Wilson–Devinney program which has a user-friendly I/O interface running under Microsoft Windows.

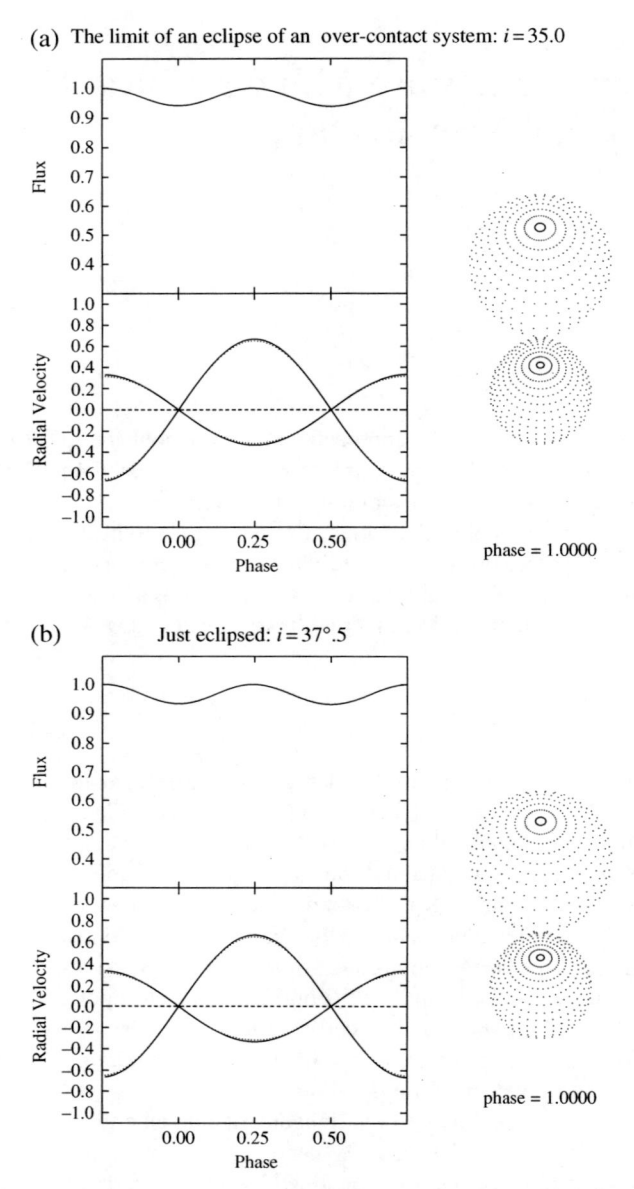

Fig. 8.1 The limit of an eclipse of an over-contact system. Created with the help of BM2 (Bradstreet 1993)

toward the observer. The surface elements may be shown in both spherical and cylindrical coordinates. One of its most interesting features is the demonstration of the effects of star spot regions on the light and radial velocity curves and on the profiles.

8.1 Binary Maker

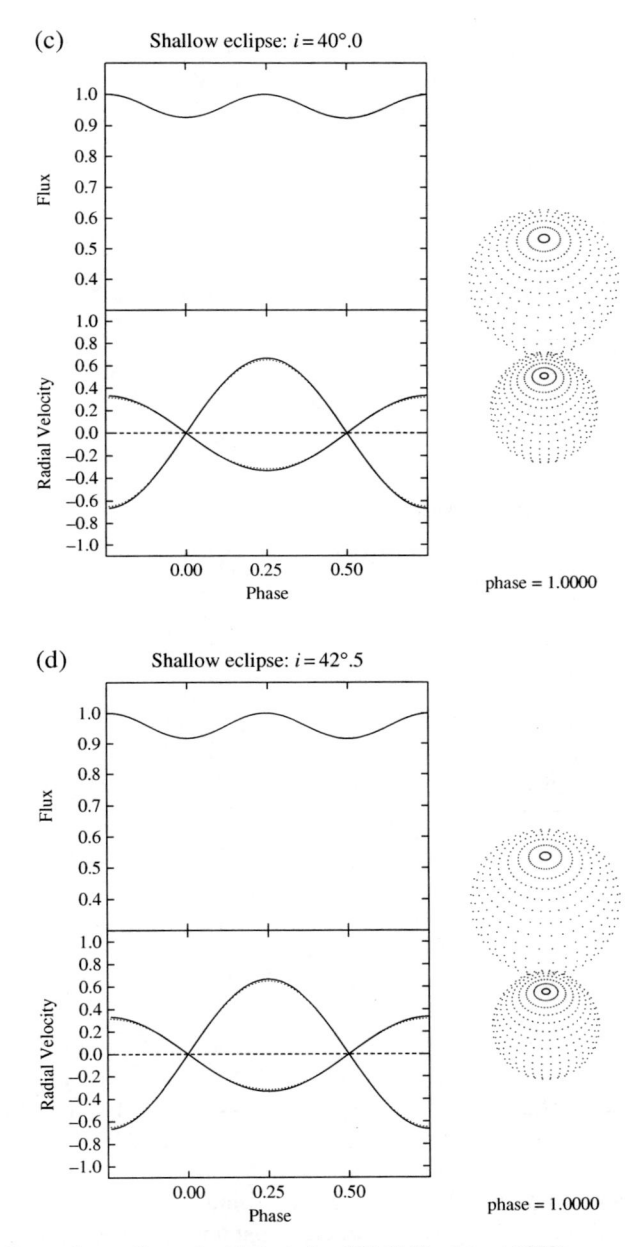

(c) Shallow eclipse: $i = 40°.0$

phase = 1.0000

(d) Shallow eclipse: $i = 42°.5$

phase = 1.0000

Fig. 8.2 Shallow eclipses. Created with the help of BM2 (Bradstreet 1993)

The present version (3.0) at the time of this writing does not make use of the 1992 WD or WD93K93 program features such as asynchronous rotation, nonlinear limb-darkening, multiple reflection, or Kurucz atmosphere calculations. However,

(e) Just eclipse: $i = 90°.0$ wuthout third light

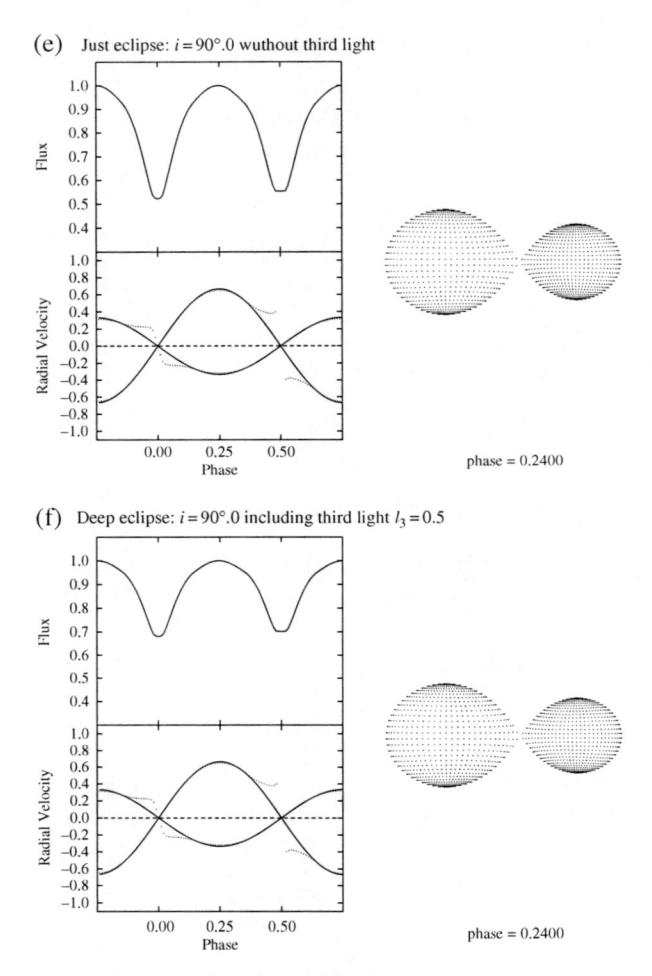

(f) Deep eclipse: $i = 90°.0$ including third light $l_3 = 0.5$

phase = 0.2400

Fig. 8.3 Deep eclipses. Created with the help of BM2 (Bradstreet 1993)

most of the features available in the pre-1979 version of the WD program are included. Unlike BM2, however, eccentric orbits are treated and BM3 runs on a variety of platforms, including Windows on PCs, Unix, and Macintosh computers.

The program has many uses for both research and teaching. It is perhaps most valuable for demonstrating the effects on the synthetic light and radial velocity curves and profiles of changes of particular parameters. The manual has been updated and defines many of the classical WD program features. We cannot recommend it strongly enough; indeed, we have illustrated many of the useful features of Binary Maker 3.0 repeatedly throughout this book, and both the previous and present editions have made use of BM2 illustrations.

8.2 PHOEBE

In Figs. 8.1–8.3 we show some pictures produced with Binary Maker 2.0 showing the light curve, radial velocity curve, and three-dimensional shape of an eclipsing binary system. They were produced to determine, for predefined geometry, the smallest inclination which leads to an eclipse.

The following parameters have been used to produce the plots:

q	$\Omega_1 = \Omega_1$	λ [nm]	T_1 [K]	$g_1 = g_2$	$x_1 = x_2$	$A_1 = A_2$	ℓ_3
0.5	2.875	550	5500	0.32	0.6	0.5	0

The normalization phase has been set to 0.25 and the phase increment was 0.01. For the first four figures we used the latitude and longitude grid numbers 10 and 20, while for the last two figures we chose 15 and 30.

8.2 PHOEBE

PHOEBE (PHysics Of Eclipsing BinariEs) by Prša & Zwitter (2005b) is a modeling package for eclipsing binary stars, built on top of WD program (Wilson & Devinney 1971, Wilson 1979). The introductory paper by Prša & Zwitter (2005b) overviews most important scientific extensions (incorporating observational spectra of eclipsing binaries into the solution-seeking process, extracting individual temperatures from observed color indices, main sequence constraining and proper treatment of the reddening), numerical innovations (suggested improvements to WD's Differential Corrections method, the new Nelder & Mead's downhill Simplex method), and technical aspects (back-end scripter structure, graphical user interface). While PHOEBE retains 100% WD compatibility, its add-ons are a powerful way to enhance WD by encompassing even more physics and solution reliability. The operability of all these extensions is demonstrated on a synthetic main sequence test binary; applications to real data will be published in follow-up papers. PHOEBE is released under the GNU General Public License, which guaranties it to be free, open to anyone interested to join in on future development.

PHOEBE started as a wrapper, but the authors characterize it now as a standalone *EB* modeling suite *based* on the WD model with several physical enhancements among them:

- color indices as indicators of individual temperatures (color constraints);
- spectral energy distribution (SED) as independent data source;
- main sequence constraints; and
- interstellar and atmospheric extinction.

PHOEBE is built in three layers: The lowermost layer is the modeling engine, currently employing WD. On top of it is the extension layer, where all scientific, numerical, and technical extensions are incorporated. The topmost layer is the user interface layer, which serves as a bridge between the user and the model. PHOEBE uses a scripting language especially designed for modeling eclipsing binaries and giving a lot of flexibility to the user. Although the scripter is currently being rewritten into python, we give an example of its syntax:

```
open_parameter_file ("input.phoebe")
mark_for_adjustment (phoebe_incl, 1)
set res = minimize_using_dc ()
print res
adopt_minimizer_results (res)
```

PHOEBE currently uses WD as its back-end, but it can accommodate any physical model instead or in addition to WD. It is written in ANSI C, which makes it fully portable to any platform and any compiler around. Finally, it features a full-fledged graphical user interface displayed in Fig. 8.4, which brings intuitivity and ease of clicking to the *EB* community.

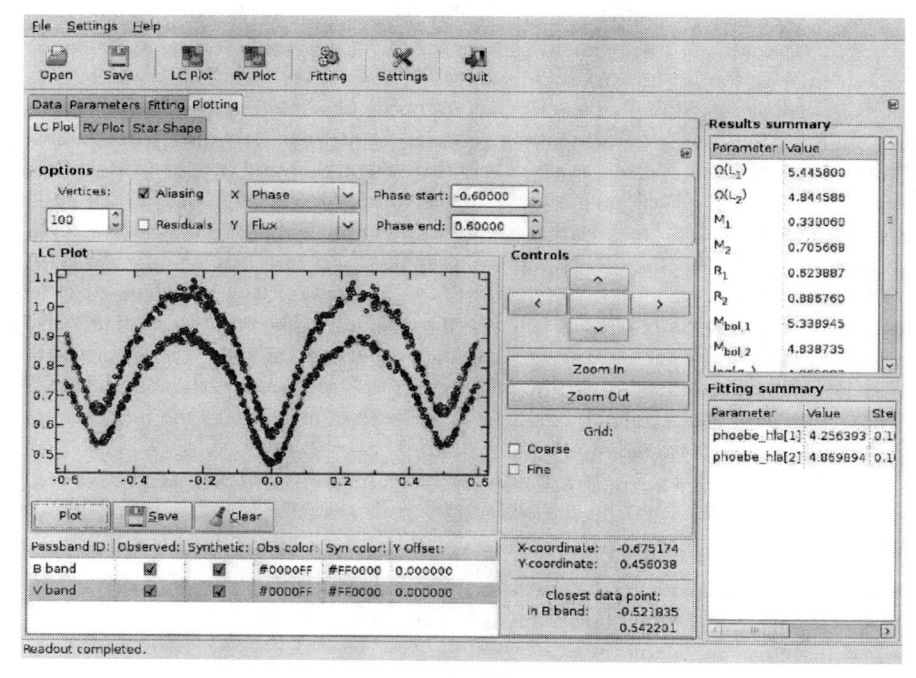

Fig. 8.4 A screen-shot of the PHOEBE graphical user interface. Courtesy Andrej Prša, Villanova University, Villanova, PA

PHOEBE uses a synthetic spectra database to test whether flattened, wavelength-calibrated spectra match synthetic spectra within a given level of significance. Originally, PHOEBE used the Zwitter et al. (2004) grid, but now the internal grid is based on Castelli & Kurucz's (2004) NEWODF SEDs. Accordingly, the scheme to compute limb-darkening coefficients was changed. The computation of stellar spectra of distorted stars is fully supported. The SED analysis is capable to find the values of physical parameters that have not usually been attainable by light- and RV-curve analyses, namely, metallicity and rotational velocity (see Terrell et al. 2003). In Sect. 5.1.2.2 SEDs are used to derive effective temperatures from color indices.

8.3 NIGHTFALL

NIGHTFALL by Wichmann (2002) is a freely available amateur code[2] for modeling eclipsing binary stars. It supports a large range of binary star configurations, including over-contact (common envelope) systems, eccentric (noncircular) orbits, mutual irradiance of both stars (reflection effect), surface spots and asynchronous rotation (stars rotating slower or faster than the orbital period), and the possible existence of a third star in the system. It allows the user to produce animated views of eclipsing binary stars, calculate synthetic light curves and radial velocity curves, and eventually determine the best-fit model for a given set of observational data of an eclipsing binary star system.

8.4 Graphics Packages

Douglas Phillips (University of Calgary) has been working with a number of useful visualization packages. The standard plotting tool for the University of Calgary workstations is XMGR, for two-dimensional plots, and AVS for three-dimensional plots. Figures 8.5 and 8.6 show high- and low-angle views of the phase and wavelength dependence of the residuals of intermediate modeling of the binary *Al Phoenicis* discussed in Sect. 7.3.7 and illustrates the power of the three-dimensional visualization techniques.

Figures 8.5 and 8.6 represent two viewings of the residuals plotted against phase for a succession of passbands. They therefore represent the spectra of the residuals for each phase, and can be used, in certain instances, to gauge the physical mechanism for the residuals if they are nonrandom. This case represents an unconverged solution and is shown only to illustrate the technique only.

---

[2] Nightfall can be downloaded from http://www.hs.uni-hamburg.de/DE/Ins/Per/Wichmann/Nightfall.html. It runs only on Linux platforms.

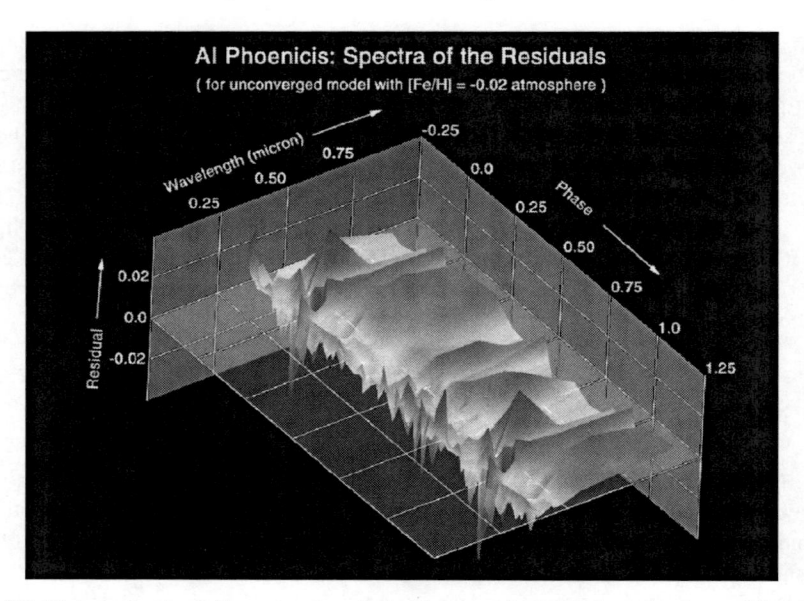

Fig. 8.5 High-angle, three-dimensional analysis of the residuals. High-angle views of the phase and wavelength dependence of residuals

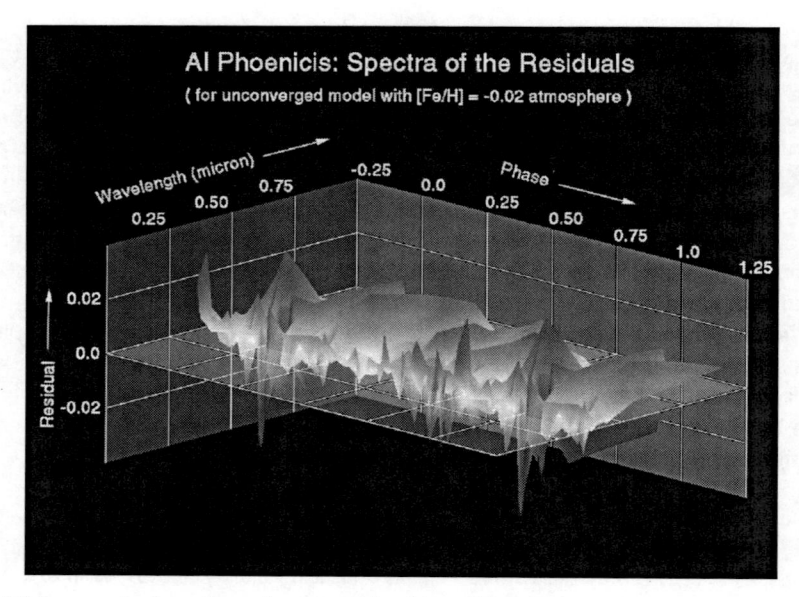

Fig. 8.6 Low-angle, three-dimensional analysis of the residuals. Low-angle views of the phase and wavelength dependence of residuals

References

Bradstreet, D. H.: 1993, Binary Maker 2.0 – An Interactive Graphical Tool for Preliminary Light Curve Analysis, in E. F. Milone (ed.), *Light Curve Modeling of Eclipsing Binary Stars*, pp. 151–166, Springer, New York

Bradstreet, D. H. & Steelmans, D. P.: 2004, *Binary Maker 3.0*, Contact Software, Norristown PA, 19087

Castelli, F. & Kurucz, R. L.: 2004, New Grids of ATLAS9 Model Atmospheres, *ArXiv Astrophysics e-prints*

Prša, A. & Zwitter, T.: 2005b, A Computational Guide to Physics of Eclipsing Binaries. I. Demonstrations and Perspectives, *ApJ* **628**, 426–438

Terrell, D., Munari, U., Zwitter, T., & Nelson, R. H.: 2003, Observational Studies of Early-Type Overcontact Binaries: TU Muscae, *AJ* **126**, 2988–2996

Wichmann, R.: 2002, *NIGHTFALL User's Guide*, Technical report, Hamburger Sternwarte, Hamburg, Germany

Wilson, R. E.: 1979, Eccentric Orbit Generalization and Simultaneous Solution of Binary Star Light and Velocity Curves, *ApJ* **234**, 1054–1066

Wilson, R. E. & Devinney, E. J.: 1971, Realization of Accurate Close-Binary Light Curves: Application to MR Cygni, *ApJ* **166**, 605–619

Zwitter, T., Castelli, F., & Munari, U.: 2004, An Extensive Library of Synthetic Spectra Covering the Far Red, RAVE and GAIA Wavelength Ranges, *A&A* **417**, 1055–1062

Chapter 9
The Structure of Light Curve Programs and the Outlook for the Future

This chapter reflects the authors' views on the structure of future light curve programs. It provides precepts for the structure of generic light curve programs and a preview of the coming decade with the goal of promoting further activities in *EB* research and light curve program development.

9.1 Structure of a General Light Curve Analysis Program

Nous avons changé tout cela   (*We have changed all that*)

Commercially available hardware and software development packages have reached a quality that they allow the components of the light curve analysis problem to be incorporated efficiently into a stable and user-friendly program. These aspects of any general light curve analysis program (hereafter, GLCAP) include the following.

- the light curve models;
- capabilities of the least-squares solvers; and
- user-friendly front-ends.

Light curve models are most appropriately provided by astronomers. Least-squares solvers could be implemented by some astronomers but more commonly by mathematicians, computer scientists, or others with a keen eye for efficiency, an extensive knowledge of numerical analysis, and programming skills. The development of a front-end typically requires the expertise of software engineers rather than astronomers or mathematicians. Of course, many gifted individuals are capable of fulfilling more than one of these rôles.

The three components of any GLCAP need to be linked appropriately. Although such a program does not yet exist, a GLCAP can be expected to emerge in the future. In order to maintain an open structure, it is likely that various experts will need to contribute to the physical model and to the means to achieve mathematical optimization. In such a distribution of labor, a framework and a number of well-defined interfaces will be needed. In the following subsections we suggest how the framework and the interfaces might be structured.

GLCAP itself should consist of several modules and major subroutines:

- light curve models (LCI) ;
- module (DETPAR) controlling and invoking least-squares solvers;
- interface (F) to least-squares solvers;
- interface (DF) to compute derivatives required by the least-squares solvers;
- interface (FEI) passing the input from user-friendly front-end to vectors **s** containing all system parameters and **CP**, a vector containing all control parameters; and
- interface (FEO) passing the output back to front-end.

9.1.1 Framework of the Light Curve Models

GLCAP may contain several light curve models of different complexity. The analysis of eclipsing binary light curve data may start with a simple model assuming circular orbits and spherical stars. Now, when computer power is not the limiting factor, we might argue that the simple models have almost nothing to offer. Nevertheless, at least for a while, simple models are likely to be used in the context of teaching and by amateur astronomers, but also for deriving initial parameter estimates very quickly. These results can then be used in more realistic models based on Roche geometry. Each light curve model $\mathcal{L}_i$ should be coded in a module LCI which in turn is built up by several submodules. Generically, we will name it LC from now on and describe its structure, input, and output quantities.

The structure of GLCAP and also LC should be modular. That is, it should separate the computation of orbit quantities, component surfaces, radiative physics, eclipse effects, and special effects in modules. Modules are collections of subroutines, are subroutines themselves on a higher level, and they are part of the overall program GLCAP but might be stored in separate files. The computation of line profiles or the computation of the orbit might be considered as a module. Solving Kepler's equation would then be a subroutine in the orbital computation module. LC should not do any file reading or file writing. This task should be carried out by an I/O module. The input to LC would be in the form of multidimensional arrays:

- **System parameters** could be stored in the vector **s** containing all system parameters, viz., inclination, mass ratio, etc. There should be a standard mapping, defining, for instance,

    $x_1 \longleftrightarrow$ inclination i, $0° \leq i < 180°$,
    $x_2 \longleftrightarrow$ mass ratio q, $q > 0$,
    $x_3 \longleftrightarrow$ temperature T_1, $T_1 > 0$, of component 1,
    $x_4 \longleftrightarrow \ldots$.

Of course, not all elements of the vector **s** need be used in subroutine LC. Therefore LC would have a subroutine StoP which would map **s** to the vector **p** containing only the parameters needed by LC. This mapping could include scaling transforms as well as other features.

- **Control parameters CP** containing a variety of control parameters, e.g., specifying the atmospheric model to use, accuracy to stop iterations, etc.

The output consists of the set $\{L^{\text{cal}}, RV^{\text{cal}}, ...\}$ of light and radial velocity curves, and such other observables as polarization curves, spectral indices of various kinds, or even line profiles.

9.1.2 Framework to Embed Least-Squares Solvers

Several available least-squares solvers could be included, among them

- least-squares solvers based on direct search algorithms, e.g., the Simplex algorithm;
- Powell's direction method;
- differential corrections with the Levenberg–Marquardt option;
- damped Gauß–Newton algorithm for constrained least-squares problems; and
- sequential quadratic programming.

The operational implementation amid such a group of solvers would depend on the needs of the analysis. For instance, it could be advantageous to do an initial search with the Simplex algorithm and then switch to the Gauß–Newton algorithm. The switching could be automatic with control of the switch to invoke any specific least-squares solvers in the subroutine DETPAR. This subroutine would have the following major input data which are passed to the least-squares solvers:

- the number of data points to be considered in the least-squares fit, N;
- the number of adjustable parameters, M;
- the number of constraints, $MCON$;
- the vector FOBS containing all individual observed data points, e.g., all light curves and all radial velocity curves;
- the vector FWEIGH containing the weights associated with the individual data points. If the weights are lumped weights, they have to be computed in advance before calling DETPAR;
- an initial guess $\mathbf{x}_0$ for the parameter vector to be determined. The vector s would be mapped onto **x** by the use of information from a vector **KEEP**, which would specify whether or not a parameter is adjustable; and
- control parameters specifying accuracy and termination criterion.

The output should consist of the

- vector **x** containing the parameters;
- vector $\varepsilon(\mathbf{x})$ containing the statistical errors of the parameters; and
- status flag to interpret the results.

Each individual least-squares solver would access the light curve model through a subroutine F. This subroutine would require N, M, and $MCON$ as passed to DETPAR, and in addition it could find the set of phase values in a common block.

The output would consist of the

- vector **FX** = $\{L^{cal}, RV^{cal},...\}$: all computed light curves and radial velocity curves and
- status flag to interpret the results.

The subroutine F would again call subroutine LC, take the output of subroutine LC, i.e., all light curves and radial velocity curves, and store it into the vector **FX**. If several light curve models were implemented, subroutine F would find this information in a common block.

9.2 Procedural Philosophies

Experto credite   *(Believe one who knows by experience)*
Vergil (70–19 B.C.) *Aeneid*, 11, 283i

How might a *general light curve analysis program* be used? It should help the user to process and analyze his/her data (we assume that photometric data have been reduced and transformed to a standard system as per the precepts of Chap. 2 and that any other data have been suitably standardized as well). It should be able to make best use of anything that is known about the system, such as its distance, amount of interstellar extinction, and reddening.

The first step is to import the data and to process them into standard format. The format would depend on the type of data. For photometric data, it would consist of triads of JDN (and decimal thereof) or phase; differential or standard-system magnitudes, flux level, or energy; and a suitable weighting factor. The data would then be transformed into a standard form, such as triads of Phase (if the period, epoch, and any variations of these elements are available) and relative light and weight.

The data would then be analyzed according to a selected light curve model. If none is specified, a series of models could be tried, beginning with a simple one that provides some rough initial parameter values. The preliminary parameter values could then be used as initial input into a more sophisticated analysis using a model based on Roche geometry. The program might request a selection of stellar atmospheres, or information supplied earlier could be used to select appropriate stellar atmospheres.

Depending on the characteristics of the binary system, the appropriate physics would be attached to the model and the data reanalyzed. The result could be

1. geometrical parameters;
2. stellar parameters;
3. physical parameters describing gas streams, disks, or other physical objects; and
4. a statistical analysis including error limits for all parameters.

If the *general light curve analysis program* were used in this way, the results could readily be compared to those of other authors. Such an analysis program could set a standard for describing and archiving the properties of eclipsing binary stars in general.

9.3 Code Maintenance and Modification

Remis velisque   (*With oars and sails; with all one's might*)

The construction of any program requires a method by which changes can be made, implemented, and remembered. Because science is usually progressive, model improvements are desirable. A prudent modeler will plan for them. Changes are easier, faster, and safer to implement if the program is modularized.

An important issue to resolve is: *Who* will make the changes, check the code for unexpected consequences or bugs, issue new version numbers, and distribute the code? Code modifications require a central authority to take responsibility to document and upgrade improvements, and, if our precepts regarding the modularization of the code are accepted, to incorporate additional modules and subroutines as they are developed. For data-acquisition and data-reduction packages, international observatories (KPNO + CTIO = AURA and ESO) take on this role. There is, however, no natural agency for a similar role to take over light curve analysis code development. Indeed, there should be no way to prevent someone from modifying a code to accomplish some particular task. But, if this is done, it is important that all such modifications be identified, especially in publications that make use of the result. While we find it unpalatable that a superauthority police the use of a code, there is a case for a code creator to seek copyright or other protection for its use to ensure that anyone modifying the code could be obliged to identify and document changes in any publication that emerges. The modifying party could be required to embed such documentation in comments before passing her/his modified code to anyone else. This protocol could reduce confusion about code legitimacy and performance.

Another possible mechanism to ensure responsible use and modification of programs would be to invoke the moral authority of the IAU through resolutions suggesting these steps at a General Assembly: Commissions 42 (Close Binary Stars), 26 (Double Stars), 27 (Variable Stars), and 25 (Photometry and Polarimetry). Perhaps others would be likely sources of such resolutions. In any case, we recommend a naming scheme for identifying successive versions and upgrades. This would simplify future investigators' work associating solutions and program versions.

Technology has settled at least one class of questions. Such questions as how upgrades could be announced and distributed (so troubling in past decades) are now rendered trivial because of the widespread communication media of the Internet and the World Wide Web.

The question of who will be responsible for maintaining the web pages and providing public access to various versions is still unresolved. In the case of a living creator, there is presumably no doubt; nor would there be any if an "heir" were designated by the progenitor. In the case that no such heir is designated, we recommend again that the relevant IAU commissions (mainly 42 – Close Binary Stars, but also 26 – Double Stars, and perhaps others) examine the question and designate an individual or group to take on the responsibility of code maintenance, upgrade, and distribution. One possibility is the creation of a multicommission Working Group on Data Analysis.

9.4 Prospects and Expectations

Tempora mutantur, et nos mutamur in illis
(The times change, and we along with them)

Modern light curve analysis was born early in the twentieth century with the pioneering work of Henry Norris Russell. Thanks first to the theoretical investigations of Zdenek Kopal beginning in the 1940s, and second to the development of practical computer applications by a number of workers two decades later, the field underwent a dramatic revolution in the 1970s. Since then light curve analysis has taken full advantage of the remarkable progress in both computer hardware and software. The Wilson–Devinney program has become the light curve analysis tool of choice by the majority of the community, and the line of improved modeling and models extends forward toward the horizon.

It seems likely that the number of observed light curves will continue to exceed the number analyzed. Indeed, the use of CCDs of increasingly large format threaten to overwhelm us with light curve data. Thousands of light curves are beginning to emerge from large field imaging projects, such as OGLE and MACHO, which have been developed to find gravitationally lensed objects. Unfortunately, the analysis of binaries from such sources suffers from the lack of radial velocities as well as spectral and color index data. Nevertheless, as image processing codes provide reliable magnitudes for more and more stars, the number of light curves will exceed our ability to analyze them unless concomitant ways of providing fast and reliable analysis can be developed. Millions are expected by the Kepler or GAIA mission, or ground-based survey such as the LSST. Section 5.3 contains promising techniques to approach this challenge. Neural networks could be used to identify the general type of light curve, and then adequate light curve models used to attack the subtleties, and provide solutions. Neural network techniques are most advantageous when the training set of well-studied light curves is very large, because only a large database can provide an adequate training set.

The quality of light curve analyses can also be expected to improve. Accurate stellar atmospheres over expanded temperature and wavelength regimes will provide the means to model the fluxes of binary components with ever-increasing accuracy. Finally, an increasing range of types of objects and astrophysical conditions can be expected to be modeled successfully with a package of standardized programs. The phenomena of extended atmospheres, semi-transparent atmospheric clouds, variable thickness disks, and gas streams, and a plethora of planetary transits as well as (from infrared data) occultations are among such programs.

9.4 Prospects and Expectations

What does the long-term future hold? Besides the determination of orbits, stellar sizes, and masses it seems likely that the detailed physics of stellar surfaces, including those arising from activity cycles, will continue to be targets of modeling work. It also seems likely that diagnostic tools will continue to be developed within light curve codes to provide more insight into stellar astrophysics. The development of ever more accurate stellar atmospheres is the key to the successful use of the analysis codes for the exploration of the radiative properties of the stars. It is crucial in transforming synthetic light curve codes into a diagnostic tool of great power. Such developments will lead to ever better tools for the elucidation and understanding of stars.

Finally, close binary research might initiate projects involving complicated physics and requiring sophisticated mathematics or huge number crunching. Here, we mention six problems:

1. The structural–dynamic readjustment of tidally distorted stars in eccentric orbits. To what extent does stellar volume vary in response to forced nonradial oscillations?
2. Computation of binary star interiors in terms of three-dimensional structural dynamics: How is matter redistributed in a binary component that almost fills its limiting lobe and is close to the beginning of mass loss through the Lagrangian point?
3. Three-dimensional radiation–hydrodynamic problems in binary systems. This includes meridional circulation and stellar winds and would treat the radiative transfer without geometric assumptions.
4. A number of sources of spectral emission could be modeled usefully. The presence of coronal plumes in systems with solar-like and later spectral type stars is one of these. Another is the stream in morphologically defined Algol systems. A third is the extensive disks around W Serpentis systems or around white dwarfs in CV systems. Such complex structures, however, are not easily characterized by one or even two parameters, but there is software available that treats these kinds of objects.
5. Investigation of dynamically evolving configurations that result from tides in eccentric orbits, as in high-mass X-ray binaries such as *GP Velorum/Vela-X1*, *Centaurus X-3*, and *V884 Scorpii* (*HD 153919*) with an optical O6.5 supergiant and $6^m\!.5$ V magnitude.
6. Adjustable *EB* parameters that change in time: Currently, the rate dP/dt of period change and orbit rotation $d\omega/dt$ (apsidal motion) are the only parameters which are traced in time. Third bodies inducing Kozai cycles as discussed in Sect. 5.2.1 can produce significant changes in binary's eccentricity and inclination [cf. *SS Lac* studied, for instance, by Torres & Stefanik (2000), Milone et al. (2000), or Eggleton & Kiseleva-Eggleton (2001)]. Whereas the longer period in Kozai cycles related to circularization and shrinkage of the orbit is of the order of 100,000 years, the shorter cycles from large to small and again back to large eccentricities is of the order of 1,000 years. Thus, the effects are measurable over decades. Another group of adjustable parameters that would be useful for

systems such as *V781 Tau* analyzed by Kallrath et al. (2006) are curve-dependent spot parameters and allowance for differential stellar rotation and latitude migration of spot groups. The idea here is that the curves have been observed during times with different spots on the stars. This work has been done for single stars by Harmon & Crews (2000) and by the Catania astronomers.

These examples reemphasize that close binaries are not only rich in physics but the ongoing need to extract the full measure of information contained in the data, leads, in turn to progress in the mathematical and numerical methods used in astrophysics. *Vive l'astrophysique*!

References

Eggleton, P. P. & Kiseleva-Eggleton, L.: 2001, Orbital Evolution in Binary and Triple Stars, with an Application to SS Lacertae, *ApJ* **562**, 1012–1030

Harmon, R. O. & Crews, L. J.: 2000, Imaging Stellar Surfaces via Matrix Light-Curve Inversion, *AJ* **120**, 3274–3294

Kallrath, J., Milone, E. F., Breinhorst, R. A., Wilson, R. E., Schnell, A., & Purgathofer, A.: 2006, V781 Tauri: A W Ursae Majoris Binary with Decreasing Period, *Astronomy and Astropyhsics* **452**, 959–967

Milone, E. F., Schiller, S. J., Munari, U., & Kallrath, J.: 2000, Analysis of the Currently Noneclipsing Binary SS Lacertae or SS Lacertae's Eclipses, *Astron. J.* **199**, 1405–1423

Torres, G. & Stefanik, R. P.: 2000, The Cessation of Eclipses in SS Lacertae: The Mystery Solved, *AJ* **119**, 1914–1929

Part IV
Appendix

Appendix A
Brief Review of Mathematical Optimization

Ecce signum (*Behold the sign; see the proof*)

Optimization is a mathematical discipline which determines a "best" solution in a quantitatively well-defined sense. It applies to certain mathematically defined problems in, e.g., science, engineering, mathematics, economics, and commerce. Optimization theory provides algorithms to solve well-structured optimization problems along with the analysis of those algorithms. This analysis includes necessary and sufficient conditions for the existence of optimal solutions. Optimization problems are expressed in terms of variables (degrees of freedom) and the domain; objective function[1] to be optimized; and, possibly, constraints. In *unconstrained optimization*, the optimum value is sought of an objective function of several variables without any constraints. If, in addition, constraints (equations, inequalities) are present, we have a *constrained optimization* problem. Often, the optimum value of such a problem is found on the boundary of the region defined by inequality constraints. In many cases, the domain X of the unknown variables $\mathbf{x}$ is $X = \mathbb{R}^n$. For such problems the reader is referred to Fletcher (1987), Gill et al. (1981), and Dennis & Schnabel (1983). Problems in which the domain X is a discrete set, e.g., $X = \mathbb{N}$ or $X = \mathbb{Z}$, belong to the field of discrete optimization ; cf. Nemhauser & Wolsey (1988). Discrete variables might be used, e.g., to select different physical laws (for instance, different limb-darkening laws) or models which cannot be smoothly mapped to each other. In what follows, we will only consider continuous domains, i.e., $X = \mathbb{R}^n$.

In this appendix we assume that the reader is familiar with the basic concepts of calculus and linear algebra and the standard nomenclature of mathematics and theoretical physics.

A.1 Unconstrained Optimization

Let $f : X = \mathbb{R}^n \to \mathbb{R}, \mathbf{x} \to f(\mathbf{x})$ be a continuous real-valued function. The problem

---

[1] The fitting of a model with free parameters to data leads to a minimization problem in which the objective function is the sum of squares of residuals.

$$UCP : \min_{\mathbf{x}}\{f(\mathbf{x})\} \iff f_* := \min_{\mathbf{x}}\{f(\mathbf{x}) \mid \mathbf{x} \in X\}, \qquad (A.1.1)$$

is called an unconstrained minimization problem. A vector $\mathbf{x}_*$ corresponding to the scalar f_* is called a minimizing point or minimizer $\mathbf{x}_*$, $f_* = f(\mathbf{x}_*)$ and is also expressed as

$$\mathbf{x}_* = \arg\min\{f(\mathbf{x})\} := \{\mathbf{x} \mid f(\mathbf{x}) \leq f(\mathbf{x}'), \ \forall \mathbf{x}' \in X\}. \qquad (A.1.2)$$

Maximization problems can be transformed into minimization problems with the relation

$$\max_{\mathbf{x}}\{f(\mathbf{x})\} = -\min_{\mathbf{x}}\{-f(\mathbf{x})\}. \qquad (A.1.3)$$

Therefore, it is sufficient to discuss minimization problems only.

A typical difficulty associated with nonlinear optimization is the problem that in most cases it is possible to determine only a locally optimal solution, not the global optimum. Loosely speaking, the global optimum is the best of all possible solutions, whereas a local optimum is the best in a neighborhood (for instance, the local sphere introduced on page 172) of $\mathbf{x}_*$ only. The formal definition reads:

Definition A.1.1 Given a minimization problem, a point $\mathbf{x}_* \in X$ is called a local minimum with respect to a neighborhood N of $\mathbf{x}_*$ (or simply local minimum)if

$$f(\mathbf{x}_*) \leq f(\mathbf{x}), \quad \forall \mathbf{x} \in N(\mathbf{x}_*). \qquad (A.1.4)$$

If $N(\mathbf{x}_*) = X$, then $\mathbf{x}_*$ is called global minimum.

$N(\mathbf{x}_*)$ could be chosen, for instance, as the ball $B_\varepsilon(\mathbf{x}_*)$ introduced on page 172. Early methods used for minimization were ad hoc search methods merely comparing function values $f(\mathbf{x})$ at different points $\mathbf{x}$. The most successful method of this group is the Simplex method [Spendley et al. (1962), Nelder & Mead (1965)] described in Chap. 4. In this group we also find the *alternating variables method*, in which in iteration k ($k = 1, 2, ..., K$) the variable x_k alone is changed in an attempt to reduce the value of the objective function, and the other variables are kept fixed. Both methods are easy to implement. They can handle nonsmooth functions, and they do not suffer from the requirement of many derivative-based optimization procedures depending on sufficiently accurate initial solutions for convergence. However, sometimes they converge very slowly. The most efficient way to use such methods is to derive an approximate solution for the minimization problem and use it to initialize a derivative-based minimization algorithm.

Derivative-based methods need the first- and possibly second-order derivatives involving the following objects:

A.1 Unconstrained Optimization

- Gradient of a scalar, real-valued differentiable function $f(\mathbf{x})$ of a vector $\mathbf{x}$; $f : X = \mathbb{R}^n \to \mathbb{R}, \mathbf{x} \to f(\mathbf{x})$

$$\nabla f(\mathbf{x}) := \left(\frac{\partial}{\partial x_1} f(\mathbf{x}), \ldots, \frac{\partial}{\partial x_n} f(\mathbf{x}) \right) \in \mathbb{R}^n. \qquad (A.1.5)$$

- *Jacobi matrix*[2] of a vector-valued function $\mathbf{F}(\mathbf{x}) = (f_1(\mathbf{x}), \ldots, f_m(\mathbf{x}))^T$

$$\mathsf{J}(\mathbf{x}) \equiv \nabla \mathbf{F}(\mathbf{x}) := (\nabla f_1(\mathbf{x}), \ldots, \nabla f_m(\mathbf{x})) = \left(\frac{\partial}{\partial x_j} f_i(\mathbf{x}) \right) \in \mathcal{M}(m, n). \qquad (A.1.6)$$

- *Hessian matrix*[3] of a real-valued function $f(\mathbf{x})$

$$\mathsf{H}(\mathbf{x}) \equiv \nabla^2 f(\mathbf{x}) := \frac{\partial}{\partial x_i} \left(\frac{\partial}{\partial x_j} f(\mathbf{x}) \right) = \left(\frac{\partial^2}{\partial x_i \partial x_j} f(\mathbf{x}) \right) \in \mathcal{M}(n, n). \qquad (A.1.7)$$

Note that the Hessian matrix of the scalar function $f(\mathbf{x})$ is the Jacobian matrix of the gradient, i.e., vector function $\nabla f(\mathbf{x})$. Derivative-based minimization algorithms are local minimizers and may optionally involve the Jacobian J and the Hessian H. In any case, we assume that $f(\mathbf{x})$ is sufficiently smooth so that the derivatives are continuous.

As in the one-dimensional case, a local minimizer $\mathbf{x}_*$ satisfies

$$\nabla f(\mathbf{x}_*) = \mathbf{0}, \quad \nabla f(\mathbf{x}_*) \in \mathbb{R}^n. \qquad (A.1.8)$$

A Taylor series expansion about $\mathbf{x}_*$ can be used to prove the following theorem on sufficient conditions:

Theorem A.1.2 *Sufficient conditions for a strict and isolated local minimizer $\mathbf{x}_*$ are that (A.1.8) is satisfied and that the Hessian $\mathsf{H}_* := \mathsf{H}(\mathbf{x}_*)$ is positive definite, that is,*

$$\mathbf{s}^T \mathsf{H}_* \mathbf{s} > 0, \quad \forall \mathbf{s} \neq \mathbf{0}, \quad \mathbf{s} \in \mathbb{R}^n. \qquad (A.1.9)$$

Here $\mathbf{s}$ denotes any vector different from zero. The widespread approach to solve the minimization problem (A.1.1) numerically is to apply *line search algorithms*. Given a solution $\mathbf{x}_k$ of the kth iteration, in the next step $\mathbf{x}_{k+1}$ is computed according to the following scheme:

---

[2] So called after Carl Gustav Jacob Jacobi (1804–1851).
[3] So called after the German mathematician Ludwig Otto Hesse (1811–1874). In the German the matrix is correctly called the "Hesse matrix," but in English, the not quite correct term "Hessian." The proper spelling would be "Hesseian" or "Hessean." $\mathcal{M}(m, n)$ denotes the set of all matrices with m rows and n columns.

- determine a search direction s_k;
- solve the *line search subproblem*, i.e., compute an appropriate damping factor $\alpha_k > 0$, which minimizes[4] $f(\mathbf{x}_k + \alpha \mathbf{s}_k)$; and
- define $\mathbf{x}_{k+1} := \mathbf{x}_k + \alpha_k \mathbf{s}_k$.

Different algorithms correspond to different ways of computing the search direction $\mathbf{s}_k$. For *damped methods* there are several procedures to compute the damping factor α_k. *Undamped methods* do not solve the line search problem and just put $\alpha_k = 1$. If $f(\mathbf{x}_k + \alpha \mathbf{s}_k)$ is exactly minimized, in the new point the gradient $\nabla f(\mathbf{x}_{k+1})$ is orthogonal to the current search direction $\mathbf{s}_k$, i.e., $\mathbf{s}_k^T \nabla f(\mathbf{x}_{k+1}) = 0$. This follows from the necessary condition

$$0 = \frac{df}{d\alpha}(f(\mathbf{x}_k + \alpha_k \mathbf{s}_k)) = (\nabla f(\mathbf{x}_k + \alpha_k \mathbf{s}_k))^T \mathbf{s}_k. \quad (A.1.10)$$

With this concept in mind several methods can be classified. *Descent methods* are line search methods in which the search direction $\mathbf{s}_k$ satisfies the descent property

$$\nabla f(\mathbf{x}_k)^T \mathbf{s}_k < 0. \quad (A.1.11)$$

The *steepest descent method* results from $\mathbf{s}_k = -\nabla f(\mathbf{x}_k)$. Obviously, (A.1.11) is fulfilled in this case. The gradient may be calculated analytically or may be approximated by finite differences with a small $h > 0$, that is, e.g., forward differences

$$\nabla_i f(\mathbf{x}) = \frac{\partial f}{\partial x_i}(\mathbf{x}) \approx \frac{f(\mathbf{x} + h\mathbf{e}_i) - f(\mathbf{x})}{h} + f''(\mathbf{x})h \quad (A.1.12)$$

or central differences

$$\nabla_i f(\mathbf{x}) = \frac{\partial f}{\partial x_i}(\mathbf{x}) \approx \frac{f(\mathbf{x} + \frac{1}{2}h\mathbf{e}_i) - f(\mathbf{x} - \frac{1}{2}h\mathbf{e}_i)}{h} + \frac{1}{24}f'''(\mathbf{x})h^2. \quad (A.1.13)$$

Here $\mathbf{e}_i$ denotes the unit vector along the ith coordinate axis. From a numerical point of view, central differences are preferred because they are more accurate and have an error which is only of the order of h^2.

Another approach to compute the search direction $\mathbf{s}_k$ is to expand the function $f(\mathbf{x})$ about $\mathbf{x}_k$ in a Taylor series up to second order. So we approximate $f(\mathbf{x})$ locally by a quadratic model, i.e., $\mathbf{x} = \mathbf{x}_k + \mathbf{s}_k$, $f(\mathbf{x}) = f(\mathbf{x}_k + \mathbf{s}_k)$, and

$$f(\mathbf{x}_k + \mathbf{s}_k) \approx f(\mathbf{x}_k) + \nabla f(\mathbf{x}_k)^T \mathbf{s}_k + \tfrac{1}{2}\mathbf{s}_k^T \mathbf{H}_k \mathbf{s}_k. \quad (A.1.14)$$

---

[4] Usually, $f(\mathbf{x}_k + \alpha \mathbf{s}_k)$ is not exactly minimized with respect to α. One possible heuristic is to evaluate f for $\alpha_m = 2^{-m}$ for $m = 0, 1, 2, \ldots$ and stop when $f(\mathbf{x}_k + \alpha_m \mathbf{s}_k) \leq f(\mathbf{x}_k)$.

A.1 Unconstrained Optimization

If both the gradient $\nabla f(\mathbf{x})$ and the Hessian matrix H are known analytically, the *classical Newton method* can be derived from (A.1.14). Applying the necessary condition (A.1.8) to (A.1.14) leads to

$$\mathsf{H}_k \mathbf{s}_k = -\nabla f(\mathbf{x}_k), \quad (A.1.15)$$

from which $\mathbf{s}_k$ is computed according to

$$\mathbf{s}_k = -\mathsf{H}_k^{-1} \nabla f(\mathbf{x}_k). \quad (A.1.16)$$

It can be shown [see, for instance, Bock (1987)] that the damped Newton method (Newton method with line search) converges globally, if H_k is positive definite. The case where H_k is not positive definite is treated only after a modification to the Hessian.

In most practical problems, the matrix of the second derivatives, i.e., the Hessian matrix H is not available. If, however, $\nabla f(\mathbf{x})$ is given analytically or can be approximated accurately enough, a *quasi-Newton method* can be applied. There are two approaches to quasi-Newton methods. Either the finite difference approximation $\bar{\mathsf{H}}_k$ of H_k is symmetrized and H_k in (A.1.16) is replaced by

$$\tfrac{1}{2}\left(\bar{\mathsf{H}}_k + \bar{\mathsf{H}}_k^T\right), \quad (A.1.17)$$

or, more efficiently, the inverse matrix H_k^{-1} is approximated by a symmetric, positive definite matrix $\tilde{\mathsf{H}}_k$. The initial matrix $\tilde{\mathsf{H}}_0$ can be any positive definite matrix. Usually, due to the absence of any better estimate, the identity matrix $\mathbb{1}$, or, if specific information of the problem can be exploited, a diagonal matrix is used to initialize $\tilde{\mathsf{H}}_0$. These methods are sometimes called *variable metric methods*. The efficiency of quasi-Newton methods depends on the formula to update $\tilde{\mathsf{H}}_{k+1}$ from $\tilde{\mathsf{H}}_k$. The construction of such formulas is still an active research field; some updates rules operate directly on the inverse H_k^{-1} and generate H_{k+1}^{-1}.

Furthermore, *conjugate direction methods* (CDMs) can be applied efficiently to quadratic optimization problems such as (A.1.14), as in that case CDMs terminate after at most n steps (or iterations). CDMs are based on the concept of the *conjugacy* of a set of nonzero vectors $\{\mathbf{s}_1, \mathbf{s}_2, \ldots \mathbf{s}_n\}$ to a given positive matrix H, that is,

$$\mathbf{s}_i^T \mathsf{H} \mathbf{s}_j = 0, \quad \forall\, i \neq j. \quad (A.1.18)$$

When applied to a quadratic function with Hessian H, CDMs generate these directions $\{\mathbf{s}_1, \mathbf{s}_2, \ldots \mathbf{s}_n\}$.

A special situation (see Appendix A.3) arises when $f(\mathbf{x})$ is built up from sums of squares, i.e., results from a least-squares problem. In that case, the Hessian H is specially linked to $f(\mathbf{x})$.

A.2 Constrained Optimization

A.2.1 Foundations and Some Theorems

For historical reasons, nonlinear constrained optimization is also referred to as nonlinear programming. A general nonlinear programming (hereafter referred to only as "NLP") problem with n variables, and n_2 equations, and n_3 inequalities is defined as **Problem NLP**

$$\begin{aligned}\text{Minimize:} \quad & f(\mathbf{x}), \quad \mathbf{x} \in \mathbb{R}^n, \\ \text{Subject to:} \quad & \mathbf{F}_2(\mathbf{x}) = 0, \; \mathbf{F}_2 : \mathbb{R}^n \to \mathbb{R}^{n_2}, \\ & \mathbf{F}_3(\mathbf{x}) \geq 0, \; \mathbf{F}_3 : \mathbb{R}^n \to \mathbb{R}^{n_3}. \end{aligned} \quad (\text{A.2.1})$$

The functions $f(\mathbf{x})$, $\mathbf{F}_2(\mathbf{x})$, and $\mathbf{F}_3(\mathbf{x})$ are assumed to be continuously differentiable on the whole vector space $\mathbb{R}^n$. The vector inequality, $\mathbf{F}_3(\mathbf{x}) \geq 0$, used for brevity, represents the n_3 inequalities $F_{3k}(\mathbf{x}) \geq 0$, $1 \leq k \leq n_3$. The set of all feasible solutions, i.e.,

$$\mathcal{S} := \left\{ \mathbf{x} \in \mathbb{R}^n \mid \mathbf{F}_2(\mathbf{x}) = 0 \wedge \mathbf{F}_3(\mathbf{x}) \geq 0 \right\} \quad (\text{A.2.2})$$

is called the feasible region $\mathcal{S}$. The binding or active constraints with respect to $\mathbf{x} \in \mathcal{S}$ are characterized by the index set

$$\mathcal{I}(\mathbf{x}) := \{ i \mid F_{3i}(\mathbf{x}) = 0, \; i = 1, \dots, n_3 \}, \quad (\text{A.2.3})$$

which is sometimes called the *active set*. In the early 1950s Kuhn & Tucker (1951) extended the theory of Lagrange multipliers, used for solving equality constrained optimization problems, to include the NLP problem (formulated as a maximum problem in the original work) with both equality and inequality constraints. This theory is based on the definition of a Lagrangian function

$$L(\mathbf{x}, \boldsymbol{\lambda}, \boldsymbol{\mu}) := f(\mathbf{x}) - \boldsymbol{\lambda}^T \mathbf{F}_2(\mathbf{x}) - \boldsymbol{\mu}^T \mathbf{F}_3(\mathbf{x}), \quad (\text{A.2.4})$$

that links the objective function $f(\mathbf{x})$ to the constraints $\mathbf{F}_2(\mathbf{x})$ and $\mathbf{F}_3(\mathbf{x})$. The vector variables $\boldsymbol{\lambda} \in \mathbb{R}^{n_2}$ and $\boldsymbol{\mu} \in \mathbb{R}^{n_3}$ are called Lagrange multipliers. They are additional unknowns of the problem.

Let $\mathbf{J}_2$ and $\mathbf{J}_3$ denote the Jacobian matrix of $\mathbf{F}_2$ and $\mathbf{F}_3$. Below we list some theorems and results in order to present a brief survey of the foundations of nonlinear optimization theory. Necessary conditions for an optimal solution to NLP problems are given by the following theorem [see, for instance, Ravindran et al. (1987)]:

Theorem A.2.1 *If $\mathbf{x}_*$ is a solution to an NLP problem, and the functions $f(\mathbf{x})$, $\mathbf{F}_2(\mathbf{x})$, and $\mathbf{F}_3(\mathbf{x})$ are differentiable, then there exists a set of vectors $\boldsymbol{\mu}_*$ and $\boldsymbol{\lambda}_*$ such that $\mathbf{x}_*$, $\boldsymbol{\mu}_*$, and $\boldsymbol{\lambda}_*$ satisfy the relations:*

A.2 Constrained Optimization

$$\text{ensure feasibility } \mathbf{F}_2(\mathbf{x}) = 0, \quad (A.2.5)$$
$$\text{ensure feasibility } \mathbf{F}_3(\mathbf{x}) \geq 0, \quad (A.2.6)$$
$$\text{complementary slackness } \boldsymbol{\mu}^T \mathbf{F}_3(x) = 0, \quad (A.2.7)$$
$$\boldsymbol{\mu} \geq 0, \quad (A.2.8)$$
$$\nabla f(\mathbf{x}) - \boldsymbol{\lambda}^T \mathbf{J}_2(\mathbf{x}) - \boldsymbol{\mu}^T \mathbf{J}_3(\mathbf{x}) = 0. \quad (A.2.9)$$

This theorem has been derived by Kuhn & Tucker (1951). Proofs are also found in Collatz & Wetterling (1971), or Fletcher (1987). Equations (A.2.5) through (A.2.9) are known as *(Karush[5]-) Kuhn–Tucker conditions*; they are also referred to as *first-order (necessary) conditions*. A point $(\mathbf{x}_*, \boldsymbol{\mu}_*, \boldsymbol{\lambda}_*)$ satisfying these conditions is called a *Kuhn–Tucker point* .

Introducing some restrictions on the functions $f(\mathbf{x})$, $\mathbf{F}_2(\mathbf{x})$, and $\mathbf{F}_3(\mathbf{x})$, called *constraint qualification* by Kuhn and Tucker, certain irregularity conditions can be excluded – particularly at a stationary point. Different constraint qualifications are discussed for instance in Bomze & Grossmann (1993) and by Gill et al. (1981). A rather general formulation is given by Bock (1987).

Definition A.2.2 Let $\mathcal{I}(\mathbf{x}')$ denote the set of active inequality constraints at $\mathbf{x}'$. Let $\tilde{\mathbf{F}}_3$ be the function composed of all F_{3i} with $i \in \mathcal{I}(\mathbf{x}')$; $\tilde{\mathbf{J}}_3$ is the associated Jacobian. Let $\mathbf{u}^T := (\mathbf{F}_2^T, \tilde{\mathbf{F}}_3^T) : \mathbb{R}^n \to \mathbb{R}^N$, $N := m + |\mathcal{I}|$. Let $L(\mathbf{x}, \boldsymbol{\mu}, \boldsymbol{\lambda})$ be the Lagrangian function of **NLP** defined in (A.2.4).

Definition A.2.3 Let $\mathbf{J}^N = \mathbf{J}^N(\mathbf{x}) = \dfrac{\partial \mathbf{u}}{\partial x}$ be the Jacobian of $\mathbf{u}$ associated with the equations and active inequalities. A feasible point $\mathbf{x}'$ is called regular if $\text{rank}(\mathbf{J}^N(\mathbf{x}')) = N$ [Bock (1987, p. 48)].

Theorem A.2.4 *Let $\mathbf{x}_* \in \mathbb{R}^n$ be a regular point and a local minimizer of problem NLP. Then there exist vectors $\boldsymbol{\mu}_*$ and $\boldsymbol{\lambda}_*$ such that $\mathbf{x}_*$, $\boldsymbol{\mu}_*$ and $\boldsymbol{\lambda}_*$ satisfy the (Karush–) Kuhn–Tucker conditions [Eqs. (A.2.5) through (A.2.9)].*

Note that the difference between Theorems A.2.6 and A.2.1 is in the assumptions, i.e., constraint qualifications. If we further define the set of directions

$$\mathcal{T}(\mathbf{x}_*) := \left\{ \mathbf{p} \neq 0 \, \middle| \, \begin{array}{l} \mathbf{J}_2(\mathbf{x}_*)\mathbf{p} = 0, \\ \tilde{\mathbf{J}}_3(\mathbf{x}_*)\mathbf{p} \geq 0, \end{array} \tilde{\mu}_{i*} \tilde{\mathbf{J}}_3(\mathbf{x}_*)\mathbf{p} = 0, \ \forall \, i \in \mathcal{I}(\mathbf{x}_*) \right\}, \quad (A.2.10)$$

and the Hessian

$$\mathbf{H}(\mathbf{x}_*, \boldsymbol{\mu}_*, \boldsymbol{\lambda}_*) := \frac{\partial^2}{\partial \mathbf{x}^2} L(\mathbf{x}_*, \boldsymbol{\mu}_*, \boldsymbol{\lambda}_*), \quad (A.2.11)$$

---

[5] It was later discovered that Karush (1939) had proven the same result in his 1939 master thesis at the University of Chicago. A survey paper by Kuhn (1976) gives a historical overview of the development of inequality-constrained optimization.

by extending the assumptions in Theorem A.2.1, the following results can be derived:

Theorem A.2.5 *(Second-Order Necessary Conditions).* *If the functions $f(\mathbf{x})$, $\mathbf{F}_2(\mathbf{x})$, and $\mathbf{F}_3(\mathbf{x})$ are twice differentiable, the second-order necessary conditions for the Kuhn–Tucker point $(\mathbf{x}_*, \mu_*, \lambda_*)$, being a local minimizer, is*

$$\mathbf{p}^T H(\mathbf{x}_*, \mu_*, \lambda_*)\mathbf{p} \geq 0, \quad \forall\, \mathbf{p} \in \mathcal{T}(\mathbf{x}_*). \tag{A.2.12}$$

The interpretation of this theorem is that the Hessian of the Lagrangian function is positive semi-definite for all directions $\mathbf{p} \in \mathcal{T}(\mathbf{x}_*)$.

Theorem A.2.6 *(Second-Order Sufficient Conditions).* *Let $(\mathbf{x}_*, \mu_*, \lambda_*)$ be a Kuhn–Tucker point of NLP, and for all directions $\mathbf{p} \in \mathcal{T}(\mathbf{x}_*)$, let the Hessian of the Lagrangian function be positive definite, i.e.,*

$$\mathbf{p}^T H(\mathbf{x}_*, \mu_*, \lambda_*)\mathbf{p} > 0, \quad \forall\, \mathbf{p} \in \mathcal{T}(\mathbf{x}_*). \tag{A.2.13}$$

Then $\mathbf{x}_*$ is a strict local minimum of NLP.

A proof of this theorem, further discussion of second-order conditions, and an even less restrictive formulation of the constraint qualification (based on a local characterization by linearized constraints) are given by Fletcher (1987).

For a special class of problems, namely convex problems, the following theorem based on the formulation in Ravindran et al. (1987) is useful:

Theorem A.2.7 *(Kuhn–Tucker Sufficiency Theorem)* . *Consider the nonlinear programming problem, NLP. Let the objective function $f(\mathbf{x})$ be convex, let $\mathbf{F}_2(\mathbf{x})$ be linear, and let $\mathbf{F}_3(\mathbf{x})$ be concave. If there exists a solution $(\mathbf{x}_*, \mu_*, \lambda_*)$ satisfying the Kuhn–Tucker conditions (Eqs. (A.2.5) through (A.2.9)), then $\mathbf{x}_*$ is an optimal solution to NLP.*

A proof of this theorem can be found, e.g., in Kuhn & Tucker (1951), Collatz and Wetterling (1971), and Bomze & Grossmann (1993).

If the functions $f(\mathbf{x})$, $\mathbf{F}_2(\mathbf{x})$, and $\mathbf{F}_3(\mathbf{x})$ satisfy the assumptions[6] of Theorem A.2.7 the optimization problem is called a *convex optimization problem*. This class of problems has the property, cf. Papadimitriou & Steiglitz (1982), that local optimality (see Appendix A.1.1) implies global optimality, i.e., every local minimum of **NLP** is a global one, if the problem is convex.

Algorithms to solve (A.2.1) are found, for instance, in Gill et al. (1981) or Fletcher (1987). Most are based on linearization techniques. Inequalities are included, for instance, by applying active set methods. The most powerful nonlinear optimization algorithms are the *Generalized Reduced Gradient algorithm* (*GRG*) and *Sequential Quadratic Programming* (*SQP*) *methods* and *Interior Point Methods*

---

[6] These assumptions guarantee that the feasible region is a convex set and that the objective function is a convex function.

A.2 Constrained Optimization

(IPM) [see, for instance, Bazaraa et al. (1993) or Wright (1996)] for problems involving many inequalities. The *GRG* algorithm was first developed by Abadie & Carpenter (1969); further information is contained in Abadie (1978), Lasdon et al. (1978), Lasdon & Waren (1978), and Gill et al. (1981, Sect. 6.3)]. It is frequently used to solve nonlinear constrained optimization problems, it is rarely used to solve least-squares problems. A similar remark holds for IPM. A special class of this method includes inequalities by adding logarithmic penalties terms to the objective function. Then the problem can solved as a nonlinear optimization problem with possible equations but no inequalities. SQP methods are described in Appendix A.2.2 and they are applied to constrained least-squares problems (Schittkowski 1988). Finally, generalized Gauß–Newton methods are an efficient approach to solve constrained least-squares problems.

A.2.2 Sequential Quadratic Algorithms

SQP methods belong to the most powerful and frequently used nonlinear optimization algorithms (Stoer 1985) to solve problem (A.2.1). The basic idea is to solve (A.2.1) by a sequence of quadratic programming subproblems. The subproblem in iteration k appears as

$$\min_{\Delta \mathbf{x}} \left\{ \tfrac{1}{2} \Delta \mathbf{x}^T \mathbf{H}_k \Delta \mathbf{x} + \nabla f(\mathbf{x}_k)^T \Delta \mathbf{x} \right\}, \quad \Delta \mathbf{x} \in \mathbb{R}^n \quad \text{(A.2.14)}$$

subject to

$$\mathbf{J}_2(\mathbf{x}_k)^T \Delta \mathbf{x} + \mathbf{F}_2(\mathbf{x}_k) = 0,$$
$$\mathbf{J}_3(\mathbf{x}_k)^T \Delta \mathbf{x} + \mathbf{F}_3(\mathbf{x}_k) \geq 0,$$

where the subscript k refers to quantities known prior to iteration k, and $\Delta \mathbf{x}$ is the correction vector to be determined. This subproblem [cf. Gill et al. (1981, Sect. 6.5.3)] is obtained by linearizing the constraints and terminating the Taylor serious expansion of the objective function of (A.2.1) after the quadratic term; the constant term $f(\mathbf{x}_k)$ has been dropped. The necessary condition derived from the Lagrangian function associated with (A.2.14) is

$$\mathbf{H}_k \Delta \mathbf{x} + \nabla f(\mathbf{x}_k) - \mathbf{J}_2(\mathbf{x}_k) \tilde{\lambda}_{k+1} - \mathbf{J}_3(\mathbf{x}_k) \tilde{\mu}_{k+1} = \mathbf{0}. \quad \text{(A.2.15)}$$

If, furthermore, λ_k denotes the vector of Lagrange multipliers (for convenience we do not distinguish between λ and μ for equations and inequalities) known prior to iteration k, and $\Delta \mathbf{x}_k$, $\tilde{\lambda}_k$, and $\tilde{\mu}_k$ represent the solution of (A.2.14), then the next iteration follows as

$$\begin{pmatrix} \mathbf{x}_{k+1} \\ \lambda_{k+1} \\ \mu_{k+1} \end{pmatrix} = \begin{pmatrix} \mathbf{x}_k \\ \lambda_k \\ \mu_k \end{pmatrix} + \alpha_k \begin{pmatrix} \Delta \mathbf{x}_k \\ \Delta \lambda_k \\ \Delta \mu_k \end{pmatrix}, \quad \begin{pmatrix} \Delta \lambda_k = \tilde{\lambda}_k - \lambda_k \\ \Delta \mu_k = \tilde{\mu}_k - \mu_k \end{pmatrix}, \quad \text{(A.2.16)}$$

where $\alpha_k \geq 0$ is a damping factor.

For the solution of the quadratic subproblems the reader is referred to Gill et al. (1981, Sect. 5.3.2) or Fletcher (1987, Chap. 10).

A.3 Unconstrained Least-Squares Procedures

A special case of unconstrained minimization arises when the objective function is of the form[7] maximum likelihood estimator (Brandt, 1976, Chap. 7) for the unknown parameter vector **x**. This objective function dates back to Gauß (1809) and in the mathematical literature the problem is synonymously called a least-squares or ℓ_2 approximation problem.

$$f(\mathbf{x}) = \|\mathbf{R}(\mathbf{x})\|_2^2 = \mathbf{R}(\mathbf{x})^T \mathbf{R}(\mathbf{x}) = \sum_{\nu=1}^{N} [R_\nu(\mathbf{x})]^2, \quad \mathbf{R}(\mathbf{x}) \in \mathbb{R}^N. \tag{A.3.1}$$

This structure may arise either from a nonlinear over-determined system of equations

$$R_\nu(\mathbf{x}) = 0, \quad \nu = 1, ..., N, \quad N > n, \tag{A.3.2}$$

or from a data-fitting problem, e.g., the one described in Chap. 4 [formula (4.1.11)] with N given data points $(t_\nu, \tilde{Y}_\nu)$ and variances σ_ν, a model function $\tilde{F}(t, \mathbf{x})$, and n adjustable parameters **x**:

$$R_\nu := R_\nu(\mathbf{x}) = Y_\nu - F_\nu(\mathbf{x}) = \sqrt{w_\nu}\left[\tilde{Y}_\nu - \tilde{F}(t_\nu, \mathbf{x})\right]. \tag{A.3.3}$$

The weights w_ν are related to the variances σ_ν by

$$w_\nu := \beta/\sigma_\nu^2. \tag{A.3.4}$$

Traditionally, the weights are scaled to a variance of unit weights. The factor β is chosen so as to make the weights come out in a convenient range. Sometimes, if variances are not known, they may be estimated by other considerations as described in Sect. 4.1.1.5. In short vector notation we get

$$\mathbf{R} := \mathbf{Y} - \mathbf{F}(\mathbf{x}) = [R_1(\mathbf{x}), ..., R_N(\mathbf{x})]^T, \quad \mathbf{F}(\mathbf{x}), \mathbf{Y} \in \mathbb{R}^N. \tag{A.3.5}$$

Our least-squares problem requires us to provide the following input:

---

[7] The minimization of this functional, i.e., the minimization of the sum of weighted quadratic residuals, under the assumption that the statistical errors follow a Gaußian distribution with variances as in (A.3.4), provides a

A.3 Unconstrained Least-Squares Procedures

1. model;
2. data;
3. variances associated with the data; and
4. measure of goodness of fit, e.g., the Euclidean norm.

In many practical applications, unfortunately, less attention is paid to the third point. It is also very important to point out that the fourth point requires preinformation related to the problem and statistical properties of the data.

Before we treat the general case of nonlinear models in Appendix A.3.3, we discuss the linear case first, i.e., $\mathbf{F}(\mathbf{x}) = \mathbf{A}\mathbf{x}$ with a constant matrix A.

A.3.1 Linear Case: Normal Equations

With $\mathbf{y} = \sqrt{\mathbf{W}}\tilde{\mathbf{y}}$ and $\mathbf{A}\mathbf{x} = \sqrt{\mathbf{W}}\tilde{\mathbf{A}}$ the weighted residual vector

$$\mathbf{R}(\mathbf{x}) = \mathbf{y} - \mathbf{A}\mathbf{x} = \sqrt{\mathbf{W}}(\tilde{\mathbf{y}} - \tilde{\mathbf{A}}\mathbf{x}) \tag{A.3.6}$$

is linear in $\mathbf{x}$ and leads to the linear least-squares problem

$$\min \|\mathbf{y} - \mathbf{A}\mathbf{x}\|_2^2, \quad \mathbf{x} \in \mathbb{R}^n, \quad \mathbf{y} \in \mathbb{R}^N, \quad \mathbf{A} \in \mathcal{M}(N, n), \tag{A.3.7}$$

with a constant matrix A. It can be shown that the linear least-squares problem has at least one solution $\mathbf{x}_*$. The solution may not be unique. If $\mathbf{x}'_*$ denotes another solution, the relation $\mathbf{A}\mathbf{x}'_* = \mathbf{A}\mathbf{x}_*$ holds. All solutions of (A.3.7) are solution of the normal equations

$$\mathbf{A}^T\mathbf{A}\mathbf{x} = \mathbf{A}^T\mathbf{y}, \tag{A.3.8}$$

and vice versa: All solutions of the normal equations are solutions of (A.3.7). Thus, the normal equations represent the necessary and sufficient conditions for the existence and determination of the least-squares solution.

The modulus $R_* = |\mathbf{R}_*|$ of the residual vector $\mathbf{R}_*$ at the solution is uniquely determined by

$$\min \|\mathbf{y} - \mathbf{A}\mathbf{x}\|_2^2 = R_* = |\mathbf{R}_*|, \quad \mathbf{R}_* = \mathbf{y} - \mathbf{A}\mathbf{x}_*. \tag{A.3.9}$$

However, only if A has full rank, the problem (A.3.7) has a unique solution and there exists a unique solution for $\mathbf{R}_*$ which can be obtained, for instance, as the solution of the linear system of equations

$$\mathbf{A}^T\mathbf{R} = \mathbf{0}. \tag{A.3.10}$$

In this case the symmetric matrix $\mathbf{A}^T\mathbf{A}$ has full rank

$$\text{rank } \mathbf{A}^T\mathbf{A} = n. \tag{A.3.11}$$

From a numerical point of view the normal equation approach for solving least-squares problems should, if used at all, be applied with great caution for the following reasons:

- the computation of $A^T A$ involves the evaluation of scalar products (numerical analysts usually try to avoid the evaluation of scalar products because this often leads to loss of significant digits due to adding positive and negative terms of similar size); and
- strong propagation of errors on the right-hand side term $A^T y$ may occur when solving (A.3.8) as the propagation depends on the condition number $\kappa_2(A^T A) = \kappa_2^2(A)$.

As shown in Deuflhard & Hohmann (1993, p. 74), during solving least-squares problems, for large residual problems the error propagation related to perturbations of the matrix A is approximately given by $\kappa_2^2(A)$ whereas for small residual problems it is better approximated by $\kappa_2(A)$. In that case solving (A.3.7) by the normal equation approach and, for instance, the Cholesky–Banachchiewicz algorithm, is not recommended. Note that in practical problems A itself may already be badly conditioned. Condition numbers of 10^2–10^3 are common which leads to $\kappa_2^2(A)$ of the order of 10^4–10^6. Therefore, it is recommended to use methods which use only the matrix A itself. Such an algorithm is described in Sect. A.3.2.

A.3.2 The Linear Case: An Orthogonalization Method

Numerical problems due to bad condition numbers of A can be limited if the least-squares problem is solved using only A. Orthogonalization methods involve orthogonal transformations P to solve linear least-squares problems. Orthogonal transformations leave the Euclidean norm of matrices invariant and lead to stable linear least-squares solvers. Householder transformations are special types of orthogonal transformations. The matrix A is transformed in such a way that

1. the solution of the problem is not changed;
2. the condition number $\kappa_2(PA)$ of the transformed matrix is not larger than $\kappa_2(A)$; and
3. the transformed matrix PA has a triangular structure very suitable for numerical computations.

Let $P_k, k \in \mathbb{N}$, be a sequence of Householder transformations. Householder transformations are special matrices of the form[8]

$$P := \mathbb{1} - 2vv^H, \qquad (A.3.12)$$

---

[8] Although for our current case it is not necessary to use complex vectors, we describe the general case, for reasons of consistency, with most of the mathematical literature. Unitary matrices in complex vector spaces correspond to orthogonal matrices in real vector spaces.

A.3 Unconstrained Least-Squares Procedures

where $\mathbb{1}$ is the (n, n) identity matrix and $\mathbf{v}$ is an arbitrary n-dimensional vector. Householder transformations $\mathbf{x} \to P\mathbf{x}$ are reflections of the vector space $\mathbb{C}^n$ with respect to the orthogonal complement

$$\{\mathbf{v}\} := \{\mathbf{y} \in \mathbb{C}^n \mid \mathbf{v}^H \mathbf{y} = 0\}. \tag{A.3.13}$$

They have the property

$$P^H P = \mathbb{1}, \tag{A.3.14}$$

where P^H is the conjugate transpose, i.e., that P is *unitary*. Unitary matrices U are known to conserve the norm of vectors $\mathbf{x}$ and linear operators A, i.e.,

$$\begin{aligned} \|U\mathbf{x}\|_2 &= \|\mathbf{x}\|_2, \\ \|U\|_2 &= \|U^H\|_2 = 1, \\ \|UA\|_2 &= \|A\|_2 = \|UA\|_2. \end{aligned} \tag{A.3.15}$$

The matrix

$$P = P_n \cdots P_1 \tag{A.3.16}$$

is unitary as well, as it is a product of unitary matrices. The vector $\mathbf{w}$ is chosen such that P maps a given vector $\mathbf{x} = (x_1, \ldots, x_n)^T$ with[9] $x_1 \neq 0$ onto a multiple of the first unit vector $\mathbf{e}_1 = (1, 0, \ldots, 0)^T$, i.e.,

$$P\mathbf{x} = k\mathbf{e}_1.$$

For a given vector $\mathbf{x} \neq \mathbf{0}$ this implies the following formulas for computing the associated Householder transformation (for $\mathbf{x} = \mathbf{0}$ there is nothing to do):

$$\begin{aligned} \sigma &= \|\mathbf{x}\|_2, \\ \beta &= \sigma(\sigma + |x_1|), \\ e^{i\alpha} &= x_1 / |x_1|, \\ k &= -\sigma e^{i\alpha}, \\ \mathbf{u} &= \mathbf{x} - k\mathbf{e}_1, \\ P &= \mathbb{1} - \beta^{-1} \mathbf{u}\mathbf{u}^H. \end{aligned} \tag{A.3.17}$$

If the matrix A has n linearly independent columns $\mathbf{a}_1, \ldots, \mathbf{a}_n$, the matrix A and the vector $\mathbf{y}$ in the linear least-squares problem (A.3.7) are finally transformed by P into a upper triangular matrix $\mathcal{A}$

---

[9] If $\mathbf{x} \neq \mathbf{0}$ we can, by appropriate permutations, always achieve $x_1 \neq 0$. For $\mathbf{x} = \mathbf{0}$ there is nothing to do.

$$\mathcal{A} = \mathsf{P}\mathsf{A} = \begin{bmatrix} \mathsf{S} \\ 0 \end{bmatrix} \in \mathcal{M}(N,n), \quad \mathsf{S} = \begin{bmatrix} s_{11} & \cdots\cdots\cdots & s_{1n} \\ 0 & s_{21} & \cdots\cdots & \vdots \\ \vdots & 0 & \ddots & \vdots \\ \vdots & \vdots & & \ddots & \vdots \\ 0 & 0 & \cdots\cdots & s_{nn} \end{bmatrix} \in \mathcal{M}(n,n), \quad (A.3.18)$$

and a vector **h**

$$\mathbf{h} = \mathsf{P}\mathbf{y} = \begin{bmatrix} \mathbf{h}_1 \\ \mathbf{h}_2 \end{bmatrix}, \quad \mathbf{h}_1 \in \mathbb{R}^n, \quad \mathbf{h}_2 \in \mathbb{R}^{N-n}. \qquad (A.3.19)$$

As usually in numerical computations, for accuracy and stability, P is not computed via the matrix multiplication (A.3.16) but is established by successive Householder transformations, i.e., successive modifications of A.

The original problem (A.3.7) now takes the form

$$\min \|\mathbf{y} - \mathcal{A}\mathbf{x}\|_2^2 = \min \|\mathsf{P}(\mathbf{y} - \mathsf{A}\mathbf{x})\|_2^2 = \min \left\| \begin{bmatrix} \mathsf{S}\mathbf{x} - \mathbf{h}_1 \\ 0 - \mathbf{h}_2 \end{bmatrix} \right\|_2^2. \qquad (A.3.20)$$

As we are using the Euclidean norm and as P is unitary we get

$$\min \|\mathbf{y} - \mathcal{A}\mathbf{x}\|_2^2 = \min \|\mathsf{S}\mathbf{x} - \mathbf{h}_1\|_2^2 + \|\mathbf{h}_2\|_2^2. \qquad (A.3.21)$$

Because $\mathbf{h}_2$ is a constant vector, $\|\mathbf{y} - \mathcal{A}\mathbf{x}\|_2^2$ takes its minimum when the unknown vector **x** is the solution of the linear system of equations

$$\mathsf{S}\mathbf{x} = \mathbf{h}_1. \qquad (A.3.22)$$

Thus, the solution of $\mathsf{S}\mathbf{x} = \mathbf{h}_1$ solves our problem (A.3.21). The upper triangular matrix S has a unique inverse matrix if and only if $S_{ii} \neq 0$ for all i. As P is regular, regularity of S is equivalent to regularity of A.

A.3.3 Nonlinear Case: A Gauß–Newton Method

In order to solve the nonlinear problem (A.3.1) we can treat it as an unconstrained optimization problem by computing the gradient $\nabla f(\mathbf{x})$ as well as the Hessian matrix H and proceeding as described in Appendix A.1. This approach derives the necessary conditions, linearizes them, and ends up with the normal equations. From a numerical point of view, this approach is not recommended. Nevertheless, for didactical reasons, we will describe it and discuss its structure. In addition we also sketch an equivalent approach avoiding the numerical problems associated with the solution of the normal equation.

A.3 Unconstrained Least-Squares Procedures

The gradient, $\nabla f(\mathbf{x})$, of $f(\mathbf{x})$ takes the simple form

$$\nabla f(\mathbf{x}) = 2\mathsf{J}^T \mathbf{R} = 2\nabla \mathbf{R}(\mathbf{x})^T \mathbf{R}(\mathbf{x}) \in \mathbb{R}^n \qquad (A.3.23)$$

with the Jacobian matrix J of $\mathbf{R}$

$$\mathsf{J}_{ij} := \frac{\partial R_i}{\partial x_j} \quad \Longleftrightarrow \quad \mathsf{J}(\mathbf{x}) := \nabla \mathbf{R}(\mathbf{x}) = \left[\nabla R_1^T, \nabla R_2^T, \dots, \nabla R_N^T\right] \; . \qquad (A.3.24)$$

The Hessian H of $f(\mathbf{x})$ follows as

$$\mathsf{H} = 2\mathsf{J}^T \mathsf{J} + 2\mathsf{B}(\mathbf{x}), \quad \mathsf{H} \in \mathcal{M}(n,n), \qquad (A.3.25)$$

with

$$\mathsf{B}(\mathbf{x}) := \sum_{i=1}^{N} \left[R_i \nabla^2 R_i\right]. \qquad (A.3.26)$$

If the second derivatives $\nabla^2 R_i$ are readily available, then (A.3.25) can be used in the quasi-Newton method. However, in most practical cases it is possible to utilize a typical property of least-squares problems. The residuals R_i are expected to be small at a solution point $\mathbf{x}_*$, and H might, under the "small residual assumption," be sufficiently well approximated by

$$\mathsf{H} \approx 2\mathsf{J}^T \mathsf{J}. \qquad (A.3.27)$$

This approximation of the Hessian matrix is also achieved if the residuals R_i are taken up to linear order. Note that by this approximation the second derivative method, the Hessian matrix, H, requires only first derivative information. This is typical for least-squares problems, and this special variant of Newton's method is called the *Gauß–Newton method*. The damped Gauß–Newton method including a line search iterates the solution of $\mathbf{x}_k$ of the kth iteration to $\mathbf{x}_{k+1}$ according to the following scheme:

- determination of a search direction $\mathbf{s}_k$ by solving the linear system

$$\mathsf{J}_k^T \mathsf{J}_k \mathbf{s}_k = -\mathsf{J}_k^T \mathbf{R}_k \qquad (A.3.28)$$

derived from (A.1.15);
- solving the *line search subproblem*, i.e., computing the damping factor

$$\alpha_k = \arg\min_{\alpha}\{f(\mathbf{x}_k + \alpha \mathbf{s}_k) \mid 0 < \alpha \leq 1\}; \qquad (A.3.29)$$

- defining

$$\mathbf{x}_{k+1} = \mathbf{x}_k + \alpha_k \mathbf{s}_k. \tag{A.3.30}$$

The Gauß–Newton method and its convergence properties depend strongly on the approximation of the Hessian matrix. In large residual problems, B(**x**) in formula (A.3.25) is not negligible when compared to $\mathsf{J}^\mathsf{T}\mathsf{J}$, and the rate of convergence becomes poor. In fact, for B(**x**) $\neq$ 0 and, sufficiently close to an optimal solution, the Gauß–Newton method achieves only a linear convergence rate. Only for B(**x**) = 0 quadratic convergence rate can be achieved. Nevertheless, it represents the traditional (although not recommended) way to solve nonlinear least-squares problems; below we show how to do better.

Note that the linear equations (A.3.28) to be solved at each iteration k are the normal equations of the linear least-squares problem

$$\min \|\mathbf{y} - \mathsf{A}\mathbf{x}\|_2^2 \tag{A.3.31}$$

with

$$\mathbf{x} = \mathbf{s}_k, \quad \mathbf{y} = \mathbf{R}_k, \quad \mathsf{A} = -\mathsf{J}_k. \tag{A.3.32}$$

We are already aware that there exist numerical techniques to solve the linear least-squares problem (A.3.31) avoiding the normal equations and using only A and **y**. If we linearize the nonlinear least-squares problem slightly differently (similar to Sect. 4.2.2) we get a Gauß–Newton method avoiding the formulation of the normal equations completely. It is based on a Taylor series expansion of the residual vector **R**(**x**) to first order:

$$\min_{\mathbf{x}} f(\mathbf{x}) = \min_{\mathbf{x}} \|\mathbf{R}(\mathbf{x})\|_2^2 \doteq \min_{\mathbf{x}} \|\mathbf{R}(\mathbf{x}_k) + \mathsf{J}(\mathbf{x}_k)(\mathbf{x} - \mathbf{x}_k)\|_2^2. \tag{A.3.33}$$

Note that the necessary optimality conditions of the linear least-squares problem (A.3.33) are again the normal equations of (A.3.31) and (A.3.32). This shows that the solutions of (A.3.33) and our original problem are the same. The expansion used in (A.3.33) is a good approximation of our original problem only

- if the residual vector **R**(**x**), or equivalently B(**x**), is sufficiently small; or
- if the difference $\Delta \mathbf{x} := \mathbf{x} - \mathbf{x}_k$ is sufficiently small.

In a damped Gauß–Newton method with step-size cutting or damping parameter α, the original problem (A.3.33) is therefore replaced by

$$\min_{\Delta \mathbf{x}} \|\mathbf{R}(\mathbf{x}_k) + \mathsf{J}(\mathbf{x}_k)\Delta \mathbf{x}\|_2^2 \tag{A.3.34}$$

subject to the line search problem described in the previous section. At first, the linear least-squares problem (A.3.34), with $\mathbf{y} = \mathbf{R}(\mathbf{x}_k)$ and $\mathsf{A} = -\mathsf{J}(\mathbf{x}_k)$, is solved with the orthogonalization method described in Appendix A.3.2, yielding the search direction $\Delta \mathbf{x}$. Then, we put

$$\mathbf{x}_{k+1} = \mathbf{x}_k + \alpha_k \Delta\mathbf{x}, \quad \alpha_k := \arg\min_{\alpha}\{f(\mathbf{x}_k + \alpha\Delta\mathbf{x}) \mid 0 < \alpha \leq 1\}, \quad (A.3.35)$$

where the damping factor α_k is determined by solving a *line search subproblem* or by exploiting *natural level functions* as in Bock (1987).

A.4 Constrained Least-Squares Procedures

While many special purpose solvers for unconstrained least-squares problems are available as public domain or commercial software, the situation is different for constrained least-squares problems of the type

$$\textbf{NLCLS} \qquad \min f(\mathbf{x}) = \min \|\mathbf{R}(\mathbf{x})\|_2^2 \quad , \quad \mathbf{x} \in \mathbb{R}^n,$$

subject to

$$\mathbf{F}_2(\mathbf{x}) = 0, \qquad (A.4.1)$$
$$\mathbf{F}_3(\mathbf{x}) \geq 0. \qquad (A.4.2)$$

One basic technique to solve nonlinear constrained optimization problems is the SQP method described in Appendix A.2.2. However, as most nonlinear least-squares problems are ill conditioned, it is not recommended solving the *NLCLS* problem directly by a nonlinear programming method. Instead, we might use the transformation described by Schittkowski (1988). This transformation and the subsequent solution of the problem by SQP methods retain typical features of a special purpose code and eliminate the need to take care of any negative eigenvalues of an approximated Hessian matrix.

Another promising approach is the generalized Gauß–Newton methods (see Sect. 4.2.2) developed by Bock (1987) and coworkers. This group provides several subroutine libraries to support the solution of least-squares problems based on models involving equations and/or differential equations. In certain cases inequalities are also supported. However, concerning the incorporation of inequalities into least-squares problems great care is required regarding the interpretation of these inequalities. If the model reflects the physics well enough, and the data are of good quality, we might argue that inequalities should not be necessary at all. However, inequalities might be used to prevent the solver, for numeric reasons, entering certain regions in parameter space while iterating. Consider, for example, a physical parameter p limited to $0 \leq p \leq 1$ and simulated data based on $p = 0.98$ contaminated with Gaussian noise. Because of the statistical noise in the data, the solver might try to evaluate a model for $p = 1.001$ in the course of iterations. If, however, some inequalities are active at the solution, a careful interpretation of the result is necessary.

A.5 Selected Bibliography

This section is intended to guide the reader to recommended books or articles related to mathematical optimization and least-squares problems.

- *Solving Least-Squares Problems* by Lawson & Hanson (1974) provides a useful introduction into the field of least squares. The book covers mostly linear least squares. A newer edition appeared in 1987.
- Branham (1990) gives an introduction to overdetermined systems and scientific data analysis on an easy-to-follow level.
- The introduction into numerical analysis by Stoer et al. (1992) provides, besides other topics, a strict and serious mathematical treatment of the linear and nonlinear least square problems.
- Readers interested in practical optimization are referred to Gill et al. (1981), a useful source on ideas and implementation issues of algorithms used in mathematical optimization.
- *Newton Methods for Nonlinear Problems* by Deuflhard (2004) gives an overview on Newton techniques to solve optimization and least-squares problems.
- *Numerical Optimization* by Nocedal & Wright (2006) is an excellent book on optimization and as such covers most of the modern methods. It covers many of the algebra-related details very well.

Appendix B
Estimation of Fitted Parameter Errors: The Details

Hoc opus, hic labor est *(Here is the work – and the labor; this is the really tough one)*

This appendix describes useful techniques regarding the estimation of errors associated with the determined parameters in *EB* analysis. The Kolmogorov–Smirnov test, a procedure which checks whether the residuals fit the normal distribution, is reviewed. We also present here the sensitivity analysis approach and the grid approach, as they provide realistic estimations of the parameter errors.

B.1 The Kolmogorov–Smirnov Test

With the Kolmogorov–Smirnov test [cf. Ostle (1963)] it is possible to check whether the residuals of a least-squares solution are normally distributed around the mean value 0. An alternative is the χ^2-test. As Linnell's program (Linnell 1989) uses the Kolmogorov–Smirnov test we prefer this method, which works as follows:

1. let $M := (x_1, x_2, ..., x_n)$ be a set of observations for which a given hypothesis should be tested;
2. let $G : x \in M \to \mathbb{R}, x \to G(x)$ be the corresponding cumulative distribution function;
3. for each observation $x \in M$ define $S_n(x) := k/n$, where k is the number of observations less than or equal to x;
4. determine the maximum $D := \max(G(x) - S_n(x) \mid x \in M)$;
5. D_{crit} denotes the maximum deviation allowed for a given significance level and a set of n elements. D_{crit} is tabulated in the literature, e.g., Ostle (1963, Appendix 2, p. 560) ; and
6. if $D < D_{\text{crit}}$, the hypothesis is accepted.

In our case the hypothesis is "The residuals $x_\nu := l_\nu^o - l_\nu^c$ are normally distributed around the mean value 0." Therefore, the cumulative distribution function $G(x)$ takes the form

$$\sqrt{2\pi}G(x) = \int_{-\infty}^{x} g(z)dz = \int_{-\infty}^{-x_0} g(z)dz + \int_{-x_0}^{x} g(z)dz, \quad g(z) := e^{-\frac{1}{2}z^2}. \quad (B.1.1)$$

In a typical light curve analysis, the absolute values of the residuals are usually smaller than $x_0 = 0.025$ light units assuming unit light at maximum. If we take $G(-x_0) = 0.490$ from a table, it is no problem to compute the second part of the integral numerically. Unfortunately, in many cases the errors in EB observations do not follow the normal distribution.

B.2 Sensitivity Analysis and the Use of Simulated Light Curves

Let us assume that a light curve solution $\mathbf{x}_*$ has been derived, and the corresponding calculated light curve $\mathcal{O}^{cal}$ is available. If we add some noise Δl_ν to the computed light $l_\nu^c(\mathbf{x}_*)$ we get the *simulated light curve* $\mathcal{O}^{sim}$. The values Δl_ν follow from (4.1.21):

$$\Delta l_\nu = \sigma \varepsilon l_\nu^b, \quad (B.2.1)$$

where σ is a characteristic measure for the noise of the data. Photoelectric light curves of high quality may have $\sigma = 0.005$ light units assuming unit light at maximum, less good observations rather have $\sigma \approx 0.01$ light units. The variable ε denotes a normalized random variable obeying a normal distribution; indeed, observations usually produce residuals ε which follow a normal distribution around a mean value $\bar{\varepsilon} = 0$. A set of normally distributed values ε can be generated as follows (Brandt 1976, p. 57):

1. Assume that the mean of the distribution function is a and that the standard deviation is σ. In addition, we ask for the biased values

$$a - 5\sigma \leq \varepsilon \leq a + 5\sigma. \quad (B.2.2)$$

The reason for this bias is that in light curve analysis we usually do not observe outliers beyond a range of 5σ.

2. Let $\rho(i)$ be a function that produces uniformly distributed random numbers within the interval [0, 1]. The generation of such functions is usually part of any Fortran compiler; i is an arbitrary number.

3. Let $g(z) = \dfrac{1}{\sqrt{2\pi}\sigma} e^{-\frac{(z-a)^2}{2\sigma}}$ be the Gaussian function. In addition, $g_{max} = g(a)$ denotes the amplitude of the Gaussian function.

4. By means of uniform random numbers $x = \rho(i)$, $0 \leq x \leq 1$, a test value $\varepsilon = a + 5\sigma(2x - 1)$ and $g(\varepsilon)$ is computed. Furthermore, a value $\varepsilon_t = \rho(i)g_{max}$ is computed to compare it with ε.

5. If $g(\varepsilon) \leq \varepsilon$, then ε is accepted as an additional random number. Otherwise, go back to 4. Due to this transformation the set of elements ε is normally distributed and has the desired properties. In our case, of course, we have $a = 0$ and $\sigma = 1$.

Now the simulated light curve $\mathcal{O}^{sim}$

$$\mathcal{O}^{sim} := \mathcal{O}^{sim}(\mathbf{x}_*) = \{l_\nu^s \mid l_\nu^s = l_\nu^c(\mathbf{x}_*) + \Delta l_\nu(\mathbf{x}_*)\} \quad (B.2.3)$$

can be reanalyzed and the parameters derived. The results will show what effect errors in the observational data will have on the uncertainty of the parameters. For sufficiently small noise, in well-behaved regions of the parameter space, the original parameter vector $\mathbf{x}_*$ is recovered within small error bounds. But if noise increases, uniqueness problems may arise, and the recovery of the parameters be jeopardized. Note that this kind of analysis is a local one. It holds only for the parameter set of interest.

Although analysis can be used to investigate parameter uncertainties, we should bear in mind that (in many cases) the residuals in *EB* observations may not follow a normal distribution.

B.3 Deriving Bounds for Parameters: The Grid Approach

The *grid approach* is very useful if some of the parameters are correlated or can only be determined with great uncertainty. Usually the mass ratio q or the inclination i are such parameters. A parameter x may be constrained by or likely be located in the interval

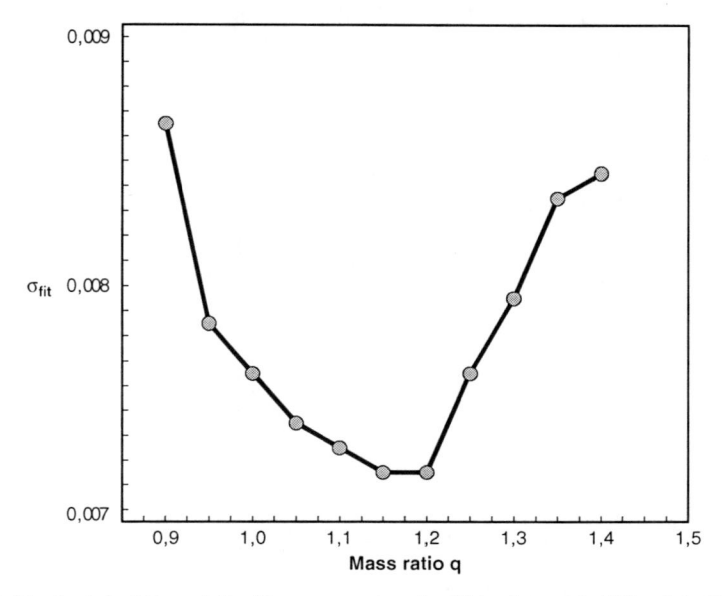

Fig. B.1 Standard deviation of the fit versus mass ratio. This plot, part of Fig. 4 in Kallrath & Kämper (1992), shows the standard deviation of the fit, σ^{fit}, versus the mass ratio, q

$$X^- \leq x \leq X^+. \tag{B.3.1}$$

Now, for some equidistant value x_i in this interval, x is fixed to x_i and the inverse problem is solved yielding the other light curve parameters and $\sigma^{\text{fit}} = \sigma^{\text{fit}}(x_i)$. If σ^{fit} is plotted versus x_i as shown in Fig. B.1, very often the result is a curve with a flat region as in Kallrath & Kämper (1992). The limits x^- and x^+ of that flat region yield realistic bounds on the uncertainty of the parameter x:

$$X^- \leq x^- \leq x \leq x^+ \leq X^+. \tag{B.3.2}$$

Appendix C
Geometry and Coordinate Systems

η ισότης η γεωμετρικη μέγα δύναται

(The geometrical identity has great meaning)

This part of the appendix contains some additional material on coordinate systems and geometry.

C.1 Rotation of Coordinate Systems

Consider two right-handed Cartesian coordinate systems with the same origin. Let the second system with coordinates (x', y', z') be generated by a counterclockwise rotation of the first one with coordinates (x, y, z) around its z-axis by an angle α. This situation is demonstrated in Fig. C.1. Then a point (x, y, z) in the first system has the coordinates (x', y', z') in the second system, and they can be computed according to

$$\begin{pmatrix} x' \\ y' \\ z' \end{pmatrix} = R_z(\alpha) \begin{pmatrix} x \\ y \\ z \end{pmatrix}, \quad R_z(\alpha) := \begin{pmatrix} \cos\alpha & \sin\alpha & 0 \\ -\sin\alpha & \cos\alpha & 0 \\ 0 & 0 & 1 \end{pmatrix}, \quad (C.1.1)$$

with

$$R_z(\alpha) := \begin{pmatrix} \cos\alpha & \sin\alpha & 0 \\ -\sin\alpha & \cos\alpha & 0 \\ 0 & 0 & 1 \end{pmatrix}. \quad (C.1.2)$$

The explicit formula reads

$$\begin{pmatrix} x' \\ y' \\ z' \end{pmatrix} = \begin{pmatrix} x\cos\alpha + y\sin\alpha \\ -x\sin\alpha + y\cos\alpha \\ z \end{pmatrix}. \quad (C.1.3)$$

The matrix $R_z(\alpha)$ is an orthogonal matrix, i.e.,

$$R_z(\alpha) R_z^{-1}(\alpha) = R_z^{-1}(\alpha) R_z(\alpha) = \mathbb{1}. \quad (C.1.4)$$

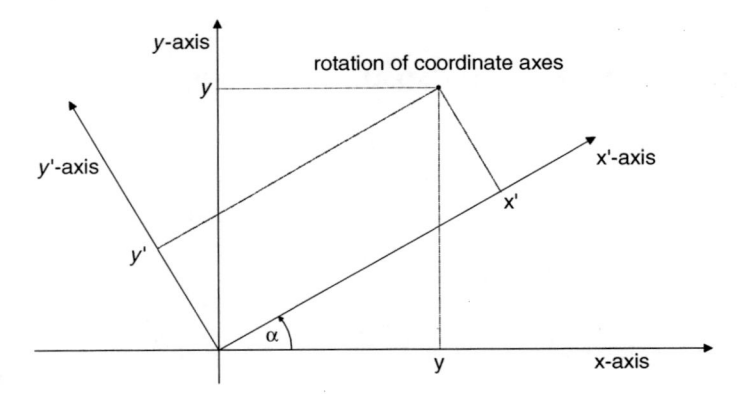

Fig. C.1 Rotation of Cartesian coordinate systems

Its inverse is the matrix

$$R_z^{-1}(\alpha) := \begin{pmatrix} \cos\alpha & -\sin\alpha & 0 \\ \sin\alpha & \cos\alpha & 0 \\ 0 & 0 & 1 \end{pmatrix}, \qquad (C.1.5)$$

which may be used to compute (x, y, z) as a function of (x', y', z').

Some caution is needed when coordinate systems are connected by several rotations around different axes. As commutativity is violated in most cases, i.e.,

$$R_x(\alpha)R_y(\beta) \neq R_y(\beta)R_x(\alpha), \qquad (C.1.6)$$

the sequence of rotations has to be considered very carefully.

C.2 Volume and Surface Elements in Spherical Coordinates

Let us start with the derivation of the volume element. The derivation of the volume element in spherical coordinates is based on the n-dimensional substitution rule in calculus:

$$\int_{\mathbf{g}(I)} f(\mathbf{u})d\mathbf{u} = \int_I f(\mathbf{g}(\mathbf{v})) \left|\det \mathbf{g}'(\mathbf{v})\right| d\mathbf{v}, \qquad (C.2.1)$$

where $\mathbf{g}(\mathbf{v})$ describes the coordinate transformation from coordinates $\mathbf{v}$ to $\mathbf{u}$, I is the domain of integration in the coordinates $\mathbf{v}$, and $\mathbf{g}'(\mathbf{v})$ is the Jacobian matrix associated with $\mathbf{g}$. In our case, the vectors $\mathbf{u}$ and $\mathbf{v}$ represent the Cartesian and spherical polar coordinates θ and φ introduced in Sect. 3.1.1, i.e.,

C.2 Volume and Surface Elements in Spherical Coordinates

$$\mathbf{u} = \begin{pmatrix} x \\ y \\ z \end{pmatrix} = \mathbf{g}(r, \theta, \varphi), \quad \mathbf{g}(r, \theta, \varphi) = r \begin{pmatrix} \cos\varphi \sin\theta \\ \sin\varphi \sin\theta \\ \cos\theta \end{pmatrix}, \tag{C.2.2}$$

and

$$I = I_r \times I_\theta \times I_\varphi = [0, \infty) \times [0, 180°] \times [0, 360°]. \tag{C.2.3}$$

The Jacobian matrix is

$$\mathbf{g}'(\mathbf{v}) := \frac{\partial \mathbf{g}(\mathbf{v})}{\partial \mathbf{v}} = \begin{pmatrix} \cos\varphi \sin\theta & r\cos\varphi \cos\theta & -r\sin\varphi \sin\theta \\ \sin\varphi \sin\theta & r\sin\varphi \cos\theta & r\cos\varphi \sin\theta \\ \cos\theta & -r\sin\theta & 0 \end{pmatrix}, \tag{C.2.4}$$

and the determinant is

$$\det \mathbf{g}'(\mathbf{v}) = r^2 \sin\theta. \tag{C.2.5}$$

Note that due to $0 \leq \theta \leq 180°$ the expression $r^2 \sin\theta$ is always nonnegative. Therefore, the volume element in spherical polar coordinates follows as

$$dV = |\det \mathbf{g}'(\mathbf{v})| \, d\mathbf{v} = r^2 \sin\theta \, dr \, d\theta \, d\varphi. \tag{C.2.6}$$

The derivation of the surface element in the form (3.1.5) requires some formulas from vector calculus and differential geometry. At first we review that for three-dimensional vectors $\mathbf{a}$, $\mathbf{b}$, and $\mathbf{c}$ the scalar product $\mathbf{a} \cdot (\mathbf{b} \times \mathbf{c})$ can be represented by a determinant according to

$$\mathbf{a} \cdot (\mathbf{b} \times \mathbf{c}) = \det \begin{pmatrix} a_1 & b_1 & c_1 \\ a_2 & b_2 & c_2 \\ a_3 & b_3 & c_3 \end{pmatrix}, \tag{C.2.7}$$

where $\mathbf{b} \times \mathbf{c}$ denotes the *cross product* or *outer product* defined as

$$\mathbf{b} \times \mathbf{c} = (b_2 c_3 - b_3 c_2, b_3 c_1 - b_1 c_3, b_1 c_2 - b_2 c_1)^\mathrm{T}. \tag{C.2.8}$$

As the determinant of a matrix with linearly dependent columns vanishes we further have

$$\mathbf{a} \times \mathbf{a} = 0, \tag{C.2.9}$$

and

$$\mathbf{a} \cdot (\mathbf{a} \times \mathbf{c}) = 0. \tag{C.2.10}$$

Finally, we need the distributive law for the cross product, i.e.,

$$(\mathbf{v}_1 + \mathbf{v}_2) \times (\mathbf{v}_3 + \mathbf{v}_4) = \mathbf{v}_1 \times \mathbf{v}_3 + \mathbf{v}_1 \times \mathbf{v}_4 + \mathbf{v}_2 \times \mathbf{v}_3 + \mathbf{v}_2 \times \mathbf{v}_4. \quad \text{(C.2.11)}$$

Let us now consider a surface embedded in a three-dimensional vector space parametrized by two coordinates θ and φ, i.e.,

$$\mathbf{r} = \mathbf{r}(\theta, \varphi) = \begin{pmatrix} x(\theta, \varphi) \\ y(\theta, \varphi) \\ z(\theta, \varphi) \end{pmatrix}. \quad \text{(C.2.12)}$$

The normal vector, $\mathbf{n}$, at a surface point, $\mathbf{r}$, is defined as

$$\mathbf{n} := \frac{\mathbf{r}_\theta \times \mathbf{r}_\varphi}{|\mathbf{r}_\theta \times \mathbf{r}_\varphi|}, \quad \text{(C.2.13)}$$

where $\mathbf{r}_\theta$ and $\mathbf{r}_\varphi$ denote the partial derivatives of $\mathbf{r}$ w.r.t. θ and φ. If the surface encloses a finite region of the three-dimensional space, the normal vector $\mathbf{n}$ points into the region external to this volume, and the vectors $\mathbf{r}_\theta$, $\mathbf{r}_\varphi$, and $\mathbf{n}$ establish the unit axis of a right-handed coordinate system. The surface element, $d\sigma$, is defined as

$$d\sigma := |\mathbf{r}_\theta \times \mathbf{r}_\varphi| d\theta d\varphi. \quad \text{(C.2.14)}$$

If we multiply (C.2.13) by $\mathbf{e}_r$, we can eliminate $|\mathbf{r}_\theta \times \mathbf{r}_\varphi|$ from (C.2.14) by using $\cos \beta$ in definition (3.1.3), i.e.,

$$\cos \beta := \frac{\mathbf{r}}{r} \cdot \mathbf{n} = \mathbf{e}_r \cdot \mathbf{n}. \quad \text{(C.2.15)}$$

If we do so, we get the expression

$$d\sigma := \frac{1}{\cos \beta} \mathbf{e}_r \cdot (\mathbf{r}_\theta \times \mathbf{r}_\varphi) d\theta d\varphi. \quad \text{(C.2.16)}$$

We now need to evaluate the term

$$\mathbf{e}_r \cdot (\mathbf{r}_\theta \times \mathbf{r}_\varphi) \quad \text{(C.2.17)}$$

for the special case of interest

$$\mathbf{r} = \mathbf{r}(\theta, \varphi) = r(\theta, \varphi)\mathbf{e}_r = r(\theta, \varphi) \begin{pmatrix} \cos \varphi \sin \theta \\ \sin \varphi \sin \theta \\ \cos \theta \end{pmatrix}. \quad \text{(C.2.18)}$$

C.2 Volume and Surface Elements in Spherical Coordinates

In this case we have

$$\mathbf{r}_\theta = r_\theta \mathbf{e}_r + \mathbf{r}_\theta^S, \quad \mathbf{r}_\varphi = r_\varphi \mathbf{e}_r + \mathbf{r}_\varphi^S \tag{C.2.19}$$

where $\mathbf{r}_\theta^S$ and $\mathbf{r}_\varphi^S$ denote the derivatives for the unit-sphere case, i.e.,

$$\mathbf{r}_\theta^S = r(\theta, \varphi) \begin{pmatrix} \cos\varphi \cos\theta \\ \sin\varphi \cos\theta \\ \cos\theta \end{pmatrix}, \quad \mathbf{r}_\varphi^S = r(\theta, \varphi) \begin{pmatrix} -\sin\varphi \sin\theta \\ \cos\varphi \sin\theta \\ \cos\theta \end{pmatrix}.$$

If we substitute (C.2.19) into (C.2.17) and apply (C.2.11), we get

$$\mathbf{e}_r \cdot (\mathbf{r}_\theta \times \mathbf{r}_\varphi) = \mathbf{e}_r \cdot (\mathbf{r}_\theta^S \times \mathbf{r}_\varphi^S). \tag{C.2.20}$$

The other three terms vanish due to (C.2.7) and (C.2.9). If we evaluate (C.2.20) we get

$$\mathbf{e}_r \cdot (\mathbf{r}_\theta^S \times \mathbf{r}_\varphi^S) = \det \begin{pmatrix} \cos\varphi \sin\theta & r\cos\varphi \cos\theta & -r\sin\varphi \sin\theta \\ \sin\varphi \sin\theta & r\sin\varphi \cos\theta & r\cos\varphi \sin\theta \\ \cos\theta & -r\sin\theta & r\cos\theta \end{pmatrix} = r^2 \sin\theta. \tag{C.2.21}$$

So, finally, we get the surface element

$$d\sigma = \frac{1}{\cos\beta} r^2 \sin\theta \, d\theta \, d\varphi. \tag{C.2.22}$$

For readers who prefer graphical demonstrations, we also provide the following: we can apply the sine rule to the infinitesimal small triangle ABC (left part of Fig. C.2):

$$\frac{ds}{\sin d\theta} = \frac{r}{\sin(180° - d\theta - 90° - \beta)}. \tag{C.2.23}$$

Considering that $d\theta$ is an infinitesimal small angle, i.e., $\sin d\theta \approx d\theta$ and $\cos(\beta + d\theta) \approx \cos\beta$, we get

$$\frac{ds}{d\theta} = \frac{r}{\cos\beta} \quad \Leftrightarrow \quad ds = \frac{r}{\cos\beta} d\theta. \tag{C.2.24}$$

The right part of Fig. C.2 shows the projection of r onto the x–y plane, yielding

$$R = r\cos(90° - \theta) = r\sin\theta, \tag{C.2.25}$$

and an infinitesimal small triangle containing the angle $d\varphi$ and the arc dl. The arc dl follows as

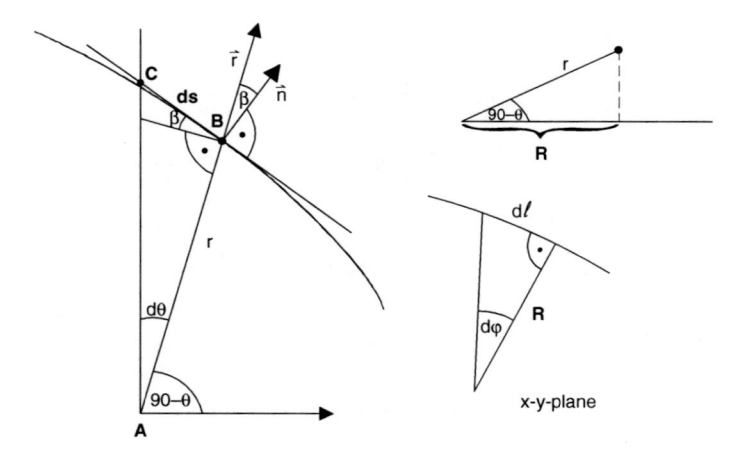

Fig. C.2 Derivation of the surface element

$$\frac{\mathrm{d}l}{R} = \tan\mathrm{d}\varphi \quad \Leftrightarrow \quad \mathrm{d}l \approx R\mathrm{d}\varphi = r\sin\theta\mathrm{d}\varphi. \tag{C.2.26}$$

As θ and φ establish an orthogonal coordinate system we get

$$\mathrm{d}\sigma = \mathrm{d}s\mathrm{d}l = \frac{1}{\cos\beta}r^2\sin\theta\mathrm{d}\varphi\mathrm{d}\theta \tag{C.2.27}$$

as before.

C.3 Roche Coordinates

Although the Roche potential is frequently used among binary astronomers, the Roche coordinates are less known. Roche coordinates, as investigated by Kitamura (1970) and Kopal (1970, 1971), try to establish an (orthogonal) coordinate system (u, v, w) associated with the Roche potential Ω in the circular case. The centers of both components are singularities of this coordinate system. The first coordinate is just equal to the Roche potential, i.e., $u = \Omega$. Unlike polar coordinates or other more frequently used coordinate systems which are related to the Cartesian coordinates by some explicit formulas, Roche coordinates cannot be described by closed analytical expression but can only be evaluated numerically. This property limits the practical use. Kopal & Ali (1971) and again Hadrava (1987) showed that it is not possible to establish a system of three orthogonal coordinates based on the Roche potential. The requirement that such coordinates exist [Cayley's (1872a, b) problem] imposes a necessary condition [the Cayley–Darboux equation (Darboux 1898)] which must be satisfied by the function Ω. Hadrava (1987) defines generalized Roche coordinates in asynchronous rotation binaries with eccentric orbits and calculates them in the form of power series of the potential (3.1.77). His definition abandons the orthogonality of the coordinates v and w. As an example for the application of Roche coordinates we mention Hadrava's (1992) investigation of the radiative transfer in an atmosphere or in a rotationally oblated star.

Finally, the reader more interested in the mathematical background of coordinate systems and the classification of the Roche coordinates is referred to Neutsch (1995).

C.4 Solving Kepler's Equation

Standard and improved techniques for solving Kepler's equation are described, e.g., in Neutsch & Scherer (1992) – a rich resource book strongly recommended for celestial mechanics in general. A less well-known technique to solve Kepler's equation is the following iterative scheme based on Padé approximants. The roots of the function

$$f(x) := x - e \sin x - M \qquad (C.4.1)$$

can be computed according to

$$x_{i+1} = x_i + \Delta x_i, \quad \Delta x_i = -\frac{ff'}{f'^2 - \frac{1}{2}ff''}, \qquad (C.4.2)$$

with the correction, Δx_i,

$$\begin{aligned}\Delta x_i &= -\frac{ff'}{f'^2 - \frac{1}{2}ff''} \\ &= \frac{(x_i - e \sin x_i - M)(1 - e \cos x_i)}{(1 - e \cos x_i)^2 - \frac{1}{2}(x - e \sin x_i - M)e \sin x_i}.\end{aligned} \qquad (C.4.3)$$

This method is also known as Halley's method. The derivation based on the formal concept of Padé approximants is described in Kallrath (1995). The advantage of this method is that it has a large convergence region; it does not depend too much on the initial value x_0.

Appendix D
The Russell–Merrill Model

Locus poenitentiae   *(Opportunity for repentence)*

D.1 Ellipticity Correction in the Russell–Merrill Model

This material gives further details on the Russell–Merrill model already discussed in Sect. 6.2.1.

The expression for the ellipticity correction was developed as follows (Russell & Merrill 1952, p. 43). The basic model for the unrectified system consists of two similar triaxial ellipsoids with equal limb-darkening and gravity (or, sometimes, "gravity brightening") coefficients. For synchronous rotation, the ellipticity of star 1 is assumed to be representable by equations of the kind

$$\frac{a_1 - b_1}{\bar{r}_1} = \frac{3m_2}{2m_1}(1 + 2K_1)\bar{r}_1^3 \qquad (D.1.1)$$

and

$$\frac{b_1 - c_1}{\bar{r}_1} = \frac{m_1 + m_2}{2m_1}(1 + 2K_1)\bar{r}_1^3, \qquad (D.1.2)$$

where a, $b = a\sqrt{1 - \eta^2}$, and $c = a\sqrt{1 - \zeta^2}$ are the radii (a) in the line through the stellar centers, (b) in the direction perpendicular to this line but in the orbital plane, and (c) in the polar direction, respectively. Note that b and c are the radii "seen" during eclipses. The quantity K depends on the central condensation and lies in the range 0.0018–0.018. The surface brightness[1] in a given passband is assumed to obey

$$J = J_0(1 - x + x\cos\gamma)[1 - y(1 - g/\bar{g})], \qquad (D.1.3)$$

---

[1] Note that J is the usual symbol for mean intensity in radiative transfer theory. Russell used J instead of I for ordinary intensity.

where J_0 is the mid-disk surface brightness, x is the limb-darkening coefficient, g is the (local) gravitational acceleration, $\bar{g}$ is the average of g over the surface, and y is given by

$$y = \frac{c_2}{4\lambda T} \frac{e^{\frac{c_2}{\lambda T}}}{e^{\frac{c_2}{\lambda T}} - 1} \quad , \quad c_2 = \frac{hc}{k}. \tag{D.1.4}$$

The second expression in (D.1.4) is a gray-body approximation. A correction for different-sized components made use of a theorem demonstrated by Russell (1948) that the light distribution as given by (D.1.3) is the same for a uniformly bright star of surface brightness $J_0(1 - \frac{1}{3}x)$ and having axes $1 + Na$, $1 + Nb$, and $1 + Nc$, where

$$N = \frac{(15 + x)(1 - \frac{K-2}{4K+2}y)}{15 - 5x} \approx \frac{(15 + x)(1 + y)}{15 - 5x} \tag{D.1.5}$$

and where the factor multiplying y is in the range -0.996 to -0.957 for the expected range of K (0.0018–0.018, respectively).

The effect of the ellipticity on the light curve is to provide an observed variation in system flux:

$$\ell_{comp} = \left(1 - \tfrac{1}{2}Nz\cos^2\theta\right)(\ell_{max} - \ell_{1max} f_1 - \ell_{2max} f_2), \tag{D.1.6}$$

where $z = \eta^2 \sin^2 i$, and the observed flux is

$$\ell = A_0 + A_2 \cos 2\theta. \tag{D.1.7}$$

The coefficient A_2 is intrinsically negative, and Russell & Merrill (1952, p. 43) express the peak light as $\ell_{max} = A_0 - A_2$, and $Nz = -4\dfrac{A_2}{A_0 - A_2}$. The rectified light then becomes

$$\ell_{rect} = \frac{(A_0 - A_2)\ell_{comp}}{A_0 + A_2 \cos 2\theta} = \ell_{max}(1 - L_1 f_1 - L_2 f_2), \tag{D.1.8}$$

where $L_1 = \dfrac{\ell_{1max}}{\ell_{max}}$ and $L_2 = \dfrac{\ell_{2max}}{\ell_{max}}$. A rectification in phase is also required because of the ellipticity of the components; the rectified phase is computed from

$$\sin^2 \Theta = \frac{\sin^2 \theta}{1 - z\cos^2 \theta}. \tag{D.1.9}$$

D.1 Ellipticity Correction in the Russell–Merrill Model

This completes the rectification due to the deviation of the stars from sphericity. The effects of gravity brightening and reflection must be treated now. The effect of centrifugal forces is to widen the equatorial axes at the expense of the polar one and causes the star to brighten toward the rotation poles; the effect of tidal distortion (increasing the star's diameter along the line of centers) is to darken it at the extremes of the axis a. The net center-to-limb variation across an ellipsoid is given by (D.1.3). Express the coordinates in the directions of the ellipsoid axes as X, Y, and Z and take the gravity-darkening relative to the Y-axis, $a = b(1+u)$, $c = b(1-v)$, so that (D.1.3) may be written as

$$J = J_0(1 - x + x\cos\gamma)\left[1 + 4y\left(1 - \frac{r}{b}\right)\right]. \quad \text{(D.1.10)}$$

As $\cos\gamma$ is the direction cosine of the normal to the ellipsoid at the position X, $\cos\gamma = rX/a^2$. Setting $X = a\cos\beta$, and neglecting second-order terms in Z and u, the relative surface brightness J in (D.1.10) becomes

$$J = J_0\left\{1 - x + x\cos\beta - u\left[x\cos\beta + 4y(1-x)\cos^2\beta + (4y-1)x\cos^3\beta\right]\right\}. \quad \text{(D.1.11)}$$

The theoretical basis for the treatment of the reflection effect is that of Milne (1926) who assumed a parallel beam of incoming radiation, with corrections from Sen (1948) for penumbral effects. The expected enhancement in light from the second star "reflection" is

$$\Delta L_2 = f(\varepsilon)r_2^2 L_1, \quad \text{(D.1.12)}$$

where an approximation may be used for $f(\varepsilon)$

$$f(\varepsilon) = 0.30 + 0.40\cos\varepsilon + 0.10\cos 2\varepsilon. \quad \text{(D.1.13)}$$

Here ε is the phase angle from full phase (i.e., from mid-secondary minimum for star 2). The net result of the reflection effect alone (Russell & Merrill 1952, p. 45) is

$$\begin{aligned}\Delta L &= L_c r_h^2 f(\varepsilon_h) + L_h r_c^2 f(\varepsilon_c) \\ &= (L_c r_h^2 + L_h r_c^2)(0.30 - 0.10\cos^2 i + 0.10\sin^2 i \cos 2\theta) \\ &\quad + 0.40(L_c r_h^2 - L_h r_c^2)\sin i \cos\theta.\end{aligned} \quad \text{(D.1.14)}$$

Note that both ellipticity and reflection contribute $\cos 2\theta$ values, but with opposite sign, and the effect of reflection is to decrease the effect of ellipticity. The $\cos\theta$ contribution is due, however, to the reflection effect. Note that this is true for Russell's ellipsoid model, but not for an equipotential model.

Instead of removing the enhancement due to the reflection effect the prescription requires the addition of light at other phases. It is done this way to avoid iterative rectification. Given a situation in which the light curve outside the eclipse is well represented by a truncated Fourier series involving only terms A_{0-4}, and star 2 is heated by star 1, the total radiation reemitted by star 2 shows up in the effects on coefficients A_3 and A_4 as

$$A_3 = 0.005 L_1 r_2^2 < 0.0008 L_1, \qquad (D.1.15)$$

assuming $r_2 \leq 0.4$ and

$$A_4 = (0.002 - 0.050 r_2^2 - 0.006 r_1^2) L_1 r_2^2 < 0.0012 L_1 \qquad (D.1.16)$$

(Russell & Merrill 1952, p. 52). For cases, where the distortion of the figures of the stars is large, Russell and Merrill cite Kopal (1946, Eqs. (210) and (220.1) on pp. 135 and 139):

$$A_3 = -0.59 (L_2 m_1 / m_2) r_2^4 \qquad (D.1.17)$$

and

$$A_4 = +0.27 (L_2 m_1 / m_2) r_2^5. \qquad (D.1.18)$$

A crude form of the mass–luminosity relation ($m \propto L^{0.26}$) was then used to evaluate the magnitude of the expressions. As $A_3 \leq 0.33 r_2^4$ and thus $0.001 \leq A_3 \leq 0.01$ for $0.23 \leq r_2 \leq 0.41$ and $A_4 \leq 0.16 r_2^5$, so that $A_4 \leq 0.001$ for $r_2 \leq 0.36$.

For cases where both sine and cosine terms must be used to represent the maxima, the general prescription is to apply "empirical corrections" (Russell & Merrill 1952, pp. 53–54). They recommended dividing by the expression

$$A_0 + B_1 \sin\theta + B_2 \sin 2\theta \ldots, \qquad (D.1.19)$$

if the light perturbations are thought to affect both stars in proportion to their brightnesses, but otherwise producing no shape or surface brightness asymmetry. If star 1 alone were affected, they recommended subtraction of the sine terms from all phases but those at which star 1 was eclipsed, and during the eclipse of star 1, the subtraction of $(1 - f)(B_1 \sin\theta + B_2 \sin 2\theta)$, where f is the fraction of that star's light which is obscured at any particular phase.

Appendix E
Subroutines of the Wilson–Devinney Program

Finis coronat opus   *(The end crowns the work)*

The WD program is a practical expression of the WD model and there have been many more publications about the model than about the program. Even the best programs ordinarily have much shorter lifetimes than good ideas, so the model has been kept conceptually separate from its software implementation. Thus papers on the model essentially never mention the names of subroutines or main programs – only papers specifically about the program [e.g., Wilson (1993)] use those names. This appendix tries to provide some of the details which might improve the reader's understanding of the program and to connect the subroutines to the relevant features in the model.

    The program's history has been one of various special-purpose versions that were developed for particular problems, followed by absorption of their capabilities into the general program. Overall, the idea has been to have one "direct problem" program (LC) and one "inverse problem" program (DC), each with multiple capabilities, rather than a library of special purpose programs. For example, the 1998 version computed light curves, radial velocity curves, spectral line profiles, and images. Versions that compute polarization curves and X-ray pulse arrival times were created and still exist separately and may eventually be absorbed. The WD 2007 version incorporated improved atmosphere simulation, parameters of eclipses by clouds in extensive stellar atmospheres of hot stars, and third-body parameters. Generalizations that a user need not worry about are embedded invisibly wherever practical. For example, computational shortcuts for many special case situations speed execution without compromising more intricate cases.

    The overall structure of the Wilson–Devinney program and its main routines LC and DC is shown in Figs. E.2 and E.3. The purpose of some subroutines is explained in Wilson (1993) and in Wilson's program manuals. Other subroutines lack a detailed description of the mathematics. Although the overall ideas of computing *EB* light curves usually can be presented in an elegant and systematic manner, the implementation of these ideas in practice requires a substantial amount of numerical analysis. An example is the problem of computing accurate derivatives and providing them with a derivative-based least-squares solver. This topic is related to the accuracy of representing the model in the computer (both the discretization of the surface of the components represented by a finite grid and the finiteness of the number representation in the machine).

A subroutine having an asterisk in its name is part of the 2009, but not of the 2008 version of the Wilson–Devinney program.

E.1 ATM – Interfacing Stellar Model ATMospheres

This subroutine was the interface to a stellar model atmosphere. The original Wilson–Devinney program uses the Carbon–Gingerich model atmospheres. In WDx2007, this subroutine is replaced by a routine written by C. Stagg (Milone et al. 1992b), implementing the Kurucz atmospheres. In both cases it returns the ratio between the flux based on the atmosphere and the blackbody law. The version of Jan. 23, 2004, and later versions incorporate the models of Kurucz (1993) in an use of external tables of metallicity, temperature, and log g to provide modern atmosphere simulations. The tables are automatically read in the 2007 version.

E.2 ATMx – Interfacing Stellar Model ATMospheres

The WD subroutine interfacing to Kurucz atmosphere is for the ATMx. It performs a four-point interpolation in log g and then an m-point Lagrangian interpolation. This routine exploits an a priori computed file that contains a block for each of the 19 compositions, each block listing the temperature limits and Legendre coefficients for every band, log g, and temperature sub-interval. With 11 log g's, 25 bands, 19 compositions, 4 temperature subintervals, and 10 Legendre coefficients with 2 temperature limits per subinterval, the data file contains 250,800 numbers.

Where grid elements require elements outside the grid, a tranfer is made to a Planckian through a "ramp" function that provides smooth continuity. That way Van Hamme & Wilson (2003) established an atmosphere to blackbody transition region in T_{eff} and in log g and avoid any discontinuity. If the T_{eff}, log g combination is outside the range of atmosphere applicability, the program smoothly connects atmosphere model intensities to bandpass blackbody intensities over built-in ranges in log g and T_{eff} whose limits can easily be changed. Similarly to the files above, a Legendre blackbody file spans 500–500,000 K and is very small, as the only dimensions are bandpass and temperature sub-interval. Note that the files can be incorporated into other binary star programs to calculate model and blackbody intensities.

E.3 BBL – Basic BLock

Subroutine BBL is the so-called Basic BLock of the WD program (see Fig. E.1). It keeps the computation of numerical derivatives as simple as possible and at the same time tries to avoid redundant computations.

E.7 CLOUD – Atmospheric Eclipse Parameters

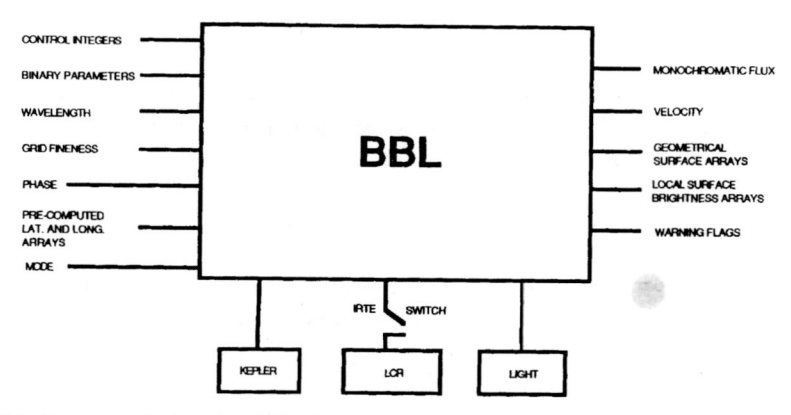

Fig. E.1 Structure of subroutine BBL. Courtesy R. E. Wilson

E.4 BinNum – A Search and Binning Utility

BinNum is a tool to find the bin in which a number is located. It is essentially similar to LOCATE in the *Numerical Recipes* by Press et al. (1992), which searches an ordered table by bisection.

E.5 BOLO – Bolometric Corrections

Subroutine BOLO uses Harris's (1963) calibration from $T = 3,970$ K to $5,800$ K, Morton & Adams (1968) from $T = 5,800$ to $37,500$ K, and the blackbody law (3.2.21) below $T = 3,970$ K and above $T = 37,500$ K to compute bolometric corrections. These corrections are needed to compute the ratio of bolometric luminosities involved in the reflection effect in Sect. 3.2.5.

E.6 CofPrep – Limb-Darkening Coefficient Preparation

Subroutine COFPREP computes the coefficients used in the limb-darkening interpolation scheme. It reads the numbers from the table selected based on input metallicity into an array which is then available to LIMDARK.

E.7 CLOUD – Atmospheric Eclipse Parameters

Added in the 2002 version of the WD program, subroutine CLOUD computes atmospheric eclipse parameters as described in Sect. 3.4.4.5.

E.8 CONJPH – Conjunction Phases

This subroutine computes the phases of superior and inferior conjunctions. For eccentric orbits this becomes a relevant and subtle issue as described on page 88.

E.9 DGMPRD – Matrix–Vector Multiplication

Subroutine DGMPRD performs matrix multiplication and returns the resulting vector **r** = A**b**, where matrix A ist stored in a one-dimensional chain and **b** is the input vector.

E.10 DMINV – Matrix Inversion

Subroutine DMINV computes the inverse of a n by n matrix A stored in a one-dimensional chain.

E.11 DURA – Constraint on X-Ray Eclipse Duration

This subroutine puts an explicit constraint on the size of a star based on the duration of an X-ray eclipse. Such constraints may be considered when X-ray eclipses of neutron stars, black-holes, or white dwarfs occur, as described in Wilson (1979) where the full mathematics is given. The basic observable is the semi-duration Θ_e of an X-ray eclipse. For circular orbits and synchronous rotation a relation

$$\Theta_e = \Theta_e(i, \Omega, q) \tag{E.11.1}$$

has been derived by Chanan et al. (1977) which relates the inclination i, the Roche potential Ω, the mass ratio q, and Θ_e. The more general eccentric, nonsynchronous case is in Wilson (1979).

E.12 ELLONE – Lagrangian Points and Critical Potentials

This subroutine computes the x-coordinates of the equilibrium points L_1^p and L_2^p and the associated critical Roche potentials Ω_1^{crit} and Ω_2^{crit}. The name *equilibrium points* is used here as a generalization of Lagrangian points, as used for the synchronous, circular case. The required input quantities are the mass ratio $q = \mathcal{M}_2/\mathcal{M}_1$, the ratio $F_1 = \omega_1/\omega$, and the distance d between the stars in units of the semi-major axis a of the relative orbit. The x-coordinate of the equilibrium points follows from the condition

E.12 ELLONE – Lagrangian Points and Critical Potentials

$$\Omega_x := \frac{\partial \Omega}{\partial x} = 0. \tag{E.12.1}$$

As the equilibrium points are located on the line connecting the centers of the components in (6.3.2), the distances of a point x on that line take a special form. The star centers are at positions $x_1 = 0$ and $x_2 = d$. The coordinate $x_{L_1^p}$ of the equilibrium point L_1^p fulfills the relation

$$0 \leq x_{L_1^p} \leq d, \tag{E.12.2}$$

so that its distances to the component centers are $x_{L_1^p}$ and $d - x_{L_1^p}$. Therefore, for $x_{L_1^p}$, the potential (6.3.2) takes the form

$$\Omega(\mathbf{r} = (x, 0, 0); q, F, d) = \frac{1}{x} + q\left[\frac{1}{\sqrt{d^2 - 2dx + x^2}} - \frac{x}{d^2}\right] + \frac{q+1}{2} F_1^2 x^2. \tag{E.12.3}$$

From (E.12.3), condition (E.12.1) takes the form

$$f(x) := \frac{\partial \Omega}{\partial x} = -\frac{1}{x^2} - q\frac{x-d}{|d-x|^3} + F_1^2(q+1)x - \frac{q}{d^2} = 0. \tag{E.12.4}$$

This nonlinear equation (E.12.4) is solved with the Newton–Raphson algorithm. Therefore, the derivative $f'(x)$ is needed:

$$f'(x) = \frac{2}{x^3} + \frac{2q}{|d-x|^3} + F_1^2(q+1). \tag{E.12.5}$$

Now the Newton–Raphson procedure proceeds as

$$x^{(n+1)} = x^{(n)} + \Delta x, \quad \Delta x = -f(x^{(n)})/f'(x^{(n)}) \tag{E.12.6}$$

with the initial value $x^{(0)} = d/2$. If sufficient accuracy is achieved, viz., if $|\Delta x| \leq 10^{-6}$, the iteration is halted and $x_{L_1^p}$ is set to

$$x_{L_1^p} := x^{(n+1)}. \tag{E.12.7}$$

With known $x_{L_1^p}$ it is easy to compute Ω_1^{crit} as

$$\Omega_1^{\text{crit}} = \frac{1}{x_{L_1^p}} + q\left(\frac{1}{|d - x_{L_1^p}|} - \frac{x_{L_1^p}}{d^2}\right) + \frac{q+1}{2} F_1^2 x_{L_1^p}^2. \tag{E.12.8}$$

The case to compute $x_{L_2^p}$ and Ω_2^{crit} for L_2^p is somewhat more difficult. The computation of $x_{L_2^p}$ and Ω_2^{crit} is only valid for $F_1 = F_2 = 1$ and $d = 1$. For nonsynchronous or noncircular cases these quantities are not needed. In the valid cases, the inequality

$$d \leq x_{L_2^p} \tag{E.12.9}$$

holds; Wilson uses the same function $f(x)$ as defined in (E.12.4) but computation is performed after explicit setting of $F_j = 1$ and $d = 1$. Furthermore, this time, the initial value

$$x^{(0)} = 1 + \mu^{1/3} + \tfrac{1}{3}\mu^{2/3} + \tfrac{1}{9}\mu^{3/3} + \ldots, \tag{E.12.10}$$

with

$$\mu := \tfrac{1}{3}\frac{Q}{Q+1}, \quad Q := \begin{cases} q, & 0 \leq q \leq 1, \\ q^{-1}, & q \geq 1. \end{cases} \tag{E.12.11}$$

is used. If convergence is completed, $x_{L_2^p}$ follows from

$$x_{L_2^p} := x^{(n+1)}. \tag{E.12.12}$$

For completeness we note for $d = 1$ and $F_1 = 1$

$$\Omega_2^{\text{crit}} = \frac{1}{x} + q\left(\frac{1}{\sqrt{1 - 2x + x^2}} - x\right) + \frac{q+1}{2}x^2 \tag{E.12.13}$$

with

$$x := \begin{cases} x_{L_2^p}, & 0 \leq q \leq 1, \\ 1 - x_{L_2^p}, & q \geq 1. \end{cases} \tag{E.12.14}$$

This subroutine now works also for very extreme mass ratios, e.g., $q = 10^{-6}$.

E.13 FOUR – Representing Eclipse Horizon

FOUR computes a Fourier series for the representation of the boundaries of the eclipsed regions. In the 1998 and later versions this subroutine is replaced by FOURLS.

E.14 FOURLS – Representing Eclipse Horizon

FOURLS computes the Fourier coefficients by solving a least-squares problem. It replaces subroutine FOUR present in versions older than 1998. The new subroutine fits the horizon points by least squares and avoids the Fourier approach. It is more accurate and the sorting routine is no longer needed.

E.15 GABS – Polar Gravity Acceleration

GABS computes the polar acceleration due to effective gravity in cm^2/s.

E.16 JDPH – Conversion of Julian Day Number and Phase

Subroutine JDPH allows to convert phase into Julian Day Number (JDN) but also JDN into phase. It computes a phase (phout) based on an input JD (xjdin), reference epoch (t0), period (p0), and dP/dt (dpdt). It also computes a JD (xjdout) from an input phase (phin) and the same ephemeris.

E.17 KEPLER – Solving the Kepler Equation

This subroutine solves Kepler's equation (3.1.28) with the Newton–Raphson scheme (see Appendix E.12) with initial value $E^{(0)} = M$. Iterations are stopped when $|\Delta E| \leq 10^{-10}$. Eventually, the true anomaly is computed according to (3.1.27). If $\upsilon < 0$, then 2π is added to ensure that $0 \leq \upsilon < 2\pi$.

E.18 LC and DC – The Main Programs

These are the main programs of the Wilson–Devinney program. LC solves the direct problem: From a given set of parameters, phased light and radial velocity curves, spectral line profiles, star dimensions, or sky coordinates for producing images are computed. DC solves the inverse problem: The parameters are derived from observations. The structure of the main programs is illustrated in Figs. E.2 and E.3.

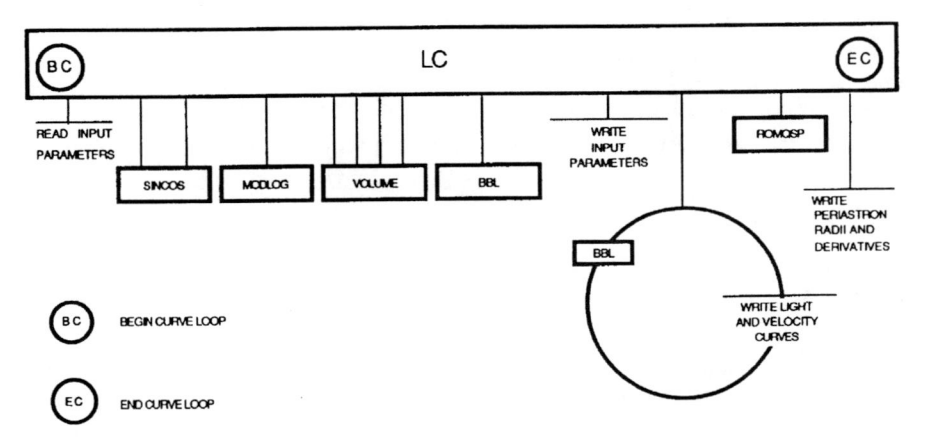

Fig. E.2 The WD main program: LC. Courtesy R. E. Wilson

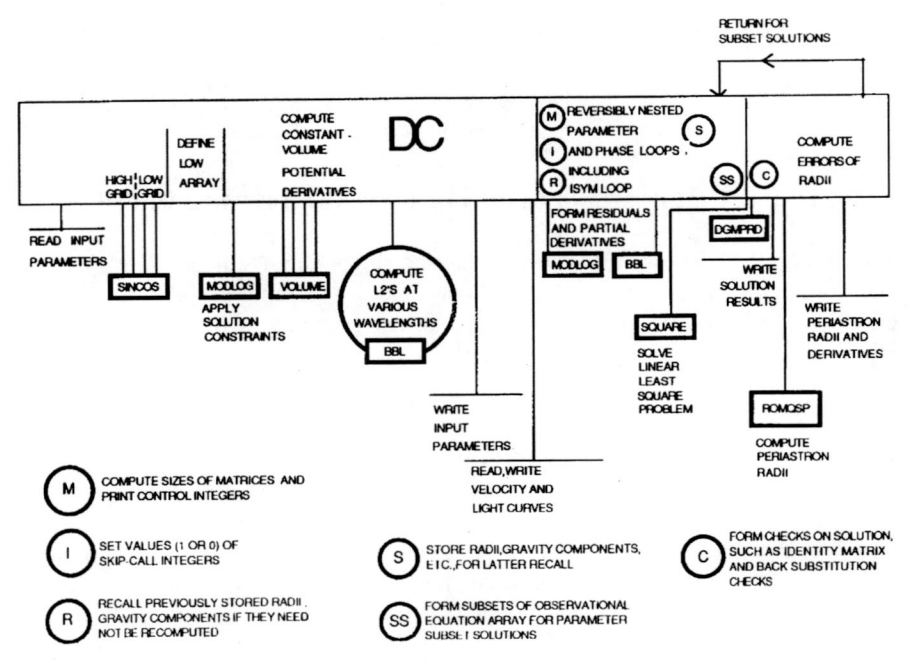

Fig. E.3 The WD main program: DC. Courtesy R. E. Wilson

E.19 LCR – Aspect Independent Surface Computations

This subroutine (see Fig. E.4) oversees all aspect-independent computations of the stellar surfaces by calling other subroutines in proper sequence. This includes the shapes of the components and the potential gradient on both stars. It updates those quantities if they change in an eccentric orbit. In that case, LCR computes the volume from the potential Ω_p at periastron and finds the phase-specific potential for that volume. Then, the polar temperature is computed by (3.2.19) from the average surface effective temperature. Eventually, LCR calls subroutine LUM, and either OLUMP or LUMP.

E.20 LEGENDRE – Legendre Polynomials

Based on the recursive relationship this subroutine evaluates the Legendre polynomials used in the atmosphere calculation. LEGENDRE reconstructs values for intensity from the precomputed coefficients stored in *atmcof.dat* and *atmcofplanck.dat*.

E.21 LIGHT – Projections, Horizon, and Eclipse Effects

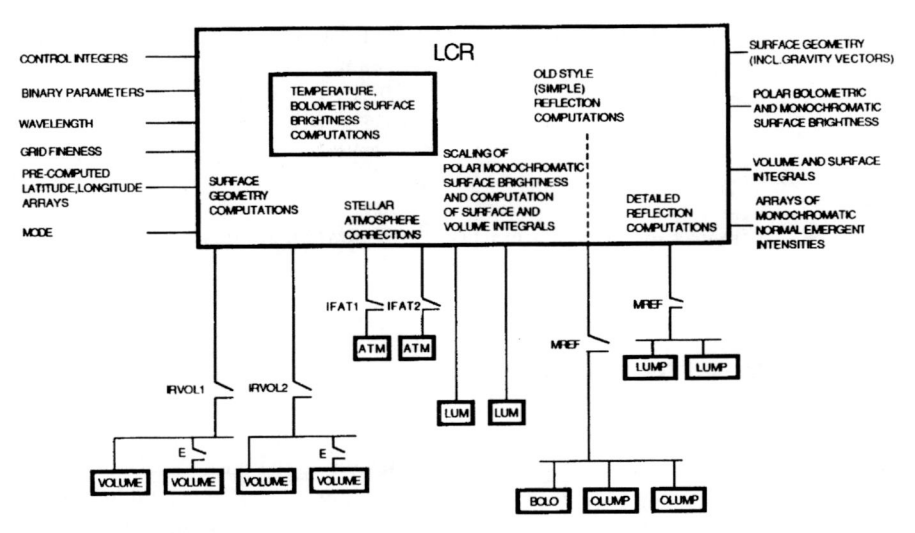

Fig. E.4 Subroutine LCR. Courtesy R. E. Wilson

E.21 LIGHT – Projections, Horizon, and Eclipse Effects

Subroutine LIGHT performs the aspect computations involving the projections, horizon, and eclipse effects and summation over the visible surface S'' of each star. The WD program uses a normal vector, **n**, pointing inward and a line-of-sight vector, **s**, pointing from the binary to the observer. In the first part of subroutine light there is a test which decides which star is in front. If the phase Φ is close to 0.5,

$$(\Phi - 0.5)^2 \leq 0.0625 = 0.25^2, \tag{E.21.1}$$

then the primary is in front, otherwise the secondary is in front.

For the star in front, subroutine LIGHT checks all grid points for

$$\cos \gamma < 0. \tag{E.21.2}$$

For the orientation of **n** and **s** reviewed above, the grid point is visible if the condition is fulfilled. The whole horizon is then represented by an array (θ^H, ρ^H), the nearest points to the horizon. These points are identified when integrating the star in front and detecting a change in the sign of $\cos \gamma$. LIGHT fits a Fourier representation $\rho(\theta)$ of the horizon by least-squares in subroutine FOURLS.

The star in the background, possibly eclipsed by the star in front, is treated similarly. Visible grid points are identified (E.21.2). A quick test is then applied to ascertain if a visible point is eclipsed. This test is followed by a fine-tuned test using $\rho(\theta)$ which is improved by a fractional area correction if the boundary lies between adjacent grid points.

The related program `WD93K93`, developed in Calgary, integrates the flux by Simpson integration for points visible on the "distant" star.

E.22 LimbDark – Limb Darkening

`LimbDark` supports the computation of local limb darkening by interpolating in Van Hamme's (1993) band-specific limb-darkening tables.

E.23 LinPro – Line Profiles

Added in the 1998 version of the WD program, subroutine `LinPro` computes spectral line profiles of absorption and emission lines are generated for MPAGE=3 in LC (not DC). The profiles are for rotation only, although other broadening mechanisms may be added later. Blending is incorporated, including blending of mixed absorption and emission lines. Lines can originate either from an entire star or from designated sub-areas of the surface, as explained below. Spectra are formed by binning, with the spectra of the two stars formed separately. The user can add them (weighted by observable flux) if spectra of the binary are needed.

E.24 LUM – Scaling of Polar Normal Intensity

The computation of the flux from each surface element is based on the local intensity. In order to integrate star brightness (in subroutine `LIGHT`) we need to know the normal intensity at a reference point on the surface, ordinarily the pole. However, the input parameter to the WD program is the luminosity in units determined by the user. Subroutine LUM accomplishes the necessary inversion such that luminosity becomes input and reference intensity is output. It uses (6.3.7) to compute the polar normal intensity I_j required to yield the relative monochromatic luminosity when the local fluxes are suitably integrated over the surface. For both components LUM also computes and stores the local bolometric and monochromatic ratios $G_j(\mathbf{r}_s)$ of normal intensities at all local points $\mathbf{r}_s$ to that at the pole according to (6.3.8). LUM also implements model atmosphere corrections.

E.25 LUMP – Modeling Multiple Reflection

As the multiple reflection effect involves many iterative computations it is strongly recommended to pay some attention to the structure and logic of computations. As in Wilson (1990) we now put the formulas in a more symmetrical form and label the stars A and B. The effective irradiance fluxes are denoted by primes (F'), while the

E.25 LUMP – Modeling Multiple Reflection

"intrinsic" fluxes (those which would exist in the absence of the reflection effect) are unprimed.

Let us start by computing the irradiance flux F_B'' from component B received at a surface point on component A. Combining (6.3.10) and (3.2.47) we get

$$F_B'' = I_B \sum_\varphi \sum_\theta \left\{ R_B \frac{\cos \gamma_A}{\rho^2} G_B D(\gamma_B) \frac{\cos \gamma_B}{\cos \beta_B} r_B^2 \sin \theta_B \Delta\varphi_B \Delta\theta_B \right\}, \quad (E.25.1)$$

where γ denotes, at a given surface point, the angle between the local surface normal and the line-of-sight to a given surface element on the other star, and ρ is the distance between that point and the surface element. If, in the common coordinate system, the surface point of component A and the surface element of the other star have the coordinates $\mathbf{r}_A$ and $\mathbf{r}_B$, $\cos \gamma_A$ follows simply as

$$\cos \gamma_A = \mathbf{n}_A \cdot \frac{\mathbf{r}_B - \mathbf{r}_A}{\rho}, \quad \rho = |\mathbf{r}_B - \mathbf{r}_A|. \quad (E.25.2)$$

The effective irradiance flux considers the local bolometric albedo, A_A, and gives us

$$F_B' = A_A F_B''. \quad (E.25.3)$$

The intrinsic flux, F_A, is given by

$$F_A = \mathcal{D}_A I_B G_A,$$

where $\mathcal{D}$ is the effective bolometric limb-darkening factor introduced on page 122.

Wilson (1990) expresses the bolometric flux ratio F_B'/F_A in the kth iteration as

$$\frac{F_B'}{F_A} = \frac{F_B'}{F_A}\left(\mathbf{R}_A^{(k)}\right) = \frac{C_B}{G_B} \sum_\varphi \sum_\theta G_A K_A R_A^{(k)} Q_{AB} D(\gamma_A), \quad (E.25.4)$$

with C_B, which is constant for a given binary star system and surface grid

$$C_B = \frac{I_A A_B \Delta\theta_A}{I_B \mathcal{D}_B}, \quad (E.25.5)$$

and K_A, which (as well as G_A) does not change in the course of iterations

$$K_A = r_A^2 \frac{\sin \theta_A \Delta\varphi_A}{\cos \beta_A}, \quad (E.25.6)$$

and, finally, Q_{AB},

$$Q_{AB} = \frac{\cos \gamma_A \cos \gamma_B}{\rho^2}. \quad (E.25.7)$$

The iterative procedure (note that $\mathbf{R}_A^{(k)}$ really represents a vector of reflection factors $R_A^{(k)}$ over the whole surface) is then defined by

$$R_A^{(k+1)} = 1 + \frac{F'_B}{F_A}\left(\mathbf{R}_A^{(k)}\right), \quad T_A^{(k+1)} = \sqrt[4]{R_A^{(k)}} T_A^{(k)}. \qquad (E.25.8)$$

The formulas for the other components are just achieved by interchanging the subscripts A and B. For further details and the logic of the implementation the reader is referred to Wilson (1990).

E.26 MLRG – Computing Absolute Dimensions

MLRG stands for mass, luminosity, radius, and gravity. This subroutine computes absolute dimensions and other quantities for the stars of a binary star system. This includes the masses, radii, and absolute bolometric luminosities of the stars in solar units as well as the logarithm to base 10 of the mean surface acceleration (effective gravity) of both components.

E.27 MODLOG – Handling Constraints Efficiently

This subroutine controls some of the geometrical constraints. If, for instance, a contact binary should be modeled, then this subroutine enforces the equations $A_2 = A_1$, $g_2 = g_1$, and $\Omega_2 = \Omega_1$.

E.28 NEKMIN – Connecting Surface of Over-Contact Binaries

This subroutine is only called for contact binaries (modes 1 and 3). A plane through the connecting neck defines the star boundaries. The ("vertical") plane is essentially at the neck minimum, so not exactly at the L_1^p point. Subroutine NEKMIN computes the x-coordinate of that plane as described by Wilson & Biermann (1976).

E.29 OLUMP – Modeling the Reflection Effect

In order to understand the meaning of the bolometric albedos used in the Wilson–Devinney model to describe the reflection effect, it is useful to have a detailed knowledge of how the reflection effect is modeled. We first describe subroutine OLUMP, which means "old LUMP." OLUMP is used for eccentric orbit calculations. It is a bit less accurate but much faster than subroutine LUMP.

E.29 OLUMP – Modeling the Reflection Effect

As shown in Fig. E.5, d denotes the distance between the star centers in unit a. For a circular orbit $d \equiv 1$. The index 2 refers to the secondary component. Let

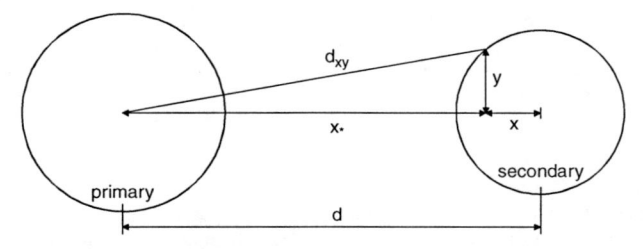

Fig. E.5 Geometry of the reflection effect 1. The basic geometry

$$\mathbf{r}_j = (x_j, y_j, z_j) \leftrightarrow (r_j, \theta_j, \varphi_j), \quad j = 1, 2, \quad (\text{E.29.1})$$

be the coordinates with the origin in the center of component j. The coordinate systems can be transformed into each other by

$$x_2 = d - x_1, \quad y_2 = -y_1, \quad z_2 = z_1. \quad (\text{E.29.2})$$

Spherical coordinate systems, $\mathcal{C}_1$ and $\mathcal{C}_2$, introduced in Sect. 3.1.1, also are used. All quantities indexed by * refer to the coordinate system with center in the origin of the irradiating star. If two signs are given, as in (E.29.3), the upper one refers to the case that the primary component is the irradiated star. The coordinates x, y, and z always refer to the irradiated component:

$$x_1 = \begin{Bmatrix} d \\ 0 \end{Bmatrix} \pm x, \quad y_1 = \pm y, \quad z_1 = z, \quad x_* = x_1 - \begin{Bmatrix} d \\ 0 \end{Bmatrix}. \quad (\text{E.29.3})$$

According to Fig. E.5 the distance from a point of the irradiated star to the center of the irradiating star is given by

$$d = \sqrt{x_*^2 + y_1^2 + z_1^2}, \quad d_{xy} = \sqrt{x_*^2 + y_1^2}. \quad (\text{E.29.4})$$

The transformation

$$\mathbf{r} = (r, \theta, \varphi) \to (r_*, \theta_*, \varphi_*) = (d, \theta_*, \varphi_*) \quad (\text{E.29.5})$$

is, as shown in Fig. E.6, realized by

$$\cot \theta_* = \frac{z}{x_*} \quad (\text{E.29.6})$$

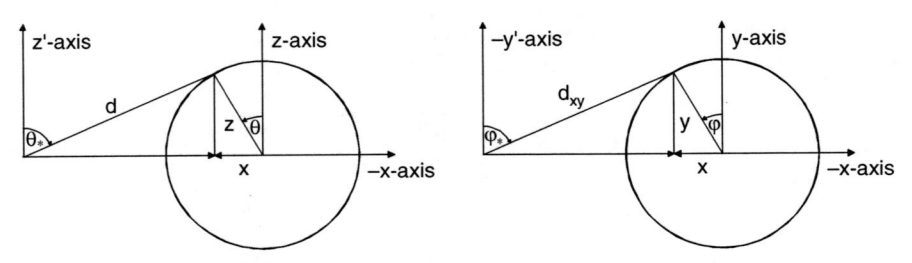

Fig. E.6 Geometry of the reflection effect 2. The relevant angles are shown. Quantities indexed by * refer to the coordinate system with center in the origin of the irradiating star. Although, for simplicity, the figure shows circles it should not give the impression that the irradiated star is modeled as a sphere, which is not the case

and

$$\cos^2 \varphi_* = \frac{1}{1 + y^2/x_*^2} = \left[\frac{x_*}{d_{xy}}\right]^2. \quad (E.29.7)$$

Furthermore, $\cos^2 \varphi_*$ is needed in the following computations.

The ellipticity correction used in (3.2.45) is the ratio between the flux $F_E(\mathbf{r})$ in an arbitrary point outside of an ellipsoidal star with axes $a = r_{\text{point}}$, $b = r_{\text{pole}}$, and $c = r_{\text{side}}$, and the flux $F_S(\mathbf{r})$ in the same point which would result from a spherical star of the same luminosity:

$$E = E(\mathbf{r}) = \frac{F_E(\mathbf{r})}{F_S(\mathbf{r})} = \frac{G}{S}. \quad (E.29.8)$$

Here G is the incident flux at F. In the following, only those values $G = G(\mathbf{r})$ are of interest for which x is a surface point of the irradiated star. The nomenclature is similar to that in Russell & Merrill (1952). $i(= \theta_*)$ and $\theta(= \varphi_*)$ are spherical polar coordinates with center at the origin of the radiating star. Conservation of flux, and the observation that for a star with unit luminosity we have $S = 1/4\pi$, yields the normalization condition:

$$\int G(i, \theta) d\Omega = \frac{1}{4\pi} \int E(i, \theta) d\Omega = 1, \quad d\Omega = \sin i \, di \, d\theta. \quad (E.29.9)$$

According to Russell & Merrill (1952, p. 33), it follows that

$$E(i, \theta) = C\left[1 - \frac{Nz}{2}\cos^2\theta\right], \quad (E.29.10)$$

E.29 OLUMP – Modeling the Reflection Effect

where N describes limb darkening and gravity darkening according to Russell and Merrill, or (D.1.5), and

$$\frac{1}{C} = \frac{1}{4\pi} \int \left(1 - \frac{Nz}{2} \cos^2 \theta\right) d\Omega = 1 - \frac{N}{8\pi} \int z \cos^2 \theta \, d\Omega. \quad (E.29.11)$$

The quantity z follows from Fig. E.9 and the definition of a, b, and c therein and is given by

$$z = \frac{1 - b^2/a^2}{1 + b^2/c^2 \tan^2 i}. \quad (E.29.12)$$

Wilson approximates the integral in expression (E.29.11) by

$$\int_0^{2\pi} \int_0^{\pi} z \cos^2 \theta \sin i \, di \, d\theta = 2\pi (1 - b^2/a^2) \int_0^{\pi/2} \frac{\sin i}{1 + \dfrac{b^2}{c^2 \cdot \tan^2 i}} di, \quad (E.29.13)$$

where z has been eliminated using (E.29.12) and eventually gets

$$\frac{1}{C} = 1 - \frac{N}{4} \left(1 - \frac{b^2}{a^2}\right) \left(0.9675 - 0.3008 \frac{b}{c}\right). \quad (E.29.14)$$

The required quantities $\cot i = \cot \theta_*$ and $\cos^2 \theta = \cos^2 \varphi_*$ have already been computed in (E.29.6) and (E.29.7).

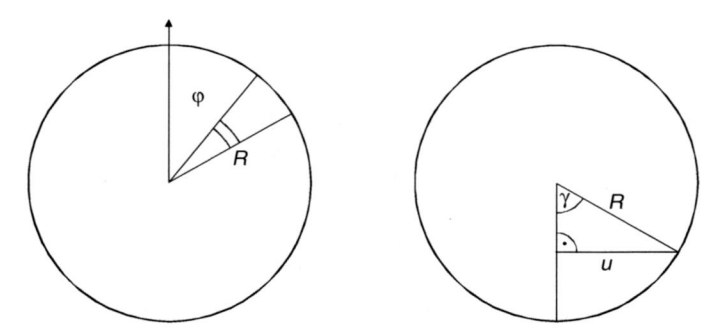

Fig. E.7 Geometry of the reflection effect 3. Here we see the quantities r_c and r_H involved in describing the visible part of the irradiating star

The bolometric flux $F_1(a)$ received at a point a is, after application of the correction for limb darkening, modified by a multiplicative correction factor $E(\theta_*, \varphi_*)$

$$F_1(\mathbf{r}) = E(\theta_*, \varphi_*) F_1(\mathbf{r}, \text{sphere}). \quad (E.29.15)$$

The step from an irradiating point source to an extended source star requires an explicit treatment of penumbral regions. Modeling penumbral effects forces us also to include limb darkening. Derived in Chapter 3 for a linear limb-darkening law, formula (3.2.31) gives the flux received from a unit disk with unit intensity at disk center

$$F = \pi \left(1 - \frac{x}{3}\right). \quad (E.29.16)$$

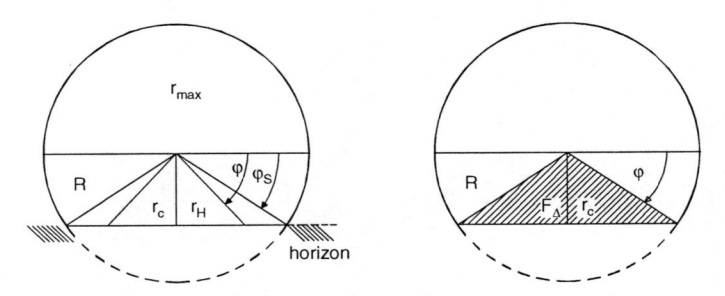

Fig. E.8 Geometry of the reflection effect 4. The local coordinates on the stellar disk are illustrated

Figure E.7 illustrates the geometry of the visible part of the irradiating star seen from a point x of the irradiated star. The horizon is assumed to be a straight line. In the first step, $\varphi_s = \varphi_s(r_c, R)$ is computed according to

$$\rho := \sin \varphi_s = \frac{r_c}{R} \leftrightarrow \varphi_s = \arcsin \frac{r_c}{R} = \arcsin \rho. \quad (E.29.17)$$

for given values r_c and R. To each value φ we can assign a value r_H

$$r_H = r_H(\varphi) = \frac{r_c}{\sin \varphi}. \quad (E.29.18)$$

As illustrated in Fig. E.8, we can compute $\sin \gamma_{\max}$ as a function of r_H

$$\sin \gamma_{\max} = \frac{r_H}{R} \quad (E.29.19)$$

and, with $R \equiv 1$, we get

$$\gamma_{\max} = \gamma_{\max}[r_H(\varphi)] = \arcsin \frac{r_c}{\sin \varphi}. \quad (E.29.20)$$

The contribution F_Δ of the large triangle area in Fig. E.7 is

E.29 OLUMP – Modeling the Reflection Effect

$$F_\Delta = 2 \int_{\varphi_s}^{\pi/2} \int_0^{\gamma_{max}} I(r) r \, dr \, d\varphi = F_1 + F_2 + F_3, \tag{E.29.21}$$

where

$$F_1 = (1-x)\rho\sqrt{1-\rho^2}, \tag{E.29.22}$$

$$F_2 = \tfrac{2}{3}\left(\frac{\pi}{2} - \arcsin\rho\right), \tag{E.29.23}$$

$$F_3 = -\tfrac{2}{3}x \int_{\varphi_s}^{\pi/2} \left(1 - \rho^2/\sin^2\varphi\right)^{3/2} d\varphi. \tag{E.29.24}$$

Wilson computes F_3 with a Gauß integration as $\sum_{i=1}^{3} w_i f_i$. Based on the computed quantities, it is now possible to define the penumbra function $P = P(r_c, R)$

$$P = P(r_c, R) = \begin{cases} 1, & \text{case 1}: \rho \geq 1, \\ P_1, & \text{case 2}: 0 \leq \rho < 1, \\ P_2, & \text{case 3}: -1 < \rho < 0, \\ 0, & \text{case 4}: \rho \leq -1. \end{cases} \tag{E.29.25}$$

Cases 1–4 represent four geometric possibilities:

1. the irradiating star is completely above the local horizon;
2. it is more than half above the local horizon;
3. it is less than half above the local horizon; or
4. it is completely below the horizon.

The quantities P_1 and P_2 are defined by

$$P_1 = \frac{F_{\sec} + F_\Delta}{F}, \quad P_2 = \frac{F - (F_{\sec} + F_\Delta)}{F} = 1 - P_1. \tag{E.29.26}$$

The contribution $F_{\sec}$ is

$$F_{\sec} = \tfrac{1}{2}F + 2\int_0^{\varphi_s}\int_0^{\pi/2} dF(\gamma) = \tfrac{1}{2}F + \frac{1}{\pi}\varphi_s F \tag{E.29.27}$$

and eventually

$$F_{\sec} = \left(\frac{\pi}{2} + \arcsin\rho\right)\left(1 - \frac{x}{3}\right). \tag{E.29.28}$$

Now the dependence of the factor P on r_c and R according to (E.29.22), (E.29.23), (E.29.24), and (E.29.28) is transformed into a dependence on ρ defined in (E.29.17). As can be seen in Fig. E.9, it is also possible to define the fractional radius ρ above or below the horizon as

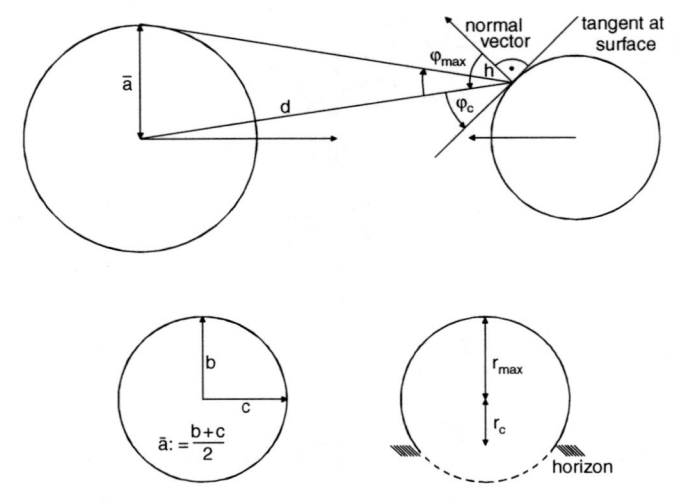

Fig. E.9 Geometry of the reflection effect 5. The relation among geometrical quantities describing the visible part of the irradiating star

$$\rho := \frac{\rho_c}{\rho_{max}} = \frac{\frac{\pi}{2} - h}{\arcsin\left(\frac{b+c}{2d}\right)}, \quad \rho_c = \frac{\pi}{2} - h, \quad (E.29.29)$$

where additional auxiliary quantities

$$\rho_{max} = \arcsin\left(\frac{\bar{a}}{d}\right), \quad \bar{a} := \frac{b+c}{2}, \quad (E.29.30)$$

and

$$\cos h = -\frac{\mathbf{n} \cdot \mathbf{r}_*}{|\mathbf{r}_*|}, \quad (E.29.31)$$

appear. The replacement of the approximation (E.29.17) by (E.29.29) is only useful for $D \gg \bar{a}$. In contact systems we have $D \cong 3\bar{a}$; in detached systems the approximation is even better.

If (3.2.45) is to be used instead of (3.2.44), $\cos \varepsilon$ is replaced by $\cos \bar{\varepsilon}$. In the case of a point source the ratio F_1/F_2 depends on the angle defined in (E.29.31), i.e., on $\cos \varepsilon = \cos h$. As shown in Fig. E.10, $\bar{\varepsilon}$ can be defined as

E.29 OLUMP – Modeling the Reflection Effect

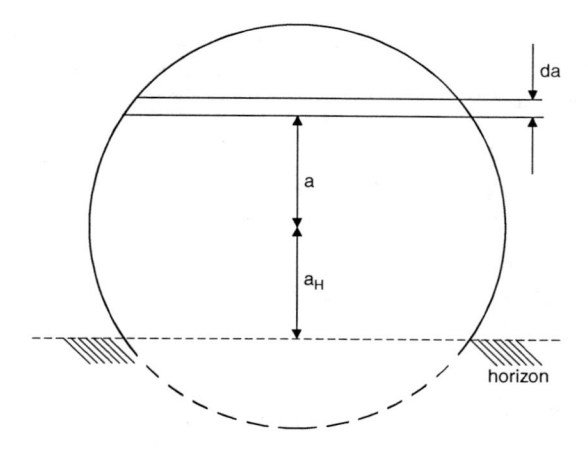

Fig. E.10 Geometry of the reflection effect 6. The geometry at the horizon

$$\bar{\varepsilon} = \frac{\pi}{2} - \varepsilon_{\text{mean}}, \qquad (E.29.32)$$

where $\varepsilon_{\text{mean}}$ can be interpreted as the mean height above the horizon. However, it is easier to define a mean radius or mean (linear) height $\bar{a}$. From

$$\varepsilon_{\text{mean}} = \bar{a}\rho_{\text{max}} \qquad (E.29.33)$$

the angle, $\varepsilon_{\text{mean}}$ can be computed. Then, $\bar{a}$ follows as

$$\bar{a} = \frac{\int_{a_H}^{1} a\sqrt{1-a_H^2}\, da}{\int_{a_H}^{1} \sqrt{1-a_H^2}\, da} = \frac{N}{D}, \qquad (E.29.34)$$

with

$$N := \tfrac{1}{2}\left[a\sqrt{1-a^2} + \arcsin a\right]_{a_H}^{1}, \qquad (E.29.35)$$

and

$$D := -\tfrac{1}{3}\left[a(1-a^2)^{3/2}\right]_{a_H}^{1} \qquad (E.29.36)$$

we eventually find

$$\bar{a} = \frac{\tfrac{1}{3}[1-a_H^2]^{3/2}}{\tfrac{\pi}{4} - \tfrac{1}{2}[a_H(1-a_H^2)^{1/2} + \arcsin a_H]}. \qquad (E.29.37)$$

The height a_H above the horizon is identical with ρ computed from (E.29.29).

More detailed computations of the reflection effect and also multiple reflection are done in subroutine LUMP (see Appendix E.25).

E.30 OMEGA* – Computing $\Omega(r)$

This subroutine, similar to Wilson's subroutine VOLUME, computes the Roche potential Ω for a given (dimensionless) mean radius r_* by solving the equation

$$\bar{r}^V(\Omega) = r_*, \tag{E.30.1}$$

where $\bar{r}^V(\Omega)$ is calculated according to (6.3.21). For given values q and Ω, a modified version of Wilson's subroutine SURFAS yields not only the surface points but also the volume and thus the mean radius $\bar{r}^V$ according to (6.3.21). Thus, the implicit condition (E.30.1) can be solved w.r.t. Ω by applying Newton's method, i.e.,

$$\Omega^{(n+1)} = \Omega^{(n)} - \frac{\bar{r}^V\left[\Omega^{(n)}\right] - r_*}{\dfrac{d\bar{r}}{d\Omega}\left[\Omega^{(n)}\right]}. \tag{E.30.2}$$

The derivative in the denominator of (E.30.2) can be approximated by the finite difference expression

$$\frac{d\bar{r}}{d\Omega}[\Omega] \approx \frac{\bar{r}^V(\Omega + \Delta\Omega) - \bar{r}^V(\Omega)}{\Delta\Omega}, \quad \Delta\Omega = 0.01. \tag{E.30.3}$$

The initial guess $\Omega^{(0)}$ is computed by the approximations

$$\Omega_1^{(0)} = \frac{1}{r_1} + q, \quad \Omega_2^{(0)} = q\left(\frac{1}{r_2} + \frac{1}{q}\right) + \frac{1}{2}(1-q), \tag{E.30.4}$$

and

$$\Omega_2^{(0)} = q\left(\frac{1}{r_2} + \frac{1}{q}\right) + \frac{1}{2}(1-q) = \frac{3}{2} + \left(\frac{1}{r_2} - \frac{1}{2}\right)q, \tag{E.30.5}$$

based on Kopal's (1959, p. 129, formula 2-3) approximation

$$r = \frac{1}{\Omega - q} \tag{E.30.6}$$

for component 1 and the transformation (6.3.3) for changing into the coordinate frame of component 2. Iterations are stopped if

$$\left|\bar{r}^V(\Omega) - r_*\right| < \varepsilon = 0.00001. \tag{E.30.7}$$

E.31 PLANCKINT – Planck Intensity

This subroutine returns the Planck intensity of the disk center over the range of temperature 500 $K \leq T \leq 500,300$ K. This is the integral of the Planck function (or blackbody function) folded by the response function of a particular bandpass over the whole star.

E.32 READLC* – Reading Program Control Parameters

This subroutine reads all control parameters needed in WD95 in order to run the least-squares solvers or to produce graphics.

E.33 RING – The Interface Ring of an Over-Contact Binary

This subroutine is called by SURFAS and is related to the construction of the surface of over-contact binaries. It computes the area fraction of surface elements intersecting the ring (resp. the plane), separating the components of over-contact binaries. Finally, subroutine RING, computes a Fourier representation of the ring.

E.34 RanGau – Generation of Gaussian Random Numbers

Subroutine RanGau generates pseudo random numbers with Gaussian probability in the range $[-\infty, +\infty]$.

E.35 RanUni – Generation of Uniform Random Numbers

Subroutine RanUni generates pseudo random numbers with uniform probability in the range $[-1, +1]$. The input number S_n, from which both output numbers are generated, should be larger than the modulus 10^8 and smaller than twice the modulus. The returned number smod will be in that range and can be used as the input S_{n+1} on the next call.

E.36 ROMQ – Distance Computation of Surface Points

This subroutine replaces the older subroutine ROMQSP, which was programmed in single precision.

E.37 ROMQSP – Distance Computation of Surface Points

For a single point (θ, φ) on the surface of a component, this subroutine computes the distance r to the center of that component

$$(\theta, \varphi; \Omega, F, q) \rightarrow r(\theta, \varphi; \Omega, F, q). \tag{E.37.1}$$

In addition, the following derivatives are computed:

$$\frac{dr}{dq}, \quad \frac{dr}{d\Omega}, \quad \frac{d\Omega}{dr}. \tag{E.37.2}$$

ROMQSP provides a convenient means to generate tables of dimensions and useful derivatives for output of the main program. The subroutine was replaced with ROMQ in the 1998 and later versions.

E.38 SIMPLEX* – Simplex Algorithm

This subroutine contains the Simplex algorithm. It calls subroutine SSR to compute light curves for a parameter vector **x** suggested by the Simplex algorithm.

E.39 SinCos – Surface Grid Sine and Cosines

This subroutine computes and stores the sine and cosine values for all surface grid points. This save some computing time.

E.40 SQUARE – Building and Solving the Normal Equations

This subroutine builds the normal equation, inverts the left-hand side of the normal equations, and determines the parameter corrections. Furthermore, the correlation matrix and parameter standard deviations are computed. The inversion of the normal equations is performed by subroutine DMINV.

E.41 SPOT – Modeling Spots

This subroutine checks whether a surface point (θ, φ) lies within any of n specified spots and corrects the local temperature. If a surface point is in more than one spot, this subroutine adopts the product of the spot temperature factors. Note that WD uses North polar "latitudes," running from 0° at the "North" pole to 180° at the other. In addition, θ_s^c and φ_s refer to the coordinates on stars of a particular spot of radius ρ_s. The angular distance Δ_s of the point (θ, φ) from the center of spot s

follows from (3.4.3), and the spot-free local temperature T_l at (θ, φ) is modified by the temperature factor t_f according to (3.4.5). The effects of star spots are treated as part of the aspect computations. The reason is that only a quarter of the surface points are stored, so as to save on memory needs. Note that spots break the up–down and right–left symmetry of the model star.

E.42 SSR* – Computation of Curves and Residuals

This subroutine receives the adjustable parameters from the Simplex algorithm, and the Levenberg–Marquardt algorithm picks up the all other input parameters from common blocks, prepares the total set of parameters and other data to the WD program, and invokes subroutine DC to compute all light and radial velocity curves for all observed phase values and residuals. SSR is used within the context of the Simplex algorithm and the Levenberg–Marquardt-type damped differential corrections algorithm. The standard deviation of the fit and errors of the parameters are returned to the calling subroutines.

E.43 SURFAS – Generating the Surfaces of the Components

This subroutine generates the spherical and rectangular coordinates of the surface elements of each component, computes the rectangular components of the surface potential gradient and other quantities which only depend on the surface elements. The grid spacing and the motivation for it has been described in Sect. 4.5.3.

E.44 VOLUME – Keeping Stellar Volume Constant

For eccentric orbits, this subroutine computes the phase-dependent Roche potential such that the volumes of the stars are kept constant.

References

Abadie, J.: 1978, The GRG Method for Nonlinear Programming, in H. J. Greenberg (ed.), *Design and Implementation of Optimization Software*, pp. 335–363, Sijthoff and Noordhoff, The Netherlands

Abadie, J. & Carpenter, J.: 1969, Generalization of the Wolfe Reduced Gradient Method to the Case of Nonlinear Constraints, in R. Fletcher (ed.), *Optimization*, pp. 37–47, Academic Press, New York

Bazaraa, M., Sheraldi, H. D., & Shetly, C. M.: 1993, *Nonlinear Programming*, Wiley, Chichester, UK, 2nd edition

Bock, H. G.: 1987, *Randwertproblemmethoden zur Parameteridentifizierung in Systemen nichtlinearer Differentialgleichungen*, Preprint 142, Universität Heidelberg, SFB 123, Institut für Angewandte Mathematik, 69120 Heidelberg

Bomze, I. M. & Grossmann, W.: 1993, *Optimierung – Theorie und Algorithmen*, Wissenschaftsverlag, Mannheim

Brandt, S.: 1976, *Statistical and Computational Methods in Data Analysis*, North-Holland, Amsterdam, 2nd edition
Branham, R. L.: 1990, *Scientific Data Analysis: An Introduction to Overdetermined Systems*, Springer, New York
Cayley, A.: 1872a, Sur les Surfaces Orthogonales, *Compt. Rend. Acad. Sci. Paris* **75**, 324–330
Cayley, A.: 1872b, Sur les Surfaces Orthogonales (suite), *Compt. Rend. Acad. Sci. Paris* **75**, 381–385
Chanan, G. A., Middleditch, J., & Nelson, J. E.: 1977, The Geometry of the Eclipse of a Point-Like Star by a Roche-Lobe-Filling Companion, *ApJ* **208**, 512–517
Collatz, L. & Wetterling, W.: 1971, *Optimierungsaufgaben*, Springer, Berlin, Germany, 2nd edition
Darboux, J. G.: 1898, *Lecons sur les Systèmes Orthogonaux et les Coordonnées Curvilignes*, Gauthier-Villars, Paris
Dennis, J. E. & Schnabel, R. B.: 1983, *Numerical Methods for Unconstrained Optimisation and Nonlinear Equations*, Prentice Hall, Englewood Cliffs, NJ
Deuflhard, P.: 2004, *Newton Methods for Nonlinear Problems. Affine Invariance and Adaptive Algorithms*, Springer, Berlin
Deuflhard, P. & Hohmann, A.: 1993, *Numerische Mathematik – Eine algorithmische orientierte Einführung*, Walter de Gruyter, Berlin, Germany, 2nd edition
Fletcher, R.: 1987, *Practical Methods of Optimization*, Wiley, Chichester, UK, 2nd edition
Gill, P. E., Murray, W., & Wright, M. H.: 1981, *Practical Optimization*, Academic Press, London
Hadrava, P.: 1987, The Roche Coordinates in Non-Synchronous Binaries, *Ap. Sp. Sci.* **138**, 61–69
Hadrava, P.: 1992, Radiative Transfer in Rotating Stars, *A&A* **256**, 519–524
Harris, D. L.: 1963, The Stellar Temperature Scale and Bolometric Corrections, in K. A. Strand (ed.), *Basic Astronomical Data*, pp. 263–272, University of Chicago Press, Chicago, IL
Kallrath, J.: 1995, Rational Function Techniques and Padé Approximants, in J. Hagel (ed.), *Nonlinear Perturbation Methods with Emphasis to Celestial Mechanics*, pp. 79–107, Madeira University Press, Madeira, Portugal
Kallrath, J. & Kämper, B.-C.: 1992, Another Look at the Early-Type Eclipsing Binary BF Aurigae, *A&A* **265**, 613–625
Karush, W.: 1939, Minima of Functions of Several Variables with Inequalities as Side Conditions, Master thesis, Department of Mathematics, University of Chicago, Chicago, IL
Kitamura, M.: 1970, The Geometry of the Roche Coordinates, *Ap. Sp. Sci.* **7**, 272–358
Kopal, Z.: 1946, An Introduction to the Study of Eclipsing Variables, *Harvard Observatory Monograph* **6**, 1–220
Kopal, Z.: 1959, Close Binary Systems, Chapman & Hall, London
Kopal, Z.: 1970, The Roche Coordinates in Three Dimensions and their Application to Problems of Double-Star Astronomy, *Ap. Sp. Sci.* **8**, 149–171
Kopal, Z.: 1971, The Roche Coordinates for the Rotational Problem, *Ap. Sp. Sci.* **10**, 328–331
Kopal, Z. & Ali, A. K. M. S.: 1971, On the Integrability of the Roche Coordinates, *Ap. Sp. Sci.* **11**, 423–429
Kuhn, H.: 1976, Nonlinear Programming: A Historical View, in R. Cottle & C. Lemke (eds.), *Nonlinear Programming*, Vol. 9 of *SIAM-AMS Proceedings*, pp. 1–26, American Mathematical Society, Providence, RI
Kuhn, H. W. & Tucker, A. W.: 1951, Nonlinear Programming, in J. Neumann (ed.), *Proceedings Second Berkeley Symposium on Mathematical Statistics and Probability*, pp. 481–492, University of California, Berkeley, CA
Kurucz, R. L.: 1993, New Atmospheres for Modelling Binaries and Disks, in E. F. Milone (ed.), *Light Curve Modeling of Eclipsing Binary Stars*, pp. 93–102, Springer, New York
Lasdon, L. S. & Waren, A. D.: 1978, Generalized Reduced Gradient Method for Linearly and Nonlinearly Constrained Programming, in H. J. Greenberg (ed.), *Design and Implementation of Optimization Software*, pp. 363–397, Sijthoff and Noordhoff, The Netherlands
Lasdon, L. S., Waren, A. D., Jain, A., & Ratner, M.: 1978, Design and Testing of a Generalized Reduced Gradient Code for Nonlinear Programming, *ACM Trans. Math. Software* **4**, 34–50
Lawson, C. L. & Hanson, R. J.: 1974, *Solving least-squares Problems*, Prentice Hall, Englewood Cliffs, NJ

References

Linnell, A. P.: 1989, A Light Synthesis Program for Binary Stars. III. Differential Corrections, *ApJ* **342**, 449–462

Milne, E. A.: 1926, The Reflection Effect in Eclipsing Binaries, *MNRAS* **87**, 43–55

Milone, E. F., Stagg, C. R., & Kurucz, R. L.: 1992, The Eclipsing Binary AI Phoenicis: New Results Based on an Improved Light Curve Analysis Program, *ApJ Suppl.* **79**, 123–137

Morton, D. C. & Adams, T. F.: 1968, Effective Temperatures and Bolometric Corrections of Early-Type Stars, *ApJ* **151**, 611–621

Nelder, J. A. & Mead, R.: 1965, A Simplex Method for Function Minimization, *Comp. J.* **7**, 308–313

Nemhauser, G. L. & Wolsey, L. A.: 1988, *Integer and Combinatorial Optimization*, Wiley, New York

Neutsch, W.: 1995, *Koordinaten*, Spektrum, Heidelberg

Neutsch, W. & Scherer, K.: 1992, *Celestial Mechanics – An Introduction to Classical and Contemporary Methods*, Wissenschaftsverlag, Mannheim, Germany

Nocedal, J. & Wright, S. J.: 2006, *Numerical Optimization*, Springer Series in Operations Research, Springer, New York, 2nd edition

Ostle, B.: 1963, *Statistics in Research*, Iowa State University Press, Ames, IA

Papadimitriou, C. H. & Steiglitz, K.: 1982, *Combinatorial Optimization: Algorithms and Complexity*, Prentice Hall, Englewood Cliffs, NJ

Press, W. H., Flannery, B. P., Teukolsky, S. A., & Vetterling, W. T.: 1992, *Numerical Recipes – The Art of Scientific Computing*, Cambridge University Press, Cambridge, UK, 2nd edition

Ravindran, A., Phillips, D. T., & Solberg, J. J.: 1987, *Operations Research. Principles and Practice*, Wiley, New York

Russell, H. N.: 1948, Idealized Models and Rectified Light Curves for Eclipsing Variables, *ApJ* **108**, 388–412

Russell, H. N. & Merrill, J. E.: 1952, The Determination of the Elements of Eclipsing Binary Stars, *Princeton. Obs. Contr.* **26**, 1–96

Schittkowski, K.: 1988, Solving Nonlinear Least-Squares Problems by a General Purpose SQP-Method, in K.-H. Hoffmann, J.-B. Hiriat-Urruty, C. Lemarechal, & J. Zowe (eds.), *Trends in Mathematical Optimization*, No. 84 in International Series of Numerical Mathematics, pp. 195–309, Birkhäuser, Basel, Switzerland

Sen, H. K.: 1948, Reflection Effect in Eclipsing Binaries for a Point-Source of Light, *Proc. Nat. Acad. Sci.* **34**, 311–317

Spendley, W., Hext, G. R., & Himsworth, F. R.: 1962, Sequential Application of Simplex Designs in Optimisation and Evolutionary Operation, *Technometrics* **4**, 441–461

Stoer, J.: 1985, Foundations of Recursive Quadratic Programming Methods for Solving Nonlinear Programs, in K. Schittkowski (ed.), *Computational Mathematical Programming*, No. 15 in NATO ASI Series, Springer, Heidelberg, Germany

Stoer, J., Bulirsch, R., Bartels, R., Gautschi, W., & Witzgall, P.: 1992, *Introduction to Numerical Analysis*, Texts in Applied Mathematics, Springer, Heidelberg, 2nd edition

Van Hamme, W. & Wilson, R. E.: 2003, Stellar Atmospheres in Eclipsing Binary Models, in U. Munari (ed.), *GAIA Spectroscopy: Science and Technology*, Vol. 298 of *Astronomical Society of the Pacific Conference Series*, pp. 323–328, San Francisco

Wilson, R. E.: 1979, Eccentric Orbit Generalization and Simultaneous Solution of Binary Star Light and Velocity Curves, *ApJ* **234**, 1054–1066

Wilson, R. E.: 1990, Accuracy and Efficiency in the Binary Star Reflection Effect, *ApJ* **356**, 613–622

Wilson, R. E.: 1993, Computation Methods and Organization for Close Binary Observables, in J. C. Leung & I.-S. Nha (eds.), *New Frontiers in Binary Star Research*, Vol. 38 of ASP Conference Series, pp. 91–126, Astronomical Society of the Pacific, San Francisco, CA

Wilson, R. E. & Biermann, P.: 1976, TX Cancri – Which Component Is Hotter?, *A&A* **48**, 349–357

Wright, S.: 1996, *Primal-Dual Interior-Point Methods*, Society for Industrial and Applied Mathematics, Philadelphia, PA

Appendix F
Glossary of Symbols

This table gives the page numbers on which the symbol occurs or gives an equation (in parentheses) in which it is defined or used.

A	matrix in linear least-squares functional	(A.3.7)
$\mathcal{A}$	surface of a star in physical units	(4.4.19)
a	isothermal sound speed	140
a	semi-major axis of the relative orbit, in units of $\mathcal{R}_\odot$	11
a_j	semi-major axis of the absolute orbit of component j, in units of $R_\odot$	(4.4.17)
B	apparent magnitude in blue passband of Johnson system	38
B(**x**)	second derivative term in the Hessian of a nonlinear least-squares functional	(A.3.25)
$B_\nu(T)$	Planck function	(3.2.21)
b	exponent used to compute the flux-dependent weight w^{flux}	(4.1.22)
C	number of type of observables used	169
$\hat{C}$	covariance matrix	(4.3.12)
c	speed of light; $c = 2.9979 \cdot 10^8$ m $\cdot$ s^{-1}	105
c	apparent color index	45
c_0	color index outside atmosphere	45
c	vector of calculated observables	(4.1.5)
D	telescope aperture	12
D	distance of the binary	156
d	separation $d(\phi)$ between binaries in eccentric orbits	(3.1.36)
dV	differential volume element	(3.1.4)
dσ	differential surface element	(3.1.5)
d	vector of unweighted residuals; **d** := **o** − **c**	(4.1.6)
E	eccentric anomaly	87
E_0	initial epoch	42
e	the orbital eccentricity = separation of foci / $2a$	83
F	rotation parameter	100
F_j	rotation parameter for star j; $F_j = \omega_j/\omega$	173
F	force per unit mass	(3.1.58)

$\mathbf{F}_2(\mathbf{x})$ vector-valued function representing equations in a constrained optimization problem (A.2.1)
$\mathbf{F}_3(\mathbf{x})$ vector-valued function representing inequalities in a constrained optimization problem (A.2.1)
f mass function $f(M_1, M_2, i)$ (4.4.33)
f focal length of the spectrograph camera 50
f fill-out parameter (3.1.101)
f_j fill-out factor of component j 112
$f(\mathbf{x})$ continuous real-valued objective function in an optimization problem (A.1.1)
f mass function $f(M_1, M_2, i)$ (4.4.33)
f_* optimal objective function value of a nonlinear optimization problem (4.1.12)
G gravity constant; $G = 6.673 \cdot 10^{-11}$ m^{3}kg^{-1}s^{-2} 96
g exponent in gravity-brightening law as expressed in terms of bolometric flux (3.2.13)
g surface gravity acceleration (3.2.11)
$\mathbf{g}$ vector of surface gravity acceleration (3.2.10)
$g_\odot$ solar surface acceleration; $g_\odot = 2.74 \cdot 10^2$ m · s^{-2} (4.4.21)
$H(\mathbf{x})$ Hessian matrix of a real-valued function (A.1.7)
h Planck's constant; $h = 6.62608 \cdot 10^{-34}$ J · s 55
I local monochromatic intensity 119
i orbital inclination; angle between orbital plane and plane-of-sky (angular degree) 83
$J(\mathbf{x})$ Jacobian matrix of a vector-valued function (A.1.6)
j referring to component j of the binary system if used as subscript xxxv
k Boltzmann's constant; $k = 1.3807 \cdot 10^{-23}$ J · K^{-1} 120
k ratio of radii, $k = r_s/r_g$ in Russell–Merrill notation 91
k refering to iteration k in an iterative algorithm if used as subscript 354
l_{jp} normalized monochromatic luminosity of component j in passband p, $p \in U, B, V, u, b, v, y$, $l_{jp} = L_{jp}/(L_{1p} + L_{2p})$ 24
ℓ system light received (observable radiation flux in a given passband in energy/time/area), often normalized in some way (3.2.50)
ℓ_D system light received by observer at distance D (3.2.51)
ℓ_j light received from component j (3.2.48)
ℓ_3 third light, sum of all contributions to ℓ from any systems parts beyond the binary pair (usually assumed to be constant) 132
$\mathcal{L}$ bolometric luminosity (radiant power in Watts or $L_\odot$, over 4π steradians, units could be W/micron or $L_\odot$) (4.4.24)
L_j monochromatic luminosity (in a specified passband) of component j over 4π steradians 286
L_1^p inner Lagrangian point 109
L_2^p outer Lagrangian point 110
$L_\odot$ solar luminosity; $L_\odot = 3.82 \cdot 10^{26}$ W (4.4.26)
M mean anomaly (3.1.29)
M binary system mass (in units of $M_\odot$); $M = M_1 + M_2$ (4.4.17)
M_j mass of component j (in units of $M_\odot$) (4.4.17)

F Glossary of Symbols

M_V	absolute visual magnitude	(4.4.32)
$M_\odot$	solar mass, $M_\odot = 1.9891 \cdot 10^{30}$ kg	
M^{bol}	absolute bolometric magnitude	(4.4.26)
m	apparent magnitude	42
m	number of adjustable parameters in a least-squares problem	169
n	number of data points in a least-squares problem	(4.1.7)
n	polytropic index	93
n_e	electron number density (m^{-3})	141
n_2	number of equations in a constrained optimization problem	(A.2.1)
n_3	number of inequalities in a constrained optimization problem	(A.2.1)
$\mathcal{O}_c$	set of observed or calculated data points of observable or curve type c	169
o	vector of observed observables	(4.1.4)
P	Householder matrix	(A.3.12)
P	binary orbital period	42
P	fractional polarization	(2.3.1)
$\mathcal{P}$	set of adjustable parameters	190
$\mathbf{p}^G$	vector containing the geometrical parameter involved in the Wilson–Devinney model	285
p	pressure	99
Q	Stokes quantity (parameter)	(2.3.2)
q	binary system mass ratio: $q = \mathcal{M}_2/\mathcal{M}_1$	(3.5.1)
q_{ph}	photometric mass ratio	173
q_{sp}	spectroscopic mass ratio	173
R	universal gas constant; $R = 8.31451$ Jmol^{-1}K^{-1}	(3.2.9)
R	spectral resolution	(2.2.1)
$\mathcal{R}$	mean radius of a star in physical units	(4.4.20)
$\mathbf{R}(\mathbf{x})$	weighted residual vector	(4.1.9)
R_j	mean radius of component j; usually the "equal volume radius" in units of $R_\odot$	24
$R_\odot$	solar radius, $R_\odot = 6.971 \cdot 10^8$ m	(4.4.22)
R(α)	rotation matrix, rotation by angle α	(C.1.2)
r	modulus of the radius vector, $r = |\mathbf{r}|$	(3.1.1)
$\bar{r}$	mean radius in units of a	206
$\bar{r}_j$	mean radius of component j in units of a	206
$r_{s,g}$	mean radius of smaller and greater star, respectively, in the Russell–Merrill model in units of a	287
r^{back}	distance from star center to surface along the line of centers in opposite direction to the other component in units of a	287
r^{point}	distance from star center to surface measured along the line of centers toward the other star in units of a	287
r^{pole}	distance from star center to surface measured perpendicular to the orbital plane in units of a	287
r^{side}	distance from star center to surface measured 90° from the line of centers in the orbital plane in units of a	287
r	radius vector	(3.1.1)

$\mathcal{S}$	feasible region (feasible set)	(A.2.2)
S	upper triangle matrix in Householder's algorithm	(A.3.18)
S	line-of-sight direction, $\mathbf{S} = (S_x, S_y, S_z)^T$	81
s	search direction	354
s	normalized line-of-sight direction, $\mathbf{s} = (s_x, s_y, s_z)^T$	(3.1.15)
σ_e	Thomson scattering cross-section per electron	(3.4.10)
T_{eff}	mean effective temperature	118
T_j	mean effective temperature of component j	174
T_p	polar effective temperature	(3.2.16)
T_{pj}	polar effective temperature of component j	(6.3.8)
T_l	local effective surface temperature of a surface element	125
$T_\odot$	effective temperature of the sun, $T_\odot = 5770$ K	(4.4.25)
t	time	42
$t_{\text{I}}, t_{\text{II}}$	times of successive primary and secondary minima	(3.1.22)
U	unitary matrix	(A.3.15)
U	Stokes quantity (parameter)	(2.3.2)
U	apparent magnitude in ultraviolet passband of Johnson system	38
V	apparent magnitude in visual passband of Johnson system	38
v	reflection vector to generate Householder matrix	(A.3.12)
W	weight matrix in least-squares problem	(4.1.8)
w^c	curve-dependent weight of a data point	(4.1.20)
w^{flux}	flux-dependent weight of a data point	(4.1.20)
w^{intr}	intrinsic weight of a data point	(4.1.20)
w	weight vector	170
X	airmass	45
x	component-centered coordinates, x-coordinate	(3.1.1)
x^s	plane-of-sky coordinates, x-coordinate	79
x_c	x-coordinate of the center of mass	(3.1.57)
$\mathbf{x}_0$	initial point or guess in nonlinear algorithms	172
$\mathbf{x}_*$	solution of a nonlinear algorithms	(4.1.12)
y	component-centered coordinates, y-coordinate	(3.1.1)
y^s	plane-of-sky coordinates, y-coordinate	79
z	component-centered coordinates, z-coordinate	(3.1.1)
z^s	plane-of-sky coordinates, z-coordinate	79
α, α_k	damping factor in Newton-type algorithms	354
β	angle between radius vector and surface normal	(3.1.3)
β	exponent in gravity-brightening law as expressed in terms of effective temperature	(3.2.16)
γ	radial velocity of the center-of-mass of a binary system	149
γ	angle between line-of-sight and surface normal	(3.1.16)
Δ	minimum angular separation in radians	12
δ	plane-of-sky distance between the center of components	(3.1.9)
δ	ratio of radiative of gravitational force	(3.1.85)
$\Delta\lambda$	wavelength resolution element	(2.2.1)
ΔT	temperature difference; $\Delta T := T_1 - T_2$	24

F Glossary of Symbols

δt	heliocentric correction	42
Θ_a	duration of eclipse at apastron	(3.1.26)
Θ_e	semi-duration of X-ray eclipses	(4.1.18)
Θ_p	duration of eclipse at periastron	(3.1.26)
θ	(true) phase angle or "geometrical phase"	83
θ	colatitude (zero at "North" pole)	(3.1.1)
λ	wavelength	12
λ	direction cosine (x-component)	(3.1.1)
λ, λ_k	damping factor in the Levenberg–Marquardt method	191
λ	vector of Lagrange multipliers associated with the equations in a constrained optimization problem	(A.2.4)
μ	direction cosine (y-component)	(3.1.1)
μ	vector of Lagrange multipliers associated with the inequalities in a constrained optimization problem	(A.2.4)
ν	true longitude in orbit	83
ν	direction cosine (z-component)	(3.1.1)
π	parallax, $\pi = \mathrm{const}/D$ with distance D	76
ρ	mass density	93
σ	constant in Boltzmann's law; $\sigma = 5.6705 \cdot 10^{-8}$ Jm^{-2}s^{-1}K^{-4}	(3.2.17)
σ^{data}	inner noise of the data	196
σ^{fit}	standard deviation of the fit	(4.1.13)
τ	optical depth	116)
τ	Heliocentric Julian Date	(2.1.1)
υ	true anomaly	83
Φ	photometric phase, orbital cycles from a reference phase	(2.1.1)
Φ_s	constant offset to phase	42
φ	longitude (zero in the direction toward the companion star, increasing counterclockwise)	(3.1.1)
Ψ	Roche potential in physical units	(3.1.59)
Ψ^{eff}	effective Roche potential (in physical units) for circular orbits and asynchronous rotation	(3.1.71)
ψ	ψ-function in the Russell–Merrill model	(6.2.5)
ω	argument of periastron	83
ω	(time-averaged) angular velocity of the orbital motion in radians per second	(3.1.62)
ω_j	angular rotation velocity of component j in radians per second	98
$\tilde{\omega}_j$	$\tilde{\omega}_j := \omega_j - \omega$, $j = 1, 2$	(3.1.71)
Ω	longitude of the ascending node (measured in the plane-of-sky)	83
Ω	Roche potential	(3.1.64)
Ω_j	Roche potential of component j	172
$\Omega^{\mathrm{I,O}}$	Roche potential of inner and outer Langrangian surfaces, respectively	110

Index

A

Abbreviations, xxxiii
Absolute dimensions, 22, 173, 233, 247, 288, 311, 396
    computation of, 205–210
Absorption, 105, 135, 144–146, 159, 250, 252, 290–292, 324, 326
Accretion disk, 137, 140, **142**, 161, 272, 293
    magnetism, 161
Active set, **356**, 358
Activity cycle, 246, 249, 346
Adaptive optics, 12
Advice to observers, 60–62
Airy disk, 12
Albedo, 175, 177, 215, 315, 317, 396
    bolometric, 26, 124–125, 127, 214, 276, 395–396
Algol paradox, 26–27, **137**–139
Algorithm, 25, 143, 157, 172, 174–175, 177, 183–184, 186, 197, 245, 281, 289, 343, 354, 358–389, 391
    Cholesky–Banachiewicz, 362
    Gauß–Newton, 184–185, 187, 198, 343, 359, 364–367, **364**
    generalized reduced gradient, 358–359
    genetic, 198
    interior point methods, 358–359
    Levenberg–Marquardt, *see* Levenberg–Marquardt algorithm
    maximum entropy, 297
    minimization, 352–353
    Newton–Raphson, 389, 391
    Price, 198
    sequential quadratic, 343, **359**
    Simplex, 25, **192**–196, 198, 202, 279, 324, 343, 406–407
Analysis
    Fourier, 23, 51–52, 54, 59, 182, 271, 276, 295–296, **296**, 384, 390, 393, 405
    least-squares, *see* least-squares
    light curve, 175, 189, 341
    line-profile, 54–55, **55**, 135, 279
    simultaneous least-squares, 224, 311
    spectral, 48, 51
    time domain, 296
Analytic derivatives, 156–157, 189, 210, **214**, 216, 223, 353
Anomaly
    eccentric, 87–89, **87, 88**, 237
    mean, 88–89, **88**, 156, 237
    true, 83–84, **87**, 89, 134, 156, 237
Apastron, 85–86, 201
Apsidal motion, 29, 75, 84, 88, **132**, 183, 235–236, 238
Apsides, 85, 132
Astronomical Almanac, 53
Astroseismology, 38
Atmospheres
    Carbon–Gingerich, 284, 386
    convective, 317
    empirical corrections, 119, 198, 320, 384
    extended, 136, 289–292, 325, 346
    gray, 115
    Kurucz, 26, 119, 123, 158, 279–280, 284, 287, 306–309, 318–326, 386
    Mihalas, 119
    nonemissive, 289
    radiative, 287–288, 315
Attenuation, 144–145, 159–160, 208, 228, 326

B

Balmer lines, 137, 144
Barycenter, 29, 78, 82, 235–237
Best fit, 76, 169, 243, 255, 306, 318, 337
Biaxial ellipsoid, 254
Binaries
    AM Herculis systems, 60
    astrometric, 11, 247

bizarre, 61
classification of, 109, 252
close, 6, 15, 18, **20**, 22, 92–95, 98, 117, 124, 131–132, 139–140, 146, 160, 235, 258, 310, 347
contact, 6, 17, 95, 110–111, 396
detached, **20**, 112
double-contact, 102, 113, 138
double-lined spectroscopic, 14, 209, 234
early-spectral type, 103
eclipsing, 3, 11, 14–15, 17–18, 22
evolution, 47, 65, 140, **236**
evolution in clusters, 315
in clusters, 305
magnetic, 60
morphology of, 109
over-contact, 6, 14–15, 17, 19–21, **20**, 23, 26, 95, 111–112, 396, 405
protostar, 13
semi-detached, 5, 9, 17, 19–20, **20**, 23, 27, 110, 112–113, 172, 174, 242, 287
single-lined spectroscopic, **14**, 28, 209, 245, 251
spectral–visual, 11
spectroscopic, 10–11, **13**, 54, 152
visual, 11, 54–55, 131
Wolf–Rayet, 108, 144, 289
X-ray, 22, 62, 108, 112, 140, 149, 161, 293, 347
Binary Maker, 331–335
Binnendijk model, 278
Blackbody
  assumptions, 120
  computations, 279, 288, 320, 326
  law, 199, 231, 387
  radiation, 115, 119, 123
Blaze angle, 50
Blue stragglers, 48, **314**
Bolometric correction, 207–208, 387
Bolometric flux, 115–118, 121, 125, 215, 395, 399
Boltzmann equation, 139
Boltzmann's constant, 120, 256
Boltzmann–Saha equation, 153
Bouguer extinction method, 44
Bremsstrahlung, 57
  non-thermal, 57
  thermal, 57

C
Ca II H&K lines, 132, 157–158
Cataclysmic variables, 9, 14, 20, 60, 101, 112, 140, 254

Cayley's problem, 378
Cayley–Darboux equation, 378
CCD, xxxiii, 38, **43**, 62, 146, 228
CCD detectors, 38
CCF, xxxiii
CDM, xxxiii
Center-to-limb variation, xxxiii, 91, **120**, 123, 292, 320, 383
Centrifugal limit, 113, 138
Characteristic function, 128–129, 214
Chemical composition, 22, 77, 119, 288, 321
Cherepashchuk model, 289–294
Chi-square-test, 369
Chromospheres, 120, 123, 136, 157–158, 246, 324
Circularization, 85, 347
Circumstellar envelope, 152
Circumstellar flows, 76, 144
Circumstellar matter, 50, 77, 131, 136–138, 140, 144, 153
Clouds, 10, 27, 77, 136–137, 140, 142, **144**, 144–145, 155, 228, 326
CLV, xxxiii
CMD, xxxiii
Coincidence correction, 39
Color excess, **159**, 208, 227, 231
Color index, 7, 13, 45, **45**, 159, 199, 224–228, 233, 272, 279, 282, 346
Color-magnitude diagram, 47, 225, 314, 318
Common envelope
  binaries, 14, 17, 110–112, 114
  evolution, 98, 101
Companion eclipse function, 129
Component
  primary, 21, **65**, 90, 274
  secondary, **65**, 274
Condition number, 175, 189–190, 362
Conjugacy, 355
Conjugate direction method, 196, 355
Conjunction, 83, 88, 134, 224, 288, 291
  primary, 84–85
  superior, 84, 236–237
Constant volume assumption, **102**, 115
Constraint, 171–172, 183, 185, 227, 230, 285, 335, 343, **351**, 388, 396
  active, 356–357
  eclipse duration, 179
  geometrical, 396
  main sequence, xxxiii, 233–234
  qualification, 357–358
  use of, 178
Contact discontinuity, 144
Contact parameter, 20, **110**, 111–112, 315

Index

Convection, 26, 125, 132, 251
Convective envelopes, 26, 117, **125**, 158, 315, 317
Convergence, 26, 107, 184, 186, 189–190, 198–199, 202–203, 352, 366, 379
Cool Algol, 273
Coordinates
  barycentric, 89
  Cartesian, 77–78, 143
  cylindrical, 278, 297
  plane-of-sky, *see* plane-of-sky
  Roche, 97, 378
  rotation of, 373
  spherical, 78
Coronae, 57, 157
Coronal emission, 120, 157, 347
Correlation, 174
Cosmic rays, 43–44
Covariance, 186, **188**, 204–205
Covariance matrix, 188, 204–205
Cross-correlation, xxxiii, **51**, 51–54, 57, 146, 278
Curve of growth, 56

D

Damped least-squares method, *see* Levenberg–Marquardt algorithm
Damping factor, 186, 191, 203, 354, 360, 365, 367
Data
  binning, 179, 189, 317, 387, 394
  fundamental stellar, 3, 51, 321
  photoelectric, 41, 44, 180, 315, 370
  polarization, 37, 75, 152, 154, 183
  spectrophotometric, 37, 39, 53–54, 278
  spectroscopic, 14–15, 25, 76, 138, 297
Data fitting, 160, 257, 360
Data reduction, 30, 42–44, **44**, 66, 213, 241, 345
Dead time correction, 39
Degree of contact, 110–111, 174
Depth relation, 270
Differences
  central, 354
  finite, 175, 189, 210, 216, 354–355, 404
  forward, 354
Differential Corrections, 25, 156, 175, **187**, 190, 198, 202, 205, 214, 279, 282, 324, 343, 407
Direct problem, 76, **170**, 211, 385, 391
Disks, 77, **142**
  opaque, 291
  semi-transparent, 289
  thick, 24, 27, 138, 272–273, 293
Displacement of minima, 85
Distance estimation, 28, 222, 230
  direct, 160, 221, **231**, 326
  inverse, 326
Distance modulus, 208, **231**
Disturbances, 132, 138, 140, 142, 235
Doppler
  imaging, 135
  profile analysis, 37
  profile mapping, **60**
  shift, 8, 50, 54, 56, 142, 146, 249

E

EB, xxxiii
Eccentricity, 15, 83, 91–92, 94, 172–173, 209, 235–236, 244, 251, 281–282, 311
Echelle spectra, 146, 158, 321
Eclipse
  atmospheric, **136**, 136–137, 289, 325, 387
  duration of, 199–201, 311, 388
  horizon, 129, 279, 281, 390
  moment of the, 295
  occultation, 92, 199–200, 225, 252, 254, 267, 269–272
  partial, 199, 201, 269, 291–292, 297, 314
  self-, **129**, 129–130, 281
  total, 199–200, 225, 267–268, 291–292, 321
  transit, **250**, 254, 268–270
  X-ray, 179, 183, 285, 311, 388
Eclipsing binaries
  classification, **15**, 17–18, 20
  data analysis, 15, 22, 75, **169**, 173, 221, 271
  definition, 14, 65, 90
  importance, 3, 22
  large number of light curves, 27, **241**
  modeling, **75**, 124, 128, 254
  observable curve, 75
Ellipsoidal effect, 126
Ellipsoidal variation, **6**, 17, 124, 174
Ellipticity, 92, 94, 133, **136**, 270, 327, 381–383, 398
Ellipticity effect, 23, 136
Ephemeris phase correction, 274
Epoch, 28, 42, 75, 238, 242, 257, 273, 288, 324, 344, 391
Equation of condition, 44–45, 170, 223, 267
Equilibrium
  convective, 117, **125**
  points, **102**, 388

points (computation), 389
radiative, 105, 116–117, **125**
Equipotential surfaces, 3, **19**, 77, 82, 95, 99, 103, 105–106, 109, 112, 114
Errors
  estimation of, 179, 198, 276, 369
  in the presence of correlations, 205, 278
  interpretation of, 30, 204–205
  numerical, 95, 204
  probable, 188, 205, 318, 325
  sources of, 37–38, 204, 214, 223, 274
  sources of photometric, 181–182
  statistical, 202–205, 343
  systematic, 61, 174, 204, 233, 282
ESPaDOnS, 325
Euclidean norm, 361–362, 364
Extinction, 41, 45, 47, 61, 64, 231–232
  atmospheric, 39, 41, 43, 53, 181, 335
  circumstellar, 144–145
  differential, 53
Extrasolar planets, 245ff

F
Faculae, 123, 132–133
Feasible region, **171**, 356, 358
Fibonacci line search, 192
Fill-out factor, 20, **110**, 110–112, 172, 315, 331
Finite differences, 175, 189, 210, 216, 354–355, 404
Flat-fielding, 43
Flocculi, 132
Flux
  local bolometric, 116–118, 125
  monochromatic, 37, 121–122, 125, **128**, 214–215
  total emitted, 114–115, 122, 125, 128–129, 214, 277, 281
Force
  centrifugal, 18, 77, 96, 99, 102, 108, 383
  Coriolis, 99–100
  field, 101, 104, 108
  radiation, 104
  tidal, 18, 77, 85, 92, 102, 138, 278
  time-independent, 87
Fourier transforms, 51–52, 54, 295
FOTEL, 278
Frequency domain method, 296–297

G
GAIA mission, ix, **64**, 64–65, 221, 248, 321, 326, 346
Gas streams, 24, 26, 57, 77, 123, 137–141, **139**, 160, 346

Gauß–Newton algorithm, 184–185, 187, 198, 343, 359, **364**, 364–367
Gauß–Legendre quadrature, 279
Gauß quadrature, 274, 277
Generalized inverse, 186
General light curve analysis programs, 341ff
GDDSYN, 297
GENSYN, 297
Gliese 229b, 252
Gliese 876b, 248
Gradient, 96, 98, 101–103, 106, 108, 116, 124, 139–141, 188, 242, 353–355, 364–365, 407
Graphical user interface, ix, 305, **331**
Graphics packages, 337
Gravitational radiation, 112
Gravitational waves, 26
Gravity brightening, **115**, 115–116, 381, 383
Gravity darkening, 115–118, 125, 177, 214, 266, 284, 315, 317, 381, 383
Grid approach, 205, 369, 371

H
Hα line, 132–133, 157
Hα line profiles, 143
Hβ-index, 140
Halley's method, 379
Hankel transforms, 295
Hardie extinction method, 45
Heisenberg Uncertainty Principle, 55
Heliocentric correction, 42
Hessian matrix, 187, 189, 198, **353**, 355, 357–358, 364–367
Hill model, 279
Hipparcos mission, 11, 64, 76, 156, 245, 250, 310, 321, 326
Horizon, 107, 126, **129**, 131, 279–281, 297, 346, 390, 393, 400–404
Householder transformation, 362–364
HST, xxxiii
Hubble Space Telescope, 12, 248–250, 255, 307, 319
Hydrodynamics, 17, 117, 139, 142–144, 155, 256, 326, 347

I
IAU, xxxiii
Ideal gas law, 115
Image processing, 43
Image processing packages
  DAOGROW, 44
  DAOPHOT, 44
  DOPHOT, 44
  IRAF, 44

Index

MIDAS, 44
ROMAFOT, 44
Inclination, 11, 83
Initial parameters, 169, 185, **198**, 198–199, 205, 242, 352, 379, 389–391, 404
Initial point, 172
Intensity, 104
Interferometers, 131
Interferometry
    aperture synthesis, 12
    intensity, 13
    long-base, 13
    phase, 12
    speckle, 13
Intermediate orbits, 266
International Astronomical Union, 47, 53, 61–62, 138, 345
Interpolation, 210–213, 232, 242–243, 268–269, 280–282, 288, 306, 309, 386–387
    cubic splines, 213
Interstellar
    dust, 57
    extinction, **159**, 228–229, 232–233, 326, 335, 344
    reddening, **159**, 199, 208, 344
Intrinsic variability, 7, 14, 222, 234
Inverse problem, 76, **169**, 169–171, 174, 183, 211, 372, 385, 391
Ionization potential, 153
Isoplantic patch, 12

J

Jacobian matrix, 175, 186, 191, **353**, 356–357, 365, 374–375
Julian Day Number, 42, 155–156, 344, 391

K

Kepler mission, ix, 221, 241, 346
Kepler's equation, 88–89, 101, 237
    numerical solution of, 379, 391
Kepler's law, 11, 15, 28, 97, 257
Keplerian problem, 87
Kolmogorov–Smirnov test, 369–370
Kozai cycles, **236**, 347
Kuhn–Tucker
    conditions, 186, 357–358
    point, 357–358
    sufficiency theorem, 357–358
Kwee effect, 135, *see* O'Connell effect

L

Lagrange multipliers, 172, 186, **356**, 359
Lagrangian
    function, 356
    inner L. point, 19, 102, 105, **109**, 109–110, 112, 137, 140
    outer L. point, 110
    points, 19, 102, 107, **109**, 140, 331, 388
Least-squares, 123, 183, 186, 202, 204–205, 214, 221–223, 228, 230–231, 233, 238, 266, 276, 278, 282, 293, 296, 306, 310, 322, 341–343, 359–367, 369, 385, 393, 405
    constrained, 359
    damped, 184
    equality constrained, 185
    linear, 361
    nonlinear, 75, 172, 360–361, **364**, 364–366
    unconstrained, 360
Legendre polynomials, 93, 309–310, 392
Leibniz's rule, 215
Levenberg–Marquardt algorithm, 26, 184–187, 190, 198, 203, 407
Light curve solution, 169, 171
Light curves, 14, 42
    asymmetries in, 4, 271, 314
    distortions of, 135, 140, 145
    EA-type, 15, 17
    EB-type, 15, 17
    EW-type, 15, 17
    far-ultraviolet, 13, 120, 157, 271, 307, 321
    Fourier analysis of, 23, 182, 271, 276
    Fourier fit, 384
    infrared, 17, 61
    maxima, 135, 271, 384
    monochromatic, 280
    perturbations of, 23, 135, 182, 271, 296, 384
    simulated, 254, 319, 370
    synthetic, 15, 24–25, 202, **266**, 278, 306, 331, 337, 347
    types of, 15
Light loss, 92, 140, 199, 267, 269, 274, 290, 295
Light-time effect, 132, 235–236, 278, 288
Limb eclipse effect, 153
Limb-darkening, 91, 115
    bolometric, 123, 395, 399
    far-ultraviolet, 123, 321
    linear, 120–121, 275
    logarithmic, 123, 284
    nonlinear, 122, 321
    polynomial, 122
    quadratic, 122, 284
    solar, 120, 123, 320
    square-root, 123, 284

Limiting lobes, 287
Line blending, 53, 57
Line broadening, 55–57, 150, 324
 collisional, 55
 Doppler, 56, 325
 natural, 55
 rotational, 53, 57
 Stark, 56
Line profiles, 49, 56, 75, 142, 144, **150**, 152, 324
Line-of-sight, 56, 80–82, 85, 120, 129, 136, 145, 276, 324, 393
Linear dispersion, 50
Linnell model, 279–282
Local Group galaxies, 27
LSST, **ix**, 221, 241, 346
Luminosity
 bolometric, 87, 118, 207, **208**, 126, 275, 286, 387
 monochromatic, **208**, 286
Lunar occultations, 12

M
MACHO, xxxiii, 63–64, 241, 346
Magnetic breaking, 112
Magnetic fields, 9, 46, 50, 56–57, 59, 152
Magnetohydrodynamics, 161
Magnetometry, 37, **59**, 314
Magnitude
 absolute, 7, 208, 225, 229, 233, 244, 315
 apparent, 42, 231
 bolometric, 207
Magnitude–flux relation, 42
Mass function, 14, **209**, 209–210, 294
Mass loss, 22, 84, 112, 138–139, 347
Mass motions, 56, 99
Mass transfer, 65, 84, 112, 137–140, 161
 large scale, 137
 nonconservative, 139
Mass-luminosity relation, 209, 384
Massive Compact Halo Object, 63–64, 241, 346
Matching approach, **242**, 306
Matrix inversion, 204, 265, 388
Maximum likelihood, 171, **360**
Mean free path, 139
Meridional circulation, 117, 347
Metallicity, 64, 159, 225, 228, 321, 337, 386
Method
 alternating variables, 352
 autocorrelation, 12
 Bessel, 212
 classical Newton, 355
 conjugate direction, 196, 355
 damped, 354
 derivative-based, 352
 descent, 354
 frequency domain, 266, **295**, 296
 Gauß–Newton, 184–185, 187, 198, 343, 359, **364**, 364–367
 Gauß–Legendre, 279
 generalized reduced gradient, 358
 Levenberg–Marquardt, *see* Levenberg–Marquardt algorithm
 line-search, 353
 of multiple subsets, 177–178, **190**
 orthogonalization, 362
 Powell Direction Set, 192
 quasi-Newton, 355, **355**, 365
 Simplex, 54, **192**, 194–196, 198, 278–279, 343, 352
 Simulated Annealing, 197, 244, 306
 steepest descent, 192, 354
 undamped, 354
 variable metric, 355
Microlensing, **120**, 132, 136
Microturbulent velocities, 150, 307, 324
Minimum
 global, **172**, 172–173, 352, 358
 local, **172**, 172–173, 198–199, 352, 358
 primary, 42, **65**, 84, 92, 112, 176, 200
 secondary, 211, 276
Model
 Binnendijk, 278
 Budding, 295
 Cherepashchuk, 289–293
 ellipsoidal, 23–24, 91–92, **92**, 95, 242, 383
 extended Roche, 102
 geometrical, 77
 Hadrava, 278
 Hill, 279
 Lin and Pringle, 140
 Linnell, 279–282
 Mochnacki, 297
 NDE, 89, 273–274
 physical, 77
 rectification, 274
 Roche, 95
 Russell–Merrill, 89
 semi-classical, 291
 spherical, 23, **90**
 versus code, 265
 Wilson–Devinney, 282–287
 Wood, 92, 95, 277, **277**
Modes, 286
Moment equation, 11

Index

MSC, xxxiii
Multiple systems, 10

N

Natural level functions, 186, 367
Neural networks, 242, **245**, 346
NIGHTFALL, 337
Nonlinear programming, 356
Normal distribution, 369
Normal equations, 361
Normal points, **179**, 189, 281, 292
Novae, 22, 101

O

O'Connell effect, **6**, 135, 271, 314–315, 318
O–C curve, 132, 235
Objective function, 351
Oblateness, 17, 94
Oblateness effect, **17**
Observables, **75**, 170
  systemic, **76**, 156
Observatories
  Asiago Observatory, 321
  Cracow Observatory, 322–323
  Dominion Astrophysical Observatory, 14
  Harvard College Observatory, 39, 250
  Haute Provence Observatory, 322
  McDonald Observatory, 41
  Mt. Laguna Observatory, 47
  Narrabri, 13
  Ondřejov Observatory, 278
  Princeton Observatory, 39
  Rothney Astrophysical Observatory, 39, 62
  U.S. Naval Observatory, 12
Occultation, 92, 112, 176, 199–200, 315
OGLE, xxxiii, 62, 241, 245, 250, 256–257, 346
Optical depth, 6, 17, 65, 91, 103, 116–117
Optical Gravitational Lensing Experiment, xxxiii, 62, 241, 245, 250, 256–257, 346
Optimization, 351
  constrained, 351, **356**
  convex, 358
  derivative-free, 191
  discrete, 351
  inequality constrained, **356**
  inequality nonlinear, 356, 358
  unconstrained, 351–352, **352**
Optimum
  global, 352
  local, 352
Orbital elements, 83–84
Orbital period, 209
Orbits
  absolute, 82
  circular, 82
  eccentric, 77, **82**
  edge-on, 14, **83**
  escape, 141
  relative, 82
  trajectories, 141
Oscillations
  nonradial, 87, 95, 101–102

P

Padé approximants, 379
Parallax, 76, 156–157
Parallax measurements, 11
Particle trajectories, 139
Passband, 37
Penumbra function, 401
Penumbral effect, 126
Periastron, 83, 86, 92
Period change, 178, 235, **238**, 242, 257
Phase
  eccentric orbit, 88
  geometrical, 83
  orbital, 42
  photometric, 14, **42**
Phase shift, 85
PHOEBE, 335–337
Photographic plate, 43
Photometer
  multichannel, 181
  polarizing, 39
  pulse-counting, 39, 41
  RADS, xxxiii, **41**, 181
  two-channel, 39
  Walraven, 39
Photometric systems
  Cousins, 38
  DDO, 38
  Geneva, 38
  intermediate-band, 48
  Johnson, 38
  Johnson–Cousins, 38
  narrow-band, 48
  Strömgen, 38
  transformation, 38, 41, 45–47, 233, 344, 347
  Vilnius, 38
  Walraven, 38
  Washington, 38
Photometry
  astronomical, 38
  background-limited, 38
  broadband, 48
  CCD, 38, 41, 44–45, **45**, 47

cluster, 47
   infrared, **47**, 271
   photoelectric, 37–47
   photon-limited, 37
   synthetic, 281
Photomultiplier, 38
Plages, 132
Planck function, 119
Planck law, 280
Planck's constant, 120
Plane-of-sky, 79–83
   coordinates, 131
   projected distance, **81**, 85, 91, 130, 290
Planets (individual)
   Gliese 229b, 252
   Gliese 876b, 248
   HAT P-1b, 256
   HD 209458b, 247, **255**
   OGLE-TR-56b, 250, **257**
   TR-113b, 250
   TR-132b, 250
   WASP-1b, 256
Point mass, 95
Polar gravity acceleration, 391
Polarimetry, **57**, 57–60, 62, 66
Polarization, 75–76, 144, **152**, 152–155, 327, 385
   circular, 56
   circumstellar, 152, 155
   fractional, 57
   limb, 152
   photospheric, 144, 152–153
   sources of, 57, 152
Polytropic
   gas spheres, 77
   index, 93–94
Position angle, 37, 57
Positive definite, 353
Potential, 87, 93
   at periastron, 115
   centrifugal, 95, **97**, 100
   effective, 87, 99–101
   energy, 77, 95
   function, 77
   gradient, 27, 77, 116
   gravitational, **97**
   modified Roche, **97**
   Roche, *see* Roche potential
Precession, 29, 235
Pressure, 99
Prominences, 123
Proper motion, 11
Proximity effects, 146

Pulsating variables, 51, **234**
Pulse arrival times, 76, 311

R

Radial velocity, 14, 42, 50–51, **51**
   amplitude, 147
   curve, **51**, 51–52, 54
Radiation hydrodynamics, 139
Radiation pressure, **103**, 104–105, 107–108
   inner, 105
   outer, 106
Radiative envelopes, 103, 116–117, 315, 317
Radiative transfer, 104, 117, 143–144, 347, 379, 381
Rapid alternate detection system (RADS), **39**, 181
Rectification, 76, 95, **270**
Reflection effect, 5, 17, 115, 123, **125**, **128**, 315, 387, 394, 396, 404
   multiple, 124, 127, 321, **394**, 394–395, 404
Regularization, 186
   Tikhonov, 289
Relativistic effects, 132, 235
Resolution
   angular, 12
   spatial, 11
   spectral, **37**, **48**, 53, 62, 314, 320
   telescopic, 11
   time, 41
Resolving power, 12
Restricted three-body problem, **82**, 141
Rings, 137
Robust estimation, **171**, 203
Robustness, 203
Roche
   coordinates, 97, 378
   geometry, 6, 24, 26, 95, 103, 108, 136–137, 142, 198, 254, 342, 344
   lobes, 5, 19–20, 102, 105, 107, 110–113, 137–138, 140, 178, 201, 293
   model, 25, 27, 77, 94–95, **95**, 100, 102, 114, 131, 211, 277–287
   potential, 20, 28–29, 79, 95–97, 103, 107, 110, 114–115, 159, 172, 201, 224, 228, 243, 257, 331, 378, 388, 404, 407
Rossiter effect, 56, **146–147**
Rotation, 94
   asynchronous, 19–20, **95**, 98, 100–102, 111, 113, 149, 279, 333, 337, 378
   axial, 92, 97, 99
   fast, 98, 113
   matrices, 80
   orbital, 29, 85, 92, 94, 140, 238, 288, 347

Index   425

stellar, 56, 136, 150, 246, 249, 347
subsynchronous, 98
synchronous, 18, 94, 96, 102, 111, 114,
   179, 201, 287, 381, 388
Rotation effect, 146
Royal road of eclipses, **23**, 266
RS CVn-like phenomena, 135
Rucinsky model, 282
Russian school, **289**, 291, 294

S

Saha equation, 153
SB1, 14
SB2, 14
Scattering
   anisotropic, 57
   electron, 58
   multiple, 141
   Rayleigh, 144, 152
   Thomson, 144, 152
Scintillation, 180
SED, xxxiii, 158
Self-eclipse function, 129
Sensitivity analysis, 369–370
Sequential quadratic programming, 343, **359**
Shape relation, 270
SHELLSPEC, 327
Shot noise, 37, **180**
Signal-to-noise ratio, 37, **53**
Simpson integration, 281, 394
Simultaneous fitting, **182**, 278, 311, 315
Sky background, 43
Skylab mission, 157
Small residual assumption, 188
Smoothed particle hydrodynamics, 143
Solar data, 411, 415
Solar prominences, 123
Solution
   light curve, 169, 171
   locally optimal, 352
   nonunique, 361
   optimal, 351–352
   system, 169, 171
   unique, 186
Sound speed, 140
Spectra disentangling, 54
Spectral classification, 49
Spectral energy distribution, **158**
Spectral line strength, 146
Spectral resolution, 48
Spectrophotometry, 39, 48, **54**, 59–60, 62
Spectroscopy, 48–50
Spectroscopy packages

KOREL, 54
REDUCE, 52
SPSYN, 54
VCROSS, 52
Spectrum synthesis, 280
Spots, **132**, 136, 279, 281–282, 295
   accuracy problems, 385
   bright, 60, 115
   dark, 60
   elliptic, 134, 279, 327
   latitude, 133
   longitude, 133
   radius, 133
   temperature factor, 133
SQP, xxxiii
Standard candles, 7, 27
Standard deviation, 171
Star, 66
   age, 22, 47
   comparison, **39**, **47**
   composition, 22
   irradiated, 125
   irradiating, 126
   primary, **65**, 92
   program, 41
   reference, 12
   standard, 41
Star clusters, 22–23, 27, 47
   age of, 47
   color-magnitude diagram, 47, 225,
      314, 318
   globular, 7, 47–48, 305
   open, 305, 315
Star clusters (individual)
   Hyades, 209
   NGC 104 (47 Tuc), 225–226, 319
   NGC 752, 209, 311, 315–316
   NGC 2301, 4
   NGC 5466, 175–176, 310, 312–315
   NGC 6838 (M71), 310, 317–319
   TW Hydrae, 249
Stars (individual) (by constellation)
   AB Andromedae, 147
   DS Andromedae, 209
   QX Andromedae (H235 in NGC 752),
      311, 315
   ε Aurigae, 15, 26
   ζ Aurigae, 136
   BF Aurigae, 124, 176
   44i Bootis, 131
   TY Bootis, 21, 23–24, 176
   S Cancri, 324
   α Canis Majoris (Sirius), 11

CX Canis Majoris, 50
RZ Cassiopeiae, 324
SX Cassiopeiae, 144
TV Cassiopeiae, 324
YZ Cassiopeiae, 20
V664 Cassiopeiae, 124
U Cephei, 139, 324
DQ Cephei, 137
Z Chamaeleontis, 140
RW Comae Berenices, 135
TU Crucis, 50
KU Cygni, 27
V444 Cygni, 136–137, 289, 292–293
V836 Cygni, 285
EK Draconis, 60
HP Draconis, 321, 323
DY Herculis, 41
HZ Herculis, 124
V728 Herculis, 317
RT Lacertae, 271–273, 298
SS Lacertae, 327
AR Lacertae, 298
δ Librae, 146
EH Librae, 52
GQ Lupi, 249
α Lyrae (Vega), 251
β Lyrae, 5–6, 27, 113, 137, 146
AQ Monocerotis, 50
AX Monocerotis, 27, 144–145, 238, 324
TU Muscae, 21, 117, 288
V566 Ophiuchi, 297
VV Orionis, 132, 138, 140–142
51 Pegasi, 246
II Pegasi, 298
β Persei (Algol), 5, 17–18, 21, 25, 153
RW Persei, 98
AI Phoenicis, 52, 311, 320–321, 337
β Pictoris, 251
SZ Piscium, 298
V356 Sagittarii, 27
V536 Sagittarii, 113
V884 Scorpii, 347
RZ Scuti, 98, 112, 144
RW Tauri, 137
V781 Tauri, 347
AW Ursae Majoris, 297
GP Velorum, 310, 347
Stars (individual) (other designations)
    Algol, 5, 17–18, 21, 25, 153
    Centaurus X-3, 347
    E1114+182, 60
    H235 (Heinemann 235, in NGC 752), 311, 315

HD 77581, 311
HD 152270, 144
HD 153919, 347
HD 209458, 247, 250, 254–255
HD 27130, 209
HR 6469, 85
HR 8799, 249
M71 eclipsing binaries, 317
NGC 5466 eclipsing binaries, 314
OGLE-TR-56a, 257
Sirius B, 11
SS 433, 293–294
Vega, 251
Vela X-1, 311
Stars (types), 7–8
    β Lyrae (EB), 6
    δ Scuti stars, 42
    Algols, 5, 15–17, 27, 113
    asymptotic giant branch stars, 8
    Cepheids, 7
    contact systems, 6
    eclipsing binary, 5, 14–17
    eruptive variables, 4, 8–10
    flare stars, 8
    over-contact systems, 6
    Population I, II, 7
    pulsating variables, 7–8
    RS CVn-like stars, 59, 135
    SX Phoenicis stars, 7, 42
    UX UMa variables, 101
    W Serpentis stars, 138
    W UMa variables, 17, 112, 310
    Wolf–Rayet stars, 22, 144, 292–293
Stefan–Boltzmann law, 116, 118, **118**, 125
Stellar
    atmospheres, 119, **307**, 307–310, 386
    evolution, vii, 8, 22–23, 208, 234
    evolutionary models, 246
    spectra, 55
    surface, 77
    surface imaging, 120, 132, 136
    tomography, 54
    winds, 103, 109, 136–137, **144**
Stokes quantities, 58
Stopping criteria, 188, **196**, **202**
Sun, 120, 132, 157
Sunspots, 59
Superhumps, 140
Surface
    brightness, **90**, 274–278, 282, 384
    critical, 19
    element, 79
    equipotential, 105

Index

gravity acceleration, 116, 118
grid size, 213
inner Lagrangian, 110–111
limiting, 109
normal vector, 79
of constant density, 99
of constant pressure, 99
outer Lagrangian, 111–112
stellar, 102
visible, 128
Symbols, xxxv, 411
mathematical, xxxv
Symmetry, 211
exploiting, 211
on the surface, 211
orbital, 211

T

Telescopes
8-m class, 140
automatic photometric, xxxiii, 61–62
Hubble Space Telescopes, 12
Temperature, 77
local effective, 284
mean effective, 118
modified effective, 125
polar effective, **118**, 284
Termination criteria, 202–203
Thermal contact, 287
Third body, 29, 65, 131–132, **235**, 235–236, 238, 278, 288–289, 327
Third light, 91, 128, **131**, 131–132, 214, 232–233, 235, 274–275
Tidal evolution, 85
Time scale, 3
dynamic, 3
nuclear, 3, 137
thermal, 3, 139
hydrostatic, 95
Times of minima, **42**, 132
Transit, **92**, 112, 176, 199, 200, 315
Triaxial ellipsoids, 23, 92, 94–95, 266, 277, 298, 381
True longitude, 83
True phase angle, 83

U

Ultraviolet emission, 157
Ultraviolet resonance lines, 103
Uniqueness, 171, **176**, 176–178, 191, 202
Unitary, 363

V

Variable stars, 14, 47, 351
cataclysmic, **9**, 20, **101**, 112, 142
eclipsing, 5
eruptive, 8, 14
in clusters, 47
magnetic, 59
pulsating, **7**
rotating, 14
Vector
angular velocity, 98
line-of-sight, 82
residual, 185
surface normal, 79
transposed, xxxv
Volume element, **79**, 374
von Zeipel theorem, 105–106, **117**, 117–118

W

WD (Wilson Devinney model/program), xxxiii, 282ff, 305ff, 385ff
Weights, 170, 179, **179**, 181–182, 185, 278, 293, 311, 360
curve-dependent, 179, 181
intrinsic, 179
light-dependent, 179
White dwarfs, 101
WINK, 277
Wood model, 277

X

X-ray
eclipse duration, 179, 183, 311, **388**
pulsar, **155**, 155–156, 310–311
pulse arrival times, 62, **155**, 385

Z

Zeeman splitting, **50**, 56
longitudinal, 56
transverse, 56
Zero phase, 29, 42

ASTRONOMY AND ASTROPHYSICS LIBRARY

Series Editors: G. Börner · A. Burkert · W. B. Burton · A. Coustenis
M. A. Dopita · A. Eckart · E. K. Grebel · B. Leibundgut
A. Maeder · V. Trimble

The Stars By E. L. Schatzman and F. Praderie

Modern Astrometry 2nd Edition
By J. Kovalevsky

The Physics and Dynamics of Planetary Nebulae By G. A. Gurzadyan

Galaxies and Cosmology By F. Combes, P. Boissé, A. Mazure and A. Blanchard

Observational Astrophysics 2nd Edition
By P. Léna, F. Lebrun and F. Mignard

Physics of Planetary Rings Celestial Mechanics of Continuous Media
By A. M. Fridman and N. N. Gorkavyi

Tools of Radio Astronomy 4th Edition, Corr. 2nd printing
By K. Rohlfs and T. L. Wilson

Tools of Radio Astronomy Problems and Solutions 1st Edition, Corr. 2nd printing
By T. L. Wilson and S. Hüttemeister

Astrophysical Formulae 3rd Edition (2 volumes)
Volume I: Radiation, Gas Processes and High Energy Astrophysics
Volume II: Space, Time, Matter and Cosmology
By K. R. Lang

Galaxy Formation 2nd Edition
By M. S. Longair

Astrophysical Concepts 4th Edition
By M. Harwit

Astrometry of Fundamental Catalogues
The Evolution from Optical to Radio Reference Frames
By H. G. Walter and O. J. Sovers

Compact Stars. Nuclear Physics, Particle Physics and General Relativity 2nd Edition
By N. K. Glendenning

The Sun from Space By K. R. Lang

Stellar Physics (2 volumes)
Volume 1: Fundamental Concepts and Stellar Equilibrium
By G. S. Bisnovatyi-Kogan

Stellar Physics (2 volumes)
Volume 2: Stellar Evolution and Stability
By G. S. Bisnovatyi-Kogan

Theory of Orbits (2 volumes)
Volume 1: Integrable Systems and Non-perturbative Methods
Volume 2: Perturbative and Geometrical Methods
By D. Boccaletti and G. Pucacco

Black Hole Gravitohydromagnetics
By B. Punsly

Stellar Structure and Evolution
By R. Kippenhahn and A. Weigert

Gravitational Lenses By P. Schneider, J. Ehlers and E. E. Falco

Reflecting Telescope Optics (2 volumes)
Volume I: Basic Design Theory and its Historical Development. 2nd Edition
Volume II: Manufacture, Testing, Alignment, Modern Techniques
By R. N. Wilson

Interplanetary Dust
By E. Grün, B. Å. S. Gustafson, S. Dermott and H. Fechtig (Eds.)

The Universe in Gamma Rays
By V. Schönfelder

Astrophysics. A New Approach 2nd Edition
By W. Kundt

Cosmic Ray Astrophysics
By R. Schlickeiser

Astrophysics of the Diffuse Universe
By M. A. Dopita and R. S. Sutherland

The Sun An Introduction. 2nd Edition
By M. Stix

Order and Chaos in Dynamical Astronomy
By G. J. Contopoulos

Astronomical Image and Data Analysis
2nd Edition By J.-L. Starck and F. Murtagh

The Early Universe Facts and Fiction
4th Edition By G. Börner

ASTRONOMY AND ASTROPHYSICS LIBRARY

Series Editors: G. Börner · A. Burkert · W. B. Burton · A. Coustenis
M. A. Dopita · A. Eckart · E. K. Grebel · B. Leibundgut
A. Maeder · V. Trimble

The Design and Construction of Large Optical Telescopes By P. Y. Bely

The Solar System 4th Edition
By T. Encrenaz, J.-P. Bibring, M. Blanc, M. A. Barucci, F. Roques, Ph. Zarka

General Relativity, Astrophysics, and Cosmology By A. K. Raychaudhuri, S. Banerji, and A. Banerjee

Stellar Interiors Physical Principles, Structure, and Evolution 2nd Edition
By C. J. Hansen, S. D. Kawaler, and V. Trimble

Asymptotic Giant Branch Stars
By H. J. Habing and H. Olofsson

The Interstellar Medium
By J. Lequeux

Methods of Celestial Mechanics (2 volumes)
Volume I: Physical, Mathematical, and Numerical Principles
Volume II: Application to Planetary System, Geodynamics and Satellite Geodesy
By G. Beutler

Solar-Type Activity in Main-Sequence Stars
By R. E. Gershberg

Relativistic Astrophysics and Cosmology
A Primer By P. Hoyng

Magneto-Fluid Dynamics
Fundamentals and Case Studies
By P. Lorrain

Compact Objects in Astrophysics
White Dwarfs, Neutron Stars and Black Holes
By Max Camenzind

Special and General Relativity
With Applications to White Dwarfs, Neutron Stars and Black Holes
By Norman K. Glendenning

Planetary Systems
Detection, Formation and Habitability of Extrasolar Planets
By M. Ollivier, T. Encrenaz, F. Roques F. Selsis and F. Casoli

The Sun from Space 2nd Edition
By Kenneth R. Lang

Tools of Radio Astronomy 5th Edition
By Thomas L. Wilson, Kristen Rohlfs and Susanne Hüttemeister

Astronomical Optics and Elasticity Theory
Active Optics Methods
By Gérard René Lemaitre

High-Redshift Galaxies
Light from the Early Universe
By I. Appenzeller

Eclipsing Binary Stars: Modeling and Analysis
By Josef Kallrath, Eugene F. Milone

Physics, Formation and Evolution of Rotating Stars By Maeder, André

Solar System Astrophysics
Background Science and the Inner Solar System & Planetary Atmospheres and the Outer Solar System
By Milone, Eugene F., Wilson, William

Solar System Astrophysics
Planetary Atmospheres and the Outer Solar System
By Milone, Eugene F., Wilson, William

Solar System Astrophysics
Background Science and the Inner Solar System
By Milone, Eugene F., Wilson, William J.F.

The Universe in X-Rays
By Trümper, Joachim E.; Hasinger, Günther (Eds.)